Structure-Function Relationships of Human Pathogenic Viruses

Structure-Function Relationships of Human Pathogenic Viruses

Edited by

Andreas Holzenburg

Texas A&M University
Microscopy and Imaging Center
Department of Biology
Department of Biochemistry and Biophysics
College Station, Texas

and

Elke Bogner

Friedrich-Alexander Universität Erlangen-Nürnberg
Institut für Klinische und Molekulare Virologie
Erlangen, Germany

Kluwer Academic / Plenum Publishers
New York, Boston, Dordrecht, London, Moscow

ISBN 0-306-46768-2

©2002 Kluwer Academic / Plenum Publishers, New York
233 Spring Street, New York, New York 10013

http://www.wkap.nl/

10 9 8 7 6 5 4 3 2 1

A C.I.P. record for this book is available from the Library of Congress

To Our Parents

Contributors

Dr. Stephan Becker   Institut für Virology, Robert-Koch-Str.17, 35037 Marburg, Germany

Dr. Anne Bellon   Aventis Behring GmbH, Postfach 1230, 35002 Marburg, Germany

Dr. Elke Bogner   Institut für Molekulare und Klinische Virologie, Universität Erlangen-Nürnberg, Schloßgarten 4, 91054 Erlangen, Germany

Dr. Klaus Boller   Paul-Ehrlich-Institut, Paul-Ehrlich-Str. 5159, 63225 Langen, Germany

Dr. Jay C. Brown   Department of Microbiology and Cancer Center, University of Virginia Health System, Charlottsville, Virginia 22908-0734, USA

Dr. Armin Ensser   Institut für Molekulare und Klinische Virologie, Universität Erlangen-Nürnberg, Schloßgarten 4, 91054 Erlangen, Germany

Dr. Heinz Feldmann   Canadian Science Centre for Human and Animal Health, 1015 Arlington Street, Winnipeg, Manitoba R3E 3R2, Canada

Dr. Anke Feldmann   Institut für Virology, Robert-Koch-Str.17, 35037 Marburg, Germany

Dr. Helmut Fickenscher   Abteilung Virologie, Hygiene-Institut, Ruprecht-Karls-Universität Heidelberg, Im Neuenheimer Feld 324, 69120 Heidelberg, Germany

Dr. Wolfgang Garten   Institut für Virology, Robert-Koch-Str.17, 35037 Marburg, Germany

Dr. Hans Gelderblom   Robert-Koch-Institut, Nordufer 20, 13353 Berlin, Germany

Dr. Charles Grose   Department of Internal Medicine, University of Iowa, Iowa City, Iowa 52242, USA

Dr. Anne Halenius   Robert Koch-Institut, Division of Viral Infections, Nordufer 20, 13353 Berlin, Germany

Dr. Hartmut Hengel   Robert Koch-Institut, Division of Viral Infections, Nordufer 20, 13353 Berlin, Germany

Dr. Andreas Holzenburg   Texas A&M University, Microscopy and Imaging Center, Department of Biology, Department of Biochemistry and Biophysics, College Station, TX 77843-2257, U.S.A.

Dr. Fred Homa   Infectious Disease Research, Pharmacia Corp., Kalamazoo, Michigan 49001 USA

Dr. Lindsey Hutt-Fletcher   School of Biological Sciences, University of Missouri-Kansas City, 5007 Rockhill Road, Kansas City, MO, USA

Dr. Joachim Jäger   Astbury Centre of Structural Molecular Biology, School of Biological Sciences, University of Leeds, Leeds, LS2 9JT, UK

Dr. Hans-Dieter Klenk   Institut für Virologie, Philipps-Universität Marburg, Robert-Koch-Str. 17, 35037 Marburg, Germany

Dr. Andrea Maisner   Institut für Virologie, Philipps-Universität Marburg, Robert-Koch-Str. 17, 35037 Marburg, Germany

Dr. Michael A. McVoy   Department of Pediatrics, Medical College of Virginia campus of Virginia Commonwealth University, Richmond, Virginia 23298, USA

Dr. Elke Mühlberger   Institut für Virologie, Philipps-Universität Marburg, Robert-Koch-Str. 17, 35037 Marburg, Germany

Dr. Frank Neipel   Institut für Molekulare und Klinische Virologie, Universität Erlangen-Nürnberg, Schloßgarten 4, 91054 Erlangen, Germany

Dr. Janice D. Pata   Department of Molecular Biophysics and Biochemistry, Bass Center, Yale University, New Haven, CT 06520-8114, USA.

Dr. Herbert Pfister   Department of Virology, University of Cologne, Fürst-Pückler-Str. 56, Cologne, Germany

Dr. Stephan Pleschka   Institut für Virologie, Justus-Liebig-Universität Giessen, Frankfurter Str. 107, 35392 Giessen, Germany

Dr. Sigrun Smola-Hess   Department of Virology, University of Cologne, Fürst-Pückler-Str. 56, Cologne, Germany

Yoon-Jae Song   Department of Microbiology, School of Medicine, University of Iowa, Iowa City, Iowa 52242, USA

Dr. Mark F. Stinski   Department of Microbiology, School of Medicine, University of Iowa, Iowa City, Iowa 52242, USA

Dr. Martin Vey   Aventis Behring GmbH, Postfach 1230, 35002 Marburg, Germany

Dr. Viktor Volchkov   Ecole-Normale-Superiure-de Lyon, 46, Allee d'Italie, 69007 Lyon, France

Dr. Ralf Wagner   Institut für Virology, Robert-Koch-Str.17, 35037 Marburg, Germany

Dr. Winfried Weissenhorn   European Molecular Biology Laboratory (EMBL), B.P.181, 6 rue Jules Horowitz, 38042 Grenoble, France

Dr. Thorsten Wolff   Robert-Koch-Insitut, Division of Viral Infections, Nordufer 20, 13353 Berlin, Germany

Dr. Albert Zimmermann   Robert Koch-Institut, Division of Viral Infections, Nordufer 20, 13353 Berlin, Germany

Preface

Per aspera ad astra   *(Lucius Annaeus Seneca Jr., 4BC – 65 AD)*

The idea for this book crystallised during stimulating discussions of viral structure-function relationships at the 1998 International Herpesvirus Workshop in York (UK). Over the last ten years, the amalgamation of biochemical and biophysical methods has proven extremely valuable for answering present and emerging questions in Virology. In line with this are recent breakthroughs in the area of Structural Biology with the structure determination of individual proteins contributing as much to the field as the characterisation of key virus-host interactions. Recent progress allows us now to understand viral "life" cycles in more detail than ever before.

The purpose of this book is to illustrate fundamental processes involved in viral life cycles using well known and well characterised examples with the view to endow the reader with a better understanding of viral maturation and defense mechanisms by combining a number of cutting-edge developments into a single volume.

Bearing in mind that each virus system exhibits specific sets of maturation patterns and/or strategies against the host's immune response, contributions were chosen to cover the entire spectrum including a chapter on viroids. It is the declared aim of this book to illustrate the complexity of viral pathogenesis in a comprehensive yet concise way. The topics covered include (i) structural requirements for entering the host, (ii) viral replication, (iii) viral determinants for capsid formation and DNA packaging, (iv) determinants for viral maturation, (v) pathogenesis, (vi) viral oncogenesis and (vii) defense mechanisms.

The first section deals with virus entry and egress, especially how membrane proteins could be linked to vesicular trafficking in the cell. The crystal structure of HIV gp41 and HA2 shows for the first time how virus-host membrane fusions are actually accomplished. In the second section insights into the structure-function relationships of polymerases and HIV reverse transcriptase are provided with atomic details and the chapter following on from this describes how Marburg virus reproduction depends on four essential nucleocapsid-associated proteins. The third section reports on the progress in our understanding and appreciation of the roles of virus-encoded proteins involved in DNA packaging and encapsidation in herpesviruses. In part four, the envelope protein-dependent infection pathway in measles virus is unravelled, the endocytosis and trafficking motif-bearing gE/gI complex is identified as mediating cell-to-cell spread and determining virulence in varizella zoster, and the RNA editing mechanisms of the GP gene family in Ebola virus as well as its role in cytotoxicity are highlighted.

The section on pathogenesis deals with enigmatic proteins that take a deadly turn: prions replicating without a genome. Two more chapters are devoted to wolves in sheeps' clothing: the pathogenesis of HIV and influenza HA's proteolytic activation and tissue tropism. Section six describes how papillomavirus oncogenesis depends on the interaction of E5-E7 with cellular proteins and provides a comparative overview of rhadinovirus pathogenesis in man and relevant animal models. Section seven provides information on CMV's smart immune evasion strategies like stealth mechanisms that downregulate MHC-I-dependent antigen presentations and proteins that play a role in cell cycle arrest.

We hope that this book will prove particularly useful for research and teaching staff in medicine, microbiology, and human biology whether in an academic or industrial environment. We also anticipate this book to become a favourite with medicine and microbiology students at the advanced undergraduate and (post)graduate level.

Andreas Holzenburg
Elke Bogner

Acknowledgments

Elke Bogner and Andreas Holzenburg would like to thank the Alexander von Humboldt-Foundation (Feodor-Lynen Program) for laying the foundations of our long-term collaborative research activities and enabling us to seed ideas for this book. We are especially grateful to all our colleagues who submitted timely data in a timely fashion and are indebted to Roche Diagnostics GmbH (Penzberg), Aventis Behring GmbH (Marburg) and Pharmacia GmbH (Erlangen) for sponsoring the colour figures. We would also like to thank the German Research Foundation (DFG), Dr. Bernhard Fleckenstein (Erlangen) and the Office of the Vice President for Research (Texas A&M University, College Station) for their support.

Contents

3. Viral determinants for capsid formation and packaging

4. Determinants for viral maturation

5. Pathogenesis

6. Viral oncogenesis

7. Defense mechanisms

Structure-Function Relationships of Human Pathogenic Viruses

1. Cell entry and egress

Chapter 1.1

Epstein Barr Virus Glycoproteins and their Roles in Virus Entry and Egress

LINDSEY HUTT-FLETCHER
School of Biological Sciences, University of Missouri-Kansas City, 5007 Rockhill Road, Kansas City, MO, USA

1.   INTRODUCTION

Epstein-Barr virus (EBV) is carried by more than ninety percent of the adult population of the world. Most infections occur in childhood and are subclinical. However a primary infection in adolescence or older often manifests as acute infectious mononucleosis. Long term persistence of the virus is also implicated in the development of immunoblastic lymphoma, Burkitt's lymphoma, Hodgkin's disease, nasopharyngeal carcinoma, oral hairy leukoplakia and gastric carcinoma, diseases which reflect the predominant tropism of the virus for two distinct cell types, B lymphocytes and epithelial cells (Rickinson and Kieff, 1996).

The structure of the EBV virion is typical of all herpesviruses (Kieff, 1996). The icosahedral capsid which has a diameter of 100 nm protects a protein core and a double stranded linear DNA genome of approximately 184 kilobases. The capsid is surrounded by a tegument of amorphous proteins and is enclosed by a lipid membrane that contains multiple unique glycoprotein species. To date, there is evidence for expression of twelve EBV encoded membrane proteins (Table 1), many of which are known to be present in the virion envelope. These include the molecules that play an essential role in virus attachment and penetration as infection of the cell is initiated, and play an equally important part in the correct assembly and egress of new virions for spread within and between hosts.

Structure-Function Relationships of Human Pathogenic Viruses, Edited by
Holzenburg and Bogner, Kluwer Academic/Plenum Publishers, New York, 2002

Table 1. Expressed EBV ORFs known or predicted to encode membrane proteins

ORF	Protein	Conserved Homolog	Function
BLLF1a/b	gp350/220	none	attachment to B cell receptor CR2
BXLF2	gp85	gH	attachment to epithelial cell receptor/coreceptor; fusion; oligomerizes with gL and gp42
BKRF2	gp25	gL	chaperone for gH
BZLF2	gp42	none	interaction with B cell coreceptor HLA class II
BLRF1	gp15	gN	complex of gN and gM important to acquisition and loss of envelope
BBRF3	gp84/113	gM	
BALF4	gp110	gB	assembly and exit from nucleus
BDLF3	gp150	none	?
BILF2	gp78	none	?
BILF1	gp64	none	?
BFRF1	p38	none	primary envelopment?
BMRF2	?	none	?

2.   VIRUS ENTRY

Entry of EBV into either B cells or epithelial cells requires two events that are quite distinct, attachment to a specific receptor on the cell surface and subsequent fusion of the virus envelope with the cell membrane,. Studies conducted over the last several years have revealed that both processes are surprisingly different for B cells and epithelial cells. Although some of the same virus proteins are used for both cells they are used in different and in some cases opposing ways that can influence the directional spread of virus.

2.1 Attachment

2.1.2 Attachment to B lymphocytes

Early work done on attachment of EBV to a B cell first identified not the virus protein involved but a candidate virus receptor on the cell surface. It began with the serendipitous observation that EBV and C3d, a fragment of the third component of complement, bound to closely associated or identical molecules (Einhorn *et al.*, 1978; Jondal *et al.*, 1976; Yefenof *et al.*, 1976; Yefenof, Klein, and Kvarnung, 1977) The EBV receptor and the C3d receptor exhibited the same cellular distribution, they co-capped and could be reciprocally blocked by their respective ligands. Further studies with monoclonal and polyclonal antibodies provided formal proof that EBV attaches to the second complement receptor, CR2 or CD21 on cells (Fingeroth *et al.*, 1984; Frade *et al.*, 1985; Nemerow *et al.*, 1985) and can bind to the isolated molecule (Nemerow, Siaw, and Cooper, 1986) with an affinity several orders of magnitude higher than that of the natural ligand (Moore *et al.*, 1989). Cloning of CR2 (Moore *et al.*, 1987; Weis *et al.*, 1988) also allowed demonstration that a soluble form of the receptor can block B cell infection (Nemerow *et al.*, 1990).

CR2 is a type I membrane protein of approximately 145 kilodaltons. It is a member of a large family of membrane proteins involved in tissue repair, inflammation and the immune response, characterized by structural modules known as short consensus repeats (SCRs). The tandomly repeated SCRs of CR2, which comprise the entire extracellular domain, are 60-75 amino acids in length, each forming discrete structural units (Moore *et al.*, 1989) that probably provide some segmental flexibility to the molecule (Weisman *et al.*, 1990). Each SCR consists of both variable and conserved sequences and includes four invariant cysteine residues that are disulfide bonded in a consistent pattern. CR2 exists in at least two alternatively spliced forms, one of 16 and the other of 15 SCRs, followed in both cases by a predicted 24 amino acid transmembrane domain and a C-terminal cytoplasmic tail of 34 amino acids (Moore *et al.*, 1987; Weis *et al.*, 1988). The amino terminal SCR-1 and SCR-2, found in both forms of the protein, are responsible for EBV binding and the important interaction points of the conformationally complex binding site have been mapped (Martin *et al.*, 1991). Although the binding site of C3d and EBV overlap and there is a monoclonal antibody,

OKB7, that blocks binding of both to CR2 (Nemerow *et al.*, 1985), it has proven possible to distinguish the two and explain in molecular terms how it is that although human C3d can bind to both mouse and human CR2, EBV interacts exclusively with the human receptor.

CR2 is part of a signal transduction complex on B cells (Fearon and Carter, 1995) and the interaction with CR2 may do more than simply tether the virus to the cell surface. It apparently triggers endocytosis of the virion into smooth walled vesicles (Nemerow and Cooper, 1984; Tanner *et al.*, 1987) and several observations have suggested that it may alter the phenotype of the B cell and facilitate expression of virus genes (Bohnsack and Cooper, 1988; Hutt-Fletcher, 1987; Masucci *et al.*, 1987; Sinclair and Farrell, 1995). EBV binding to CR2 on a resting B cell also activates the NF-B transcription factor which mediates activation of the viral promoter used first during latency (Sugano *et al.*, 1997). Thus, although signaling via CR2 is not essential for entry into a lymphoblastoid cell (Martin, Marlowe, and Ahearn, 1994), it is likely that it plays an important role in establishing infection in a resting B cell.

The attachment protein in the virus envelope that binds to CR2 is an abundant glycoprotein called gp350/220 because it is present in two alternatively spliced forms with masses of approximately 350 and 220 kilodaltons (Beisel *et al.*, 1985; Hummel, Thorley-Lawson, and Kieff, 1984). The splice, which maintains same reading frame, results in the loss of residues 500 to 757 of the full length 907 amino acid protein. Antibodies to gp350/220 block virus binding and infection, recombinant gp350 binds to CR2 expressing cells and to isolated CR2 and soluble gp350/220 blocks virus binding and infection (Nemerow *et al.*, 1987; Tanner *et al.*, 1987; Tanner *et al.*, 1988). The proteins are the product of the BLLF1 (Baer *et al.*, 1984) open reading frame (ORF) (Table 1) and are extensively modified by both N and O-linked sugars (Beisel *et al.*, 1985). Attachment is, however, the result of a protein-protein interaction with CR2. The binding site on gp350/220 is thought to include a short sequence (EDPGFFNVE) 21 amino acids from the N-terminus of both forms of the attachment molecule (Nemerow *et al.*, 1989) and deletion of the valine and glutamic acid residues abrogates binding (Tanner *et al.*, 1988). The sequence has striking similarity to a sequence of C3d (EDPGKQLNVE) which has been proposed to be responsible for complement binding to CR2 (Lambris *et al.*, 1985), although there still remains some controversy with respect to this latter point (Diefenbach and Isenman, 1995; Nagar *et al.*, 1998). There is no known reason for the expression of two forms of the EBV attachment protein, but the initial interaction with the elongated CR2 apparently separates the virus envelope and the cell membrane by approximately 50 nm (Nemerow and Cooper, 1984), a considerable distance, so one possibility is that exchange of the larger for the smaller form might bring the virus in a little closer. The region of gp350 that is lost as a result of the splice includes three repeats of a 21 amino acid motif with amphipathic characteristics and it has also been suggested that this may be membrane interactive (Tanner *et al.*, 1988). However, what advantage, if any, might accrue by deletion of such a structure is unclear.

2.1.3   Attachment to epithelial cells

The identity of the epithelial cell receptor for EBV is much less certain. It was originally suggested that that CR2 or a CR2-like molecule was expressed on epithelial cells (Corso, Eversole, and Hutt-Fletcher, 1989; Sixbey *et al.*, 1987; Thomas and Crawford, 1989; Young *et al.*, 1986). However these data were obtained with monoclonal antibodies which probably cross reacted with unrelated molecules on epithelial tissues (Young *et al.*, 1989). Thus there remains no conclusive evidence for epithelial cell expression of CR2 *in vivo* and although some epithelial lines may express low levels of CR2 (Fingeroth *et al.*, 1999) and stable transfection of a cDNA clone of CR2 into an epithelial cell line can render a significant proportion of the cells permissive to infection (Li *et al.*, 1992) it is unclear whether this is a biologically relevant event. Later, very provocative studies proposed that dimeric IgA, specific for EBV, carries the virus into epithelial cells via the polymeric IgA receptor (Sixbey and Yao, 1992). Epithelial polarization *in vitro* influences the outcome of IgA-mediated entry. In polarized cells virus is transported intact from the basal to the apical surface, in non-polarized cells immediate early and early virus proteins are expressed (Gan *et al.*, 1997). Thus infection with EBV that involves production of IgA might predispose to productive infection of epithelial cells that have lost polarity as a result of some unrelated cytopathic event. This is of particular relevance to the development of nasopharyngeal carcinoma as individuals with this disease have long been know to have unusually high titers of secretory IgA (Zeng *et al.*, 1985).

It is now, however, also clear that certain gastric carcinoma cells can be infected with EBV in a CR2-independent manner (Borza and Hutt-Fletcher, 1998; Imai, Nishikawa, and Takada, 1998; Yoshiyama *et al.*, 1997), without involvement of gp350/220 (Janz *et al.*, 2000). Although the identity of this novel receptor has not yet been reported, the EBV glycoprotein that serves as its ligand is known to be glycoprotein gp85 or gH (Molesworth *et al.*, 2000; Oda *et al.*, 2000). Interestingly, this is not only a protein important to attachment of virus to epithelial cells, but also a molecule that is critical for penetration of both epithelial cells and B cells.

2.2   Penetration

2.2.1   Penetration into B lymphocytes

Penetration of any enveloped virus into a cell involves fusion of the virion envelope with the membrane of the cell. This can occur either at the cell surface or after endocytosis. Although EBV has been shown to fuse at the cell surface of the Burkitt lymphoma-derived cell line Raji (Seigneurin *et al.*, 1977), gp350/220, as indicated above, stimulates endocytosis into thin-walled non-clathrin-coated vesicles in normal B cells (Nemerow and Cooper, 1984) and lymphoblastoid cell lines (LCLs). Treatment with agents that interfere with endocytosis also inhibits fusion and since the reduction in pH to which endocytosed virus is exposed is not required, this suggests that there may be differences in the endosomal and plasma membranes that are important to the event (Miller and Hutt-Fletcher, 1992).

Probably the best understood paradigm of virus cell fusion is provided by the RNA viruses such as influenza virus and the human immunodeficiency virus. In these models a single type 1 membrane glycoprotein encoded by the virus is cleaved during processing to create two species that re-associate and mediate both binding and fusion. The original amino terminal segment mediates attachment and, at least in the case of the human immunodeficiency virus, also interacts with a coreceptor on susceptible cells. A hydrophobic sequence, the "fusion peptide" at the newly created amino terminus of the segment containing the transmembrane domain is exposed by pH- or receptor-induced conformational changes and mediates disruption and mixing of the lipid bilayers of cell and virus (Hernandez *et al.*, 1996; Wyatt and Sodroski, 1998). However, no such paradigm has yet been identified for any herpesvirus and the involvement of many unique protein species in the process, none of which include readily identifiable "fusion peptides" has made understanding it that much more difficult.

At least three EBV glycoproteins are required for fusion to take place. These three are glycoproteins gp85, gp42 and gp25, which together form a detergent-stable, non-covalently linked complex in the virion (Li, Turk, and Hutt-Fletcher, 1995). Liposomes that incorporate EBV envelope proteins can bind to and fuse with appropriate receptor positive cells, but if the gp85 complex is removed they can only bind (Haddad and Hutt-Fletcher, 1989).

The largest species in the complex, gp85, is a 708 amino acid protein with a relatively hydrophobic sequence overall. It includes 5 potential N-linked glycosylation sites and the mature protein carries approximately 10 kilodaltons of N-linked sugar. Glycoprotein gp85 is the product of the BXLF2 ORF (Heineman *et al.*, 1988; Oba and Hutt-Fletcher, 1988) and is the homolog of a highly conserved glycoprotein found throughout the herpesvirus family and named gH after the prototype in herpes simplex virus. In every virus so far studied gH is a critical player in virus cell fusion (Forghani, Ni, and Grose, 1994; Forrester *et al.*, 1992; Fuller, Santos, and Spear, 1989; Gompels and Minson, 1986; Keller *et al.*, 1987; Liu *et al.*, 1993a; Peeters *et al.*, 1992). All of the homologs are predicted to be type 1 membrane proteins and the EBV gH is no exception. The gH family shares little sequence homology, but if aligned at a conserved potential N-linked glycosylation site at the carboxyl terminus the members show a colinearity of cysteine residues that suggests a conservation of secondary structure (Klupp and Mettenleiter, 1991).

Each member of the gH family of proteins is also dependent on a second glycoprotein, gL, for folding and transport throughout the cell. In the absence of gL, gH is misfolded and fails to exit the endoplasmic reticulum (Duus, Hatfield, and Grose, 1995; Hutchinson *et al.*, 1992; Kaye, Gompels, and Minson, 1992; Liu *et al.*, 1993b; Spaete *et al.*, 1993; Yaswen *et al.*, 1993). The smallest member of the EBV complex, gp25, a 137 amino acid protein with three potential N-linked glycosylation sites, is the EBV gL homolog. There is even less sequence homology between members of this second conserved family of glycoproteins. The identity of the EBV gL was first suggested by the positional homology of its gene, BKRF2, with those of the gL glycoproteins of herpes simplex virus and cytomegalovirus (McGeoch *et al.*, 1988), before it was experimentally confirmed to function as a chaperone for gH (Li, Turk, and Hutt-Fletcher, 1995; Pulford, Lowrey, and Morgan, 1995; Yaswen *et al.*, 1993).

Typically the gL homologs, which are always smaller than their gH partners, have amino terminal hydrophobic domains with the characteristics of a signal sequence and no other hydrophobic domains predicted to be membrane spanning. In the absence of gH the gL proteins are either secreted, or, as is the case for EBV gL, expressed as type 2 membrane proteins (unpublished data). And despite the absence of sequence homology between different herpesvirus gL homologs the conservation of function within the family is so marked that different members are able to substitute for each other, at least in terms of transport functions. Thus the human cytomegalovirus gL can substitute for the gL of the related betaherpesvirus human herpesvirus 6 (Anderson, Liu, and Gompels, 1996), and, even more remarkably, the EBV gL and the gL homolog of the more distantly related alphaherpesvirus varicella zoster virus have chaperone functions in common (Li *et al.*, 1997a). No determinations have yet been made of the interactive domains of EBV gH and gL, but in herpes simplex virus only the 323 amino terminal residues of gH are required for association with gL (Peng *et al.*, 1998) and in human herpesvirus 6 only the amino terminal 230 residues are necessary (Anderson, Liu, and Gompels, 1996). A current model of gH-gL complexes in herpes simplex virus proposes a dimeric structure with the amino terminal domain of gL hidden within the amino terminus of gH (Peng *et al.*, 1998). Although the strong functional conservation between gL homologs of different viruses supports the idea that such a model may be widely applicable, if the EBV gL is anchored in the virus and cell membrane via an uncleaved signal sequence, a somewhat different structure appears likely for this case at least.

The EBV gH-gL complex required for penetration into the B lymphocyte also differs from that of many other herpesviruses in at least one other important respect, namely the inclusion of the third protein, gp42 which has weak homology with a C-type lectin (Spriggs *et al.*, 1996). Glycoprotein gp42 is the product of the BZLF2 ORF (Li, Turk, and Hutt-Fletcher, 1995) and although a third glycoprotein has been shown to associate with the human cytomegalovirus gH-gL complex (Huber and Compton, 1997; Li, Nelson, and Britt, 1997), gp42 only has obvious viral homologs in certain of the other gammaherpesviruses (Ensser, Pflanz, and Fleckenstein, 1997; Telford *et al.*, 1995). It is a 223 residue protein with four potential N-linked glycosylation sites and like both gH and gL carries approximately 10 kDa of N-linked sugar (Li, Turk, andHutt-Fletcher, 1995). Its orientation in the membrane is similar to that of gL in that it is anchored by an uncleaved signal sequence and the region responsible for interaction with the gH-gL oligomer lies between residues 34 and 58 (Wang *et al.*, 1998). Between residues 122 and 223 there is a second region that binds to the polymorphic β_1 domain of the HLA-DR chain on the B cell surface (Spriggs *et al.*, 1996) in an allele-specific manner (Haan and Longnecker, 2000). Several observations indicate that this interaction is critical to infection. First, there is the behavior of a monoclonal antibody called F-2-1 neutralizes B cell infection by blocking virus-cell fusion (Miller and Hutt-Fletcher, 1988). This antibody was originally thought to interact with gH because of its ability to immunoprecipitate gH and its associated proteins, but was later found to bind to an epitope on gp42. The same F-2-1 antibody blocks the ability of a soluble form of gp42, gp42.Fc, to bind to HLA class II (Li *et al.*, 1997b). Second, a monoclonal antibody to HLA-DR that blocks binding of gp42.Fc to HLA class II also blocks virus infection. Third, the soluble gp42.Fc protein can inhibit B cell infection (Li *et al.*, 1997b). Finally, B cells lacking HLA class II can only be infected if class II expression is restored (Haan *et al.*, 2000; Li *et al.*, 1997b). These results have been interpreted to mean that HLA class II can serve as a coreceptor for EBV on the B cell surface.

More recently a recombinant EBV which lacks gp42 has been shown to be unable to infect B cells unless cells and attached virus are treated with an exogenous fusogen such as polyethylene glycol (Wang and Hutt-Fletcher,

1998), or unless a three part gH-gL-gp42-like complex is reformed by addition of exogenous gp42.Fc (Wang *et al.*, 1998). Thus a minimal model (Figure 1) of EBV infection of the B cell proposes that: 1) virus attaches to CR2 by means of gp350/220, initially at a considerable distance from the cell surface; 2) virus comes closer to the membrane, perhaps as a result of the flexibility of CR2 and the sequential use of gp350 and gp220, 3) gp42 binds to HLA class II; 4) gH and gL mediate fusion. There is no direct evidence for step 4, but since infection by virus lacking gp42 can be restored by a soluble form of the protein with no membrane interactive domains (Wang *et al.*, 1998), it is very unlikely that gp42 plays any direct role in fusion. By analogy with other herpes viruses and by virtue of the fact that it is the most hydrophobic member of the complex it would appear that gH is the most likely protein to be directly involved. In its role as chaperone gL may be responsible for maintaining gH in a non-fusogenic state until some interaction at the cell membrane, perhaps that between gp42 and HLA class II, provides a trigger for change (Haan and Longnecker, 2000)

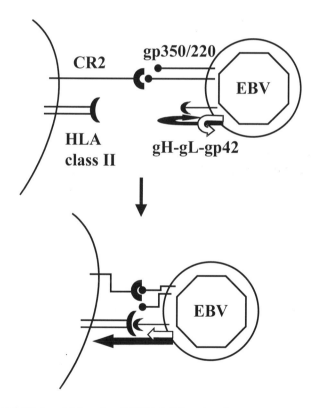

Figure 1. A minimalist model for entry of EBV into B lymphocytes.

2.2.2   Penetration into epithelial cells

Just as there are clear differences in attachment of EBV to B cells and epithelial cells, there are also clear differences in the penetration process (Figure 2). In the first place no interaction with HLA class II appears to be required and glycoprotein gp42 is dispensable (Wang *et al.*, 1998). Instead, a series of experiments suggest that there is a direct interaction between a two part gH-gL complex and the epithelial cell surface. First, a monoclonal antibody called E1D1, which reacts with a conformationally complex epitope on gH (Li, Turk, and Hutt-Fletcher, 1995; Pulford, Lowrey, and Morgan, 1995), has no effect on B cell transformation by EBV but very efficiently neutralizes infection of both epithelial cells transfected with CR2 (Li, Turk, and Hutt-Fletcher, 1995) and the more biologically relevant gastric carcinoma cell lines (Molesworth *et al.*, 2000). Second, if the recombinant virus that lacks gp42 is treated with soluble gp42.Fc to reform a three part gH-gL-gp42.Fc complex the virus loses its ability to infect epithelial cells. If a form of gp42.Fc is used in which the gH-gL interactive domain between residues 34 and 58 has been deleted but the HLA class II reactive domain between residues 122 and 223 has been retained, only B cell infection and not epithelial cell infection is inhibited. Third, biochemical analysis indicates that wild type virus contains both three part gH-gL-gp42 and two part gH-gL complexes (Wang *et al.*, 1998). This implies the existence of two distinct complexes, one of only gH and gL and one of gH-gL and gp42.

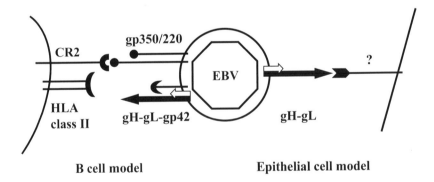

B cell model   **Epithelial cell model**

Figure 2. A comparison of the models for B cell and epithelial cell entry.

If only gH and gL are required for infection of epithelial cells, why does addition of a third component matter? The most likely explanation would appear to be that the presence of gp42 inhibits an interaction between gH-gL and a novel entity that is present on epithelial cells, but not on B lymphocytes. This novel molecule must serve the same function for epithelial cell entry as does HLA class II for B cell entry. One of the most interesting questions now remaining is whether or not the primary receptor and coreceptor for epithelial cells are one and the same or whether gH, which is involved in both attachment and penetration of epithelial cells interacts sequentially with two independent species.

2.2   Implications for tropism and pathogenesis

The nature and distribution of the receptors and coreceptors used by viruses to infect cells have obvious impact on tropism and pathogenesis. The fact that gp42 can interact with some, but not all alleles of HLA class II suggests that some individuals may be more or less susceptible to infection (Haan and Longnecker, 2000; Li *et al.*, 1997b). Although it is unlikely in an outbred human population that any individual expresses only HLA alleles that fail to interact with gp42, the number of interactive alleles expressed might influence viral load and hence the potential outcome of persistent infection. Likewise, if the as yet unknown receptor or coreceptor on epithelial cells should prove to be polymorphic, its distribution might explain some of the genetic predispositions to epithelial tumors (Rickinson and Kieff, 1996). In addition the observation that different receptors are used for different cell types suggests that fitness of virus made in one cell type might be different for another. Interactions of newly synthesized viral glycoproteins with cellular receptors have been shown to mediate interference and block infection by retroviruses and herpesviruses (Coffin, 1996; Geraghty, Jogger, and Spear, 2000). In the same way, interaction of gp350/220 and gp42 with their receptors CR2 and HLA class II within the B cell or gH and its receptor/coreceptor in epithelial cells might compromise the ability of the virus to reinfect a cell of the same type as that in which it is made. These and other issues should be fertile ground for further study.

3.   VIRUS ASSEMBLY AND EGRESS

3.1   Pathways of egress

Historically, two models have been proposed for egress of herpesviruses from cells. One, originally proposed for egress of herpes simplex virus, suggested that tegumented nucleocapsids bud through the inner nuclear membrane into the endoplasmic reticulum and in doing so acquire their final envelopes (Johnson and Spear, 1982). This model has now been challenged by several observations (Browne *et al.* 1996; Gershon *et al.*, 1994; Granzow *et al.*, 1997; Radsak *et al.*, 1966; Whealy *et al.*, 1991; Whitely *et al.*, 1999; Zhu *et al.*, 1995) that instead suggest that although virus first buds through the inner nuclear membrane it subsequently fuses back out into the cytoplasm, perhaps to acquire or exchange with additional tegument proteins (Klupp, Granzow, and Mettenleiter, 2000; Figure 3). A second and final envelope is then acquired by budding back into the *trans* Golgi region. In both cases virus is finally delivered to the extracellular space via exocytosis.

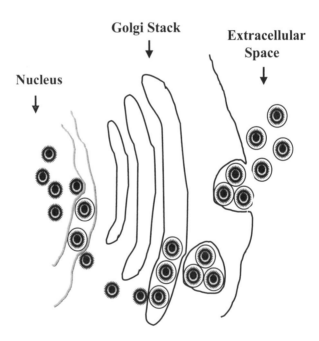

Figure 3. The current envelopment-deenvelopment-reenvelopment model for egress of herpesviruses.

Although the issue has not been extensively studied in EBV, the available evidence, which is based on the reciprocal distribution of two glycoproteins, suggests that EBV conforms to the second of the two models. Glycoprotein gp110, the product of the BALF4 ORF, is found predominantly in the membranes of the nucleus and endoplasmic reticulum and carries mainly high mannose sugars, consistent with a lack of processing in the Golgi apparatus (Gong and Kieff, 1990; Lee, 1999; Papworth *et al.*, 1997). A string of four arginine residues in the the cytoplasmic tail of gp110 apparently function as an endoplasmic reticulum retention motif (Lee, 1999). In contrast gp350/220, which carries large amounts of Golgi-dependent O-linked sugars, is found primarily in the plasma membrane and Golgi. The virion envelope contains abundant amounts of gp350/220 but only very small amounts of a differentially glycosylated form of gp110, gp125, that carries some Golgi-modified sugars (Emini *et al.*, 1987; Gong and Kieff, 1990; Lee, 1999). These data are most consistent with the exchange of envelopes and a model in which primary envelopment is followed by deenvelopment and acquisition of a second envelope from a late Golgi membrane that contains fully processed glycoproteins.

3.2   Proteins known to be involved in egress and possible models for their behavior

At least three glycoproteins are confirmed to be involved in assembly and egress of EBV. Glycoprotein gp110 was originally hypothesized to play a role in virus egress because of its unique location in the membranes of the nucleus and the endoplasmic reticulum. The glycoprotein is the homolog of the highly conserved herpesvirus glycoprotein gB (Gong *et al.*, 1987; Pellett *et al.*, 1985) which in most herpesviruses is found in abundance in the virion envelope and plays a critical role in fusion of the virus envelope with the uninfected cell membrane. Since its distribution appeared to eliminate the possibility of a similar function in EBV a role in modification of the nuclear for egress was proposed instead (Gong and Kieff, 1990). The subsequent derivation of recombinant viruses that lack gB has lent support to such a model. No nucleocapsids or enveloped viruses were detected in cells making virus that lacked gB (Lee and Longnecker, 1997). Although the extremely small numbers of virus producing cells available for electromicroscopy in this study raise some questions about the precise nature of the defect, the work is entirely consistent with a role for gB in budding through the inner nuclear membrane.

The two other glycoproteins currently known to play a role in virus assembly and egress are the products of the BLRF1 and the BBRF3 ORFs. These two form a non-covalently linked complex within which, like gH and gL, the proteins are co-dependent for expression (Lake, Molesworth, and Hutt-Fletcher, 1998). They are also the homologs of another two conserved herpesvirus glycoproteins designated, respectively, gN and gM. The EBV gN, like its counterparts in other herpesviruses, is a small type 1 membrane protein of 102 amino acids. It carries only O-linked sugar, most of which is α2,6 linked sialic acid. Expressed as a recombinant protein it is incompletely processed and lacks all of the sialic acid residues that are found on the native molecule. The native protein forms a stable association with gM and co-expression of gN and gM as recombinant proteins restores authentic processing of gN. Glycoprotein gM is predicted to be a type 3 membrane protein with multiple transmembrane domains and a 78 amino acid cytoplasmic tail which is highly charged and rich in prolines. It is expressed in virus-producing cells as at least three differentially glycosylated forms of approximately 48, 84 and 113 kilodaltons. All three of these species are also expressed as recombinant proteins in the absence of gN. However, in a recombinant virus that lacks gN, gM is not detectable at all (Lake and Hutt-Fletcher, 2000). This suggests that in the absence of gN, gM interacts with another virus protein that influences its turnover or processing.

A virus lacking a mature gN-gM complex has several significant defects (Lake and Hutt-Fletcher, 2000). First, a dramatic proportion of the recombinant virus capsids remain associated with condensed chromatin in the nucleus of virus producing cells. Second, cytoplasmic vesicles containing enveloped virus, which are clearly evident in cells making wild type virus, are scarce. Virus egress is impaired and the majority of virus that is released lacks a complete envelope. The small amount of enveloped virus that is released and can bind to cells is further impaired in infectivity at a step following fusion. These data are consistent with the hypothesis that the predicted charged, proline rich cytoplasmic tail of gM, interacts with the virion tegument and that this interaction is important both for association of capsids with cell membrane to assemble and release enveloped particles and for dissociation of the capsid from the membrane of the newly infected cell on its way to the cell nucleus. The phenotype of EBV lacking the gN-gM complex is more striking than that of most alphaherpesviruses lacking the same complex, but resembles in many respects the phenotype of pseudorabies virus lacking glycoproteins gM, gE, and gI (Brack *et al.*, 1999). Since EBV does not encode homologs for gE and gI this suggests that functions that may have some redundancy in alphaherpesviruses have been concentrated in the EBV gN-gM complex.

Although gB, gN and gM are the only glycoproteins that are currently directly implicated in virus egress, it is likely that others play a role either in assembly or intracellular trafficking. For example, although gp350/220 does not play an essential role in egress (Janz *et al.*, 2000), in polarized cells it is targeted to the basolateral surface (Chodosh *et al.*, 2000) and thus may play a role in directional movement of exocytic vesicles carrying virus. There are also several membrane proteins known to be expressed in virus-producing cells to which no phenotype has yet been attributed. Perhaps the most likely to be involved in egress is p38, the product of the BFRF1 ORF (Farina *et al.*, 2000) This protein is the homolog of the alphaherpesvirus UL34 product, a type 2 membrane protein which is involved in primary envelopment of herpes simplex and pseudorabies virus capsids (Klupp, Granzow, and Mettenleiter, 2000; Roller *et al.*, 2000; Shiba *et al.*, 2000). Other EBV membrane proteins with potential involvement in either assembly or egress include gp78, the product of the BILF2 ORF (Mackett *et al.*, 1990), gp64, the product of the BILF1 ORF (unpublished) and the BMRF2 gene product (Penaranda *et al.*, 1997). Only gp150, the product of the BDLF3 ORF (Kurilla *et al.*, 1995; Nolan and Morgan, 1995) can be eliminated as having any essential replicative function (Borza and Hutt-Fletcher, 1998), and even in this case there remains at least a possibility that a role for the protein might be identified in polarized cells

4. CONCLUSIONS

Much of the investigation of the biology of EBV has naturally and appropriately focussed on its extraordinary transforming potential. As the field matures, however, more attention can legitimately be given to the function of those proteins that are responsible for the production of new virus. This chapter has considered work on only the first and last steps in this process that involve membrane proteins, but even this limited arena has produced several biologically important discoveries and has raised perhaps even more provocative questions.

Examination of the entry process and identification of at least some of the proteins involved has revealed interactions that have, and should continue to provide important insights into tropism and pathogenesis. An understanding of the fusion process itself is still, however, almost completely lacking and is an issue that urgently needs creative investigation. Studies of virus assembly and egress are still in their infancy, but already have revealed a complex process that may provide insight not only into virus function, but also into the more fundamental processes of vesicular trafficking in the cell. In this area it seems likely that work done with herpesviruses in which it has historically been easier to study lytic replication may lead the way. It is, however, already clear that each virus has evolved in a somewhat different way to exploit its own biological niche, so surprises should be expected.

Finally, although this chapter considers only the essential replicative functions of virus membrane proteins it must be noted that many of them certainly play additional more subtle roles *in vivo*. Those roles may be more difficult to to identify, except by analysis and analogy with closely related primate viruses, but are probably nonetheless critical to explaining the extraordinary success of EBV as a universal parasite of man.

REFERENCES

Anderson, R. A., Liu, D. X., and Gompels, U. A. 1996. Definition of a human herpesvirus-6 betaherpesvirus-specific domain in glycoprotein gH that governs interaction with glycoprotein gL: substitution of human cytomegalovirus glycoproteins permits group-specific complex formation. *Virology* **217**, 517-526.

Baer, R., Bankier , A. T., Biggin, M. D., Deininger, P. L., Farrell, P. J., Gibson, T. J., Hatfull, G., Hudson, G. S., Satchwell, S. C., Seguin, C., Tuffnell, P. S., and Barrell, B. G. 1984. DNA sequence and expression of the B95-8 Epstein-Barr virus genome. *Nature* **310**, 207-211.

Beisel, C., Tanner, J., Matsuo, T., Thorley-Lawson, D., Kezdy, F., and Kieff, E. 1985. Two major outer envelope glycoproteins of Epstein-Barr virus are encoded by the same gene. *J. Virol.* **54**, 665-674.

Bohnsack, J. F., and Cooper, N. R. 1988. CR2 ligands modulate B cell activation. *J. Immunol.* **141**, 2569-2576.

Borza, C., and Hutt-Fletcher, L. M. 1998. Epstein-Barr virus recombinant lacking expression of glycoprotein gp150 infects B cells normally but is enhanced for infection of the epithelial line SVKCR2. *J. Virol.* **72**, 7577-7582.

Brack, A. R., Dijkstra, J. M., Granzow, H., Klupp, B. G., and Mettenleiter, T. C. 1999. Inhibition of virion maturation by simultaneous deletion of glycoproteins E, I, and M of pseudorabies virus. *J. Virol.* **73**, 5364-5372.

Browne, H., Bell, S., Minson, T., and Wilson, D. W. 1996. An endoplasmic reticulum-retained herpes simplex virus glycoprotein H is absent from secreted virions: evidence for reenvelopment during egress. *J. Virol.* **70**, 4311-4316.

Chodosh, J., Gan, Y., Holder, V. P., and Sixbey, J. W. 2000. Patterned entry and egress by Epstein-Barr virus in polarized CR2- positive epithelial cells. *Virology* **266**, 387-396.

Coffin, J. M. (1996). Retroviridae: The viruses and their replication. *In* "Fields Virology" (B. N. Fields, D. M. Knipe, and P. M. Howley, Eds.), Vol. 2, pp. 1767-1847. 2 vols. Lippincott-Raven, Philadelphia.

Corso, B., Eversole, L. R., and Hutt-Fletcher, L. M. 1989. Hairy leukoplakia: Epstein-Barr virus receptors on oral keratinocyte plasma membranes. *Oral Surg. Oral Med. Oral Pathol.* **67**, 416-421.

Diefenbach, R. J., and Isenman, D. E. 1995. Mutation of residues in the C3dg region of human complement component C3 corresponding to a proposed binding site for complement receptor type 2 (CR2, CD21) does not abolish binding of iC3b or C3dg to CR2. *J. Immunol.* **154**, 2303-2320.

Duus, K. M., Hatfield, C., and Grose, C. 1995. Cell surface expression and fusion by the varicella-zoster virus gH:gL glycoprotein complex: analysis by laser scanning confocal microscopy. *Virology* 210, 429-440.

Einhorn, l., Steinitz, M., Yefenof, E., Ernberg, I., Bakacs, T., and G., K. 1978. Epstein-Barr virus (EBV) receptors, complement receptors and EBV infectibility of different lymphocyte fractions of human peripheral blood. *Cell. Immunol.* 35, 43-58.

Emini, E. A., Luka, J., Armstrong, M. E., Keller, P. M., Ellis, R. W., and Pearson, G. R. 1987. Identification of an Epstein-Barr virus glycoprotein which is antigenically homologous to the varicella-zoster glycoprotein II and the herpes simplex virus glycoprotein B. *Virology* 157, 552-555.

Ensser, A., Pflanz, R., and Fleckenstein, B. 1997. Primary structure of the alcelaphine herpesvirus 1 genome. *J. Virol.* 71, 6517-6525.

Farina, A., Santarello, R., Gonnella, R., Bei, R., Muraro, R., Cardinali, G., Uccini, S., Ragona, G., Frati, L., Faggioni, A., and Angeloni, A. 2000. The *BFRF1* gene of Epstein-Barr virus encodes a novel protein. *J. Virol.* 74, 3235-3244.

Fearon, D. T., and Carter, R. H. 1995. The CD19/CR2/TAPA-1 complex of B lymphocytes: linking natural to acquired immunity. *Ann. Rev. Immunol.* 13, 127-149.

Fingeroth, J. D., Diamond, M. E., Sage, D. R., Hayman, J., and Yates, J. L. 1999. CD-21 dependent infection of an epithelial cell line, 293, by Epstein-Barr virus. *J. Virol.* 73, 2115-2125.

Fingeroth, J. D., Weis, J. J., Tedder , T. F., Strominger, J. L., Biro, P. A., and Fearon, D. T. 1984. Epstein-Barr virus receptor of human B lymphocytes is the C3d complement CR2. *Proc. Natl. Acad. Sci. USA* 81, 4510-16.

Forghani, B., Ni, L., and Grose, C. 1994. Neutralization epitope of the varicella-zoster virus gH:gL glycoprotein complex. *Virology* 199, 458-462.

Forrester, A., Farrell, H., Wilkinson, G., Kaye, J., Davis-Poynter, N., and Minson, T. 1992. Construction and properties of a mutant of herpes simplex virus type 1 with glycoprotein H coding sequences deleted. *J. Virol.* 66, 341-348.

Frade, R., Barel, M., Ehlin-Henricksson, B., and Klein, G. 1985. gp140 the C3d receptor of human B lymphocytes is also the Epstein-Barr virus receptor. *Proceedings of the National Academy of Sciences* 82, 1490-1493.

Fuller, A. O., Santos, R. E., and Spear, P. G. 1989. Neutralizing antibodies specific for glycoprotein H of herpes simplex virus permit viral attachment to cells but prevent penetration. *J. Virol.* 63, 3435-3443.

Gan, Y., Chodosh, J., Morgan, A., and Sixbey, J. W. 1997. Epithelial cell polarization is a determinant in the infectious outcome of immunoglobulin A-mediated entry by Epstein-Barr virus. *J. Virol.* 71, 519-526.

Geraghty, R. J., Jogger, C. R., and Spear, P. G. 2000. Cellular expression of alphaherpesvirus gD interferes with entry of homologous and heterologous alphaherpesviruses by blocking access to a shared gD receptor. *Virology* 268, 147-158.

Gershon, A. A., Sherman, D. L., Zhu, Z., Gabel, C. A., Ambron, R. T., and Gershon, M. D. 1994. Intracellular transport of newly synthesized varicella-zoster virus: final envelopment in the *trans*-Golgi network. *J. Virol.* 68, 6372-6390.

Gompels, U. A., and Minson, A. 1986. The properties and sequence of glycoprotein H of herpes simplex type 1. *Virology* 153, 230-247.

Gong, M., and Kieff, E. 1990. Intracellular trafficking of two major Epstein-Barr virus glycoproteins, gp350/220 and gp110. *J Virol* 64, 1507-1516.

Gong, M., Ooka, T., Matsuo, T., and Kieff, E. 1987. Epstein-Barr virus glycoprotein homologous to herpes simplex virus gB. *J. Virol.* 61, 499-508.

Granzow, H., Weiland, F., A., J., B.G., K., Karger, A., and Mettenleiter, T. C. 1997. Ultrastructural analysis of the replication cycle of pseudorabies virus in cell culture: a reassessment. *J. Virol.* 71, 2072-2082.

Haan, K. M., Kwok, W. W., Longnecker, R., and Speck, P. 2000. Epstein-Barr virus entry utilizing HLA-DP or HLA-DQ as a coreceptor. *J. Virol.* **74**, 2451-2454.

Haan, K. M., and Longnecker, R. 2000. Coreceptor restriction within the HLA-DQ locus for Epstein-Barr virus infection. *Proc. Natl. Acad. Sci. USA* **97**, 9252-9257.

Haddad, R. S., and Hutt-Fletcher, L. M. 1989. Depletion of glycoprotein gp85 from virosomes made with Epstein-Barr virus proteins abolishes their ability to fuse with virus receptor-bearing cells. *J. Virol.* **63**, 4998-5005.

Heineman, T., Gong, M., Sample, J., and Kieff, E. 1988. Identification of the Epstein-Barr virus gp85 gene. *J Virol* **62**, 1101-1107.

Hernandez, L. D., Hoffman, L. R., Wolfsberg, T. G., and White, J. M. 1996. Virus-cell and cell-cell fusion. *Ann. Rev. Cell Dev. Biol.* **12**, 627-661.

Huber, M. T., and Compton, T. 1997. Characterization of a novel third member of the human cytomegalovirus glycoprotein H-L complex. *J. Virol.* **71**, 5391-5398.

Hummel, M., Thorley-Lawson, D., and Kieff, E. 1984. An Epstein-Barr virus DNA fragment encodes messages for the two major envelope glycoproteins (gp350/300 and gp220/200). *J. Virol.* **49**, 413-417.

Hutchinson, L., Browne, H., Wargent, V., Davis-Poynter, N., Primorac, S., Goldsmith, K., Minson, A. C., and Johnson, D. C. 1992. A novel herpes simplex virus glycoprotein gL forms a complex with glycoprotein H (gH) and affects normal folding and surface expression of gH. *J. Virol.* **66**, 2240-2250.

Hutt-Fletcher, L. M. 1987. Synergistic activation of cells by Epstein-Barr virus and B-cell growth factor. *J. Virol.* **61**(3), 774-81.

Imai, S., Nishikawa, J., and Takada, K. 1998. Cell-to-cell contact as an efficient mode of Epstein-Barr virus infection of diverse human epithelial cells. *J. Virol.* **72**, 4371-4378.

Janz, A., Oezel, M., Kurzeder, C., Mautner, J., Pich, D., Kost, M., Hammerschmidt, W., and Delecluse, H. J. 2000. Infectious Epstein-Barr virus lacking major glycoprotein BLLF1 (gp350/220) demonstrates the existence of additional viral ligands. *J. Virol.* **74**, 10142-10152.

Johnson, D. C., and Spear, P. G. 1982. Monensin inhibits the processing of herpes simplex virus glycoproteins, their transport to the cell surface, and the egress of virus from infected cells. *J. Virol.* **43**, 1102-1112.

Jondal, M., Klein, G., Oldstone, M. B. A., Bokish, V., and Yefenof, E. 1976. Surface markers on human B and T lymphocytes. *Scand. J. Immunol.* **5**, 401-410.

Kaye, J. F., Gompels, U. A., and Minson, A. C. 1992. Glycoprotein H of human cytomegalovirus (HCMV) forms a stable complex with the HCMV UL115 gene product. *J. Gen. Virol.* **73**, 2693-2698.

Keller, P. M., Davison, A. J., Lowe, R. S., Riemen, M. W., and Ellis, R. W. 1987. Identification and sequence of the gene encoding gpIII, a major glycoprotein of varicella-zoster virus. *Virology* **157**, 526-533.

Kieff, E. (1996). Epstein-Barr Virus and its Replication. *In* "Fields Virology" (B. N. Fields, D. M. Knipe, and P. M. Howley, Eds.), Vol. 2, pp. 2343-2396. 2 vols. Lippincott-Raven, Philadelphia.

Klupp, B., and Mettenleiter, T. C. 1991. Sequence and expression of the glycoprotein gH gene of pseudorabies virus. *Virology* **182**, 732-741.

Klupp, B. G., Granzow, H., and Mettenleiter, T. C. 2000. Primary envelopment of pseudorabies virus at the nuclear membrane requires the UL34 gene product. *J. Virol.* **74**, 10063-10073.

Kurilla, M. G., Heineman, T., Davenport, L. C., Kieff, E., and Hutt-Fletcher, L. M. 1995. A novel Epstein-Barr virus glycoprotein gp150 expressed from the BDLF3 open reading frame. *Virology* **209**, 108-121.

Lake, C. M., and Hutt-Fletcher, L. M. 2000. Epstein-Barr virus that lacks glycoprotein gN is impaired in assembly and infection. *J. Virol.* **74**, 11162-11172.

Lake, C. M., Molesworth, S. J., and Hutt-Fletcher, L. M. 1998. The Epstein-Barr virus (EBV) gN homolog BLRF1 encodes a 15 kilodalton glycoprotein that cannot be authentically processed unless it is co-expressed with the EBV gM homolog BBRF3. *J. Virol.* **72**, 5559-5564.

Lambris, J. D., Ganu, V. S., Hirani, S., and Muller-Eberhard, H. J. 1985. Mapping of the C3d receptor (CR2) binding site and a neoantigenic site in the C3d domain of the third component of complement. *Proceedings of the National Academy of Sciences* **82**, 4235-4239.

Lee, S. K. 1999. Four consecutive arginine residues at positions 836-839 of EBV gp110 determine intracellular localization of gp110. *Virology* **264**, 350-358.

Lee, S. K., and Longnecker, R. 1997. The Epstein-Barr virus glycoprotein 110 carboxy-terminal tail domain is essential for lytic virus replication. *J. Virol.* **71**, 4092-4097.

Li, L., Nelson, J. A., and Britt, W. J. 1997. Glycoprotein H-related complexes of human cytomegalovirus: identification of a third protein in the gCIII complex. *J. Virol.* **71**, 3090-3097.

Li, Q. X., Buranathai, C., Grose, C., and Hutt-Fletcher, L. M. 1997a. Chaperone functions common to nonhomologous Epstein-Barr virus gL and Varicella-Zoster virus gL proteins. *J. Virol.* **71**(2), 1667-70.

Li, Q. X., Spriggs, M. K., Kovats, S., Turk, S. M., Comeau, M. R., Nepom, B., and Hutt-Fletcher, L. M. 1997b. Epstein-Barr virus uses HLA class II as a cofactor for infection of B lymphocytes. *J. Virol.* **71**(6), 4657-4662.

Li, Q. X., Turk, S. M., and Hutt-Fletcher, L. M. 1995. The Epstein-Barr virus (EBV) BZLF2 gene product associates with the gH and gL homologs of EBV and carries an epitope critical to infection of B cells but not of epithelial cells. *J. Virol.* **69**, 3987-3994.

Li, Q. X., Young, L. S., Niedobitek, G., Dawson, C. W., Birkenbach, M., Wang, F., and Rickinson, A. B. 1992. Epstein-Barr virus infection and replication in a human epithelial system. *Nature* **356**, 347-350.

Liu, D. X., Gompels, U. A., Foa-Tomasi, L., and Campadelli-Fiumi, G. 1993a. Human herpesvirus 6 glycoprotein H and L homologues are components of the gp100 complex and the gH external domain is the target for neutralizing monoclonal antibodies. *Virology* **197**, 12-22.

Liu, D. X., Gompels, U. A., Nicholas, J., and Lelliott, C. 1993b. Identification and expression of the human herpesvirus 6 glycoprotein H and interaction with an accessory 40K glycoprotein. *J. Gen. Virol.* **74**, 1847-1857.

Mackett, M., Conway, M. J., Arrand, J. R., Haddad, R. S., and Hutt-Fletcher, L. M. 1990. Characterization and expression of a glycoprotein encoded by the Epstein-Barr virus *Bam*HI 1 fragment. *J. Virol.* **64**, 2545-2552.

Martin, D. R., Marlowe, R. L., and Ahearn, J. M. 1994. Determination of the role for CD21 during Epstein-Barr virus infection of B lymphoblastoid cells. *J. Virol.* **68**, 4716-4726.

Martin, D. R., Yuryev, A., Kalli, K. R., Fearon, D. T., and Ahearn, J. M. 1991. Determination of the structural basis for selective binding of Epstein-Barr virus to human complement receptor type 2. *J. Exp. Med.* **174**, 1299-1311.

Masucci, M. G., Szigeti, R., Ernberg, I., Hu, C. P., Torsteindottir, S., Frade, R., and Klein, G. 1987. Activation of B lymphocytes by Epstein-Barr virus/CR2 receptor interaction. *Eur. J. Immunol.* **17**, 815-820.

McGeoch, D. J., Dalrymple, M. A., A.J., D., Dolan, A., M.C., F., McNab, D., L.J., P., Scott, J. E., and Taylor, P. 1988. The complete DNA sequence of the unique long region in the genome of herpes simplex virus type 1. *J. Gen. Virol.* **69**, 1531-1574.

Miller, N., and Hutt-Fletcher, L. M. 1988. A monoclonal antibody to glycoprotein gp85 inhibits fusion but not attachment of Epstein-Barr virus. *Journal of Virology* **62**, 2366-2372.

Miller, N., and Hutt-Fletcher, L. M. 1992. Epstein-Barr virus enters B cells and epithelial cells by different routes. *J. Virol.* **66**(6), 3409-14.

Molesworth, S. J., Lake, C. M., Borza, C. M., Turk, S. M., and Hutt-Fletcher, L. M. 2000. Epstein-Barr virus gH is essential for penetration of B cell but also plays a role in attachment of virus to epithelial cells. *J. Virol.* **74**, 6324-6332.

Moore, M. D., Cooper, N. R., Tack, B. F., and Nemerow, G. R. 1987. Molecular cloning of the cDNA encoding the Epstein-Barr virus/C3d receptor (complement receptor type 2) of human B lymphocytes. *Proceedings of the National Academy of Sciences* **84**, 9194-9198.

Moore, M. D., DiScipio, R. G., Cooper, N. R., and Nemerow, G. R. 1989. Hydrodynamic, electron microscopic and ligand binding analysis of the Epstein-Barr virus/C3dg receptor (CR2). *J. Biol. Chem.* **34**, 20576-20582.

Nagar, B., Jones, R. G., Diefenbach, R. J., Isenman, D. E., and Rini, J. M. 1998. X-ray crystal structure of C3d: a C3d fragment and ligand for complement receptor 2. *Science* **280**, 1277-1281.

Nemerow, G. R., and Cooper, N. R. 1984. Early events in the infection of human B lymphocytes by Epstein-Barr virus. *Virology* **132**, 186-198.

Nemerow, G. R., Houghton, R. A., Moore, M. D., and Cooper, N. R. 1989. Identification of the epitope in the major envelope proteins of Epstein-Barr virus that mediates viral binding to the B lymphocyte EBV receptor (CR2). *Cell* **56**, 369-377.

Nemerow, G. R., Mold, C., Keivens Schwend, V., Tollefson, V., and Cooper, N. R. 1987. Identification of gp350 as the viral glycoprotein mediating attachment of Epstein-Barr virus (EBV) to the EBV/C3d receptor of B cells: sequence homology of gp350 and C3 complement fragment C3d. *J. Virol.* **61**, 1416-1420.

Nemerow, G. R., Mullen, J. J., Dickson, P. W., and Cooper, N. R. 1990. Soluble recombinant CR2 (CD21) inhibits Epstein-Barr virus infection. *J. Virol.* **64**, 1348-1352.

Nemerow, G. R., Siaw, M. F. E., and Cooper, N. R. 1986. Purification of the Epstein-Barr virus/C3d complement   receptor CR2 of human B lymphocytes: antigenic and functional properties of the purified protein. *Journal of Virology* **58**, 709-712.

Nemerow, G. R., Wolfert, R., McNaughton, M., and Cooper, N. R. 1985. Identification and characterization of the Epstein-Barr v irus receptor on human B lymphocytes and its relationship to the C3d complement receptor (CR2). *J. Virol.* **55**, 347-51.

Nolan, L. A., and Morgan, A. J. 1995. The Epstein-Barr virus open reading frame BDLF3 codes for a 100-150 kDa glycoprotein. *J. Gen. Virol.* **76**, 1381-1392.

Oba, D. E., and Hutt-Fletcher, L. M. 1988. Induction of antibodies to the Epstein-Barr virus glycoprotein gp85 with a synthetic peptide corresponding to a sequence in the BXLF2 open reading frame. *J. Virol.* **62**, 1108-1114.

Oda, T., Imai, S., Chiba, S., and Takada, K. 2000. Epstein-Barr virus lacking glycoprotein gp85 cannot infect B cells and epithelial cells. *Virology* **276**, 52-58.

Papworth, M. A., Van Dijk, A. A., Benyon, G. R., Allen, T. D., Arrand, J. R., and Mackett, M. 1997. The processing, transport and heterologous expression of Epstein-Barr virus gp110. *J. Gen. Virol.* **78**, 2179-2189.

Peeters, B., Dewind, N., Broer, R., Gielkins, A., and Moormann, R. 1992. Glycoprotein H of pseudorabies virus is essential for entry and cell-to-cell spread of the virus. *J. Virol.* **66**, 3888-3892.

Pellett, P. E., Biggin, M. D., Barrell, B., and Roizman, B. 1985. Epstein-Barr virus may encode a protein showing significant amino acid and predicted secondary structure homology with glycoprotein B of herpes simplex virus. *J. Virol.* **56**, 807-813.

Penaranda, M. E., Lagenaur, L. A., Pierek, L. T., Berline, J. W., MacPhail, L. A., Greenspan, D., Greenspan, J., and Palefsky, J. M. 1997. Expression of Epstein-Barr virus BMRF-2 and BDLF-3 genes in hairy leukoplakia. *J. Gen. Virol.* **78**, 3361-3370.

Peng, T., Ponce de Leon, M., Novotny, M. J., Jiang, H., Lambris, J. D., Dubin, G., Spear, P. G., Cohen, G., and Eisenberg, R. J. 1998. Structural and antigenic analysis of a truncated from of the herpes simplex virus glycoprotein gH-gL complex. *J. Virol.* **72**, 6092-6103.

Pulford, D. J., Lowrey, P., and Morgan, A. J. 1995. Co-expression of the Epstein-Barr virus BXLF2 and BKRF2 genes with a recombinant baculovirus produces gp85 on the cell surface with antigenic similarity to the native protein. *J. Gen. Virol.* **76**, 3145-3152.

Radsak, K., Eickmann, M., Mockenhaupt, T., Bogner, E., Kern, H., Eis-Hubinger, A., and Reschke, M. 1966. Retrieval of human cytomegalovirus glycoprotein B from the infected cell surface for virus envelopment. *Arch. Virol.* **141**, 557-572.

Rickinson, A. B., and Kieff, E. (1996). Epstein-Barr Virus. 3 ed. *In* "Fields Virology" (B. N. Fields, D. M. Knipe, and P. M. Howley, Eds.), Vol. 2, pp. 2397-2446. 2 vols. Lippincott-Raven, Philadelphia.

Roller, R. J., Zhou, Y., Schnetzer, R., Ferguson, J., and DeSalvo, D. 2000. Herpes simplex virus type 1 $U_L 34$ gene product is required for viral envelopment. *J. Virol.* **74**, 117-129.

Seigneurin, J.-M., Villaume, M., Lenoir, G., and deThe, G. 1977. Replication of Epstein-Barr virus: ultrastructural and immunofluorescent studies of P3HR1-superinfected Raji cells. *J. Virol.* **24**, 835-845.

Shiba, C., Daikoku, T., Goshima, H., Yamauchi, Y., Koiwai, O., and Nishiyama, Y. 2000. The UL34 gene product if herpes simplex virus type 2 is a tail-anchored type II membrane protein that is significant for virus envelopment. *J. Gen. Virol.* **81**, 2397-2405.

Sinclair, A. J., and Farrell, P. J. 1995. Host cell requirements for efficient infection of quiescent primary B lymphocytes by Epstein-Barr virus. *J. Virol.* **69**, 5461-5468.

Sixbey, J. W., Davis, D. S., Young, L. S., Hutt-Fletcher, L., Tedder, T. F., and Rickinson, A. B. 1987. Human epithelial cell expression of an Epstein-Barr virus receptor. *J. Gen. Virol.* **68**, 805-11.

Sixbey, J. W., and Yao, Q.-Y. 1992. Immunoglobulin A-induced shift of Epstein-barr virus tissue tropism. *Science* **255**, 1578-1580.

Spaete, R. R., Perot, K., Scott, P. I., Nelson, J. A., Stinski, M. F., and Pachl, C. 1993. Co-expression of truncated human cytomegalovirus gH with the UL115 gene product or the truncated human fibroblast growth factor receptor results in transport of gH to the cell surface. *Virology* **193**, 853-861.

Spriggs, M. K., Armitage, R. J., Comeau, M. R., Strockbine, L., Farrah, T., MacDuff, B., Ulrich, D., Alderson, M. R., Mullberg, J., and Cohen, J. I. 1996. The extracellular domain of the Epstein-Barr virus BZLF2 protein binds the HLA-DR beta chain and inhibits antigen presentation. *J. Virol.* **70**, 5557-5563.

Sugano, N., Chen, W., Roberts, M. L., and Cooper, N. R. 1997. Epstein-Barr virus binding to CD21 activates the initial viral promoter via NFκB induction. *J. Exp. Med.* **186**, 731-737.

Tanner, J., Weis, J., Fearon, D., Whang , Y., and Kieff, E. 1987. Epstein-Barr virus gp350/220 binding to the B lymphocyte C3d receptor mediates adsorption, capping and endocytosis. *Cell* **50**, 203-213.

Tanner, J., Whang, Y., Sample, J., Sears, A., and Keiff, E. 1988. Soluble gp350/220 and deletion mutant glycoproteins block Epstein-Barr virus adsorption to lymphocytes. *J. Virol.* **62**, 4452-4464.

Telford, E. A., Watson, M. S., Aird, H. C., Perry, J., and Davison, A. J. 1995. The DNA sequence of equine herpesvirus 2. *J. Mol. Biol.* **249**, 520-528.

Thomas, J. A., and Crawford, D. H. 1989. Epstein-Barr virus/complement receptor and epithelial cells. *Lancet* **2**, 449-450.

Wang, X., and Hutt-Fletcher, L. M. 1998. Epstein-Barr virus lacking glycoprotein gp42 can bind to B cells but is not able to infect. *J. Virol.* **72**, 158-163.

Wang, X., Kenyon, W. J., Li, Q. X., Mullberg, J., and Hutt-Fletcher, L. M. 1998. Epstein-Barr virus uses different complexes of glycoproteins gH and gL to infect B lymphocytes and epithelial cells. *J. Virol.* **72**, 5552-5558.

Weis, J. J., Toothaker, L. E., Smith, J. A., Weis, J. H., and D.T., F. 1988. Structure of the human B lymphocyte receptor for C3d and the Epstein-Barr virus and relatedness to other members of the family of C3/C4 binding proteins. *J. Exp. Med.* **167**, 1047-1066.

Weisman, H. F., Bartow, T., Leppo, M. K., Marsh, H. C. J., Carson, G. R., Concino, M. F., Boyle, M. P., Roux, K. H., Weisdeldt, M. L., and Fearon, D. T. 1990. Soluble human complement receptor type 1: in vivo inhibitor of complement suppressing post-ischemic myocardial inflammation and necrosis. *Science* **249**, 146-151.

Whealy, M. E., Card, J. P., Meade, R. P., Robbins, A. K., and Enquist, L. W. 1991. Effect of brefeldin A on alphaherpes virus membrane protein glycosylation and virus egress. *J. Virol.* **65**, 1066-1081.

Whitely, A., Bruun, B., Minson, T., and Browne, H. 1999. Effects of targeting herpes simplex virus type 1 gD to the endoplasmic reticulum and *trans*-Golgi network. *J. Virol.* **73**, 9515-9520.

Wyatt, R., and Sodroski, J. 1998. The HIV-1 envelope glycoproteins: fusogens, antigens, and immunogens. *Science* **280**, 1884-1888.

Yaswen, L. R., Stephens, E. B., Davenport, L. C., and Hutt-Fletcher, L. M. 1993. Epstein-Barr virus glycoprotein gp85 associates with the BKRF2 gene product and is incompletely processed as a recombinant protein. *Virology* **195**, 387-396.

Yefenof, E., Klein, G., Jondal, M., and Oldstone, M. B. A. 1976. Surface markers on human B- and T-lymphocytes. IX, Two color immunofluorescence studies on the association between EBV receptors and complement receptors on the surface of lymphoid cell lines. *Int. J. Cancer* **17**, 6923-7000.

Yefenof, E., Klein, G., and Kvarnung, K. 1977. Relationships between complement activation, complement binding and EBV absorption by human hematopoietic cell lines. *Cell. Immunol.* **31**, 225-233.

Yoshiyama, H., Imai, S., Shimizu, N., and Takada, K. 1997. Epstein-Barr virus infection of human gastric carcinoma cells: implication of the existence of a new virus receptor different from CD21. *J. Virol.* **71**, 5688-5691.

Young, L. S., Clark, D., Sixbey, J. W., and Rickinson, A. B. 1986. Epstein-Barr virus receptors on human pharyngeal epithelium. *Lancet* **1**, 240-242.

Young, L. S., Dawson, C. W., Brown, K. W., and Rickinson, A. B. 1989. Identification of a human epithelial cell surface protein sharing an epitope with the C3d/Epstein-Barr virus receptor molecule of B lymphocytes. *Int. J. Cancer* **43**, 786-794.

Zeng, Y., Zhang, L. G., Wu, Y. C., Huang, Y. S., Huang, N., Q,, Li, J. Y., Wang, B., Jiang, M. K., Fang, Z., and Meng, N. N. 1985. Prospective studies on nasopharyngeal carcinoma in Epstein-Barr virus IgA/VCA antibody-positive persons in Wuzhou City, China. *Int. J. Cancer* **36**, 545-547.

Zhu, Z., Gershon, M. D., Hao, Y., Ambron, R. T., Gabel, C. A., and Gershon, A. A. 1995. Envelopment of varicella-zoster virus: targeting of viral glycoproteins to the *trans*-Golgi network. *J. Virol.* **69**, 7951-7959.

Chapter 1.2

Structure and Function of Viral Glycoproteins in Membrane Fusion

Winfried Weissenhorn
European Molecular Biology Laboratory (EMBL),
B.P.181, 6 rue Jules Horowitz, 38042 Grenoble, France;
e-mail: weissen@embl-grenoble.fr

1. INTRODUCTION

1.1 Virus attachment and entry

Enveloped viruses bind specific cellular receptors via their surface glycoproteins, thus directly triggering either fusion of viral and cellular membranes or endocytosis of virus particles followed by membrane fusion in the endosomal compartment. Both receptor-binding or H^+ binding events are known to induce conformational changes in the surface glycoproteins, notably in the fusion domain subunit, which initiates the membrane fusion reaction; this leads to the subsequent delivery of the genetic material into the target cell, establishing an infection. This chapter focuses mainly on the structure and function of viral fusion proteins from members of four virus families, namely, Orthomyxoviridae, Retroviridae, Filoviridae and Paramyxoviridae.

Influenza viruses (Orthomyxoviridae) bind with low affinity to cellular receptors such as sialic acid groups on glycoproteins and glycolipids via their hemagglutinin (HA) glycoproteins (Sauter *et al.*, 1989). Binding of multiple HA molecules is necessary to mediate tight association and subsequent virus endocytosis into the endosomal compartment. The receptor binding site is highly conserved among all influenza virus subtypes despite their antigenic variation (Skehel and Wiley, 2000). There is evidence that Ebola viruses (Filoviridae) follow a similar pH-dependent track (Wool-Lewis and Bates, 1999), although the nature of the cellular receptors that

Structure-Function Relationships of Human Pathogenic Viruses, Edited by
Holzenburg and Bogner, Kluwer Academic/Plenum Publishers, New York, 2002

interact with their surface glycoprotein (GP) is still unknown. Retroviruses such as HIV-1 and HTLV-1 enter their target cells at the plasma membrane in a pH-independent manner. HIV-1 employs its glycoprotein gp160 to attach to the cellular receptor CD4 (Kwong *et al.,* 1998) which then allows a second interaction with a member of the chemokine receptor family (Berger *et al.,* 1999). Both subsequent interactions are necessary for the successful establishment of an infection. (Wyatt and Sodroski, 1998; Salzwedel *et al.,* 2000). Members of the Paramyxoviridae (SV5) also use sialic acid-containing cellular receptors for hemagglutinin-neuraminidase (HN) glycoprotein mediated attachment. Interaction of HN with the fusion glycoprotein (F) is thought to transfer signals that subsequently lead to membrane fusion mediated by F at the plasma membrane in a pH-independent mechanism. However, some strains also fuse membranes independently of HN (Paterson *et al.,* 2000 and references therein).

1.2   Processing of viral glycoproteins

Glycoproteins from Orthomyxoviridae (influenza virus, HA), Filoviridae (Ebola virus, Gp), Retroviridae (HIV-1, gp160; HTLV-1, gp62) and Paramyxoviridae (SV5, F protein) are synthesized as precursor molecules which are subsequently cleaved into two subunits, a receptor binding domain and a membrane-anchored fusion domain. Cleavage of the precursor is absolutely required for membrane fusion activity of most viral glycoproteins (Klenk *et al.,* 1975; Huang *et al.,* 1981; Klenk and Garten, 1994). A common feature of all glycoproteins is either a conserved multibasic cleavage recognition sequence R-X-K/R-R or a monobasic cleavage site, depending on the virus strain. The multibasic cleavage site of influenza virus HA (Garten *et al.,* 1981), SV5 F (Paterson *et al.,* 1984) HIV-1 gp160 (McCune *et al.,* 1988) and Ebola virus GP (Volchkov *et al.,* 1998) allows efficient intracellular processing by subtilisin-like enzymes, such as furin (Hallenberger *et al.,* 1992; Klenk and Garten, 1994), that show a wide tissue distribution and are normally active in the processing of hormone and growth factor precursors. In contrast, the presence of a monobasic cleavage site in the glycoprotein (Sendai virus F, Hidaka *et al.,* 1984; influenza virus HA, Gething et al., 1980) normally leads to extracellular cleavage-activation with a more restricted tissue distribution for virus replication (Klenk and Rott, 1988; Klenk and Garten, 1994). The efficiency of HA cleavage is also influenced by sequences surrounding the recognition site (Walker and Kawaoka, 1993). Cleavage of the precursor molecules produces either covalently linked (influenza virus HA1 and HA2; Ebola virus Gp1 and Gp2, SV5 F1 and F2) or non-covalently linked glycoproteins (HIV-1, gp120 and gp41; HTLV-1 gp46 and gp21). The oligomeric state is trimeric in all cases (Russell *et al.*, 1994; Weissenhorn *et al.*, 1999 and references therein). The endoproteolytic cleavage places a hydrophobic fusion peptide to the N-terminus of the fusion domain.

2. THREE CONFORMATIONS OF INFLUENZA VIRUS HEMAGGLUTININ

2.1 The native HA structure

The structure of influenza virus hemagglutinin (BHA, solubilized form virus by bromelain cleavage; Brand and Skehel., 1972), forms a homotrimer, with each monomer containing covalently linked subunits HA1 and HA2. The HA1 subunit contains the receptor binding site and the major antigenic determinants (Wiley and Skehel, 1987). The HA1 domain folds into an eight-stranded "Swiss roll" β-sheet structure that binds sialic acid groups at the top of the molecule, in a highly conserved binding pocket independently of the antigenic variation of viruses (Figure 1; Wilson *et al.*, 1981; Weis *et al.*, 1988; Skehel and Wiley, 2000). In addition, the N- and C-termini of HA1 interact with the stem of HA2 in an extended conformation. HA2, the fusion protein, anchors HA to the viral membrane and folds into a central triple stranded coiled coil structure (Figures 1 and 2 B, helices D and C) that is followed by a loop region (segment B) and an anti-parallel helix (helix A) which extends towards the N-terminal end. This arrangement buries the hydrophobic N-terminal fusion peptide in the trimer interface (Figures 1 and 2B).

2.2 The structure of the precursor HA$_0$

In the structure of uncleaved hemagglutinin, the cleavage into HA1 and HA2 has been prevented by mutating R329 to Q329, the monobasic cleavage site. The structure of HA$_0$ shows that only 19 residues around the cleavage site are in a conformation which is different from the one seen in the native cleaved HA structure (Chen *et al.*, 1998; Wilson *et al.*, 1981). This difference entails an outwards projection of the last residues of HA1 (323 to 329) and the N-terminal residues of HA2 (1-12), thus exposing the proteolytic cleavage site (Figure 2A). Upon cleavage, HA2 residues 1 to 10 fill a mostly negatively charged surface cavity adjacent to the cleavage site

(Chen *et al.,* 1998) which leads to the sequestering of the fusion peptide as seen in BHA (Wilson *et al.,* 1981; Figures 1 and 2B). The HA_0 structure reveals new evidence for the origin of HA sensitivity to low pH. In addition, it also contributes to the understanding of the correlation of HA_0 precursor cleavage and influenza virus pathogenesis, indicating possible modes of increased protease accessibility (Bosch *et al.,* 1989; Chen *et al.,* 1998; Garten and Klenk, 1999).

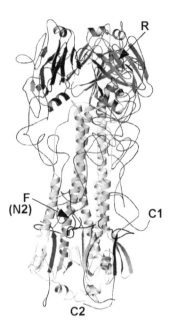

Figure 1. Ribbon diagram of bromelain solubilzed, cleaved influenza virus A/Hong Kong/68 BHA (Wilson *et al.,* 1981). N and C-termini are shown (C1 of HA1 and C2 of HA2). The position of fusion peptides (F) (N-terminus of HA2, residues 1 to 22 are coloured black; N2), buried in the trimer interface, and the receptor binding site (R) in HA1 are indicated by arrows for one monomer. Figures 1 to 3 have been prepared with the programs Molscript (Kraulis, 1991) and Raster 3D (Merrit and Bacon, 1997).

2.3   Cleavage produces metastable HA

Cleavage activates the fusion potential of HA and creates a metastable structure which undergoes a molecular switch triggered by the low pH (between pH 5 and pH 6) in endosomes (Maeda and Ohnishi, 1980; Huang *et al.,* 1981; White *et al.,* 1981; Skehel *et al.,* 1982). The same changes can be induced *in vitro* by temperature shifts and by urea treatment (Wharton et al. 1986; Ruigrok *et al.,* 1986; Carr *et al.,* 1997). The stability of the metastable conformation is also influenced by the presence of oligo-saccharides in the stem loop region of HA2 (Ohuchi *et al.,* 1997). Together, these experiments show that destabilization of the native metastable structure leads to the activation of the fusion potential. HA mutants with amino acid substitutions in the charged cavity which accommodates the fusion peptide after cleavage fuse membranes at a higher pH than wild-type HA (Daniels *et al.,* 1985; Lin *et al.,* 1997). These changes may affect the priming of HA and thus lower the threshold of activation (Skehel and Wiley, 2000). In addition, other mutations have been described that destabilize the location of the N-terminus of HA2 and may thus facilitate the extrusion of the N-terminus at higher pH, demonstrating again increased HA metastability (Daniels et al., 1985; Weis et al., 1990).

2.4 The low-pH conformation of HA

The activation of fusion of viral and endosomal membranes is accompanied by irreversible structural changes in HA (Skehel *et al.,* 1982). However, not only does membrane-anchored HA undergo structural rearrangements upon lowering the pH, but so does HA released from membranes by bromelain treatment (BHA; Brand and Skehel, 1972). Experiments *in vitro* showed that BHA aggregates through the exposure of the fusion peptides at low pH (Skehel *et al.,* 1982; Doms *et al.,* 1985). A change in structure was also detected by its increased sensitivity to proteolysis (Daniels *et al.,* 1985; Ruigrok *et al.,* 1988). This allowed the generation of a protease resistant core of HA2 by cleavage with thermolysin and trypsin resulting in the crystallizable fragment TBHA2 (HA1 residues 1 to 27; HA2 residues 38 to 175; Bullough *et al.,* 1994).

The crystal structure of TBHA2 shows three major changes compared to the neutral pH metastable structure of BHA (Figure 2B and C; Bullough *et al.,* 1994; Wilson et al. 1981). (i) A loop region (residues 55 to 76) refolds into a helix (segment B in BHA) and extends the central triple stranded coiled coil in a process that projects the fusion peptide approximately 100 Å away from its buried position in BHA. The potential of the B segment to become helical had been suggested by sequence analysis (Ward and Dopheide, 1980; Chambers *et al.,* 1990) and had been experimentally shown by peptide studies (Carr and Kim, 1993). (ii) Another dramatic change occurs between helices C and D where a short fragment unfolds to form a reverse turn which positions helix D anti-parallel against the central helices ABC. (iii) The chain reversal repositions the small β-sheet hairpin (E and F)

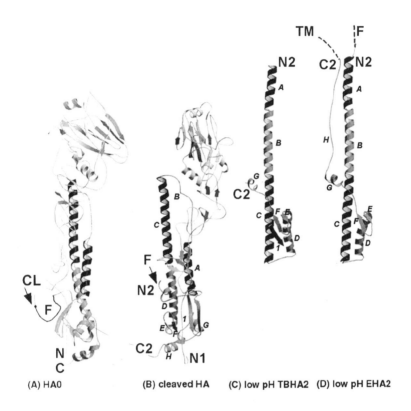

Figure 2. Ribbon diagrams of three conformations of HA. Only monomers are shown for clarity. (A) Uncleaved precursor R329Q HA$_0$ (Chen *et al.*, 1998); the position of the cleavage site is indicated by an arrow and a black circle (CL); the exposed fusion peptide sequence is colored black (F); (B) cleaved BHA (Wilson *et al.*, 1981); The position of the fusion peptide sequence (F) is indicated by an arrow. (A) is rotated by approximately 80° along the central coiled coil compared to (B) to show the exposed loop containing the furin cleavage site and the fusion peptide sequence, which is buried in the interface of the trimer in cleaved BHA (see Figure 1). (C) Low-pH conformation of thermolysin-solubilized TBHA2 (Bullough *et al.*, 1994); (D) Low-pH conformation of *E. coli* expressed EHA2 (Chen *et al.*, 1999). The N- and C-termini are indicated and secondary structure segments of HA2 are labeled according to Bullough *et al.*, (1994). Dashed lines indicate a potential path of the N- and C-terminal ends (F, fusion peptide and TM, transmembrane anchor) in EHA2.

and helix G in an extended structure packed against the central coiled coil. The TBHA2 structure also contains a short β-strand derived from HA1 (residues 10 to 17) that is covalently linked to β-strand F of HA2. The low-pH induced conformational change reverses the direction of this segment of HA1 from pointing up (towards the receptor-binding domain HA2 in BHA;

Wilson et al. 1981) to pointing into the opposite direction in TBHA2 (Figure 2C, β-strand labeled 1; Bullough *et al.,* 1994). Therefore, the connecting regions between HA1 and HA2 may also alter conformation during low-pH treatment. However, the structures of the globular heads (HA1) do not change upon acidification (Bizebard *et al.,* 1995). It has been suggested that the structural changes in HA2 serve to project the fusion peptide towards the target membrane and bend the molecule into two halves. This then places the fusion peptide and the transmembrane region to the same end of a rod-shaped molecule, resulting in a conformation which has been proposed to be crucial for membrane fusion (Bullough *et al.,* 1994; Weissenhorn *et al.,* 1997).

Expression of a construct containing HA2 residues 38 to 175 in *E. coli* (EBHA2) resulted in a structure which was biochemically indistinguishable from TBHA2 produced by low-pH treatment and proteolysis of viral HA (Ruigrok *et al.,* 1986; Chen *et al.,* 1995). Indeed, the X-ray structure of a related molecule EHA2, (containing residues 23 to 185, produced in *E. coli*), confirmed the same structure for EBHA2 as for TBHA2 (Chen *et al.,* 1999). These studies further proved that cleaved HA is metastable and that the interactions between HA1 and HA2 trap HA2 behind a free-energy barrier in a neutral-pH conformation. The spontaneous folding of HA2 into the low-pH conformation suggests therefore that it represents the lowest-energy state of the molecule (Bullough *et al.,* 1994; Chen *et al.,* 1995; Skehel and Wiley, 2000).

The structure of EHA2 (residues 23 to 185) also shows that most of the C-terminal sequences are ordered, when compared to the TBHA2 (residues 38 to 175) structure. Here, the C-terminal region packs in anti-parallel fashion into a groove between α-helices of the central coiled coil in an extended conformation (Chen *et al.,* 1999; Figure 2D). The structure ends in a N cap (Presta and Rose, 1988; Richardson and Richardson, 1988), whereas residues 34 to 37 terminate and stabilize the central helices and make hydrogen bond contacts to the last ordered residues 174 to 178 of the C-terminal end. This arrangement places the C-terminus of EHA2 right next to the N-terminal end in the rod-like structure (Chen *et al.,* 1999; Figures 2D and 3A).

The structures of TBHA2, EBHA2 and EHA2 are extremely thermo-stable and do not bind lipid bilayers *in vitro* (Ruigrok *et al.,* 1988; Chen *et al.,* 1995; Chen *et al.,* 1998). However, engineering a hydrophilic FLAG sequence N-terminal to the fusion peptide of full-length extracellular HA2 (F185; residues 1- 185) resulted in a soluble trimeric molecule that bound to liposomes upon proteolytic removal of the FLAG peptide. The hydrophobic behaviour could be also reversed by removal of the fusion peptide through thermolysin treatment, as previously shown (Daniels *et al.,* 1985; Chen *et al.,* 1998). These results confirmed that solely the fusion peptide interacts directly with the lipid bilayer and are consistent with the finding that only the fusion peptide sequence becomes labelled by hydrophobic photolabels during low-pH HA induced membrane fusion processes (Durrer *et al.,* 1996).

3.   COMPARISON OF VIRAL FUSION PROTEIN STRUCTURES

3.1   Common structural features

The discovery that influenza virus HA2 folds spontaneously into the low-pH conformation in the absence of the receptor-binding domain (Chen et al. 1995) placed the analysis of fusion proteins from other virus families onto a rational platform. Structural similarities between viral membrane fusion proteins were initially based on the prediction of potential coiled coil regions in the membrane fusion domain adjacent to the hydrophobic fusion peptide (Gallaher *et al.,* 1989; Chambers *et al.,* 1990; Gallaher, 1996). Biochemical work, showing protease resistant thermostable α-helical core structures that fold into rod-like molecules detected by electron microscopy, provided further evidence for structural similarities between membrane fusion domains derived from HIV-1 (Weissenhorn *et al.,* 1996; 1997; Blacklow *et al.,* 1995; Lu *et al.,* 1995), Ebola virus (Weissenhorn *et al.,* 1998), Simian parainfluenza virus 5 (SV5) (Joshi *et al.,* 1998) and influenza virus HA2 (Ruigrok *et al.,* 1988; Chen *et al.,* 1995).

Structures of fusion protein domains from MoMuLv TM (Fass *et al.,* 1996), HIV-1 gp41 (Weissenhorn *et al.,* 1997; Chan *et al.,* 1997; Tan *et al.,* 1996), SIV-1 gp41 (Caffrey *et al.,* 1998; Malashkevich et al. 1998), Ebola virus GP2 (Weissenhorn *et al.,* 1998; Malashkevich et al. 1999), HTLV-1 gp21 (Kobe *et al.,* 1999) and paramyxoviruses SV5 F1 (Baker *et al.,* 1999) and human RSV F1 (Zhao *et al.,* 2000) have three major structural similarities in common with the low-pH conformation of influenza virus HA2 (Bullough *et al.,* 1994; Chen *et al.,* 1999; Figure 3): (i) The central part of the rod-like structures is formed by a triple stranded coiled coil domain which varies in length among the different virus families. (ii) The chain reverses at the end of the rod and the C-terminal sequences pack in anti-parallel fashion with respect to the central coiled coil structure. (iii) This arrangement leads to the positioning of the N-terminus (fusion peptide) right next to the C-terminus (membrane anchor) at one end of the rod.   The structures suggest how the fusion molecules can be anchored in both the target and the viral membrane at a pre-fusion state by placing the rod-like structures in parallel orientations in between two membranes. This has been suggested to be instrumental for the close apposition of membranes and may also provide the driving force for bilayer fusion (Weissenhorn *et al.*, 1997).

Interestingly, the structure of a core SNARE (soluble NSF attachment receptor) complex involved in synaptic vesicle fusion consists of a four-helical bundle with membrane anchors of the v-SNARE (vesicle) and the t-SNARE (target membrane) at the same end of the rod (Figure 3G; Sutton *et al.*, 1998). There, it has been shown that this core complex is fusion-active *in vitro* (Nickel *et al.*, 1999; Parlati *et al.*, 1999). The similarity of the coiled coil structures and especially the positioning of both membrane anchors at the same end of the rod suggests that viral fusion molecules and SNARE assemblies may mediate the fusion of lipid bilayers in a similar way (Skehel and Wiley, 2000).

3.2   Structure of HIV-1 gp41

The core gp41 structure (as determined by Weissenhorn et al. 1997) contains a central triple stranded coiled coil with 15 layers of homotrimeric interactions followed by a linker region (which is not present in the structure) and subsequent 6 layers of pseudo-heterotrimeric helical interactions packing in anti-parallel fashion into conserved grooves made up by two central helices. The structure lacks 31 residues towards the N-terminal end and 18 residues towards the membrane anchor (Figure 3B). The importance of the placement of the fusion peptide in close proximity to the membrane anchor was first recognized in the HIV-1 gp41 structure. There, the theoretical distance which could be covered by the 18 missing residues in the direction of the transmembrane anchor was observed to be too short to overcome the necessary distance of 75 Å which could then place the membrane anchor at the opposite end of the fusion peptide position (Weissenhorn *et al.*, 1997). The HIV-1 gp41 structure also lacks 39 residues connecting a central helix to an outer layer which would contain a short conserved disulfide linked loop region and two potential carbohydrate sites. A structure of SIV gp41, solved by NMR, showed that the loop region, missing in all HIV-1 gp41 crystal structures, extends at the end of the rod, opposite to the fusion peptide position (Caffrey *et al.*, 1998). This further substantiated the assumption that the C-terminal end of gp41 extends into the same direction as the hydrophobic fusion peptide. There is biochemical evidence that the loop region interacts non-covalently with the receptor binding domain, gp120 (Cao *et al.*, 1993; Moore *et al.*, 1994; Wyatt *et al.*, 1997). This attachment could provide a path for signal transduction from gp120 to gp41 upon receptor binding. Differential antibody reactivity provided indirect evidence that this loop region may change its conformation

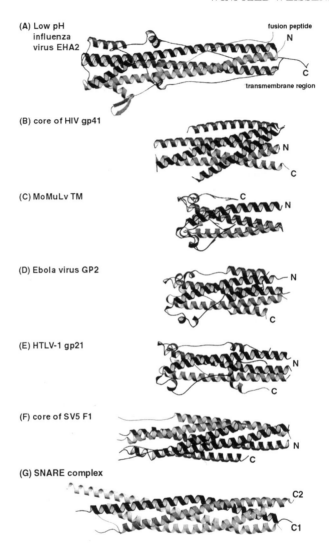

Figure 3. Core structures of viral membrane fusion proteins. The positions of the N- and C-termini at the same end of the trimeric helical hairpin structures are indicated by N (likely positions of fusion peptide sequences) and C (suggested positions of transmembrane anchors). (A) low pH influenza virus EHA2 (Chen *et al.,* 1999); (B) protease resistant core of HIV-1 gp41 (Weissenhorn *et al.,* 1997 and references therein); (C) recombinant fragment of Moloney murine leukemia virus TM (Fass *et al.,* 1996); (D) recombinant fragment of Ebola virus GP2 (Weissenhorn *et al.,* 1998); (E) recombinant fragment of HTLV-1 gp21 (Kobe *et al.,* 1999); (F) recombinant fragment of a protease resistant core of SV5 F1 (Baker *et al.,* 1999). (G) Recombinant synaptic fusion complex containing the v-SNARE synaptobrevin and the t-SNAREs syntaxin and SNAP-25B. The C-terminal anchors of syntaxin 1A (C1) and synaptobrevin (C2) are indicated (Sutton *et al.,* 1998).

in a proposed transition from a native gp41 conformation in complex with gp120 to the rod-like conformation observed in the crystal structure (Weissenhorn *et al.,* 1996).

Further evidence for two conformations of gp41, analogous to the two conformations of influenza virus HA2, came from the following findings. (i) A number of mutagenesis studies showed that residue changes within the central coiled coil structure affect infectivity and membrane fusion, but do not alter processing and cell surface expression of gp120/g41 complexes (Dubay *et al.,* 1992; Cao *et al.,* 1993; Chen *et al.,* 1993). (ii) Receptor-binding has been shown to increase the exposure of gp41 epitopes (Sattentau and Moore, 1991). (iii) Peptides derived from gp41 sequences, corresponding either partly or completely to the inner triple stranded coiled coil or to the outer helix, have been shown to have potent anti-viral activity (Jiang *et al.,* 1992; Wild *et al.,* 1992; 1994). The gp41 structure provided further evidence that the peptides exert their action during a conformational change (Chen *et al.,* 1995). Indeed it has been shown that a C-terminal peptide DP-178 (Wild *et al.,* 1992) does not interact with native gp41 (in complex with gp120) but associates with gp41 after the induction of receptor-mediated conformational changes (Furuta et al. 1998). Together, these data provide strong evidence that the crystal structure of gp41 represents a conformation similar to the low-pH fusion conformation of influenza virus HA2 but different from a yet unknown "native" gp41 conformation.

3.2   Structures from MoMuLv, Ebola virus and HTLV-1

The structures of a fragment of the MoMuLV TM (Figure 3C; Fass *et al.,* 1996), Ebola virus GP2 (Figure 3D; Weissenhorn *et al.,* 1998) and HTLV-1 gp21 (Figure 3E; Kobe *et al.,* 1999) show a high degree of resemblance, despite their low sequence similarity. They contain the same loop structure that brings about the chain reversal at one end of the rod-like structure, in all cases characterized by a conserved CX_6C motif. This motif is thought to provide a linker to the receptor-binding domain either covalently or non-covalently. Therefore, the loop structure has been proposed to not only reverse the chain direction but also to transduce potential signal(s) from the receptor-binding domain to the fusion domain upon receptor–mediated conformational changes (Weissenhorn *et al.,* 1998). The core structures of MoMuLV TM, Ebola Gp2 and HTLV-1 gp21 contain a central coiled coil containing 8, 10 and 12 layers of homotrimeric interactions, respectively. They also differ in the length of their C-terminal anti-parallel helical segments that form the outer layers (missing in the MoMuLv TM structure). In addition, the N- and C-terminal regions which link the coiled coil to the fusion peptides and transmembrane anchors (which are missing in the structures), also have different lengths.

Another similarity between the structures of MoMuLv TM, Ebola GP2 and HTLV-1 gp21 is the presence of a conserved chloride ion in the central coiled coil. This feature of an ionic interaction in the coiled coil packing is comparable to polar interactions formed by a layer of glutamines in HIV-1 gp41 (Weissenhorn *et al.,* 1997; Chan *et al.,* 1997). The polar interaction has been suggested to be important in MoMuLv TM (Fass *et al.,* 1996), like other polar interactions in coiled coil structures that promote specificity in their folding (Oakley *et al.,* 1998). It is therefore conceivable that the polar interactions observed in the crystal structures of viral fusion proteins could be important for their folding or even their refolding triggered by receptor-binding induced conformational changes.

Endoproteolytic cleavage of the glycoprotein precursors places the fusion peptides from HTLV-1 and MoMuLv molecules at the N-terminal end and that of Ebola virus GP to an internal position, probably similar to the one present in the Avian sarcoma virus (ALSV) glycoprotein (Gallaher *et al.,* 1996). It has been suggested that the N-terminal part of Ebola GP2 (that is missing in the structure) may fold back towards the central coiled coil structure via a disulfide bond between a cysteine positioned close the to endoproteolytic cleavage site and one which is accessible at the beginning of the core coiled coil (Weissenhorn *et al.,* 1998). The formation of this loop-like structure driven by a conserved proline residue (Hernandez and White, 1998; Delos and White, 2000) could again thus place the fusion peptide at the end of the rod (Weissenhorn *et al.,* 1998). The mode of action of the Ebola virus fusion peptide may be similar to that of other known internal fusion peptides including those from rabies virus and VSV (Durrer *et al.,* 1995) and from flavivirus (Rey *et al.,* 1996).

3.4   F1 structure from SV5

The structure of the SV5 F1 core fragment contains a long (17 layers) central triple stranded coiled coil followed by an anti-parallel stretch of residues in an extended conformation, as well as 8 helical turns packing into a groove made up by two central helices (Figure 3F; Baker et al. 1999). The structure represents a fusion domain core and peptides thereof are potent membrane fusion inhibitors *in vitro* (Rapaport *et al.,* 1995; Yao and Compans, 1996; Joshi *et al.,* 1998), whose inhibiton mechanism may be comparable to those derived from HIV-1 gp41 (Doms and Moore, 2000). The F1 core structure also contains potential ions buried in the hydrophobic core, possibly coordinated by conserved asparagine residues as seen in the structures of MoMuLV TM, Ebola GP2 and HTLV gp21 (Baker *et al.*, 1999). However, the main difference to the other fusion protein structures is the length of the intervening loop region connecting the inner core and the outer layers. In SV5 F the "loop" region constitutes a separate domain of 250 residues, which is missing in the structure and has a yet unknown function. Another difference is that in F1 α-helices extend into the fusion peptide sequence at the N-terminus and all the way to the C-terminus, leaving only seven potentially unstructured residues leading towards the transmembrane anchor. The absence of major flexible linkers at the N- and C-termini in this case has been suggested to affect the membrane fusion process mediated by F1 in such a way as to result in a different mode of fusion, where bending of the fusion molecule brings the membranes into close apposition. This was proposed to induce bilayer fusion and subsequent final refolding of the molecule into the post-fusion conformation, as observed by crystallography (Baker *et al.*, 1999).

4. Implications for membrane fusion

4.1 Model for membrane fusion

A general model for membrane fusion (Weissenhorn *et al.*, 1997) has been suggested based on the structure of HIV-1 gp41, its overall similarity with the low-pH conformation of influenza virus HA2 (Bullough *et al.*, 1994), and the discovery of a spontaneous low-pH conformation fold of HA2 when expressed without the receptor-binding domain (Chen *et al.*, 1995). Viral glycoproteins project their receptor-binding domains towards the cellular membrane and interact with their respective receptors (Figure 4A). This leads to conformational changes in the glycoprotein either through direct receptor interactions (HIV-1; SV5; HTLV-1) or upon endocytosis followed by acidification in the endosomal compartment (influenza virus; Ebola virus). The conformational changes expose and extend the N-terminal fusion peptides that may then interact with cellular membranes (Figure 4B). Further refolding in the fusion proteins results in a conformation where the fusion peptide and the transmembrane anchors are juxtaposed at the same end of the rod-shaped molecule (Figure 4C). It is not known whether the two changes occur simultaneously or in stepwise fashion. However, it is conceivable, that as a consequence of the conformational rearrangements, the two membranes are pulled together and brought into close proximity with the fusion peptides anchored in the target membrane and the transmembrane regions in the viral membrane (Figure 4C). This process may require the recruitment of several trimers at the site of fusion (Blumenthal *et al.*, 1996;

Danieli *et al.,* 1996; Hernandez *et al.,* 1996; Kanaseki et al. 1997). The clustering of the hydrophobic ends of the trimeric molecules may provide points of initial membrane mixing and fusion pore formation (Spruce *et al.,* 1989; 1991). The distance between the two membranes could be as close as 15 Å, which could help to overcome the hydration force (Leikin *et al.,* 1993), a barrier to membrane fusion, and allow lipid mixing, pore formation and finally content mixing (Hernandez *et al.,* 1996). The conformational change may also move the covalently linked receptor binding domains out of the way by orienting them into the opposite direction of the hydrophobic anchors as seen in the TBHA2 structure (Figure 2C; Bullough *et al.,* 1994), a step which has been also proposed to take place in case of Ebola virus GP2-mediated fusion (Weissenhorn *et al.,* 1998). Non-coavlently linked receptor binding domains (HIV-1 gp120) may be therefore simply released at the site of fusion. At the end of the fusion process, the fusion protein is anchored in the same membrane by its fusion peptide and its transmembrane anchor, in a final inactive conformation (Figure 4F; Wharton *et al.,* 1995).

4.2   Function of fusion peptides

Viral fusion peptides are hydrophobic glycine-rich sequences that are essential for membrane fusion. The fusion peptide sequence is generally 20 to 30 amino acids long and is conserved within a virus family (Wiley and Skehel., 1987; Bosch *et al.,* 1989; Gallaher *et al.,* 1996). The central function of the fusion peptide is to attach and anchor the fusion protein to the target membrane. A secondary role may be the initial destabilization of lipid bilayers to promote lipid mixing. Importantly, it has been shown that only the fusion peptide sequence inserts into the target lipid bilayer during the influenza virus HA-mediated fusion process (Weber *et al.,* 1994; Durrer *et al.,* 1996).

Fusion peptide sequences contain glycines approximately spaced by four positions. The N-terminal glycine residue in the fusion peptide of HA2 seems to be of central importance, which can be only changed to alanine without loss of function (Steinhauer *et al.,* 1995; Qiao *et al.,* 1999). Otherwise, changes of glycine residues within the HA fusion peptide (except for that at position eight) had no apparent effect on the fusion activity (Steinhauer *et al.,* 1995). Deletion mutants also showed that the length of the fusion peptide is important for HA fusion activity (Steinhauer *et al.,* 1995) while the mutagenesis of charged residues within the fusion peptide can be tolerated (Gething *et al.,* 1986; Steinhauer *et al.,* 1995). In addition, glycine to alanine substitutions in the fusion peptide of SV5 F at positions three and seven dramatically increased the fusion activity (Horvath and Lamb, 1992).

Figure 4. Model for viral membrane fusion (modified from Weissenhorn *et al.,* 1997).

(A) Viral glycoproteins project their receptor binding domains towards cellular receptors.

(B) This interaction or subsequent exposure to low pH in the endosomes triggers a conformational change in the viral glycoprotein that extends the N-terminal fusion peptide (F) towards the target membrane.

(C) Simultaneously or subsequently to the fusion peptide exposure, the N-terminal parts and the C-terminal regions assemble into the core structures, as observed by crystallography. At this stage the fusion peptide (F) is anchored in the cellular membrane near the position of the viral transmembrane anchor (A). This arrangement leads to the close apposition of cellular and viral membranes. Flexible linkers at their N- and C-termini may account for the orientation of the helical hairpin structures between two bilayers and several trimers may aggregate at the initial fusion site (Blumenthal *et al.,* 1996; Danieli *et al.,* 1996; Hernandez *et al.,* 1996).

(D) The refolding procedure may be also accompanied by the induction of membrane curvature and stalk formation characterized by lipid mixing of the inner leaflet (Siegel 1993; Chernomordik *et al.,* 1995).

(E) This intermediate step may be followed by a hemifusion stage (Chernomordik *et al.,* 1998) which then leads to the final fusion pore opening and to complete membrane fusion (Melikyan *et al.,* 1997; Chernomordik *et al.,* 1998).

(F) At the post fusion stage, viral membrane fusion proteins are anchored in the same membrane with their membrane anchors (A) and their fusion peptides (F) (Skehel *et al.,* 1995; Wharton *et al.,* 1995).

(A)

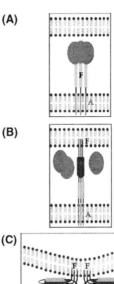

(B)

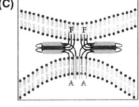

(C)

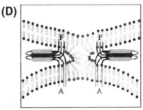

(D)

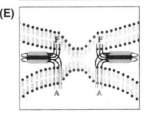

(E)

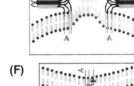

(F)

Changes within the fusion peptide may not only have an effect on its structure but may also alter the association of fusion peptides within a lipid bilayer during fusion (Baker *et al.,* 1999).

Synthetic fusion peptides have been shown to fuse liposomes *in vitro* (Lear and DeGrado, 1987; Wharton *et al.,* 1988; Rafalski *et al.,* 1990). However, the action of isolated peptides may differ from that of fusion peptides in context of the complete fusion molecule, as a number of un-related amphiphilic peptides fuse membranes *in vitro* efficiently. Fusion peptides derived from HA and HIV-1 assume flexible random conformations in solution and seem to become structured upon insertion into a membrane, with evidence for amphiphilic α-helices (Durell *et al.,* 1997; Gray *et al.,* 1996) and membrane insertion may occur in an oblique fashion (Bradshaw *et al.,* 2000).

Although it is generally thought that fusion peptides insert into the target membrane, their association with the viral membrane during fusion has been also proposed to be of functional importance (Bentz *et al.,* 1990; Gaudin *et al.,* 1995; Kozlov and Chernomordik, 1998). In addition, exposure of the fusion peptide by activation of the fusion potential without providing appropriate target membranes in close proximity leads to insertion of the fusion peptide into the viral membrane and thus to inactivation of the fusion potential.

4.3   Role of C-terminal linkers

The C-terminal ends of most viral fusion proteins have been suggested to be flexible because of their susceptibility to proteolysis or their disorder in protein crystals. The flexibility of this part was proposed to be of functional importance in orienting a fusion molecule core structure in between two membranes (Weissenhorn *et al.,* 1998). A conserved tryptophan-rich motif in HIV gp41 and the length of the linker towards the transmembrane region has been found to be important for membrane fusion (Salzwedel *et al.,* 1999; Munoz-Barroso et a., 1999), which is consistent with the proposed functional role. Similar results were derived from SNARE-mediated membrane fusion experiments where the linker region between the core structure and the membrane anchors of the SNAREs have been found to be absolutely crucial for successful membrane fusion *in vitro* (McNew *et al.,* 2000). In contrary, up to 4 residues can be deleted between the C-terminal anchor and the ectodomain of SV5 F1 with no effect on fusion, implying differences in the fusion mechanism (Baker *et al.,* 1999).

Flexible C-terminal linker regions may become ordered in full-length fusion proteins or by extension of their N-terminal ends. This becomes evident by comparison of the structure of TBHA2 (residues 38 to 175), where residues beyond 153 were mostly disordered (Bullough *et al.,* 1994) with a longer construct, EHA2, (residues 23 to 185; Figure 2 C, D). In EHA2, the C-terminal region is ordered and shows multiple interactions with the N-terminal core (Chen *et al.,* 1999). This could imply that the linkers might be initially flexible but become ordered structures during the fusion process or only in a final post-fusion stage (Weissenhorn *et al.,* 1998; Skehel and Wiley, 2000). Therefore, both the N- and C-termini of viral fusion proteins may interact with each other at a post-fusion stage and their assembly from flexibly linked ends to stable structures might be part of the force driving the fusion process.

4.4　Intermediates in membrane fusion

Artificial liposomes can be fused by polyethylene glycol (PEG) *in vitro* in a process which suggests two intermediate stages during membrane fusion: stalk formation and a hemifusion stage (Kozlov *et al.,* 1989; Siegel, 1993; Lee and Lentz, 1997). Stalk formation (Figure 4D) is characterized by local lipid mixing of the inner leaflet with negative membrane curvatures between monolayers (Chernomordik *et al.,* 1995). This step was suggested to be followed by a hemifusion stage, characterized by the merging of both outer leaflets of two opposing bilayers but no exchange of contents (Figure 4E; Chernomordik and Zimmerberg, 1995). Replacement of the HA transmembrane region by a phosphatidylinositol-glycan anchor has been show to block fusion at the hemifusion stage (Kemble *et al.,* 1994) and a similar fusion defect has been observed for GPI anchored HIV-1 gp160 (Salzwedel *et al.,* 1993). In addition, sub-optimal activation of HA led to the observation of unrestricted lipid mixing and transient pore flickering but not of complete membrane fusion after reapplying optimal fusion conditions, which suggested the formation of an unrestricted hemifusion diaphragm (Chernomordik *et al.,* 1998). Likewise, inhibitory peptides derived from HIV gp41 were shown to block membrane fusion at the hemifusion stage (Munoz-Barroso *et al.,* 1998). Viral membrane fusion proteins may participate in the stalk formation through their fusion peptide bilayer interactions, which may lead to the intermediate hemifusion stage (Colotto and Epand, 1997). Secondly, a ring-like fence formed by the aggregation of multiple low-pH activated fusion proteins (Blumenthal *et al.,* 1996; Danieli *et al.,* 1996) may promote the extension of the hemifusion diaphragm to the opening of a lipidic pore (Chernomordik *et al.,* 1998). The latter step therefore absolutely requires functional transmembrane regions (Melikyan *et al.,* 1995; Kemble *et al.,* 1994; Melikyan *et al.,* 1997). The formation of a physical pore has been observed electrophysiologically during HA-mediated membrane fusion processes (Spruce *et al.*, 1989; 1991; Zimmerberg *et al.*, 1994; Melikyan *et al.*, 1993). The initially formed pores showed similar flickering properties as have been measured during protein-free membrane fusion processes (Chanturiya *et al.*, 1997), indicating that the pore can open and close repeatedly before it becomes committed to full dilation.

The lipid composition of the bilayer can also influence the fusion reaction (Chernomordik *et al.*, 1995). Inverted cone-shaped lipids like lysophosphatidyl-choline in the outer monolayer form stalks inefficiently while their presence in the inner monolayer favourably promotes stalk and pore formation (Chernomordik *et al.*, 1995; 1998). Lysophosphatidyl-choline stabilizes positively curved monolayers and inhibits HA-driven fusion reactions (Chernomordik *et al.*, 1998) as well as rabies virus-induced membrane fusion *in vitro* (Gaudin, 2000), implying an important role for membrane curvature in the fusion mechanism.

4.5   Involvement of cytoplasmic domains

The cytoplasmic domain of HIV gp41 is not required for fusion (Gabuzda *et al.*, 1992; Wilk *et al.*, 1992). Similarly, cell fusion studies using chimeric HA molecules with exchanged transmembrane regions as well as engineered cytoplasmic tails had either no effect on the fusion activity (Schroth-Diez *et al.*, 1998) or some limited effect on the enlargement of the fusion pore (Kozerski *et al.*, 2000). In addition, changes in acylation of the carboxy-terminal end of HA had no dramatic effect on fusion although it seems to affect syncytia formation (Fischer *et al.*, 1998) and cause a restriction of virus replication (Jin *et al.*, 1996). In contrary, elongation and deletions in the cytoplasmic tail of hemagglutinin of fowl plaque virus affected its fusion activity in a virus system (Ohuchi *et al.*, 1998).

4.6   Energy provided for membrane fusion

The refolding of viral fusion molecules in general, as anticipated from the conformational transitions observed between native HA2 and HA2 in the low-pH conformation (Wilson et al. 1981; Bullough *et al.*, 1994), releases free energy which could be used at several steps of the membrane fusion process. (i) The refolding mechanism brings the viral and cellular membranes into close proximity by overcoming repelling forces (Rand, 1981). (ii) Interactions between fusion peptides and target membranes may participate in the bending of the outer monolayer, a process required for the formation of a negatively curved membrane structure (Razinkov *et al.,* 1998; Chernomordik *et al.,* 1999). (iii) The aggregation of multiple fusion proteins may provide additional energy to restrict fusion to the site of complex formation and to complete pore formation. (iv) The fusion peptides may disturb the lipid bilayer and thus lower its rupture tension. (v) Flexible linkers at the N-termini and at the C-termini may help to position the fusion molecule between two lipid bilayers; however, they may also become ordered at final stages of the fusion/refolding reaction and provide additional energy (Skehel and Wiley, 2000).

4.7 Alternative models

A number of alternative models for membrane fusion place the fusion protein structures at a post-fusion stage. There, intermediate conformations of the fusion proteins have been proposed to drive the fusion reaction which might be closely coupled to the major refolding transition (Kozlov and Chernomordik, 1998; Chernomordik *et al.,* 1998; 1999; Körte *et al.,* 1999; Melikyan *et al.,* 2000; Bentz, 2000; Bonnafous and Stegmann, 2000).

However, the resemblance of coiled coiled structures from viral fusion proteins and their cellular SNARE counterparts (Figure 3) indicates major similarities in their respective fusion reactions. In both cases, the helical coiled coil structures evolved to (i) bring two membranes into close proximity while (ii) providing the necessary energy for lipid mixing and fusion through their folding process. This strongly suggests a common membrane fusion mechanism as summarized here and elsewhere (Weissen-horn *et al.,* 1999; Skehel and Wiley, 1998; 2000; Jahn and Südhof, 2000). In intracellular fusion processes, the assembly of the correct SNARE complex is absolutely necessary for fusion (McNew *et al.,* 2000) and it has been shown that during neurotransmitter exocytosis SNARE complexes are assembled before $Ca^{(2+)}$-triggered membrane fusion occurs, indicating an important role for the core structure in fusion (Xu *et al.,* 1999). The final confirmation for the functionality of viral fusion protein structures in membrane fusion, however, needs to await the development of *in vitro* fusion assays performed with membrane anchored fusion proteins in the fusion-conformation as observed by crystallography.

4.8 Membrane fusion inhibitors

Peptides corresponding to an N-terminal region of gp41 (like DP-107) and to a C-terminal region (like DP-178 and T-20) were described as having potent anti-viral activity (Jiang *et al.*, 1992; Wild *et al.*, 1992; 1994). The early suggestion that they exert their activity during a receptor-induced conformational change (Chen *et al.*, 1995) and their ability to interact with leucine zipper structures (Wild *et al.*, 1995) is consistent with the structure of HIV-1 gp41 (Weissenhorn *et al.*, 1997; Chan *et al.*, 1997) for several reasons. (i) The assembled complex containing N- and C-terminal helices has no anti-viral activity (Lu *et al.*, 1995). (ii) gp41 is highly thermostable with a temperature dependent denaturation at approximately 80°C (Blacklow *et al.*, 1994; Lu *et al.*, 1995; Weissenhorn *et al.*, 1996; 1997), which makes it unlikely that the complex could come apart when pre-assembled. (iii) The individual peptides interact with gp41 only after receptor-induced conformation changes and show no binding to native gp41 (in complex with gp120) (Furuta *et al.*, 1998). Therefore it has been proposed that these peptides interfere with the conformational change in a dominant negative manner by binding to an intermediate "open" conformation of gp41 (see Figure 4B) and consequently blocking the formation of the helical hairpin structure (Figures 3B and 4C) (Chan *et al.*, 1997; Weissenhorn *et al.*, 1997; 1999). It is noteworthy that DP178 is still active when added after mixing of the target cells (Monoz-Barroso *et al.*, 1998) which may be reflected by the relatively slow kinetics of the induction of receptor-mediated conformational changes (Jones *et al.*, 1998). T-20, which blocks cell entry and fusion at a concentration of less than 2 ng/ml *in vitro*, has also been shown to reduce the viral load in HIV-1 patients significantly, thus providing proof of principle (Kilby *et al.*, 1998).

Efforts are currently underway to develop specific inhibitors targeted to a prominent hydrophobic cavity within the N-terminal coiled coil region (Chan *et al.*, 1997). A cyclic D-peptide has been shown to interact with the proposed conserved cavity and block fusion *in vitro* at micromolar concentrations (Eckert *et al.*, 1999). Screening of a completely synthetic combinatorial library resulted in potential molecules which were also targeted to the same hydrophobic site in the N-terminal core, substituting binding of the C-terminal helix residues Trp117, Trp120 and Ile 124 (Ferrer *et al.*, 1999; Zhou *et al.*, 2000).

The gp41 sequence and especially the sequence forming the N-terminal triple stranded coiled coil, which is the target for peptides like DP-178 and T-20, is highly conserved among different virus clades (Myers *et al.*, 1995). The proposed drug target cavity in the N-terminal part of gp41 is conserved among all clades including HIV-2 and SIV (Chan *et al.*, 1997). Although it has been shown that viruses can develop resistance towards DP-178 *in vitro* (Rimsky *et al.*, 1998), the high sequence conservation of gp41 and the inhibitory effect at the virus cell entry level makes gp41 a prominent and promising new drug target to treat HIV-1 infections.

The principle of membrane fusion inhibition by peptides corresponding to the coiled coil structures of the fusion proteins has been extended to other virus families. These include peptides derived from the glycoproteins from HTLV-1 (Sagara *et al.,* 1996), and from paramyxoviruses SV5 (Rapaport *et al.,* 1995; Lamb *et al.,* 1999; Joshi *et al.,* 1998), human respiratory syncytial virus (Lambert *et al.,* 1996) and measles virus (Wild and Buckland, 1997). The proposed similar mode of action of these inhibitory peptides further substantiates the fact that enveloped viruses with a similar molecular architecture of their fusion protein domains induce membrane fusion by essentially the same mechanism (Skehel and Wiley, 1998; Weissenhorn *et al.,* 1999).

ACKNOWLEDGMENTS

The author wishes to thank Don C. Wiley for his continuous support and his stimulating and contagious enthusiasm for viral membrane fusion research during the time in his laboratory at Children's Hospital, Boston.

REFERENCES

Baker, K.A,, Dutch, R.E., Lamb, R.A., and Jardetzky TS., 1999, Structural basis for paramyxovirus-mediated membrane fusion. *Mol. Cell* 3:309–319.

Bentz, J., Ellens, H., and Alford, D., 1990, An architecture for the fusion site of influenza hemagglutinin. *FEBS Lett.* 276:1-5.

Bentz, J., 2000, Membrane fusion mediated by coiled coils: a hypothesis. *Biophys. J.* 78:886-900.

Berger, E.A., Murphy, P.M., and Farber, J.M., 1999, Chemokine receptors as HIV-1 coreceptors: roles in viral entry, tropism, and disease. *Annu. Rev. Immunol.* 17:657-700.

Bizebard, T., Gigant, B., Rigolet, P., Rasmussen, B., Diat, O., Bosecke, P., Wharton, S.A., Skehel, J.J., and Knossow, M., 1995, Structure of influenza virus haemagglutinin complexed with a neutralizing antibody. *Nature* 376:92-94.

Blacklow, S. C., Lu, M., and Kim, P. S.,1995, A trimeric subdomain of the simian immunodeficiency virus envelope glycoprotein. *Biochemistry,* 34:14955-14962.

Blumenthal, R., Sarkar, D. P., Durell, S., Howard, D. E. and Morris, S. J., 1996, Dilation of the influenza hemagglutinin fusion pore revealed by the kinetics of individual cell-cell fusion events. *J. Cell Biol.* 135:63-71.

Bonnafous, P., and Stegmann, T., 2000, Membrane perturbation and fusion pore formation in influenza hemagglutinin-mediated membrane fusion. A new model for fusion. *J. Biol. Chem.* 275:6160-6166.

Bosch, F.X., Garten, W., Klenk, H.D., and Rott, R., 1981, Proteolytic cleavage of influenza virus hemagglutinins: primary structure of the connecting peptide between HA1 and HA2 determines proteolytic cleavability and pathogenicity of Avian influenza viruses. *Virology* 113:725–735.

Bosch, M. L., Earl, P. L., Fargnoli, K., Picciafuoco, S., Giombini, F., Wong-Staal, F. and
    Franchini, G., 1989, Identification of the fusion peptide of primate immunodeficiency
    viruses. *Science* 244:694-697.
Bradshaw, J.P., Darkes, M. J., Harroun, T. A., Katsaras, J., and Epand, R.M., 2000, Oblique
    membrane insertion of viral fusion peptide probed by neutron diffraction. *Biochemistry*
    39:6581-6585.
Brand, C.M., and Skehel, J.J., 1972, Crystalline antigen from the influenza virus envelope.
    Nat New Biol. 238:145-147.
Bullough, P.A., Hughson, F.M., Skehel, J.J. and Wiley, D.C., 1994, Structure of influenza
    haemagglutinin at the pH of membrane fusion. *Nature* 371:37-43.
Bullough, P.A., Hughson, F.M., Treharne, A.C., Ruigrok, R.W., Skehel, J.J., and Wiley, D.C.,
    1994, Crystals of a fragment of influenza haemagglutinin in the low pH induced
    conformation. *J. Mol. Biol.* 236:1262-1265.
Caffrey, M., Cai, M., Kaufman, J., Stahl, S.J., Wingfield, P.T., Covell, D.G., Gronenborn,
    A.M., and Clore, G.M., 1998, Three-dimensional solution structure of the 44 kDa
    ectodomain of SIV gp41. *EMBO J.* 17:4572–4584.
Cao, J., Bergeron, L., Helseth, E., Thali, M., Repke, H. and Sodroski, J., 1993, Effects of
    amino acid changes in the extracellular domain of the human immunodeficiency virus type
    gp41 envelope glycoprotein. *J. Virol.* 67:2747-2755.
Carr, C.M. and Kim, P. S., 1993, A spring-loaded mechanism for the conformational change
    of influenza hemagglutinin. *Cell* 73:823-832.
Carr, C. M., Chaudhry, C. and Kim, P.S., 1997, Influenza hemagglutinin is spring-loaded by a
    metastable native conformation. *Proc. Natl. Acad. Sci.* 94:14306-14313.
Chambers, P., Pringle, C.R. and Easton, A.J., 1990, Heptad repeat sequences are located
    adjacent to hydrophobic regions in several types of virus fusion glycoproteins. *J. Gen.
    Virol.* 71: 3075-3080.
Chan, D.C., Fass, D., Berger, J.M. & Kim, P.S., 1997, Core structure of gp41 from the HIV
    envelope glycoprotein. *Cell* 89:263-273.
Chanturiya, A., Chernomordik, L.V., and Zimmerberg, J., 1997, Flickering fusion pores
    comparable with initial exocytotic pores occur in protein-free phospholipid bilayers. *Proc.
    Natl. Acad. Sci. USA* 94:14423–14428.
Chen, J., Wharton, S. A., Weissenhorn, W., Calder, L. J., Hughson, F. M., Skehel, J. J. and
    Wiley, D. C., 1995, A soluble domain of the membrane-anchoring chain of influenza
    virus hemagglutinin (HA2) folds in Escherichia coli into the low-pH-induced
    conformation. *Proc. Natl. Acad. Sci. USA* 92:12205-12209.
Chen, C.H., Matthews, T.J., McDanal, C.B., Bolognesi, D.P., and Greenberg, M.L., 1995, A
    molecular clasp in the human immunodeficiency virus (HIV) type 1 TM protein
    determines the anti-HIV activity of gp41 derivatives: implication for viral fusion. *J. Virol.*
    69:3771-3777.
Chen, S. S., Lee, C. N., Lee, W. R., McIntosh, K., and Lee, T. H., 1993, Mutational analysis
    of the leucine zipper-like motif of the human immunodeficiency virus type 1 envelope
    transmembrane glycoprotein. *J. Virol.*, 67:3615-1619.
Chen, J., Lee, K.-H., Steinhauer, D.A., Stevens, D.J., Skehel, J.J., and Wiley, D.C., 1998.
    Structure of the hemagglutinin precursor cleavage site, a determinant of influenza
    pathogenicity and the origin of the labile conformation. *Cell* 95:409–417.
Chen, J., Skehel, J.J., and Wiley, D.C., 1998, A polar octapeptide fused to the N-terminal
    fusion peptide solubilizes the influenza virus HA2 subunit ectodomain. *Biochemistry*
    37:13643–13649.

Chen, J., Skehel, J.J., and Wiley, D.C., 1999, N- and C-terminal residues combine in the fusion-pH influenza hemagglutinin HA(2) subunit to form an N cap that terminates the triple-stranded coiled coil. *Proc. Natl. Acad. Sci. USA* **96**:8967–8972.

Chernomordik, L., Kozlov, M.M., and Zimmerberg, J., 1995, Lipids in biological membrane fusion. *J. Membr. Biol.* **146**:1–14.

Chernomordik, L., Chanturiya, A., Green, J., and Zimmerberg, J., 1995, The hemifusion intermediate and its conversion to complete fusion: regulation by membrane composition. *Biophys J.* **69**:922-929.

Chernomordik, L.V., and Zimmerberg, J., 1995, Bending membranes to the task: structural intermediates in bilayer fusion. *Curr Opin Struct Biol.* **5**:541-547.

Chernomordik, L., Frolov, V., Leikina, E., Bronk, P., and Zimmerberg, J., 1998, The pathway of membrane fusion catalyzed by influenza hemagglutinin: restriction of lipids, hemifusion, and lipidic fusion pore formation. *J. Cell Biol.* **140**:1369–1382.

Chernomordik, L.V., Leikina, E., Kozlov, M.M., Frolov, V.A, and Zimmerberg, J., 1999, Structural intermediates in influenza haemagglutinin-mediated fusion. *Mol. Membr. Biol.* **16**:33-42.

Colotto, A., and Epand, R.M., 1997, Structural study of the relationship between the rate of membrane fusion and the ability of the fusion peptide of influenza virus to perturb bilayers. *Biochemistry* **36**:7644–7651.

Danieli, T., Pelletier, S.L., Henis, Y.I. & White, J.M., 1996, Membrane fusion mediated by the influenza virus hemagglutinin requires the concerted action of at least three hemagglutinin trimers. *J. Cell Biol.* **133**:559-569.

Daniels, R.S., Downie, J.C., Hay, A.J., Knossow, M., Skehel, J.J., Wang, M.L., and Wiley, D.C., 1985. Fusion mutants of the influenza virus hemagglutinin glycoprotein. *Cell* **40**:431–439.

Delos, S.E., and White, J.M., 2000, Critical role for the cysteines flanking the internal fusion peptide of avian sarcoma/leukosis virus envelope glycoprotein. *J. Virol.* **74**:9738-9741.

Doms, R.W., Helenius, A., and White, J., 1985, Membrane fusion activity of the influenza virus hemagglutinin. The low pH-induced conformational change. *J. Biol. Chem.* **260**:2973-2981.

Doms, R.W., and Moore, J.P., 2000, HIV-1 membrane fusion. Targets Of opportunity. *J. Cell Biol.* **151**:F9-F14.

Dubay, J.W., Roberts, S.J., Brody, B., and Hunter, E., 1992, Mutations in the leucine zipper of the human immunodeficiency virus type 1 transmembrane glycoprotein affect fusion and infectivity. *J. Virol.* **66**:4748-4756.

Durell, S.R., Martin, I., Ruysschaert, J.M,, Shai, Y., and Blumenthal, R., 1997, What studies of fusion peptides tell us about viral envelope glycoprotein-mediated membrane fusion. *Mol. Membr. Biol.* **14**:97-112.

Durrer, P., Galli, C., Hoenke, S., Corti, C., Gluck, R., Vorherr, T., and Brunner, J., 1996, H+-induced membrane insertion of influenza virus hemagglutinin involves the HA2 amino-terminal fusion peptide but not the coiled coil region.*J. Biol. Chem.* **271**:13417–13421.

Eckert, D.M., Malashkevich, V.N., Hong, L.H., Carr, P.A., and Kim, P.S., 1999, Inhibiting HIV-1 entry: discovery of D-peptide inhibitors that target the gp41 coiled-coil pocket. *Cell* **99**:103-115

Fass, D., Harrison, S.C., and Kim, P.S.,1996, Retrovirus envelope domain at 1.7Å resolution. *Nat. Struct. Biol.* **3**:465-469.

Ferrer, M., Kapoor, T.M., Strassmaier, T., Weissenhorn, W., Skehel, J.J., Oprian, D., Schreiber, S.L., Wiley, D.C., and Harrison, S.C., 1999, Selection of gp41-mediated HIV-1 cell entry inhibitors from biased combinatorial libraries of non-natural binding elements. *Nat. Struct. Biol.* **6**:953-960.

Fischer, C., Schroth-Diez, B., Herrmann, A., Garten, and W., Klenk, H.D., 1998, Acylation of the influenza hemagglutinin modulates fusion activity. *Virology* **248**:284-294.

Furuta, R. A., Wild, C. T., Weng, Y., and Weiss, C. D., 1998, Capture of an early fusion-active conformation of HIV-1 gp41. *Nat. Struct. Biol.* **5**:276-279.

Gabuzda DH, Lever A, Terwilliger E, Sodroski J., 1992, Effects of deletions in the cytoplasmic domain on biological functions of human immunodeficiency virus type 1 envelope glycoproteins. *J. Virol.* **66**:3306-3315.

Gallaher, W. R., 1996, Similar structural models of the transmembrane proteins of Ebola and avian sarcoma viruses. *Cell* **85**: 477-478.

Gallaher, W. R., Ball, J. M., Garry, R. F., Griffin, M. C., and Montelaro, R. C., 1989, A general model for the transmembrane proteins of HIV and other retroviruses. *AIDS Res. Hum. Retroviruses,* **5**:431-440.

Garten, W., Bosch, F.X., Linder, D., Rott, R., and Klenk, H.D., 1981, Proteolytic activation of the influenza virus hemagglutinin: The structure of the cleavage site and the enzymes involved in cleavage. *Virology* **115**:361–734.

Garten, W., and Klenk, H.D., 1999, Understanding influenza virus pathogenicity.*Trends Microbiol.* **7**:99-100.

Gaudin, Y., Ruigrok, R. H. W. and Brunner, J., 1995, Low-pH induced conformational changes in viral fusion proteins: implications for the fusion mechanism. *J. Gen. Virol.* 76:1541-1556.

Gaudin, Y., 2000, Rabies virus-induced membrane fusion pathway. *J Cell Biol.* **150**:601-612.

Gething, M.J., Bye, J., Skehel, J., and Waterfield, M., 1980, Cloning and DNA sequence of double-stranded copies of haemagglutinin genes from H2 and H3 strains elucidates antigenic shift and drift in human influenza virus. *Nature* **287**:301-306.

Gething, M.J., Doms, R.W., York, D., and White, J., 1986., Studies on the mechanism of membrane fusion: site-specific mutagenesis of the hemagglutinin of influenza virus. *J. Cell Biol.* **102**:11–23.

Gray, C., Tatulian, S.A., Wharton, S.A., and Tamm, L.K., 1996, Effect of the N-terminal glycine on the secondary structure, orientation, and interaction of the influenza hemagglutinin fusion peptide with lipid bilayers. *Biophys. J.* **70**:2275-2286.

Hallenberger, S., Bosch, V., Angliker, H., Shaw, E., Klenk, H.D., and Garten, W., 1992, Inhibition of furin-mediated cleavage activation of HIV-1 glycoprotein gp160. *Nature* **360**:358-361.

Hernandez, L. D., Hoffman, L. R., Wolfsberg, T. G. and White, J. M., 1996, Virus-cell and cell-cell fusion. *Annu. Rev. Cell. Devel. Biol.* **12**:627-661.

Hernandez, L.D., and White, J.M., 1998, Mutational analysis of the candidate internal fusion peptide of the avian leukosis and sarcoma virus subgroup A envelope glycoprotein. *J. Virol.* **72**:3259-3267.

Hidaka, Y., Kanda, T., Iwasaki, K., Nomoto, A., Shioda, T., and Shibuta, H., 1984, Nucleotide sequence of a Sendai virus genome region covering the entire M gene and the 3' proximal 1013 nucleotides of the F gene. *Nucleic Acids Res.* **12**:7965-7973.

Horvath, C.M., and Lamb, R.A., 1992, Studies on the fusion peptide of paramyxovirus fusion glycoproteins: roles of conserved residues in cell fusion. *J. Virol.* **66**:2443-2455.

Huang, R.T., Rott, R., and Klenk, H.D., 1981, Influenza viruses cause hemolysis and fusion of cells. *Virology* **110**:243–247.

Jahn, R., and Südhof, T.C., 2000, Membrane fusion and exocytosis. *Annu. Rev. Biochem.* **68**:863-911.

Jiang, S., Lin, K., Strick, N., and Neurath, A. R., 1992, HIV-1 inhibition by a peptide. *Nature* **365**:113.

Jin, H., Subbarao, K., Bagai, S., Leser, G.P., Murphy, B.R., and Lamb, R.A., 1996, Palmitylation of the influenza virus hemagglutinin (H3) is not essential for virus assembly or infectivity. *J. Virol.* **70**:1406-1414.

Joshi, S.B., Dutch, R.E.. and Lamb, R.A., 1998, A core trimer of the Paramyxovirus fusion protein: Parallels to Influenza virus hemagglutinin and HIV-1 gp41. *Virology* **248**:20–34.

Jones, P., Korte, T., and Blumenthal, R., 1998, Conformational changes in cell surface HIV-1 envelope glycoprotein are triggered by cooperation between cell surface CD4 and coreceptors. *J. Biol. Chem.* **273**:404-409.

Kanaseki, T., Kawasaki, K., Murata, M., Ikeuchi, Y., and Ohnishi, S.-I., 1997, Structural features of membrane fusion between influenza virus and liposome as revealed by quick-freezing electron microscopy. *J. Cell Biol.* **137**:1041-1052.

Kemble, G.W., Danieli, T., and White, J.M., 1994, Lipid anchored influenza hemagglutinin promotes hemifusion, not complete fusion. *Cell* **76**:383-391.

Kilby, J.M., Hopkins, S., Venetta, T.M., DiMassimo, B., Cloud, G.A, Lee, J.Y., Alldredge, L., Hunter, E., Lambert, D., Bolognesi, D., Matthews, T., Johnson, M.R., Nowak, M.A., Shaw, G.M., and Saag, M.S., 1998, Potent suppression of HIV-1 replication in humans by T-20, a peptide inhibitor of gp41-mediated virus entry. *Nat. Med.* **4**:1302–1307.

Klenk, H.D., Rott, R., Orlich, M., and Blodorn, J., 1975, Activation of influenza A viruses by trypsin treatment. *Virology* **68**:426-439.

Klenk, H. D., and Rott, R., 1988, The molecular biology of influenza virus pathogenicity. *Adv. Virus Res.* **34**:247–281.

Klenk, H. D., and Garten, W., 1994, Host cell proteases controlling virus pathogenicity. *Trends Microbiol.* **2**:39–43.

Klenk, H. D., and Garten, W., 1994, Activation cleavage of viral spike proteins by host proteases. In *Cellular receptors for animal viruses* (Wimmer E., ed) Cold Spring Harbor Laboratory Press, pp241-280.

Kobe, B., Center, R.J., Kemp, B.E., and Poumbourios, P., 1999, Crystal structure of human T cell leukemia virus type 1 gp21 ectodomain crystallized as a maltose-binding protein chimera reveals structural evolution of retroviral transmembrane proteins.*Proc. Natl. Acad. Sci. USA* **96**:4319–4324.

Körte, T., Ludwig, K., Booy, F.P., Blumenthal, R., and Herrmann, A., 1999, Conformational intermediates and fusion activity of influenza virus hemagglutinin. *J. Virol.* **73**:4567-4574.

Kozerski, C., Ponimaskin, E., Schroth-Diez, B., Schmidt, M. F., and Herrmann, A., 2000, Modification of the cytoplasmic domain of influenza virus hemagglutinin affects enlargement of the fusion pore. *J. Virol.* **74**:7529-7537.

Kozlov, M.M., Leikin, S.L., Chernomordik, L.V., Markin, V.S., and Chizmadzhev, Y.A., 1989. Stalk mechanism of vesicle fusion. Intermixing of aqueous contents. *Eur. Biophys. J.* **17**:121-129.

Kozlov, M.M., and Chernomordik, L.V., 1998, A mechanism of protein-mediated fusion: coupling between refolding of the influenza hemagglutinin and lipid rearrangements. *Biophys. J.* **75**:1384–1396.

Kraulis, P.J., 1991, MOLSCRIPT: a program to produce both detailed and schematic plots of protein structures. *J. Appl. Crystallogr.* **24**:924–950.

Kwong, P. D., Wyatt, R., Robinson, J., Sweet, R.W., Sodroski, J. and Hendrickson, W.A., 1998, Structure of an HIV gp120 envelope glycoprotein in complex with the CD4 receptor and a neutralizing human antibody. *Nature* **393**:648-659.

Lamb, R.A., Joshi, S.B., and Dutch, R.E., 1999, The paramyxovirus fusion protein forms an extremely stable core trimer: structural parallels to influenza virus haemagglutinin and HIV-1 gp41. *Mol. Membr. Biol.* **16**:11–19.

Lambert, D.M, Barney, S., Lambert, A.L., Guthrie, K., Medinas, R., Davis, D.E., Bucy, T., Erickson, J., Merutka, G., and Petteway, S.R. Jr., 1996, Peptides from conserved regions of paramyxovirus fusion (F) proteins are potent inhibitors of viral fusion. *Proc. Natl. Acad. Sci. U S A* **93**:2186-2191.

Lear, J.D., and DeGrado, W.F., 1987, Membrane binding and conformational properties of peptides representing the NH2 terminus of influenza HA-2. *J. Biol. Chem.* **262**:6500-6505.

Lee, J., and Lentz, B.R., 1997, Outer leaflet-packing defects promote poly(ethylene glycol)-mediated fusion of large unilamellar vesicles. *Biochemistry* **36**:421-31.

Leikin, S., Parsegian, V.A., Rau, D.C., and Rand, R.P., 1993, Hydration forces. *Annu. Rev. Physical Chem.* **44**:369-395.

Lin, Y.P., Wharton, S.A., Martin, J., Skehel, J.J., and Wiley, D.C., and Steinhauer, D. A., 1997, Adaptation of egg-grown and transfectant influenza viruses for growth in mammalian cells: selection of hemagglutinin mutants with elevated pH of membrane fusion. *Virology* **233**:402–410.

Lu, M., Blacklow, S.C., and Kim, P.S., 1995, A trimeric structural domain of the HIV-1 transmembrane glycoprotein. *Nat. Struct. Biol.* **2**:1075-1082.

McCune, J.M., Rabin, L.B., Feinberg, M.B., Lieberman, M., Kosek, J.C., Reyes, G.R., Weissman, I.L., 1988, Endoproteolytic cleavage of gp160 is required for the activation of human immunodeficiency virus. *Cell* **53**:55-67.

Maeda, T., and Ohnishi, S., 1980, Activation of influenza virus by acidic media causes hemolysis and fusion of erythrocytes. *FEBS Lett.* **122**:283-287.

Malashkevich, V.N., Chan, D.C., Chutkowski, C.T., and Kim, P.S., 1998, Crystal structure of the simian immunodeficiency virus (SIV) gp41 core: conserved helical interactions underlie the broad inhibitory activity of gp41 peptides. *Proc Natl Acad Sci U S A.* **95**:9134-9139.

Malashkevich, V.N., Schneider, B.J., McNally, M.L., Milhollen, M.A., Pang, J.X., and Kim, P. S., 1999, Core structure of the envelope glycoprotein GP2 from Ebola virus at 1.9-A resolution. *Proc. Natl. Acad. Sci. USA* **96**:2662–2667.

McNew, J.A., Parlati, F., Fukuda, R., Johnston, R.J., Paz, K., Paumet, F., Sollner, T.H, and Rothman, J.E., 2000, Compartmental specificity of cellular membrane fusion encoded in SNARE proteins. *Nature* **407**:153-159.

McNew, J.A., Weber, T., Engelman, D.M., Sollner, T.H., and Rothman, J.E., 2000, The length of the flexible SNAREpin juxtamembrane region is a critical determinant of SNARE-dependent fusion. *Mol. Cell* **4**:415-421.

Melikyan, G.B., Niles, W.D., and Cohen, F.S., 1993, Influenza virus hemagglutinin-induced cell-planar bilayer fusion: quantitative dissection of fusion pore kinetics into stages *J. Gen. Physiol.* **102**:1151–1170.

Melikyan, G.B., White, J.M., and Cohen, F.S., 1995, GPI-anchored influenza hemagglutinin induces hemifusion to both red blood cell and planar bilayer membranes. *J. Cell Biol.* **131**:679–691.

Melikyan, G.B., Brener, S.A., Ok, D.C., and Cohen, F.S., 1997, Inner but not outer membrane leaflets control the transition from glycosylphosphatidylinositol-anchored influenza hemagglutinin-induced hemifusion to full fusion. *J. Cell Biol.* **136**:995–1005.

Melikyan, G.B., Markosyan, R.M., Hemmati, H., Delmedico, M.K., Lambert, D.M., Cohen, F.S., 2000, Evidence that the transition of HIV-1 gp41 into a six-helix bundle, not the bundle configuration, induces membrane fusion. *J. Cell. Biol.* **151**:413-24.

Merrit, E.A., and Bacon, D.J., 1997, Raster 3D photorealistic graphics. *Methods Enzymol.* **277**:505-524.

Myers, G., Hahn, B.H., Mellors, J.W., Henderson, L.E., Korber, B., Jeang, K.-T., McCutchan, F.E., and Pavlakis, G.N., 1995, Human Retroviruses and AIDS, published by

Theoretical Biology and Biophysics Group, Los Alamos (National Library, Los Alamos, NM, U.S.A.).

Munoz-Barroso, I., Durell, S., Sakaguchi, K., Appella, E., and Blumenthal, R., 1998, Dilation of the human immunodeficiency virus-1 enveloipe glycoprotein fusion pore revealed by the inhibitory action of a synthetic peptide from gp41. *J. Cell Biol.* **140**: 315-323.

Munoz-Barroso, I., Salzwedel, K., Hunter, E., and Blumenthal, R., 1999, Role of the membrane-proximal domain in the initial stages of human immunodeficiency virus type 1 envelope glycoprotein-mediated membrane fusion. *J. Virol.* **73**:6089-6092.

Nickel, W., Weber, T., McNew, J.A., Parlati, F., Sollner, T.H., and Rothman, J.E., 1999, Content mixing and membrane integrity during membrane fusion driven by pairing of isolated v-SNAREs and t-SNAREs. *Proc Natl Acad Sci U S A.* **96**:12571-12576.

Ohuchi, R., Ohuchi, M., Garten, W., and Klenk, H.D., 1997, Oligosaccharides in the stem region maintain the influenza virus hemagglutinin in the metastable form required for fusion activity. *J.Virol.* **71**:3719-3725.

Ohuchi, M., Fischer, C., Ohuchi, R., Herwig, A., and Klenk, H.D., 1998, Elongation of the cytoplasmic tail interferes with the fusion activity of influenza virus hemagglutinin. *J. Virol.* **72**:3554-3559.

Oakley, M.G., and Kim, P.S., 1998, A buried polar interaction can direct the relative orientation of helices in a coiled coil. *Biochemistry* **37**:12603-12610.

Paterson, R. G., Harris, T.J.R., and Lamb, R.A., 1984, Fusion protein of the paramyxovirus simian virus 5: Nucleotide sequence of mRNA predicts a highly hydrophobic glycoprotein. *Proc. Natl. Acad. Sci. USA* **81**:6706-6710.

Paterson, R. G., Russell, C. J., and Lamb, R.A., 2000, Fusion protein of the paramyxovirus SV5: destabilizing and stabilizing mutants of fusion activation. *Virology* **270**:17-30.

Parlati, F., Weber, T., McNew, J.A., Westermann, B., Sollner, T.H., and Rothman, J.E., 1999, Rapid and efficient fusion of phospholipid vesicles by the alpha-helical core of a SNARE complex in the absence of an N-terminal regulatory domain. *Proc Natl Acad Sci U S A* **96**:12565-12570.

Presta, L.G., and Rose, G.D., 1988, Helix signals in proteins. *Science* **240**:1632–1641.

Qiao, H., Armstrong, R.T., Melikyan, G.B., Cohen, F.S., and White, J.W., 1999, A specific point mutant at position 1 of the influenza hemagglutinin fusion peptide displays a hemifusion phenotype. *Mol. Biol. Cell* **10**:2759–2769.

Rand, R. P., 1981, Interacting phospholipid bilayers: measured forces and induced structural changes. *Annu. Rev. Biophys. Bioeng.* **10**:277–314.

Rapaport, D., Ovadia. M., and Shai, Y., 1995, A synthetic peptide corresponding to a conserved heptad repeat domain is a potent inhibitor of Sendai virus-cell fusion: an emerging similarity with functional domains of other viruses. *EMBO J.* **14**:5524-5531.

Rafalski, M., Lear, J. D., and DeGrado, W.F., 1990, Phospholipid interactions of synthetic peptides representing the N-terminus of HIV gp41. *Biochemistry,* **29**:7917-7922.

Razinkov, V.I., Melikyan, G.B., Epand, R.M., Epand, R.F., and Cohen, F.S., 1998, Effects of spontaneous bilayer curvature on influenza virus-mediated fusion pores. *J. Gen. Physiol.* **112**:409-422.

Rey, F.A., Heinz, F.X., Mandl, C., Kunz, C., and Harrison, S.C., 1995, The envelope glycoprotein from tick-borne encephalitis virus at 2 Å resolution. *Nature* **375**:291-298.

Richardson, J.S., and Richardson, D.C., 1988, Amino acid preferences for specific locations at the ends of alpha helices. *Science* **240**:1648–1452.

Rimsky, L.T., Shugars, D.C., and Matthews, T.J., 1998, Determinants of human immunodeficiency virus type 1 resistance to gp41-derived inhibitory peptides. *J. Virol.* **72**:986–993.

Ruigrok, R.W., Aitken, A., Calder, L.J., Martin, S.R., Skehel, J. J., Wharton, S.A., Weis, W. and Wiley, D.C., 1988, Studies on the structure of the influenza virus haemagglutinin at the pH of membrane fusion. *J. Gen. Virol.* **69**:2785-2795.

Ruigrok, R.W.H., Martin, S.R., Wharton, S.A., Skehel, J.J., Bayley, P.M., and Wiley, D.C., 1986, Conformational changes in the hemagglutinin of influenza virus which accompany heat-induced fusion of virus with liposomes. *Virology* **155**:484-497.

Russell, R., Paterson, R.G., and Lamb, R.A., 1994, Studies with cross-linking reagents on the oligomeric form of the paramyxovirus fusion protein. *Virology* **199**:160-168.

Sagara, Y., Inoue, Y., Shiraki, H., Jinno, A., Hoshino, H., and Maeda, J., 1996, Identification and mapping of functional domains on human T-cell lymphotropic virus type 1 envelope proteins by using synthetic peptides. *J. Virol.* **70**:1564–1569.

Salzwedel, K., Johnston, P.B., Roberts, S.J., Dubay, J.W., and Hunter, E., 1993, Expression and characterization of glycophospholipid-anchored human immunodeficiency virus type 1 envelope glycoproteins. *J. Virol.* **67**:5279-5288.

Salzwedel, K., West. J.T., and Hunter, E., 1999, A conserved tryptophan-rich motif in the membrane-proximal region of the human immunodeficiency virus type 1 gp41 ectodomain is important for Env-mediated fusion and virus infectivity. *J. Virol.* **73**:2469-2480.

Salzwedel, K., Smith, E.D., Dey, B., and Berger, E.A., 2000, Sequential CD4-coreceptor interactions in human immunodeficiency virus type 1 Env function: soluble CD4 activates Env for coreceptor-dependent fusion and reveals blocking activities of antibodies against cryptic conserved epitopes on gp120. *J. Virol.* **74**:326-333.

Sattentau, Q.J., and Moore, J.P., 1991, Conformational changes induced in the human immunodeficiency virus envelope glycoprotein by soluble CD4 binding. *J. Exp. Med.* **174**:407-415.

Sauter, N.K., Bednarski, M.D., Wurzburg, B.A., Hanson, J.E., Whitesides, G.M., Skehel, J.J., and Wiley, D.C., 1989, Hemagglutinins from two influenza virus variants bind to sialic acid derivatives with millimolar dissociation constants: a 500-MHz proton nuclear magnetic resonance study. *Biochemistry* **28**:8388–8396.

Schroth-Diez, B., Ponimaskin, E., Reverey, H., Schmidt, M.F., and Herrmann, A., 1998, Fusion activity of transmembrane and cytoplasmic domain chimeras of the influenza virus glycoprotein hemagglutinin. *J. Virol.* **72**:133-141.

Siegel, D.P. , 1993, Energetics of intermediates in membrane fusion: comparison of stalk and inverted micellar intermediate mechanisms. *Biophys. J.* **65**:2124–2140.

Skehel, J.J., Baylay, P.M., Brown, E.B., Martin, S.R., Waterfield, M.D., White, J.M., Wilson, I.A., and Wiley, D.C., 1982, Changes in the conformation of the influenza hemagglutinin at the pH optimum of virus mediated membrane fusion. *Proc. Natl. Acad. Sci. USA* **79**:968-972.

Skehel, J.J., Bizebard, T., Bullough, P.A., Hughson, F.M., Knossow, M., Steinhauer, D.A., Wharton, S.A. and Wiley, D.C., 1995, Membrane fusion by influenza hemagglutinin. *Cold Spring Harb. Symp. Quant. Biol.* **60**:573-580.

Skehel, J.J., and Wiley, D.C., 1998, Coiled coils in both intracellular vesicle and viral membrane fusion. *Cell* **95**:871–874.

Skehel, J. J., and Wiley, D.C., 2000, RECEPTOR BINDING AND MEMBRANE FUSION IN VIRUS ENTRY: the influenza hemagglutinin. *Annu. Rev. Biochem.* **69**:531-569.

Spruce, A.E., Iwata, A, White, J.M. and Almers, W., 1989, Patch clamp studies of single cell fusion events mediated by a viral fusion protein. *Nature* **342**:555-558.

Spruce, A.E., Iwata, A. and Almers, W., 1991, The first millisecond of the pore formed by a fusogenic viral envelope protein during membrane fusion. *Proc. Natl. Acad. Sci. USA* **88**:3623-3627.

Steinhauer, D.A., Wharton, S.A., Skehel, J.J,, and Wiley, D.C., 1995, Studies of the membrane fusion activities of fusion peptide mutants of influenza virus hemagglutinin. *J. Virol.* **69**:6643–6651.

Sutton, R.B., Fasshauer, D., Jahn, R., and Brünger, A.T., 1998, Crystal structure of a SNARE complex involved in synaptic exocytosis at 2.4 A resolution. *Nature* **395**:347–353.

Tan, K., Liu, J., Wang, J., Shen, S., and Lu, M., 1997, Atomic structure of a thermostable subdomain of HIV-1 gp41. *Proc. Natl. Acad. Sci. USA* **94**:12303-12308.

Volchkov, V.E., Feldmann, H., Volchkova, V.A., and Klenk, H.D., 1998, Processing of the Ebola virus glycoprotein by proProtein convertase furin. *Proc. Natl. Acad. Sci. USA* **95**:5762-5767.

Walker, J.A., and Kawaoka, Y., 1993, Importance of conserved amino acids at the cleavage site of the haemagglutinin of a virulent avian influenza A virus. *J. Gen. Virol.* **74**:311–314.

Ward, C.W., and Dopheide, T.A., 1980, Influenza virus haemagglutinin. Structural predictions suggest that the fibrillar appearance is due to the presence of a coiled-coil. *Aust. L. Biol. Sci.* **33**:441-447.

Weber, T., Paesold, G., Galli, C., Mischler, R., Semenza , G., and Brunner, J., 1994, Evidence for H(+)-induced insertion of influenza hemagglutinin HA2 N-terminal segment into viral membrane. *J. Biol. Chem.* **269**:18353-18357.

Weber, T., Zemelman, B. V., McNew, J. A., Westermann, B., Gmachl, M., Parlati, F., Söllner, T. H., and Rothman, J. E., 1998, SNAREpins: minimal machinery for membrane fusion. *Cell* **92**:759-772.

Weis, W., Brown, J.H., Cusack, S., Paulson, J.C., Skehel, J.J., and Wiley, D.C., 1988, Structure of the influenza virus haemagglutinin complexed with its receptor, sialic acid *Nature* **333**:426–31.

Weis, W.I., Cusack, S.C., Brown, J.H., Daniels, R.S., Skehel, J.J.,and Wiley, D.C., 1990, The structure of a membrane fusion mutant of the influenza virus haemagglutinin. *EMBO J.* **9**:17-24.

Weissenhorn, W.,Wharton, S.A., Calder, L. J., Earl, P.L., Moss, B., Aliprandis, E., Skehel, J.J., and Wiley, D.C., 1996, The ectodomain of HIV-1 env subunit gp41 forms a soluble, alpha-helical, rod-like oligomer in the absence of gp120 and the N-terminal fusion peptide. *EMBO J.* **15**:1507-1514.

Weissenhorn, W., Dessen, A, Harrison, S. C., Skehel, J.J., and Wiley, D.C. ,1997, Atomic structure of the ectodomain from HIV-1 gp41. *Nature* **387**: 426-430.

Weissenhorn, W., Calder, L.J., Dessen, A., Laue, T., Skehel, J.J., and Wiley, D.C., 1997, Assembly of a rod-shaped chimera of a trimeric GCN4 zipper and the HIV-1 gp41 ectodomain expressed in Escherichia coli. *Proc. Natl. Acad. Sci. USA* **94**:6065-6069.

Weissenhorn, W., Carfí, A., Lee, K.H., Skehel, J.J., and Wiley, D.C., 1998, Crystal structure of the Ebola virus membrane fusion subunit, GP2, from the envelope glycoprotein ectodomain. *Mol. Cell* **2**:605–616.

Weissenhorn, W., Calder, L.J. Wharton, S.A. Skehel J.J., and Wiley D.C., 1998, The central structural feature of the membrane fusion protein subunit from the Ebola virus glycoprotein is a long triple stranded coiled coil. *Proc. Natl. Acad. Sci. USA* **95**:6032-6036.

Weissenhorn, W., Dessen, A., Calder, L.J., Harrison, S.C., Skehel, J.J., and Wiley, D.C., 1999, Structural basis for membrane fusion by enveloped viruses. *Mol. Membr. Biol.* **16**:3-9.

Wharton, S.A., Skehel, J.J., and Wiley, D.C., 1986, Studies of influenza haemagglutinin-mediated membrane fusion. *Virology* **149**:27–35.

Wharton., S.A., Martin, S.R., Ruigrok, R.W.H., Skehel, J.J., and Wiley, D.C., 1988, Membrane fusion by peptide analogues of influenza virus haemagglutinin. *J. Gen. Virol.*

69:1847-1857.

Wharton, S.A., Calder, L.J., Ruigrok, R.W., Skehel, J.J., Steinhauer, D.A. and Wiley, D.C., 1995, Electron microscopy of antibody complexes of influenza virus haemagglutinin in the fusion pH conformation. *EMBO J.* **14**:240-246.

White, J., Matlin, K., and Helenius, A., 1981, Cell fusion by Semliki Forest, influenza, and vesicular stomatitis viruses. *J. Cell Biol.* **89**:674–79.

Wild C, Oas T, McDanal C, Bolognesi D, Matthews T., 1992, A synthetic peptide inhibitor of human immunodeficiency virus replication: correlation between solution structure and viral inhibition. *Proc. Natl. Acad. Sci. U S A.* **89**:10537-41.

Wild, C.T., Shugars, D.C., Greenwell, T.K., McDanal, C.B., and Matthews, T.J., 1994, Peptides corresponding to a predictive α-helical domain of human immunodeficiency virus type 1 gp41 are potent inhibitors of virus infection, *Proc. Natl. Acad. Sci. USA* **91**:9770-9774.

Wild, C., Greenwell, T., Shugars, D., Rimsky-Clarke, L. and Matthews, T., 1995, The inhibitory activity of an HIV type 1 peptide correlates with its ability to interact with a leucine zipper structure. *Aids Res. Hum. Retroviruses* **11**:323-325.

Wild, T.F., and Buckland, R., 1997, Inhibition of measles virus infection and fusion with peptides corresponding to the leucine zipper region of the fusion protein. *J. Gen. Virol.* **78**:107-111.

Wiley, D.C, and Skehel, J.J., 1987, The structure and function of the haemagglutinin membrane glycoprotein of influenza virus. *Annu. Rev. Biochem.* **56**:365-394.

Wilk, T., Pfeiffer, T., and Bosch, V., 1992, Retained in vitro infectivity and cytopathogenicity of HIV-1 despite truncation of the C-terminal tail of the env gene product. *Virology* **189**:167-77.

Wilson, I.A., Skehel, J.J., and Wiley, D.C., 1981, Structure of the haemagglutinin membrane glycoprotein of influenza virus at 3 A resolution. *Nature* **289**:366-373

Wool-Lewis, R.J, and Bates, P., 1998, Characterization of Ebola virus entry by using pseudotyped viruses: identification of receptor-deficient cell lines. *J. Virol.* **72**:3155-3160.

Wyatt, R., Desjardin, E., Olshevsky, U., Nixon, C., Binley, J., Olshevsky, V., and Sodroski, J., 1997, Analysis of the interaction of the human immunodeficiency virus type 1 gp120 envelope glycoprotein with the gp41 transmembrane glycoprotein. *J. Virol.* **71**:9722–9731.

Wyatt, R., and Sodroski, J., 1998, The HIV-1 envelope glycoproteins: fusogens, antigens, and immunogens. *Science* **280**:1884-1888.

Xu, T., Rammner, B., Margittai, M., Artalejo, A.R., Neher, E., and Jahn, R., 1999, Inhibition of SNARE complex assembly differentially affects kinetic components of exocytosis. *Cell* **99**:713-722.

Yao, Q., and Compans, R.W., 1996, Peptides corresponding to the heptad repeat sequence of human parainfluenza virus fusion protein are potent inhibitors of virus infection. *Virology* **223**:103-112.

Zimmerberg, J., Blumenthal, R., Sarkar, D.P., Curran, M., and Morris, S.J., 1994, Restricted movement of lipid and aqueous dyes through pores formed by influenza hemagglutinin during cell fusion. *J. Cell Biol.* **127**:1885–1894.

Zhao, X., Singh, M., Malashkevich, V.N., and Kim, P.S., 2000, Structural characterization of the human respiratory syncytial virus fusion protein core. *Proc. Natl. Acad. Sci. U S A* **97**:14172-7.

Zhou, G., Ferrer, M., Chopra, R., Kapoor, T.M., Strassmaier, T., Weissenhorn, W., Skehel, J.J., Oprian, D., Schreiber, S.L., Harrison, S.C., and Wiley, D.C., 2000, The structure of an HIV-1 specific cell entry inhibitor in complex with the HIV-1 gp41 trimeric core. *Bioorg. Med. Chem.* **8**:2219-2227.

2. Viral replication

Chapter 2.1

Viral RNA-directed Polymerases: Structure and Function Relationships

JOACHIM JÄGER[1] and JANICE D PATA[2]

[1]*Astbury Centre of Structural Molecular Biology, Faculty of Biological Sciences, University of Leeds, Leeds, LS2 9JT, UK;* [2]*Department of Molecular Biophysics and Biochemistry, Bass Center, Yale University, New Haven, CT 06520-8114, USA.*

1. INTRODUCTION

1.1 Background

The majority of known viruses are equipped with their own machinery to replicate the viral genome instead of relying on polymerases available in the infected host cell. Usually, replication is initiated shortly after cell entry. Many viruses, which typically are highly efficient and obligate parasites, allocate a considerable part of their genomic message to the expression and assembly of a processive replicase complex. Making such a substantial 'investment' underpins the importance of the viral replication and transcription process.

There are some interesting differences but also some striking similarities between viral polymerases. The requirements for an efficient viral polymerase or replicase complex are binding preferentially to and priming competently from the viral genome. Each virus appears to have adopted its own replication strategy depending on the genomic nature and content. Interestingly, several unique methods of initiation have been elucidated for different families of viruses. Some viruses require sequence specific initiation signals, some utilise modifications on viral peptides as primers and others are completely independent of any signals and are able to initiate replication without any primers at all. It would be wasteful for a virus if its

Structure-Function Relationships of Human Pathogenic Viruses, Edited by
Holzenburg and Bogner, Kluwer Academic/Plenum Publishers, New York, 2002

replication machinery would be able to direct synthesis from a foreign, cellular template. For example, recent studies on poliovirus 3Dpol or 3CD have revealed possible mechanisms of recruiting cognate RNA to the viral polymerase (Paul *et al.,* 1998, 2000). Furthermore, replicases need to elongate processively to synthesise the full genome in a template directed manner (Pata *et al.,* 1995; Arnold *et al.,* 1999; Oh *et al.,* 1999; Rodriguez-Wells *et al.,* 2001). Non-processive or distributive synthesis is not desirable as the replication product would be incomplete and the polymerase would have to be re-directed to the 3' end.

In contrast with the different recruitment and initiation strategies, the basic catalytic mechanism which mediates the chemical step, the phosphoryl transfer reaction, is carried out by amino acid side chains and more general structural features highly conserved amongst all polymerases. The catalytic domains of polymerases found in some dsDNA viruses are remarkably similar to those occurring in the host system. More interestingly, four of the five conserved sequence motifs, the hallmarks of the viral RNA polymerases, first identified by Poch *et al.* (1989) and Koonin *et al.* (1989, 1991), appear to have a common role – directly or indirectly – in the phosphoryl transfer reaction. Not surprisingly, these sequence motifs form the core or map closely to the active site of RNA polymerases and reverse transcriptases.

1.2   Signature sequences and conserved motifs

Sequence alignments of viral RNA-directed polymerases (reverse transcriptases and RNA-directed RNA polymerases) identified 4 conserved sequence motifs (A-D) that are shared among these enzymes (Poch *et al.,* 1989). Motifs A and C are also found in the DNA polymerases (Delarue *et al.,* 1991, Ito and Braithwaite, 1991) and contain the strictly conserved carboxylate residues that coordinate the catalytic metal ions. An additional motif (E) was identified by sequence similarity in the reverse transcriptases, but from the RNA-directed RNA polymerase structures determined, it is clear that the structural Motif E is conserved in these enzymes as well. In all cases the structural similarity is more extensive than the sequence similarity. For example, sequence analysis of the pol α family of DNA polymerases, which includes the replicative polymerases of eukaryotes and most viruses, shows that polymerases sharing less than 20% sequence identity can have virtually identical structures, with differences primarily in solvent exposed surface loops (Wang *et al.,* 1997; Hopfner *et al.,* 1999; Zhao *et al.,* 1999; Rodriguez *et al.,* 2000). The similarities are predominantly structural.

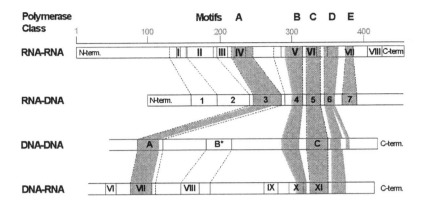

Figure 1. Overall structure-based sequence alignment of conserved motifs in the four different classes of template dependent polymerases. Note that motif E is only present in RNA-directed RNA polymerases and reverse transcriptases. (Figure adapted from Hansen *et al.*, 1997)

For current sequence alignments of the reverse transcriptase ("rvt") and RNA-directed RNA polymerase ("RNA_dep_RNA_pol") families, constructed using Hidden Markov Models, see the Pfam database (Eddy, 1996; Bateman *et al.*, 2000) at http://pfam.wustl.edu/ (rvt accession number: PF00078; RNA_dep_RNA_pol accession number: PF00680). Structure-based sequence alignments of polymerases from different families can be found in the recent literature (Hansen *et al.*, 1997; Wang *et al.*, 1997; Hopfner *et al.*, 1999).

1.3 Overall architecture of polymerases

Polymerases whose structures have been determined thus far have an overall architecture that has been likened to a right hand. Ollis *et al.*, (1985) have divided the polymerase domain in the Klenow Fragment of E. coli DNA polymerase I in fingers palm and thumb sub-domains (Ollis *et al.*, 1985; Steitz *et al.*, 1993 Joyce and Steitz, 1995; Jaeger and Pata, 1999). So far there are 3 different superfamilies of polymerase structures, with the catalytic palms being the only homologous domain within a superfamily. The majority of these polymerases have a palm domain with the same topology first found in the Klenow fragment of E. coli DNA polymerase I. This superfamily includes the DNA-directed DNA polymerases from Family A (pol I) and Family B (pol α); the monomeric DNA-directed RNA polymerases (e.g. T7 RNAP); the reverse transcriptases; and the viral RNA-directed RNA polymerases. The large UmuC/DinB family of bypass DNA polymerases (recently designated as polymerase Family Y; Ohmori *et al.,* 2001) also contains this palm domain topology (Zhou *et al.,* 2001). The Family X polymerases (e.g. pol β and poly(A) polymerase) have a palm domain with a second distinct topology, and the multimeric DNA-dependent RNA polymerases have yet a third distinct palm topology. On the basis of sequence analysis, but no structural data, the Family C (e.g. E. coli pol III) and Family D (e.g. newly identified archeal enzymes; Cann and Ishino, 1999) DNA polymerases and the cellular RNA-directed RNA polymerases (e.g. ego-1; Smardon *et al.,* 2000) involved in RNA interference may each contain still other palm domain folds. All of the polymerases discussed in this chapter belong to the first of these superfamilies and will just be referred to as the polymerases.

Most of the template dependent polymerases are multifunctional enzymes, containing more than one catalytic activity or otherwise functional domain on the same polypeptide (e.g. RT has RNase H activity and most of the DNA polymerases have 5'-3' and 3'-5' exonuclease activities; Steitz *et al.,* 1993). The fingers and thumb sub-domains are both important for correctly positioning the substrates for catalysis by the palm sub-domain. The fingers contribute to binding both the incoming NTP and the template strand (Eom *et al.,* 1996; Doublie *et al.,* 1998). Residues at the base of the fingers form one side of the nascent basepair binding pocket, which contains important structural determinants of polymerase fidelity. The thumb sub-domain binds the primer/template duplex and motif E, which forms a small hairpin structure linking the palm and thumb, binds to the primer terminus. The palm is an evolutionarily frequently recurring folding unit (Steitz *et al.,* 1994, Pelletier *et al.,* 1994, Artymiuk *et al.* 1997, Hansen *et al.,* 1997). Only two aspartate residues appear to be absolutely essential in the polymerases characterised to date. The carboxylate side chains are directly anchoring the catalytic metal ions in the polymerase active site. The chemical step is not directly mediated by protein side chain atoms but is principally catalysed by the two metal ions found within 2Å or so of the primer terminus and the incoming NTP (Doublie *et al.,* 1998). The metal ion mediated phosphoryl transfer reaction is shared by all polymerases studied thus far.

Interestingly, the palm topology is not confined just to polymerases. For example the structure of adenylate cyclase VC1 domain has the same topology as the catalytic palm and catalyses cyclisation by the same two-metal ion mechanism as the polymerases. Also, the palm has the same fold as the RRM motif (Georgiadis *et al.,* 1995; Hansen *et al.,* 1997), such as that found in U1A SNP and ribosomal protein S6.

1.4   Structure and function: polymerase motifs revisited

Sequence motifs A – D are contained within the conserved core of the polymerase palm domain. Since the conservation of secondary and tertiary structure is much more extensive than that of the primary structures, we will use the sequence motif designation in reference of the structures that contain the smaller regions of sequence similarity.

Motif A forms a β-strand that contains one of the aspartate residues responsible for binding the catalytic metal ions and a short helical turn that contains a residue (called the "steric gate" residue in the DNA polymerases; Astatke *et al.*, 1998) which contributes strongly to the discrimination between the oxy- and deoxy-ribose moiety of the incoming nucleotides. This is discussed in more detail below,   particularly in the context of HIV-1

Figure 2. Ribbon diagrams of the HIV-1 reverse transcriptase, poliovirus and phi6 dsRNA polymerase palm domains comprising sequence motifs A and C. Motif A consists of a single β-strand followed by a short turn containing a residue that discriminates between oxy- and deoxyribonucleotides.   Motif C is larger and comprises an antiparallel β-turn-β hairpin structure and two supporting helices. Figures 2 to 5 have been prepared with the programs Molscript (Kraulis, 1991).

reverse transcriptase. Motif C forms a β-turn-β structure and contains the other conserved aspartate that coordinates the catalytic metal ions. Sequences in Motif B (which is distinct from motif B of the DNA-directed polymerases; Delarue *et al.*, 1990; Hansen *et al.*, 1997) form one of the long α helices that supports the β sheet made from Motifs A and C. It also contributes to the nascent basepair binding pocket. This helix is conserved in all the polymerases, but only in the RNA-directed polymerases is it at all conserved in sequence. Motif D, like motif B is conserved in sequence only in the RNA-directed polymerases, but its structure is conserved throughout the polymerase superfamily. It forms the second long helix supporting the β-sheet of the polymerase domain, followed by a turn and a short β strand that flanks the β strand of Motif A. Motif E is also known as the "primer grip" (Jacobo-Molina *et al.*, 1993) as it is important in positioning the 3'OH of the terminal ribose (primer) appropriately for an in-line attack on the α-phosphate group of the incoming nucleotide (Beese & Steitz, 1991, Steitz *et al.*, 1993, Steitz and Steitz, 1993).

1.5 Relationships between viral polymerases and cellular polymerase structures

To date five viral polymerase structures have been determined at high resolution and a structural comparison with host polymerases reveals unprecedented insights into the evolutionary development of replication enzymes. The overall architecture and domain organisation of viral polymerases is similar to some of the host polymerase counterparts known to date. The resemblance to a right hand is preserved, but overall, the differences in the order of genomic arrangement of sub-domains prevail. The most striking differences probably occur in the fingers domain. The architecture, internal flexibility and the content of secondary structure elements vary depending on the template requirements of the individual polymerases. All DNA-directed DNA polymerases utilise α helices separating double stranded template nucleic acid, controlling the path of the template strand and the formation of Watson Crick base pairs. In reverse transcriptases, however, the template is guided by highly flexible β strands (Kohlstaedt *et al.*, 1992, Jacobo-Molina *et al.*, 1993; Huang *et al.*, 1998). In RNA dependent RNA polymerases, such as Hepatitis C virus or bacteriophage Φ6 polymerase, the fingers domain are contain α as well as β structure (Ago *et al.*, 1999, Bressanelli *et al.*, 1999; Lesburg *et al.*, 1999; Jaeger *et al.*, 2000a; Butcher *et al.*, 2001). The thumb domains are mostly α helical and, typically, two helices are positioned perpendicular to the primer:template to track the major groove of the emerging product duplex or to form multiple interactions with the ribophosphate backbone (Beese *et al.* 1993; Jacobo-Molina *et al.*, 1993; Kohlstaedt *et al.* 1992; Steitz et al, 1993). Extensions on the thumb domains appear to contribute to the processivity of the replication process (Minnick *et al.* 1996; Bedford *et al.* 1997).

In the following chapters all currently known structures of viral polymerases will be discussed in detail. HIV-1 reverse transcriptase is the first and one of the most thoroughly studied members in the subfamily of viral polymerases. Later additions to this subfamily comprise poliovirus 3Dpol and Hepatitis C virus NS5B protein. The most recent member is the polymerase of the dsRNA bacteriophage, Φ6.

2. HIV-1 REVERSE TRANSCRIPTASE

2.1 Background

Retroviruses are known to cause a wide variety of diseases in many avian and mammalian species. The discovery of a sexually transmitted human retrovirus, however, fuelled efforts in understanding (and combating) the pathogenicity and the viral life cycle in detail.   In 1981, the Human immunodeficiency virus type 1 (HIV-1, then LAV) had been linked to a fatal disease that manifests itself by unusual cancers, immunological abnormalities and, consequently, numerous opportunistic infections. Montagnier's group at the Institut Pasteur isolated a retrovirus in 1983 from lymph node cells of an AIDS patient (Chermann *et al.,* 1983). Molecular cloning, sequencing and imaging by electron microscopy confirmed that these viruses were similar to members of the lentivirus genus of the family of retroviridae.

By now, HIV-1 is one of the most thoroughly studied viruses with respect to the elucidation of life cycle, receptor binding, replication integration and maturation.   The HIV genome encodes a total of two envelope proteins, three structural proteins, various multi-function accessory proteins as well as three viral enzymes. Studies over the past decade have provided a wealth data from solution and crystallographic studies for all of the enzymes, structural proteins and envelope proteins, as well as for three of the accessory proteins.

2.2 Retroviral replication

HIV-1 reverse transcriptase is now one of the best-characterized polymerases, both structurally and functionally.   Crystal structures determined for this enzyme include the apo-enzyme, complexes of the enzyme bound to DNA substrates and complexes with nucleic acid and small molecule inhibitors.   Numerous biochemical and genetic experiments relate the function of HIV-1 reverse transcriptase to these structures (Arts and Le Grice, 1998). Reverse transcriptase has both polymerase and RNaseH activity and is responsible for converting the single stranded viral RNA genome present in the virus particle into a double-stranded DNA that is then integrated in the host cell genome.   During this process, RT uses several kinds of primer/template combinations (RNA/RNA, DNA/RNA, RNA/DNA, and DNA/DNA) and must switch templates several times (Baltimore, 1970; Temin & Mizutani, 1970). To initiate this process, RT uses the 3' end of tRNALys3 (packaged into the virus from the last host cell) to prime DNA synthesis. The tRNA is partially unwound and the last 18 nts are annealed to a complementary site (the primer-binding site, PBS) located near the 5' end of the genome. A few nucleotides can be added to the tRNA primer in the virion (probably limited by the low concentrations of nucleotides present in the virus), but otherwise replication occurs after the virus has entered the host cell cytoplasm. After initiation from the tRNA, RT continues to synthesize minus-strand DNA using the viral RNA genome as a template; meanwhile the RNase H activity of RT cleaves the viral RNA. Synthesis of the plus-sense second strand initiates from the polypurine tract (PPT), which is RNase H-resistant region near the 3' end of the genome.

HIV-1 reverse transcriptase is a heterodimer of the viral proteins p66 and p51 encoded in the pol region of the retroviral genome. The large subunit contains a polymerase domain composed of fingers, palm, and thumb sub-domains and is linked to an RNaseH domain by a connection domain. The smaller subunit is derived from the larger by a proteolytic cleavage that removes the C-terminal RNase H domain of p66. The subunits are associated in a highly asymmetric way, with the individual sub-domains being oriented differently in each subunit such that the polymerase domain adopts a much more open conformation in p66 than in p51 (Kohlstaedt *et al.*, 1992). Even though both subunits contain the protein sequences required for polymerase activity, only the open conformation is catalytically active because it is accessible to binding of primer/template nucleic acids and dNTPs and because the residues of the catalytic site are positioned in an active conformation (Kohlstaedt *et al.*, 1992; Esnouf *et al.*, 1995).

2.3   Substrate binding, mechanism of action and inhibition

Two structures of HIV-1 RT bound to DNA have been reported (Jacobo-Molina *et al.*, 1993, Huang *et al.*, 1998). The first of these, a "binary" complex, contains a dsDNA primer/template bound to RT and shows that the 18 basepair DNA duplex stretches from the polymerase active site located in the palm sub-domain, near the junction with the fingers domain to the RNase H active site. This distance is consistent with biochemical results showing that RNA in an RNA/DNA heteroduplex can be cleaved with about this same spacing. Neither the position of the incoming dNTP nor the path of template (5' to templating base) leading out of the active site was apparent from this structure. However, the structure of a "ternary" complex consisting of enzyme, primer/template DNA, and incoming dNTP elucidated both these

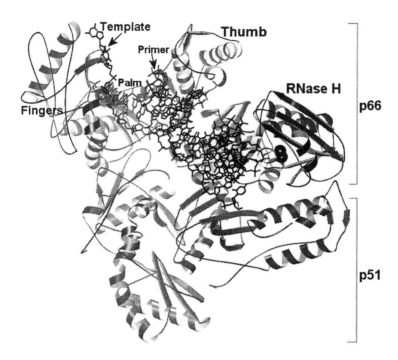

Figure 3. Ribbon diagram of a ternary HIV-1 reverse transcriptase elongation complex with bound primer:template, incoming nucleotide and divalent metal ions (Huang *et al.,* 1998). Residue Gln258 in helix H in the thumb has been mutated to Cys to provide a crosslink for the terminated, arrested primer:template. The majority of the interactions between the nucleic acid and the protein are made in the large subunit (p66). Note that the template strand enters the active site from the top over the fingers domain.

issues (Huang *et al.,* 1998). In comparison with the previous DNA complex, the ternary complex shows that when dNTP is bound, together with the primer/template and metal ions, the fingers sub-domain is in a position closer to the palm domain, forming a tight pocket around the nascent basepair. The two structures thus indicate that nucleotide binding triggers a 20° rotation of the fingers domain inwards towards the palm domain upon dNTP binding. This type of repositioning had been observed previously in DNA-dependent DNA polymerase ternary complex structures (T7 DNAP, Doublie *et al.,* 1998; and pol β, Pelletier *et al.,* 1994), and is thought to contribute to higher fidelity nucleotide incorporation, as discussed below. Additionally, the DNA template 5' to the templating base is kinked and continues alongside the fingers domain, and not between the fingers and thumb as originally proposed. This orientation of 5' template has also been seen before (T7 DNAP, Doublie *et al.,* 1998; pol β, Pelletier *et al.,* 1994), although the precise path is not well defined in any of these structures. Only the first base or two are clearly resolved in the structures, even when the template contained longer 5' extensions.

In both HIV-1 RT structures, Motif E binds to the primer strand phosphodiester backbone near the active site and was termed the "primer-grip" (Jacobo-Molina et al., 1993). A helix in the thumb domain binds in the minor groove of the primer/template DNA duplex.

The structure of a catalytic fragment of MMLV suggested that the primary selection for dNTPs over rNTPs by a DNA polymerase occurs by a "steric gating" mechanism where there is bulky residue (phenylalanine or tyrosine in the reverse transcriptases; Tyr115 in HIV-1 RT) that would clash with a 2' OH on an incoming rNTP; the HIV-1 RT ternary complex structure (Huang et al., 1998) confirms this position of the incoming dNTP.

2.4   Mechanisms of inhibition

Two classes of RT inhibitors are currently used for the treatment of HIV-1 infection: the non-nucleoside RT inhibitors (NNRTIs) and the nucleoside RT inhibitors (NRTIs). Nevirapine, one of the NNRTIs, was bound to RT in the first structure determined (Kohlstaedt et al., 1992). It was located in a hydrophobic pocket between the palm beta-sheet and the motif E primer grip. Subsequent apo enzyme structures show that this pocket does not exist prior to inhibitor binding, but instead is formed by rearrangements of amino acid side chains into alternate conformations that are stabilized by the presence of the inhibitor. This conformational change alters the position of the catalytic aspartates, and may be the primary mechanism for inhibition by the NNRTIs (Ensue et al., 1995). Additionally, NNRTI binding may also inhibit RT by reducing the flexibility of the enzyme (see below; Kohlstaedt et al., 1992).

Mutations that confer NNRTI resistance on HIV-1 RT line the inhibitor binding pocket (Smerdon et al., 1994) and appear to act by directly altering the interactions between the protein and the inhibitor. One exception to this direct interference with inhibitor binding may be the clinically important K103N mutation that appears to stabilize the closed conformation of the pocket (Arnold, 2001). Mutations that confer resistance to the nucleoside inhibitors (e.g. ddC and AZT) are not as straightforward to understand, as these mutations are not all directly in contact with the incoming nucleotide (Huang et al., 1998). See Larder and Stammers (1999) for a detailed discussion.

The structure of a small pseudoknot RNA bound to RT (Jaeger *et al.,* 1998) shows that the enzyme can also be inhibited by direct occlusion of the substrate binding site. This RNA was selected for high affinity binding and the structure shows that it partially overlaps the primer-template binding site and stabilizes the p66 polymerase domain in a closed conformation (see below).

2.5   Flexibility

The series of RT structures show that the enzyme has an extraordinary amount of flexibility. One conformational change is in the positioning of the thumb domain. The first structure of RT that had the NNI nevirapine bound showed that the thumb was in a very open position (Kohlstaedt et al., 1992; Smerdon *et al.,* 1994). Structures without any substrates bound to RT show that the thumb domain of p66 is in a position close to touching the fingers domain. In this position, there is not enough space for primer-template DNA to bind at the active site. In structures with primer/template DNA bound, the thumb is in an open position, and when NNRTI is bound, the thumb is in a still more open position. In the RT-pseudoknot inhibitor structure, the thumb is in a closed conformation (Jaeger *et al.,* 1998). Since this RNA inhibitor was selected to bind in solution to the protein in the absence of any other substrates, this suggests that the predominant (most stable) conformation of the protein in solution is one that has the thumb down. This is also suggested by the fact that the apo enzyme crystallized with the thumb down (lowest energy conformation most likely to be the one found in crystals).

A second type of flexibility exhibited by RT that is especially important is a conformational change of the fingers domain. This change is a closing in of the fingers domain that has only been seen upon binding of primer/template DNA and correct incoming dNTP (Huang *et al.,* 1998). This conformational change is important because it results in the correct positioning of 3' OH of the primer terminus, the phosphates of the incoming dNTP and the metal ions bound to the catalytic aspartates. The conformation change is induced upon binding of a dNTP that can form a correct Watson-Crick pair with the templating base. An incorrectly paired incoming nucleotide would not result in the same correct positioning of all the components required for catalysis. Kinetic characterization of the RT catalytic cycle shows that there is a rate-limiting step prior to phosphoryl transfer reaction. The conformational change of the fingers domain has been attributed to be rate limiting and is important because it contributes to the fidelity of the polymerase.

A third type of flexibility observed in HIV-1 RT is a sliding of the β-sheet containing the catalytic aspartates upon binding of NNI to a pocket formed between that β-sheet and motif E structure located between the palm and thumb domains (Rodgers *et al.,* 1994). This motion is not on nearly as pronounced as the changes in the positions of the thumb and fingers domain, but it is functionally significant nonetheless. This motion is predominantly produced by a rearrangement of two tyrosines, located on the motif C β strands, that forms the pocket in which the NNIs bind. This movement changes the position of the catalytic aspartates relative to other active site residues such that catalysis would not be efficient. This appears to be the (primary) mechanism of action of the NNI compounds.

A final type of flexibility observed by comparisons of the many RT structures determined from different crystal forms and different complexes is a swivelling of the p66 polymerase domain relative to the rest of the heterodimer (Jaeger *et al.,* 1994). The functional significance of this motion is not entirely clear, but it could be important particularly during the initiation of replication when an asymmetric tRNA molecule is used as a primer and must rotate and translate away from the polymerase domain as dNTP are added to its terminus.

3.   POLIOVIRUS 3D POLYMERASE

3.1   Background

Poliovirus, an enterovirus belonging to the Picornaviridae causes infections in humans usually by ingestion followed by viral proliferation in the gastrointestinal tract, leading to mild, if any, clinical symptoms. The (+) sense, single stranded RNA genome of poliovirus is 7.5kb long and contains a 5'-untranslated region (UTR), a single open reading frame, a short 3'UTR, and a poly(A) tail. The 5'-terminal end of the genome is 'capped' by a small peptide, VPg or 3B, via a phosphodiester bond (Lee *et al.,* 1977). The 5'UTR can be subdivided into two structural domains: firstly, a cloverleaf-like structure, which is involved (+) strand RNA synthesis (Andino *et al.,* 1990) and, secondly, the internal ribosomal entry site (IRES), which promotes translation of a polyprotein (Kitamura et al. 1981; Pelletier and Sonenberg, 1988). The viral polyprotein is subdivided into three regions, a capsid region (P1) and two nonstructural domains (P2 and P3).   The P3 region is directly involved with the process of RNA synthesis as it includes the RNA-directed RNA polymerase $3D^{pol}$. The other members of the P3 region are a small membrane-bound protein 3A, VPg (3B), and the proteinase $3C^{pro}$.

Virus replication in the infected host cell is a two-step process, carried out primarily by the viral RNA polymerase, which is thought to take place in small vesicles that are derived from the host's cellular membranes and with which the non-structural proteins of the virus are associated. First, the incoming viral RNA is transcribed into complementary (-) strands. This process is initiated by the covalent attachment of UMP to the terminal protein VPg, yielding VPgpU and VpgpUpU (Paul *et al.,* 1998). The newly synthesised (-) strand then serves as a template for progeny synthesis.

3.2   Overall structure

The structure of poliovirus 3Dpol was the first RNA-directed RNA polymerase structure determined (Hansen *et al.,* 1997) and clearly established that the RdRPs are evolutionarily related to the reverse transcriptases (discussed in more detail below). The 3Dpol structure has fingers, palm, and thumb domains, as seen in other template dependent polymerases. The core of the palm domain has the same topology as in the reverse transcriptases and contains the four conserved sequence motifs A-D. Although the top part of the fingers sub-domain is disordered in the crystals, the portion that is visible shows that the topology of the fingers domain is different from HIV-1 RT. The fingers domain is formed from two polypeptide segments that are located N-terminal to Motif A and between Motifs A and B, while the thumb domain is primarily formed from sequences that are at the C-terminus of the protein. The crystal structure also shows that Motif E is (structurally) conserved in the RNA-directed polymerases, even though the sequence is conserved only in the reverse transcriptases. Asp238 in motif A is in a position that is structurally equivalent to the "steric gate" residue Tyr115 of HIV-1 reverse transcriptase. This residue is highly conserved as an aspartate in the RNA-directed polymerases and could provide specificity for incoming rNTPs (instead of dNTPs) by hydrogen bonding to the 2'-OH of the incoming nucleotide.

Two segments of the N-terminal region are visible in the 3Dpol structure. The first segment (residues 12-37) extends from the active site cleft over the tip of the thumb and is an integral part of the thumb domain. The second segment (residues 67-97) forms an α helix at the base of the fingers domain on the side away from the active site. Since the intervening sequences are not ordered in the structure, it is not clear how the two segments are connected.

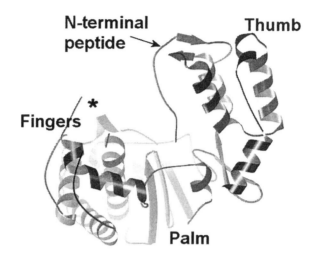

Figure 4. Ribbon diagram of poliovirus 3D polymerase (Hansen *et al.*, 1997). The crystallographically disordered fingers domain is indicated by an asterisk. The peptide bound to the helical thumb domain is donated by a symmetry related 3Dpol molecule. Site directed mutagenesis and chemical crosslinking studies corroborate the functional importance of the lattice contact and underpin the functional importance of this intermolecular polymerase interface.

The shortest distance in the crystal between the two segments that must be connected is actually between two adjacent polymerase molecules (intermolecular), rather than within the same polymerase molecule (intramolecular). The possibility of an intermolecular connection is supported by site specific cysteine mutations in the tip of the thumb domain (where the N- and C-terminal portions of the thumb are in close association) to give a disulfide crosslink that yields dimers of polymerase molecules (Hobson *et al.*, 2001). The significance of an intermolecular connection is discussed below, but a more complete structure of the poliovirus polymerase will be required to definitively resolve this issue.

3.3   Domain interactions and oligomerisation

The 3Dpol crystal lattice revealed two regions of extensive contacts between polymerase molecules. These interactions (Interfaces I and II) are much more extensive than typical crystal packing interactions (Hansen *et al.*, 1997). Interface I is formed between the front side of the thumb domain of one molecule and the back of the palm domain of an adjacent molecule. Interface II is formed between the tip of the thumb domain of one molecule and the base of the fingers domain on an adjacent molecule and involve the two ordered regions of the N-terminal domain. Interactions via interface I result in a fiber of polymerase molecules along a two-fold axis in the crystal; interactions via interface II result in the fibers coming together with a 120° crossing angle. These interactions can produce an oligomeric polymerase structure of indefinite size.

The original suggestion that polymerase-polymerase interactions are important for function came from experiments demonstrating that purified 3Dpol binds to RNA in a highly cooperative manner, and that this correlates with efficient template utilization (Pata *et al.*, 1995). Physical interactions between polymerase molecules in solution have been observed by chemical cross-linking (Pata *et al.*, 1995), and genetic interactions have been observed using the yeast two-hybrid system (Hope *et al.*, 1997).

Experiments to test the functional significance of the specific interactions observed in the crystal structure suggest that Interface I is important for RNA binding and that Interface II is important for catalytic activity. The importance of Interface I was explored by making 3Dpol that contains single and double mutations of residues involve in this interface (Hobson *et al.*, 2001). These mutations decrease RNA binding activity and concomitantly decrease template utilization efficiency. However, at high polymerase concentrations where RNA primer/template binding is observed, polymerase catalytic activity is not significantly diminished. In tissue culture, viruses containing the most disruptive interface I mutations were not viable.

Modelling of a duplex primer/template onto the structure of poliovirus polymerase, based on superimposing the homologous palm domains of the unliganded 3Dpol structure and the HIV-1 RT ternary complex, suggests that a monomer of 3Dpol would only bind to approximately one-half a turn of A-form RNA (Hansen *et al.*, 1997; Hobson *et al.*, 2001). Polymerase molecules interacting through interface I are aligned such that the adjacent polymerases could provide an extended RNA binding surface, which could explain why interface I is important for efficient RNA binding.

Interactions between the N-terminal domain and the thumb around interface II were disrupted by deletion and mutagenesis and resulted in polymerase molecules with severely decrease polymerisation activity, but RNA binding activities comparable to that of wild-type polymerase (Hobson *et al.*, 2001). If the N-terminal portion of the thumb domain is contributed by an adjacent molecule, as suggested above, then the association of two polymerase molecules via interface II is required to form a catalytically active enzyme. These experiments strongly suggest that the active form of poliovirus polymerase is an oligomer. Poliovirus RNA replication occurs in large membrane-associated complexes in the cytoplasm of infected cells. The interactions between polymerase molecules observed in the crystal lattice could contribute to the formation of these replication complexes. Sequences around both interfaces I and II are conserved among picornaviruses, suggesting that the interactions between polymerase molecules may also be conserved.

4. HEPATITIS C VIRUS POLYMERASE (NS5B)

4.1 Background

Hepatitis C virus has been identified as the main causative agent of nonA/nonB hepatitis, often establishing a persistent infection in 80% of infected (immuno-competent) individuals (Clarke, 1995, Houghton M, 1996) Infections lead to an increased propensity for the development of chronic liver disease such as cirrhosis and hepatocellular carcinoma. The virus is a significant cause of morbidity worldwide as recent estimates from WHO suggest that 3% of the total world population is persistently infected with HCV (Alter *et al.,* 1999, WHO WER-Report 2000). The current therapy, interferon-α treatment, is effective only for certain HCV genotypes, and, thus, chronic infections are a common reason for liver transplant surgery. As such it has significant economic implications for the Health Services around the world.

HCV forms a separate genus in the family Flaviviridae, the genome is a positive-stranded RNA molecule of approximately 9500 nucleotides that contains a single open reading frame encoding a polyprotein precursor of ~3100 amino acids. This precursor is cleaved co- and post-translationally presumably by a combination of cellular and viral proteases to yield a number of structural and non-structural proteins. Much work has been undertaken in an attempt to determine the function of these proteins and define their role in viral replication (Houghton, 1996). Three putative structural proteins have been identified at the N-terminus of the polyprotein. The first of these is a highly basic protein C that is likely to be the viral nucleocapsid or core antigen. The two viral glycoproteins E1 and E2 are situated downstream of core in the polyprotein and are followed by the non-structural proteins, designated NS2-NS5B.

As with other (+)-strand RNA viruses, a RNA-dependent RNA polymerase (NS5B) is involved in the initial synthesis of a complementary (-) RNA strand, which serves, in turn, as a template for the production of progeny (+)-strand RNA molecules. Sequence comparisons and multiple secondary structure predictions of NS5B, poliovirus 3Dpol and Φ6 RNA polymerase, the only known RNA-dependent RNA polymerase structures, indicate that the enzymes are functionally homologous, but, with the exception of the highly conserved fold of the palm domain, may be structurally somewhat different.

4.2   Apo structure

The RNA dependent RNA polymerase of Hepatitis C virus (NS5B) is a multi-domain protein of approximately 66kD (591 amino acid residues; Behrens *et al.* 1996). The crystallographic structure of HCV NS5B has been determined independently by several groups (Ago *et al.*, 1999; Bressanelli *et al.*, 1999, Lesburg *et al.*, 1999; Jaeger *et al.*, 2000; O'Farrell *et al.*, 2000) Before the crystallographic structures were available, however, sequence comparisons and multiple secondary structure predictions of NS5B have indicated that the domain organisation resembles that of PV 3Dpol and even HV-1 RT (Jaeger, unpublished). The analysis of these predictions also revealed that the location of secondary structure elements within motifs A and C is nearly identical to those found in previously published polymerase structures.

The four independently reported crystal forms of NS5B show a largely helical structure (41% helical and 12.7% β strand) with an elaborate arrangement of polymerase sub-domains: fingers, fingertips, palm, thumb, flap and C-terminal arm. The fingers domain carries two long extensions (L1, residues 11 – 44; L2, 138 – 161), which cross over to the thumb domain encircling the polymerase active site. Thus, the overall structure of HCV NS5B does not have the canonical U-shaped outline found in most of the classical DNA dep. DNA polymerases. The surface area buried in the interdomain contact between fingers and thumb is about 1600Å^2 (Jaeger et al 2000a). The first loop or extension (residues 11 – 44) is missing in HIV-1 RT, but the second loop occupies the same region in space as a β hairpin motif (containing residues 59-75 in HIV-1 RT). In RT, residues from this β hairpin are known to play a role in contacting the incoming nucleosidetriphosphate and the primer/template (Gosh *et al.*, 1994, Huang *et al.*, 1998). The connecting loops in the NS5B fingertips do not feature a significant amount of secondary structure and, thus, could act like flexible coils that adapt to the breathing motion of the polymerase during the catalytic cycle.

Near the N-terminus, between the base of the fingertips and the remainder of the domain there is a shallow and elongated groove that extends down into the active site. Whilst the feature is large enough to accommodate single stranded RNA, in the apo-polymerase this feature is filled with solvent molecules. The water molecules presumably mimic the oxygen atoms in the ribo-phosphate backbone.

As with all other RNA polymerases and reverse transcriptases, the palm domain forms the core of the NS5B molecule. It consists of a folding unit, which has been used frequently throughout evolution (Steitz et al., 1994; Holm and Sander, 1995; Hansen et al., 1997; Artymiuk et al. 1997). The NS5B palm consists of a three-stranded antiparallel β-sheet (β3, β6, and β7), a small helix following β3 and two α-helices (αJ and αK) supporting the antiparallel β-sheet. The tip of the β-sheet appropriately positions the trio of aspartate residues responsible for binding the two metal ions. Interestingly, in the complex of HCV polymerase (HC-J4 derived 1b genotype) with Mn2+ all three carboxylates are involved in coordination the metal ions (Jaeger, 2000). In previously reported co-crystal structures of DNA and RNA polymerases the second residue in motif C (Asp, Glu or Ser) is pointing away from the metal ions. HCV pol, however, requires all three negative charges to mediate productive binding of metal ions and incoming nucleotide (Jaeger, 2000).

The thumb domain of HCV polymerase is predominantly helical as is the case for all polymerase structures reported to date. The majority of the thumb is formed by a repeat of a two-helix motif akin to the "armadillo" (arm) repeats (Huber et al., 1997), which have been found in a number of unrelated molecules. Only the amino-terminal four a-helices in the NS5B thumb overlap with those in the poliovirus enzyme. Similar to other polymerases the surface area of the thumb facing the catalytic site is decorated with lysine and arginine residues. Replacement of theses residues by alanines virtually abolishes polymerase activity (Ferrari et al, 1999). Another highly positively charged region on the exterior face of the thumb (near the "arm repeat-like" fold) could conceivably be a site of interaction with other cellular or viral proteins (to form a functional replicase) (Bressanelli et al., 1999, O'Farrell et al., 2000). There is a distinct structural feature in the thumb that sets HCV polymerase apart from other polymerases. A 16 residue extension, termed the flap, protrudes into the active site cleft with residues being 10-12Å away from the strictly conserved active site aspartate residues. Together with residues 555-557 from the C-terminal end, the flap effectively blocks exit at the front of the polymerase domain. There is no room for nucleic acid, pyrophosphate or even water molecules to leave the polymerase active site in this direction. It seems more plausible that nucleoside triphosphates and pyrophosphate enter and leave the active site at the back of the polymerase domain. Site directed mutagenesis studies in loop2 of the fingers domain (Trowbridge and Jaeger, unpublished) and calculations of the local electrostatic potential of NS5B structures complexed with rUTP (Ago *et al.*, 1999; O'Farrell *et al.*, 2000) indicate that the back of the polymerase domain serves as an access tunnel for the incoming rNTPs.

The connection between the HCV palm and thumb forms a hairpin-like structure (motif E, residues 363 - 370) very similar to those found in poliovirus and in retroviral polymerases. In HCV NS5B, however, this segment displays fewer inter-strand hydrogen bonds and appears more flexible than in HIV-1 RT. The β-hairpin in RT (residues 225-236) has been called the "primer grip", whose proposed role is maintaining the primer terminus in an orientation appropriate for nucleophilic attack on an incoming dNTP (Jacobo-Molina *et al.*, 1993, Ghosh *et al.*, 1996). The residues within this hairpin also contribute side chains to the binding pocket for non-nucleoside RT inhibitors, such as Nevirapine and TIBO (Smerdon *et al.*, 1994, Tantillo *et al.*, 1994, Ren *et al.*, 1995). Interestingly, in HCV polymerase a hydrophobic pocket is formed by residues from motif E, the base of the thumb and the palm domain. The location of this cavity is virtually equivalent to the NNRTI binding pocket. The volume of the cavity is smaller so that Nevirapine is presumably unable to insert and prevent the complete closure of the thumb domain over the catalytic site in the holoenzyme. This movement has to occur upon substrate binding and during the translocation step in the polymerase catalytic cycle.

4.3 Binary complexes

Recently, detailed binary complexes of HCV polymerase with nucleosidetriphosphates and metal ions have been reported (Ago *et al.*, 1999, O'Farrell *et al.*, 2000). The complexes indicate how incoming nucleosidetriphosphates might access and how pyrophosphate might subsequently exit the active site without necessitating large structural changes. Furthermore, the apparently closed form of the HCV polymerase binds both ligands in productive fashion. This is inferred from numerous interactions with strictly conserved residues and a structural comparison with the terminated and arrested RT:DNA complex (Huang *et al.*, 1998).

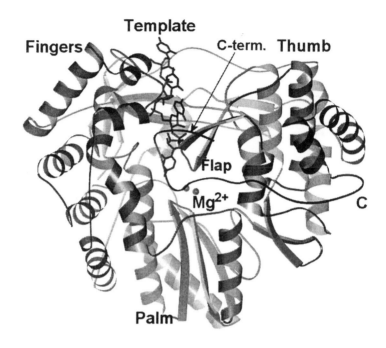

Figure 5. Ribbon diagram of HCV RNA polymerase (O'Farrell *et al.,* 2000). The metal ions are shown as spheres. The nucleic acid template between the fingers and the flap has been introduced by soaking a short oligonucleotide (rU$_5$) into a low-ionic strength crystal form of HCV (J4) polymerase in the absence of nucleosidetriphosphates to prevent turnover in the crystal lattice.

Unexpectedly, the absence of template RNA in HCV RdRp did not prevent the NTP from binding to the conserved residues in sequence motifs A, C and E. In E.coli DNA polI Klenow fragment and Taq polymerase similar soaks resulted in non-productive binding far away from the catalytic centre (Beese, *et al.,* 1993; Li *et al.,* 1998). In HCV polymerase, however, the α and β phosphates are in direct contact with both manganese ions. Interestingly, the γ phosphate of UTP points away from the active site aspartate residues and the two metal ions, instead one of the γ oxygens interacts with NH1 of Arg48. Lys51 and Arg222 are less than 3.5Å away from O2γ .The γ phosphate in the ternary RT/DNA complex overlaps more closely with the β phosphate in the NS5B/UTP complex. Thus, the pattern of interactions involving the β and γ phosphates is slightly different from those observed in other ternary polymerase complexes (T7 DNA polymerase, Taq polymerase, HIV RT, DNA polymerase β).

A complex of HCV (J4-derived) RNA polymerase with a short ssRNA (rU$_5$) intended to mimic the template reveals that the shallow groove on the fingers domain is occupied with the oligonucleotide. The RNA molecule was soaked into a low ionic strength crystal form of HCV (J4) polymerase in the absence of nucleosidetriphosphates to prevent turnover in the crystal lattice. The 3'-terminal residue of the short oligonucleotide (U5) binds within van der Waals distance of Lys141, Arg158, Ser228, Gln446, Gly448, Ser556 and Gly557. The preceding uracil residues, U4 and U3, make numerous interactions with His95, Phe162, Gln446 as well as other residues in the fingers, fingertips and the flap, which extends over from the thumb domain. This complex provides a working model for the introduction of a single stranded template RNA into the HCV polymerase active site using the shallow groove on the fingers domain as access channel. The overall conservation of the helical structure and the relative spatial location of the secondary structure elements would suggest that flavivirus and picornavirus polymerase recognise ssRNA in a similar fashion.

5. Φ6 RNA POLYMERASE

5.1 Background

Some double-stranded RNA viruses cause severe infections in humans whilst others affect economically relevant animals and plants (Fields, Knipe, Howley, 1996). Despite this breadth of host systems, most of the dsRNA viruses share a common replication strategy, whereby the virion is converted into a core particle that carries a transcriptase function. The transcriptase produces positive-sense, ssRNAs from genomic dsRNA templates. In turn, the ssRNA template is utilised for (-) strands synthesis, a process which is thought to occur inside the newly assembled core particles. The (-) strand RNA copy remains associated with the (+) strand template reconstituting the genomic dsRNA. The genome of the dsRNA bacteriophage Φ6 consists of three dsRNA segments: large (L), medium (M) and small (S) (Van Etten *et al.*, 1974). The Φ6 replicase consists of four protein species P1, P2, P4 and P7, all encoded on the L segment (Mindich *et al.*, 1988). P1 is the major structural protein assembled into a dodecahedral shell, P4 is a hexameric NTPase involved RNA packaging and P7 is a cofactor necessary for the efficient packaging. P2 (80kD, amino acids 719 to 2860) has been identified as a putative polymerase by comparative sequence alignments (Koonin 1989,

1991; Bruenn, 1991). The dsRNA polymerase is thought to be responsible for the catalysis of both (-) and (+) strand synthesis.

5.2   Overall structure of Φ6 apo-polymerase

The apo-structure of Φ6 dsRNA polymerase (P2) has been solved at 2Å resolution. The P2 protein is an 80kD protein that largely consists of α helices. The polymerase domain is very elaborate with many long loops and extension presumably forming interactions with the incoming substrates. Similar to HCV polymerase the N-terminal domain of P2 can be separated into two domains: fingers and fingertips. The fingertips, which consist of 3 long, extended loops contact the thumb domain and give the polymerase a very spherical shape. The Φ6 palm (see Figure 2; residues 316 – 331 and 397 - 478) essentially displays the canonical fold common to all polymerases (antiparallel β sheet with two supporting helices), but contains additional secondary structure elements that seem to extend the functionality of the domain (see below). The thumb is predominantly helical and, by contrast to HCV RNA polymerase, does not contain loops that extend into the active site ('flap'). The C-terminal residues (601 to 664), however, fold into a series of five helices which are closely associated with the main bulk of the polymerase domain. Helices 2 and 3 are in close contact with residues from fingers palm and thumb which form the active site cavity. However, the average temperature factor of the C-terminal domain is about twice as high as the remainder of the polymerase domain indicating a certain degree of flexibility. Overall, the structure of Φ6 polymerase resembles that of HCV polymerase and provides good evidence for an evolutionary link between dsRNA viruses and flaviviruses.

5.3   A complete initiation complex

The crystals structure of Φ6 RNA polymerase complexed with a short penta-oligonucleotide (5'-dTdTdTdCdC-3', the "DNA-form" of the 3' terminus of the Φ6 genome) shows that a narrow access tunnel in the fingers domain guides the template into the polymerase active site. In the absence of incoming nucleoside triphosphates the 3'-terminal base overshoots the active site such that it is no longer available to form an initiating Watson-Crick basepair. Instead the base inserts into a pocket formed by helices 2 and 3 in the C-terminal domain where it interacts with a distal glutamate residue.

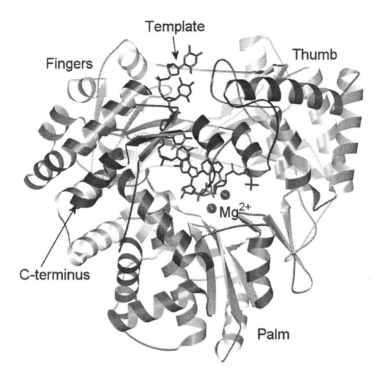

Figure 6. Ribbon diagram of Q6 RNA polymerase (Butcher *et al.*, 2001). Short soaks of this particular crystal form using GTP shows an unexpected re-arrangement of the templating bases. The 3'-terminal cytosines move away from a C-terminal binding pocket and reposition to form Watson Crick base pairs with the incoming GTP.

Short soaks of this particular crystal form using GTP shows an unexpected re-arrangement of the templating bases. The 3'-terminal cytosines move away from the C-terminal binding pocket and reposition to form Watson Crick base pairs with the incoming GTP. Upon removal of the template base the binding pocket becomes occupied by a glutamine side chain. In this initiation complex the first base in the primer strand rests flat against a tyrosine. Interestingly, the previously mentioned extensions on the palm of Φ6 polymerase provide a positively charged interaction surface and sufficient space for two GTP molecules to bind to the substrate tunnel at the back of the polymerase domain. The metal ions adjacent to the trio of conserved aspartate residues facilitate the binding of incoming nucleotides by adopting a somewhat unusual arrangement. In fact, this structure reveals a third metal binding site which has not been previously observed in any other polymerase complex. Only one of the metals binds in a position occupied in the terminated, arrested HIV-1 RT:DNA complex. The mechanism of 'overshooting' (in the absence of NTP) and 'back-ratcheting' (upon binding of NTP) to form the initiation complex in a de-novo priming scenario might be facilitated by the presence of two basic residues (Arg268 and Arg270). Both residues are functionally conserved among (+) strand and dsRNA viruses. However, the HCV polymerase C-terminal loop neither contains a strictly conserved aromatic residue equivalent to Tyr630 nor does it provide a specificity pocket that would facilitate overshooting of the template in the absence of complementary NTPs.

6.   CONCLUSIONS AND FUTURE DIRECTIONS

Many attempts have been made to establish the evolutionary relationships among the various polymerases (viral and cellular; RNA- and DNA- directed; synthesizing RNA or DNA; etc.) on the basis of sequence analysis. However, the sequence identity across the polymerase superfamily (and even within a polymerase family) is basically negligible and on its own is insufficient to determine relationships between polymerase families. Determination of polymerase structures has been essential for establishing which polymerase families share homologous domains and which do not.

The RNA-directed polymerase structures determined to date provide compelling evidence that there is an evolutionary link certainly between the different RNA dependent RNA polymerases. The relationships can be extended to RNA directed DNA polymerases if the structural conservation of the palm domains is taken into consideration. Furthermore, the overall architecture and organisation of polymerase sub-domains on the viral polyprotein is conserved. Despite the structural differences in the fingers, the functional role of this sub-domain in separating dsRNA and stem-loops, in guiding of the ssRNA template towards the active site and, perhaps most importantly, the critical role in nucleosidetriphosphate selection and fidelity (or lack thereof) is preserved. The thumb domains in all of the known RNA-directed polymerase structures are helical and play an important role in primer and duplex binding. Finally, the hairpin structure at the junction of palm and thumb (motif E) unifies RNA polymerases and reverse transcriptases and sets them apart from cellular DNA and RNA polymerases. Finally, the most universal feature of the polymerases is the two-metal-ion catalytic mechanism, being conserved throughout all polymerase classes.

Future efforts in this area will focus on larger protein:protein or protein/RNA complexes in an attempt to understanding the molecular recognition events and detailed workings of fully functional replicases. A deeper understanding of replication complexes involving cellular proteins as part of structural genomics efforts will open further avenues for the development of highly specific antivirals.

ACKNOWLEDGMENTS

The authors would like to thank Tom Steitz and Dave Rowlands for constant support, encouragement, and stimulating discussions. We gratefully acknowledge the assistance of Mark Saw, Damien O'Farrell and Rachel Trowbridge for help with illustrations and references. This manuscript was in part made possible by support from Action Research S/P/3206 and Yorkshire Cancer Research L253 to J.J. and a grant from the National Institutes of Health (J.P.).

REFERENCES

Ago, H., Adachi, T., Yoshida, A., Yamamoto, M., Habuka, N., Yatsunamj, K., Miyano, M., 1999. Crystal structure of the RNA-dependent RNA polymerase of hepatitis C virus. Structure Fold Des. 7, 1417-1426.

Agol VI, Paul AV, Wimmer E., 1999, Paradoxes of the replication of picornaviral genomes. Virus Res. Aug;62(2):129-47

Alter M. J., Kruszon-Moran D., Nainan O. V., McQuillan G. M., Gao F., Moyer L. A., Kaslow R. A., Margolis H. S., 1999, The Prevalence of Hepatitis C Virus Infection in the United States, 1988 through 1994. N Engl J Med 341:556-562.

Andino, R., E. Rieckhof, and D. Baltimore. 1990. A functional ribonucleoprotein complex forms around the 5' end of poliovirus RNA. Cell 63:369-380

Arnold JJ, Ghosh SK, Cameron CE., 1999, Poliovirus RNA-dependent RNA polymerase (3Dpol). Divalent cation modulation of primer, template, and nucleotide selection. J Biol Chem. Dec 24;274(52):37060-9.

Arts EJ, Le Grice SF., 1998, Interaction of retroviral reverse transcriptase with template-primer duplexes during replication. Prog Nucleic Acid Res Mol Biol. 58:339-93

Artymiuk PJ, Poirrette AR, Rice DW, Willett P., 1997, A polymerase I palm in adenylyl cyclase? Nature. 388(6637):33-4.

Astatke M, Ng K, Grindley ND, Joyce CM., 1998, A single side chain prevents Escherichia coli DNA polymerase I (Klenow fragment) from incorporating ribonucleotides. Proc Natl Acad Sci U S A. 95(7):3402-7.

Baltimore D., 1970, RNA-dependent DNA polymerase in virions of RNA tumour viruses. Nature. 226(252):1209-11

Bartenschlager, R., 1997, . Candidate targets for hepatitis C virus-specific antiviral therapy. Intervirology 40, 378-393.

Bateman A, Birney E, Durbin R, Eddy SR, Howe KL, Sonnhammer EL. ,2000, The Pfam protein families database. Nucleic Acids Res. 28(1):263-6)

Bedford E, Tabor S, Richardson CC., 1997, The thioredoxin binding domain of bacteriophage T7 DNA polymerase confers processivity on Escherichia coli DNA polymerase I. Proc Natl Acad Sci U S A. 94(2):479-84.

Beese LS, Derbyshire V, Steitz TA, 1993, Structure of DNA polymerase I Klenow fragment bound to duplex DNA. Science 1993 Apr 16;260(5106):352-5

Beese LS, Friedman JM, Steitz TA. , 1993, Crystal structures of the Klenow fragment of DNA polymerase I complexed with deoxynucleoside triphosphate and pyrophosphate. Biochemistry. 32(51):14095-101

Beese LS, Steitz TA, 1991, Structural basis for the 3'-5' exonuclease activity of Escherichia coli DNA polymerase I: a two metal ion mechanism. EMBO J 1991 Jan;10(1):25-33

Behrens SE, Tomei L, De Francesco R, 1996, Identification and properties of the RNA-dependent RNA polymerase of hepatitis C virus. EMBO J. 1996 Jan 2;15(1):12-22.

Brautigam CA, Steitz TA, 1998, Structural and functional insights provided by crystal structures of DNA polymerases and their substrate complexes. Curr Opin Struct Biol 1998 Feb;8(1):54-63

Brautigam CA, Steitz TA, 1998, Structural principles for the inhibition of the 3'-5' exonuclease activity of Escherichia coli DNA polymerase I by phosphorothioates. J Mol Biol 1998 Mar 27;277(2):363-77

Bressanelli, S., Tomei, L., Roussel, A., Incitti, I., Vitale, R.L., Mathieu, M., De Francesco, R., Rey, F.A., 1999. Crystal structure of the RNA-dependent RNA polymerase of hepatitis C virus. Proc. Natl. Acad. Sci. USA 96, 13034 - 13039.

Brillanti S., Miglioli M. and Barbara L., 1995, Combination antiviral therapy with ribavirin and interferon alpha in interferon alpha relapsers and non-responders: Italian experience. Journal of Hepatology 23: (Suppl. 2) 17-21

Bruenn JA., 1991, Relationships among the positive strand and double-strand RNA viruses as viewed through their RNA-dependent RNA polymerases. Nucleic Acids Res. 19(2):217-26.

Butcher SJ, Grimes JM, Makeyev EV, Bamford DH, Stuart DI. , 2001, A mechanism for initiating RNA-dependent RNA polymerization. Nature Mar 8;410(6825):235-40.

Cann IKO, Ishino Y, 1999, Archaeal DNA replication: identifying the pieces to solve a puzzle. Genetics 152:1249-1267.

Chermann JC, Barre-Sinoussi F, Dauguet C, Brun-Vezinet F, Rouzioux C, Rozenbaum W, Montagnier L., 1983, Isolation of a new retrovirus in a patient at risk for acquired immunodeficiency syndrome. Antibiot Chemother. 32:48-53

Choo, Q.-L., Kuo, G., Weiner, A.J., Overby, L.R., Bradley, D.W., Houghton, M., 1989. Isolation of a cDNA clone derived from a blood-born Non-A, Non-B viral hepatitis genome. Science 244, 359-364.

Clarke BE., 1995, Approaches to the development of novel inhibitors of hepatitis C virus replication. J Viral Hepat. 2(1):1-8.

Davis G.L., Balart L.A., Schiff E.R., Lindsay K., Bodenheimer H.C. Jr., Perrillo R.P., Carey W., Jacobson I.M., Payne J., Dienstag J.L., Van Thiel D.H., Tanburro C., Lefkowitch J., Alberth J., Meschievitz C., Ortego T.J., Gibas A. and the Hepatitis Interventional Therapy group, 1989,: Treatment of chronic Hepatitis C with recombinant interferon alpha : a multicenter randomized, controlled trial. N. Engl. J. Med. 321:1501-1506.

De Francesco R, Behrens SE, Tomei L, Altamura S, Jiricny J., 1996, RNA-dependent RNA polymerase of hepatitis C virus. Methods Enzymol. ;275:58-67.

Delarue M, Poch O, Tordo N, Moras D, Argos P, 1990, An attempt to unify the structure of polymerases. Protein Eng 3(6):461-7

Doublie S, Tabor S, Long AM, Richardson CC, Ellenberger T, 1998, Crystal structure of a bacteriophage T7 DNA replication complex at 2.2A resolution. Nature 391:252-258.

Eddy SR. , 1996, Hidden Markov models. Curr Opin Struct Biol. 6(3):361-5.

Eom SH, Wang J, Steitz TA, 1996, Structure of Taq polymerase with DNA at the polymerase active site. Nature 382:278-281.

Ensue R, Ren J, Ross C, Jones Y, Stammers D, Stuart D, 1995, Mechanism of inhibition of HIV-1 reverse transcriptase by non-nucleoside inhibitors. Nat Struct Biol. 2:303-308.

Ferrari E, Wright-Minogue J, Fang JW, Baroudy BM, Lau JY, Hong Z., 1999, . Characterization of soluble hepatitis C virus RNA-dependent RNA polymerase in Escherichia coli. J. Virol. 73, 1649-1654.

Fields BN, D.M. Knipe, P.M. Howley, R.M. Chanock, J.L. Melnick, T.P. Monath, B. Roizman, and S.E. Straus, eds., 1996, Virology. 3rd Edition. Lippincott-Raven, Philadelphia, PA.

Hansen, J.L.H., Long, A.M. & Schultz, S.C., 1997, Structure of the RNA-dependent RNA polymerase of poliovirus. Structure 5, 1109-1122.

Hobson SD, Rosenblum ES, Richards OC, Richmond K, Kirkegaard K, Schultz SC. , 2001, Oligomeric structures of poliovirus polymerase are important for function. EMBO J. 20(5):1153-63.

Holm L, Sander C., 1995, DNA polymerase beta belongs to an ancient nucleotidyltransferase superfamily.Trends Biochem Sci. 20(9):345-7.

Hope DA, Diamond SE, Kirkegaard K., 1997, Genetic dissection of interaction between poliovirus 3D polymerase and viral protein 3AB. J Virol. 71(12):9490-8.

Hope DA, Diamond SE, Kirkegaard K: Genetic dissection of interactions between poliovirus 3D pol and viral protein 3AB. J. Virol. 1997, 71:9490-9498.

Hopfner KP, Eichinger A, Engh R, Laue F, Ankenbauer W, Huber R, and Angerer B., 1999, Proc Natl Acad Sci U S A ;96(7):3600-5

Hopkins AL, Ren J, Ensue RM, Willcox BE, Jones EY, Ross C, Miyasaka T, Walker RT, Tanaka H, Stammers DK, Stuart DI., 1996, Complexes of HIV-1 reverse transcriptase with inhibitors of the HEPT series reveal conformational changes relevant to the design of potent non-nucleoside inhibitors. J Med Chem. Apr 12;39(8):1589-600.

Houghton M., 1996, Fields Virology, 3rd edition, Lippincott - Raven Publishers, Philadelphia.

Huang, H., Chopra, R., Verdine, G.L. & Harrison, S.C., 1998, . Structure of a covalently trapped catalytic complex of HIV-1 reverse transcriptase: implications for drug resistance. Science 282, 1669-1675.

Huber AH, Nelson J, and Weis WI., 1997, Three-Dimensional Structure of the Armadillo Repeat Region of -Catenin. Cell, Vol. 90, 871–882

Hwang, S.B., Park, K.-J., Kim, Y.-S., Sung, Y.C. & Lai, M.M.C. Virology 227, 439–446, 1997, .

Ishido, S., Fujita, T. & Hotta, H. Biochem. Biophys. Res. Commun. 244, 35–40

Ito, J., Braithwaite, D.K., 1991, Compilation and alignment of DNA polymerase sequences. NAR 19, 4045-4057.

Jablonski SA, Luo M, Morrow CD: Enzymatic activity of poliovirus RNA polymerase mutants with single amino acid changes in the conserved YGDD amino acid motif. J. Virol. 1991, 65:4565-4572.

Jacobo-Molina A, Ding J, Nanni RG, Clark AD Jr, Lu X, Tantillo C, Williams RL, Kamer G, Ferris AL, Clark P, et al., 1993, Crystal structure of human immunodeficiency virus type 1 reverse transcriptase complexed with double-stranded DNA at 3.0 A resolution shows bent DNA. Proc Natl Acad Sci U S A. 90(13):6320-4.

Jaeger J, 2000, BCA Spring Meeting April 2000, The British Crystallographic Association, Herriot-Watt University Edinburgh.

Jaeger J, Restle T, Steitz TA, 1998, The structure of HIV-1 reverse transcriptase complexed with an RNA pseudoknot inhibitor. EMBO J 17(15):4535-42

Jaeger J, Smerdon SJ, Wang J, Boisvert DC, Steitz TA, 1994, Comparison of three different crystal forms shows HIV-1 reverse transcriptase displays an internal swivel motion. Structure 2(9):869-76.

Jeruzalmi D, Steitz TA, 1998, Structure of T7 RNA polymerase complexed to the transcriptional inhibitor T7 lysozyme. EMBO J 17(14):4101-13

Jones, T.A., Zou, J.Y., Cowan, S.W. & Kjeldgaard, M., 1991, . Improved methods for building protein models in electron density maps and the location of errors in these models. Acta Crystallogr. A 47, 110-119.

Joyce CM, Steitz TA, 1994, Function and structure relationships in DNA polymerases. Annu Rev Biochem 1994;63:777-822

Kiefer JR, Mao C, Braman JC, Beese LS: Visualizing DNA replication in a catalytically active Bacillus DNA polymerase crystal. Nature 1998, 391:304-307.

Kiefer JR, Mao C, Hansen CJ, Basehore SL, Hofgrefe HH, Braman JC, Beese LS, 1997, Crystal structure of a thermostable Bacillus DNA polymerase I large fragment at 2.1 Å resolution. Structure 5:95-108.

Kim Y, Eom SH, Wang J, Lee DS, Suh SW, Steitz TA, 1995, Crystal structure of Thermus aquaticus DNA polymerase. Nature 1995 Aug 17;376(6541):612-6

Kitamura, N., B. L. Semler, P. G. Rothberg, G. R. Larsen, C. J. Adler, A. J. Dorner, E. A. Emini, R. Hanecak, J. J. Lee, S. van der Werf, C. W. Anderson, and E. Wimmer. 1981. Primary structure, gene organization and polypeptide expression of poliovirus RNA. Nature 291:547-553

Kohlstaedt LA, Wang J, Friedman JM, Rice PA, Steitz TA, 1992, Crystal structure at 3.5 A resolution of HIV-1 reverse transcriptase complexed with an inhibitor. Science 1992 Jun 26;256(5065):1783-90

Koonin EV, Gorbalenya AE, Chumakov KM., 1989, Tentative identification of RNA-dependent RNA polymerases of dsRNA viruses and their relationship to positive strand RNA viral polymerases. FEBS Lett. 252(1-2):42-6.

Koonin EV., 1989, The phylogeny of RNA-dependent RNA polymerases of positive-strand RNA viruses. J Gen Virol. 72 (Pt 9):2197-206.

Koonin, E. V., 1991, J. Gen. Virol. 72, 2197-2206.

Larder BA, Stammers DK., 1999, Closing in on HIV drug resistance. Nat Struct Biol Feb;6(2):103-6

Lee YF, Nomoto A, Detjen BM, Wimmer E., 1977, A protein covalently linked to poliovirus genome RNA. Proc Natl Acad Sci U S A. 74(1):59-63.

Lee, Y. F., Nomoto, A., Detjen, B. M. & Wimmer, E. A protein covalently linked to poliovirus genome RNA. Proc. Natl Acad. Sci. USA 74, 59-63 (1977).

Lesburg, C.A., Cable, M.B., Ferrari, E., Hong, Z., Mannarino, A.F., and Weber, P.C., 1999. Crystal structure of the RNA-dependent RNA polymerase from hepatitis C virus reveals a fully encircled active site. Nat. Struct. Biol. 6, 937-943.

Li Y, Kong Y, Korolev S, Waksman G., 1998, Crystal structures of the Klenow fragment of Thermus aquaticus DNA polymerase I complexed with deoxyribonucleoside triphosphates. Protein Sci. 7(5):1116-23.

Lohmann, V., Korner, F., Herian, U., Bartenschlager, R.,, 1997, . Biochemical properties of hepatitis C virus NS5B RNA-dependent RNA polymerase and identification of amino acid sequence motifs essential for enzymatic activity. J. Virol. 71, 8416-8428.

Lohmann, V., Roos, A., Komer, F., Koch, J.O., Bartenschlager, R.,, 1998, . Biochemical and kinetic analyses of NS5B RNA-dependent RNA polymerase of the hepatitis C virus. Virology 249, 108-118.

Luo, G., Hamatake, R.K., Mathis, D.M., Racela, J., Rigat, K.L., Lemm, J., Colonno, R.J.,, 2000, . De novo initiation of RNA synthesis by the RNA-dependent RNA polymerase of hepatitis C virus. J. Virol. 74, 851-863.

Marcellin P., Boyer N., Giostra E., Degott C., Courouce A.M., Degos F., Cappere H., Cales P., Couzigou P., Benhamou J.P., 1991, Recombinant human alpha interferon in patients with chronic non-A non-B hepatitis. A multicenter randomized controlled trial from france. Hepatology 13:393-397.

Matthews, B.W., 1968, Solvent content of protein crystals. J. Mol.Biol. 33, 491-497.

McDonald, J. P., V. Rapic-Otrin, J. A. Epstein, B. C. Broughton, X. Wang, A. R. Lehmann, D. J. Wolgemuth, and R. Woodgate., 1999, Novel human and mouse homologs of Saccharomyces cerevisiae DNA polymerase h. Genomics 60:20–30.

Merluzzi,V.J. et al., 1990, Inhibition of HIV-1 replication by a nonnucleoside reverse transcriptase inhibitor. Science, 250, 1411-1413.

Mindich L, Nemhauser I, Gottlieb P, Romantschuk M, Carton J, Frucht S, Strassman J, Bamford DH, Kalkkinen N., 1988, Nucleotide sequence of the large double-stranded RNA segment of bacteriophage phi 6: genes specifying the viral replicase and transcriptase. J Virol. 62(4):1180-5.

Minnick DT, Astatke M, Joyce CM, Kunkel TA., 1996, A thumb subdomain mutant of the large fragment of Escherichia coli DNA polymerase I with reduced DNA binding affinity, processivity, and frameshift fidelity. J Biol Chem. 271(40):24954-61.

Mitsuya,H., Yarchoan,R. and Broder,S., 1990, Molecular targets for AIDS therapy. Science, 249, 1533-1544.

Navaza J, Panepucci EH, Martin C., 1998, On the use of strong Patterson function signals in many-body molecular replacement. Acta Crystallogr D Biol Crystallogr. 54:817-21.

O'Farrell D, Trowbridge R, Rowlands DJ, Jaeger J, 2000, 7th International Meeting on Hepatitis C Virus and Related Viruses, Gold Coast, Brisbane.

Oh, J.W., Ito, T., Lai, M.M., 1999. A recombinant hepatitis C virus RNA-dependent RNA polymerase capable of copying the full-length viral RNA. J. Virol 73, 7694-7702.

Ohmori H, Friedberg EC, Fuchs RPP, Goodman MF, Hanaoka F, Hinkle D, Kunkel TA, Lawrence CW, Livneh Z, Nohmi T, Prakash L, Prakash S, Todo T, Walker GC, Wang Z, Woodgate R, 2001, The Y-family of DNA polymerases, Mol. Cell 8:7-8.

Ollis DL, Brick P, Hamlin R, Xuong NG, Steitz TA, 1985, Structure of large fragment of Escherichia coli DNA polymerase I complexed with dTMP Nature 1985 Feb 28-Mar 6;313(6005):762-6

Pata JD, Schultz SC, Kirkegaard K., 1995, Functional oligomerization of poliovirus RNA-dependent RNA polymerase. RNA.1(5):466-77.

Paul AV, Rieder E, Kim DW, van Boom JH, Wimmer E. , 2000, Identification of an RNA hairpin in poliovirus RNA that serves as the primary template in the in vitro uridylylation of VPg. J Virol. Nov;74(22):10359-70.

Paul AV, van Boom JH, Filippov D, Wimmer E., 1998, Protein-primed RNA synthesis by purified poliovirus RNA polymerase. Nature. 393(6682):280-4.

Pelletier H, Sawaya MR, Kumar A, Wilson SH, Kraut J, 1994, Structures of ternary complexes of rat DNA polymerase beta, a DNA template-primer and ddCTP. Science 264:1891-1903

Pelletier, J., and N. Sonnenberg. 1988. Internal initiation of translation of eukaryotic mRNA directed by a sequence derived from poliovirus RNA. Nature 334:320-325

Poch O, Sauvaget I, Delarue M, Tordo N, 1989, Identification of four conserved motifs among the RNA-dependent polymerase encoding elements. EMBO J 1989 Dec 1;8(12):3867-74

Polesky AH, Steitz TA, Grindley ND, Joyce CM, 1990, Identification of residues critical for
the polymerase activity of the Klenow fragment of DNA polymerase I from Escherichia
coli. J Biol Chem 1990 Aug 25;265(24):14579-91

Reichard O., Yun Z.B., Sonnerborg A., Weiland O., 1993, Hepatitis C viral RNA titres in
serum prior to, during, and after oral treatment with ribavirin for chronic hepatitis C. J.
Med. Virol. 41: 99-102.

Ren J, Ensue R, Garman E, Somers D, Ross C, Kirby I, Keeling J, Darby G, Jones Y, Stuart
D, et al. High resolution structures of HIV-1 RT from four RT-inhibitor complexes. Nat
Struct Biol. 1995 2:293-302.

Rice, C. M., 1996, in Virology, eds. Fields, B. N., Knipe, D. M., Howley, P. M., Chanock, R.
M., Melnick, J. L., Monath, T. P., Roizman, B. & Straus, S. E. (Lippincott, Philadelphia),
vol. 1, pp. 931-959.

Rodgers DW, Gamblin SJ, Harris BA, Ray S, Culp JS, Hellmig B, Woolf DJ, Debouck C,
Harrison SC., 1995, The structure of unliganded reverse transcriptase from the human
immunodeficiency virus type 1. Proc Natl Acad Sci U S A. 92(4):1222-6.

Rodriguez AC, Park H-W, Mao C, Beese LS, 2000, Crystal structure of a pol alpha family
DNA polymerase from the hyperthermophilic archaeon *Thermococcus sp.* 9°N-7. J. Mol.
Biol., 299: 447-462.

Rodriguez-Wells V, Plotch SJ, DeStefano JJ. , 2001, Primer-dependent synthesis by
poliovirus RNA-dependent RNA polymerase (3D(pol)). Nucleic Acids Res. 29(13):2715-
24.

Sawaya MR, Prasad R, Wilson SH, Kraut J, Pelletier H: Crystal structures of human DNA
polymerase beta complexed with gapped and nicked DNA: evidence for an induced fit
mechanism. Biochemistry 1997, 36:11205-11215.

Smardon A, Spoerke JM, Stacey SC, Klein ME, Mackin N, Maine EM. , 2000, EGO-1 is
related to RNA-directed RNA polymerase and functions in germ-line development and
RNA interference in C. elegans. Curr Biol. 10(4):169-78.

Smerdon S, Jaeger J, Wang J, Kohlstaedt,L.A., Chirino,A.J., Friedman,J.M., Rice,P.A. and
Steitz,T.A., 1994, Structure of the binding site for nonnucleoside inhibitors of the reverse
transcriptase of human immunodeficiency virus type 1. Proc. Natl Acad. Sci. USA, 91,
3911-3915.

Smerdon SJ, Jaeger J, Wang J, Kohlstaedt LA, Chirino AJ, Friedman JM, Rice PA, Steitz TA,
1994, Structure of the binding site for nonnucleoside inhibitors of the reverse transcriptase
of human immunodeficiency virus type 1. Proc Natl Acad Sci U S A 1994 Apr
26;91(9):3911-5

Steitz TA, 1998, A mechanism for all polymerases. Nature 391(6664):231-2

Steitz TA, Smerdon S, Jaeger J, Wang J, Kohlstaedt LA, Friedman JM, Beese LS, Rice, 1993,
Two DNA polymerases: HIV reverse transcriptase and the Klenow fragment of
Escherichia coli DNA polymerase I. PA Cold Spring Harb Symp Quant Biol 1993;58:495-
504

Steitz TA, Smerdon SJ, Jaeger J, Joyce CM, 1994, A unified polymerase mechanism for
nonhomologous DNA and RNA polymerases. Science 266(5193):2022-5

Steitz TA, Steitz JA., 1993, A general two-metal-ion mechanism for catalytic RNA. Proc Natl Acad Sci U S A. 90(14):6498-502.

Sun, X.-L., Johnson, R.B., Hockinan, M.A., Wang, Q.M., 2000, De novo RNA synthesis catalyzed by HCV RNA-de-pendent RNA polymerase. Biochem. Biophys. Res. Comm. 268, 798–803.

Temin HM, Mizutani S., 1970, RNA-dependent DNA polymerase in virions of Rous sarcoma virus. Nature.226(252):1211-3.

Tomei L, Vitale RL, Incitti I, Serafini S, Altamura S, Vitelli A, De Francesco R , 2000, Biochemical characterization of a hepatitis C virus RNA-dependent RNA polymerase mutant lacking the C-terminal hydrophobic sequence. J Gen Virol. 2000 Mar; 81 Pt 3:759-67.

Van Etten JL, Vidaver AK, Koski RK, Burnett JP., 1974, Base composition and hybridization studies of the three double-stranded RNA segments of bacteriophage phi 6. J Virol. 13(6):1254-62.

Wang J, Sattar AK, Wang CC, Karam JD, Konigsberg WH, Steitz TA., 1997, Crystal structure of a pol alpha family replication DNA polymerase from bacteriophage RB69. Cell. 1997 Jun 27;89(7):1087-99.

WHO, 1999, Global surveillance and control of hepatitis C. J. Viral Hepatitis 6, 35-47.

WHO, Weekly Epidemiological Record, Vol. 75, 3, 2000
http://www.who.int/emc/diseases/hepatiti

Yamashita T, Kaneko S, Shirota Y, Qin W, Nomura T, Kobayashi K, Murakami S., 1998,. RNA-dependent RNA polymerase activity of the soluble recombinant hepatitis C virus NS5B protein truncated at the C-terminal region. J. Biol. Chem. 273,15479-15486.

Zhao Y, Jeruzalmi D, Moarefi I, Leighton L, Lasken R, Kuriyan J., 1999, Crystal structure of an archaebacterial DNA polymerase. Structure 7:1189-1199.

Zhong, W., Uss, A.S., Ferrari, E., Lau, J.Y., Hong, Z., 2000, De novo initiation of RNA synthesis by hepatitis C virus nonstructural protein 5B polymerase. J. Virol. 74, 2017–2022.

Zhou B, Pata J, Steitz TA , 2001, The structure of the dinB homolog of S.solfataricus DNA bypass polymerase.

Chapter 2.2

Marburg Virus Replication and Nucleocapsid Formation: Different Jobs, Same Players

Elke Mühlberger and Stephan Becker
Institut für Virologie, Philipps-Universität Marburg, Robert-Koch-Str. 17, 35037 Marburg, Germany.

1.   INTRODUCTION

1.1   Marburg Virus Epidemics

Marburg virus (MBGV), the prototype of an emerging viral agent, is originated in Africa and causes a severe hemorrhagic disease in humans and non-human primates with high case fatality rates. The virus was first isolated in 1967 when laboratory workers in Marburg, Frankfurt (Germany), and Belgrade (Yugoslavia) handling with African green monkeys imported from Uganda became severely ill. 22% of the infected patients died despite aggressive medical management (Martini, 1971; Slenczka, 1999). Since then, only sporadic episodes of MBGV hemorrhagic fever occurred in Zimbabwe, Uganda, Kenya, and the Democratic Republic of Congo. Two accidental laboratory infections are reported from Russia (Slenczka, 1999). The last known outbreak occurred from 1998 to 2000 among miners who worked illegally in a gold mine in Durba, Democratic Republic of Congo. This outbreak was associated with a 83% case fatality rate among 75 patients (Muyembe-Tamfum *et al.*, 2000). Interestingly, sequencing of viral RNA obtained from clinical samples revealed a high degree of nucleotide diversity among the different virus isolates. Thus, several distinct MBGV lineages were associciated with the outbreak (Swanepoel *et al.*, 2000).

Structure-Function Relationships of Human Pathogenic Viruses, Edited by
Holzenburg and Bogner, Kluwer Academic/Plenum Publishers, New York, 2002

Transmission occurs by direct contact with blood, semen and other body fluids. Since there is no evidence for airborne transmission (Martini, 1971), it is presumed that MBGV is zoonotic. However, the natural host is still unknown. The only animals known to be involved in MBGV outbreaks are monkeys. Since experimentally infected monkeys died with a 100% fatality rate (Fisher-Hoch and McCormick, 1999), it is not likely that monkeys are the natural reservoir for MBGV.

1.2   The Disease

Following an incubation period of 4-10 days, the clinical symptoms of a MBGV infection begin with a sudden onset of fever and headache accompanied by diarrhoe, vomiting, hepatitis, and conjunctivitis. After the first week of illness, a characteristic maculopapular rash is noted. For about one third of the patients a severe hemorrhagic diathesis is observed leading to bleeding from puncture sites, the gastrointestinal and the urogenital tract. Death usually occurs in the second week after onset of the disease (Stille and Böhme, 1971; Martini, 1971; Gear *et al.*, 1975).

Due to its high pathogenicity, the lack of a specific treatment, and the absence of a vaccine MBGV is classified as a biosafety level 4 agent.

1.3   Pathogenesis and and Target cells

Studies describing the pathologic features of filovirus infections were either performed with human tissues collected at autopsies or by using experimentally infected animals. These studies revealed that the pathological symptoms in MBGV and Ebola virus (EBOV) infections are similar. MBGV causes a systemic infection and displays a widespread organ distribution (Geisbert and Jaax, 1998). As shown for post-mortem material, almost all organs are damaged by MBGV infection. The damage is related to necrosis observed in many internal organs including liver, spleen, kidney, and gonads (Rippey *et al.*,1984; Zaki and Goldsmith, 1999). The primary target cells, which are infected immediately after virus entry, are macrophages (Ryabchikova *et al.*, 1996). Since MBGV replicates well in macrophages (Feldmann *et al.*, 1996), it is assumed that infected and circulating mononuclear phagocytes are involved in virus spread (Ryabchikova *et al.*, 1999). Virus release from migrating macrophages results in subsequent infection of interstitial and parenchychmal cells leading to the observed pantropic effect. At late time of infection, virus production in mononuclear phagocytic cells, hepatocytes, adrenocortical cells, endothelial cells, and fibroblasts is observed (Murphy *et al.*, 1971; Ryabchikova *et al.*, 1996).

1.4   The Virus

MBGV and the closely related EBOV are the only members of the family Filoviridae which belongs together with the paramyxo-, rhabdo-, and bornaviruses to the order Mononegavirales. A common feature of all Mononegavirales is the possession of a nonsegmented negative-sense single-stranded (NNS) RNA genome. The filamentous filoviral particles are enveloped and contain a central core formed by the RNA genome and associated proteins (Figure1). The genomic RNA of MBGV is 19.1 kb in length (EMBL Nucleotide Sequence Database, accession number Z12132; Bukreyev *et al.*, 1995) with a coding capacity for 7 structural proteins (Feldmann *et al.*, 1992).

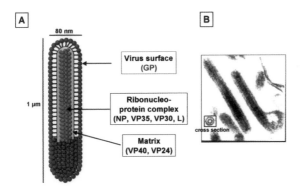

Figure 1: Structure and protein composition of MBGV virions. A. Scheme of a MBGV particle. The proteins involved in formation of the respective viral structure are parenthesized. B. Electron micrograph of MBGV particles. The inset shows a cross section through a MBGV particle (electron migrographs by courtesy of L. Kolesnikova).

1.4.1   Structural Proteins

The only envelope protein of MBGV is the glycoprotein GP (220 kDa) which is encoded by the fourth gene. It is inserted in the viral membrane and forms spikes on the surface of the virion (Figure 1; Will *et al.*, 1993). GP is highly glycosylated with over half its apparent molecular weight attributed to N- and O-linked sugar side chains (Becker *et al.*, 1996; Feldmann *et al.*, 1991). As a surface protein, GP is assumed to be responsible for receptor binding and fusion. So far, two receptors have been identified for MBGV. One is the asialoglycoprotein receptor expressed on hepatocytes (Becker, Spiess, and Klenk, 1995) which represent a main target for MBGV infection (Ryabchikova, Kolesnikova, and Netesov, 1999). However, since other cell types lacking the asialoglycoprotein receptor like endothelial cells are also susceptible for MBGV infection (Schnittler *et al.*, 1993), it has been postulated that additional receptor(s) also mediate virus entry. Very recently, the folate receptor-α has been shown to be an essential cofactor for the entry of both filoviruses into target cells (Chan et al., 2001). Due to its exposed position at the surface of both, infected cells and viral particles, GP is likely to be a main target for the immune system of the infected host.

Two putative matrix proteins with an apparent molecular weight of 38 kD (VP40) and 28 kDa (VP24) are encoded by the third and sixth gene. Both proteins are presumed to be located in the space between viral membrane and the nucleocapsid complex (Becker and Mühlberger, 1999).

The remaining four MBGV proteins are tightly associated with the viral RNA forming the ribonucleocapsid complex (Becker *et al.*, 1998). These are the nucleoprotein NP (94 kDa) encoded by the first gene (Becker *et al.*, 1994; Sanchez *et al.*, 1992), the polymerase cofactor VP35 (35 kDa) encoded by the second gene (Mühlberger *et al.*, 1998), VP30 (32 kDa) encoded by the fifth gene, and the major component of the polymerase complex L (220 kDa) encoded by the seventh gene (Mühlberger *et al.*, 1992). The nucleocapsid proteins play a dual role in the viral replication cycle: they are involved in virus morphogenesis as structural components (Kolesnikova *et al.*, 2000), and they catalyze replication and transcription of the RNA genome.

1.4.2   Genome Organization

The seven genes are arranged in a linear order on the viral RNA genome. They are either separated by short intergenic regions or they overlap (Figure 2A; Feldmann *et al.*, 1992). Each gene consists of the respective ORF and long nontranslated sequences flanking the coding regions (Figure 2B). The gene boundaries are marked by highly conserved transcriptional start and stop signals (Bukreyev *et al.*, 1995; Feldmann *et al.*, 1992; Mühlberger *et al.*, 1996). At the extreme genome ends, short extragenic regions are localized called leader and trailer (Figure 2A). These regions contain the encapsidation signals as well as the replication and transcription promoters (Mühlberger *et al.*, 1998).

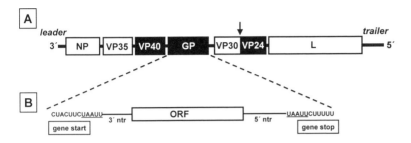

Figure 2: MBGV genome organization. A. Schematic presentation of the single-stranded, non-segmented RNA genome of MBGV. The seven genes are depicted as boxes, non-transcribed regions (leader, trailer, intergenic regions) are depicted as black lines. The overlap between the VP30 and VP24 genes is indicated by an arrow. Genes encoding nucleocapsid proteins are shown in white, genes encoding matrix or surface proteins are shown in black. B. Typical structure of a MBGV gene. Each gene is flanked by conserved transcription start and stop signals. The underlined pentamer is part of the start as well as the stop signal. The non-translated regions (ntr) are depicted as black lines, the ORF as a white box.

1.4.3 Replication Cycle

Virus entry is presumably mediated by the only surface protein GP. Whether fusion takes place at the cell membrane as described for paramyxoviruses or in endosomal vesicles is yet not known. Transcription as well as replication take place in the cytoplasm of the infected cell. During transcription the encapsidated negative-sense RNA genome is transcribed in seven monocistronic positive-sense mRNA species. Transcription starts precisely at the first nucleotide of the highly conserved transcription start signals (Figure 2B). The resulting mRNA species are polyadenylated at their 3´ ends. Since the encapsidation signals are located at the extreme genome ends within non-transcribed regions, none of the mRNA species is encapsidated. As template for polyadenylation serves a cluster of five to six U residues which is part of the transcription termination signal (Figure 2B; Feldmann *et al.*, 1992; Mühlberger *et al.*, 1996). It is assumed that generation of the polyA tail is due to a stuttering mechanism of the transcriptase complex. Viral protein synthesis occurs either at free ribosomes or in the case of GP at the rER. It is still unknown whether protein synthesis is a prerequisite for genome replication. During replication the negative sense RNA genome is transcribed in a full-length positive-sense antigenome which in turn serves as a template for production of new genomes. Both, antigenomes and genomes are encapsidated by the nucleocapsid proteins (Kolesnikova *et al.*, 2000; Mühlberger *et al.*, 1998). Assembly of the viral particles implies interaction of the preformed nucleocapsids with the matrix and the surface proteins. However, there are only few informations about the mechanisms underlying the process of viral assembly. The mature virions leave the cells by budding at the plasma membrane (Geisbert and Jahrling, 1995; Schnittler *et al.*, 1993).

2.   FUNCTION OF THE NUCLEOCAPSID PROTEINS

2.1   Reconstituted Replication and Transcription System established for Marburg virus

Approximately a decade ago, genetic engineering and genome manipulation of negative-sense RNA viruses was almost impossible. The major problem was that naked RNA genomes are not infectious. When new genomes are synthesized in the infected cell they are simultaneously packaged by the nucleocapsid proteins, and only these encapsidated forms are functional. Thus, it seemed to be impossible to manipulate negative-sense RNA genomes using recombinant DNA technology. The first approaches towards a reverse genetics system were reported by Luytjes et al. (1989) who described replication and transcription of an in vitro encapsidated RNA segment which was transfected in influenza virus-infected cells. Meanwhile, reverse genetics systems have been described for various segmented and non-segmented negative-sense RNA viruses without the need of in vitro packaging (Conzelmann, 1996). The basic system for NNS viruses was established by Pattnaik et al. (Pattnaik *et al.*, 1992) and functions as follows: The genes encoding the nucleocapsid proteins are cloned under the control of the T7 RNA polymerase promoter. Then, cells are infected with a recombinant vaccinia virus encoding the T7 RNA polymerase and thereafter transfected with the above mentioned constructs, thus leading to expression of the nucleocapsid proteins. The same cells are transfected simultaneously with a plasmid containing a virus-specific minigenomic cDNA also under the control of the T7 RNA polymerase promoter. Transcription results in a minigenomic RNA which is accepted as a template for replication and transcription by the nucleocapsid proteins.

A similar system has been established for MBGV (Figure 3A). Due to its low cytopathic effects, we used the attenuated vaccinia virus strain MVA-T7 for T7 RNA polymerase expression in cells (Sutter *et al.*, 1995). The constructed MBGV-specific minigenome consists of a reporter gene, the CAT gene, which is flanked by the short 3´ leader and 5´ trailer regions of the authentic MBGV genome (Figure 3B). Since it has been observed for other NNS viruses that the artificial minigenomes are only functional when they start precisely with the first nucleotide of the 3´ end, the sequence of the hepatitis delta ribozyme was inserted adjacent to the leader region. The ribozyme is removed via self-cleavage from the primary minigenomic RNA transcript, thus generating an exact 3´ end (Figure 3B; Pattnaik *et al.*, 1992). If the cleaved minigenome is accepted as a template for virus-specific replication and transcription, this can be monitored either by CAT gene expression or by detection of the replicated and transcribed RNA species.

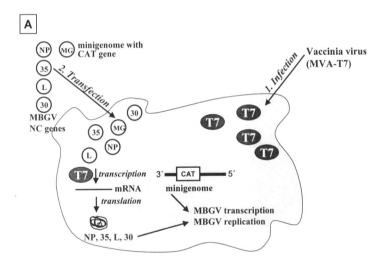

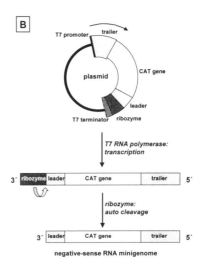

Figure 3: Model of the reconstituted MBGV replication and transcription system. A. HeLa cells are infected with MVA-T7 and subsequently transfected with plasmids encoding the four nucleocapsid proteins and the minigenome. The nucleocapsid protein genes are transcribed by the T7 RNA polymerase and translated by the cellular translation machinery. The gene coding for the minigenome is also transcribed by the T7 RNA polymerase. Finally, the synthesized RNA minigenome serves as a template for the replication and transcription complexes formed by the nucleocapsid proteins. B. Generation of a negative-sense MBGV-specific minigenome with precise 3´ and 5´ends.

2.2   Functional Analysis of the Nucleocapsid Proteins

Using the reverse genetics system it was shown that three of the four MBGV nucleocapsid proteins were essential and sufficient to support replication and transcription. These are NP, VP35, and L (Mühlberger *et al.*, 1998). To discriminate between replicated and transcribed RNA products, we took advantage of the fact that transcribed mRNA is polyadenylated, not packaged and therefore nuclease-sensitive, whereas replicated RNA species are not polyadenylated but encapsidated by the nucleocapsid proteins and hence resistant against nuclease treatment. Thus, transcribed mRNA species could be purified from lysates of MVA-T7-infected and transfected cells by oligo(dT) binding, and replicated RNA species were selected by nuclease-treatment of total RNA with following Northern blot analyses (Figure 4).

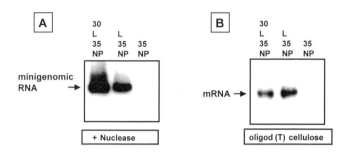

Figure 4: NP, VP35, and L are sufficient to support replication and transcription. A. Northern blot analysis of nuclease-resistant MBGV-specific RNA species indicating replication activity. B. Northern blot analysis of MBGV-specific polyadenylated mRNA species bound to oligo(dT) cellulose. The presence of these RNAs indicates transcription acitivity.

Below, it is described that NP is the major component for nucleocapsid formation and is presumed to be the encapsidating protein. Since encapsidation of the RNA genome is the prerequisite for replication and transcription, NP is essential for both processes. The other two proteins, VP35 and L, form the active polymerase complex which catalyzes synthesis of the RNA strands (Mühlberger *et al.*, 1999).

For the wide majority of other NNS viruses it is also true that three proteins are sufficient to mediate replication and transcription. This indicates that MBGV NP, VP35, and L are homologous to the nucleocapsid proteins NP/N, P, and L of rhabdo- and paramyxoviruses. Remarkably, the phosphoprotein P is highly phosphorylated among these viruses whereas MBGV VP35 is only very weakly phosphorylated (Becker and Mühlberger, 1999). The fourth MBGV nucleocapsid protein VP30, however, which is phosphorylated, was found to be dispensable for replication and/or transcription in the MBGV minigenome system.

Concerning VP30, the situation is totally different for the closely related EBOV. Here, the same three nucleocapsid proteins are sufficient for replication, however, in contrast to MBGV transcription, is strongly increased by VP30 (Mühlberger *et al.*, 1999). Thus, EBOV VP30 acts as a transcription activator.

3.   STRUCTURE OF THE NUCLEOCAPSID COMPLEX

3.1   Complexes between the different nucleocapsid proteins

Infection of target cells with MBGV induces the formation of cytoplasmic inclusion bodies which could be detected 12 h after infection (Ryabchikova *et al.*, submitted). By immunofluorescence analysis, these inclusions were shown to contain MBGV nucleocapsid proteins (Becker *et al.*, 1998), and ultrastructural analysis revealed that the inclusions consist of aggregated preformed nucleocapsids (Geisbert and Jahrling, 1995; Peters et al., 1971).

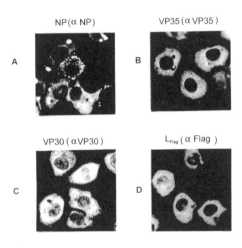

Figure 5: Distribution of singly expressed nucleocapsid proteins in HeLa cells. HeLa cells on glass cover slips were infected with vTF7-3 and subsequently transfected with plasmids encoding either NP, VP35, VP30, or L as indicated at the top of the panels. L was tagged with a FLAG epitope. At 8 h p. i., cells were fixed and probed with specific antibodies against the respective proteins. Used antibodies are given at the top of the panels.

Analyses of protein interactions among the nucleocapsid proteins which might lead to the formation of nucleocapsids and, finally, to the formation of the inclusions, suggested the presence of several complexes which were formed by two or more of the nucleocapsid proteins. The central player seems to be NP. Complexes were found between NP and VP35, NP and VP30, and between NP and NP. Additionally, NP is involved in a trimeric complex together with VP35 and L. All of these interactions resulted in the formation of intracytoplasmic inclusions which are induced by NP. This was demonstrated by the observation that only NP forms large intracytoplasmic aggregates when it is expressed alone (Figure 5, panel A) whereas the distribution of the singly expressed VP35, VP30, and L is either homogeneous (VP35 and L, panels B and D) or granular-like (VP30, panel C). Upon coexpression, the binding partners of NP (VP35, VP30) are redistributed into the NP-induced aggregates which resemble the large inclusions of nucleocapsid proteins formed during MBGV infection (Figure 6; Becker *et al.*, 1998). Beside the complexes which form inclusions, another complex was detected between VP35 and L which is homogeneously distributed in the cytosol (Becker *et al.*, 1998).

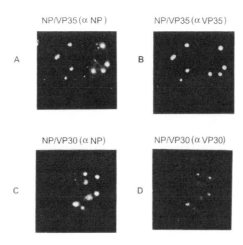

Figure 6: Interaction of VP35 and VP30 with NP. HeLa cells on glass cover slips were infected with vTF7-3 and subsequently cotransfected with plasmids encoding NP and VP35 (A and B), or NP and VP30 (C and D). At 8 h p. i., cells were fixed and probed with specific antibodies against the respective proteins. Used antibodies are given at the top of the panels.

Interactions of nucleocapsid proteins are not unique to the filoviruses. Complexes between the first and the second gene product, which parallels the interaction between NP and VP35 of MBGV, have been observed with all of the paramyxo- and rhabdoviruses (Davis *et al.*, 1986; Horikami *et al.*, 1992; Huber *et al.*, 1991; Masters and Banerjee, 1988; Ryan *et al.*, 1990; Schwemmle *et al.*, 1998). The formation of intracytoplasmic inclusions containing complexes of N (NP) and P, is also a common feature among Mononegavirales which was detected with measles virus, respiratory syncytial virus, human parainfluenza virus, and rabies virus (Chenik *et al.*, 1994; Garcia-Barreno *et al.*, 1996; Nishio *et al.*, 1996; Spehner *et al.*, 1997).

The different detected complexes between the MBGV nucleocapsid proteins are presumed to play a dual role duroing MBGV replication cycle. they are in volved in the formation of the nucleocapsid (NP-NP, NP-VP35, NP-VP30) and they catalyze the process of viral replication and transcription (see 2.2).

Of special interest is the interaction between NP and VP30. A putative homologue of this complex among the Mononegavirales can only be found in the subfamily pneumovirinae whose nucleocapsid complex is also formed by 4 proteins. Here, M2-1 accomplishes the conserved triumvirate of N, P, and L. (Garcia *et al.*, 1993). M2-1 interacts with N and is involved in the transcription of the respiratory syncytial virus genome (Collins *et al.*, 1996). However, it is not yet clear whether the interaction between N and M2-1 is necessary for the transcription elongation activity of M2.

3.2   NP as the main determinant of nucleocapsid formation

As mentioned above, NP is the central component of the nucleocapsid complex. The N-terminus of NP is hydrophobic and displays a 55% homology to the EBOV NP. The hydrophilic C-terminus is less conserved with 23% identity to EBOV Zaire NP. Sequence homologies to the nucleoproteins of other Mononegavirales can be detected only in restricted areas of the N-terminus (Sanchez *et al.*, 1992). Because of its abundance in the virion and its position in the genome, NP was presumed to be the protein responsible for the encapsidation of the RNA genome.

3.2.1   Phosphorylation of NP

Intracellular NP exists in two forms which differ in their phosphorylation state and can be distinguished by their electrophoretic mobility (nonphosphorylated NP: 92 kDa, phosphorylated NP: 94 kDa; Becker *et al.*, 1994). However, in virions only the phosphorylated form could be detected. This finding led to the assumption that phosphorylation of NP could play a modulatory role in RNA-protein and/or protein-protein interaction.

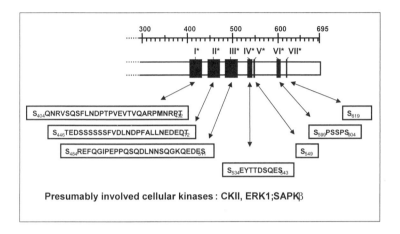

Figure 7: Schematic representation of phosphorylated regions in the C-terminus of NP (aa 300-695, shown by scale). Regions I*-VII* were determined by comparison of the NP primary structure with the information of phosphorylation state analyses. Shown is also the amino acid sequence of regions I*-VII*.

Phosphorylation state analysis of NP revealed that regardless of whether NP was isolated from virions or from transiently NP-expressing eukaryotic cell lines, the phosphoserine:phosphothreonine ratio was 85:15. Phosphotyrosine was absent. As NP displayed no autophosphorylating activity (unpublished data), these findings suggest that mainly ubiquitous cellular kinases are involved in phosphorylation of NP.

Phosphorylation of NP exclusively occurs in the hydrophilic C-terminus. Here, 7 regions can be defined containing one or more phosphoserine or - threonine residues (Figure 7).

Although the N-terminus of NP contains several potential phosphorylation acceptor sites for ubiquitous cellular protein kinases, none of them is actually used. Thus, MBGV NP shows the same topography of phosphorylation sites as phosphorylated nucleoproteins of other NNS-RNA viruses (Hsu and Kingsbury, 1982; Kawai *et al.*, 1997; Liston *et al.*, 1997; Sokol and Clark, 1973). One explanation for this phenomenon could be a similar structural organization of the nucleoproteins which is reflected in similar hydropathy profiles (Sanchez *et al.*, 1992). While the hydrophobic N-terminal half of MBGV NP might be buried in the interior of the protein, sequences in the acidic C-terminus could be accessible for protein kinases.

The amino acid context of the determined phosphorylated amino acid suggests that several cellular protein kinases are involved in the phosphorylation of NP. Region VI and VII contain phosphorylated serine residues in direct vicinity of proline residues pointing to the activity of proline-dependent kinases such as members of the mitogen-activated protein kinases (MAPK) (Davis, 1993; Lötfering *et al.*, 1999). Other phosphorylation sites suggest the activity of protein kinase CKII, PKC, and PKA.

So far, the function of phosphorylation of NP is not determined. Phosphorylation of region VI and VII did not have a significant role for the protein's function during replication and transcription (Lötfering *et al.*, 1999). The other regions have not yet been investigated. A possible functional significance of NP phosphorylation is suggested by the fact that the nucleoprotein of the closely related EBOV is phosphorylated as well (Elliott *et al.*, 1985).

So far, only for the rabies virus N a functional role of phosphorylation has so far been detected. Phosphorylation of rabies N seems to control the binding strength of the protein to the RNA template. The nonphosphorylated N displayed an enhanced binding strength for the template RNA which appeared to have a negative effect on transcription and replication. Another role for the phosphorylation of nucleoproteins has been detected by Gombart et al. who showed that the fully phosphorylated nucleoprotein of measles virus is preferentially incorporated into the nucleocapsids of measles virus (Gombart *et al.*, 1995).

3.2.2   NP forms the core of the nucleocapsid

The template for the RNA synthesis of NNS RNA viruses is the encapsidated genome. For MBGV it was hypothesized that NP is the RNA-encapsidating protein. This assumption was underlined by the observation that NP was the only protein of the nucleocapsid complex which was able to induce the formation of inclusion-like aggregates (see above).

A closer examination of the NP-induced inclusion bodies by immunoelectron microscopy revealed that these contained regularly arranged structures which appeared ring-like in cross section (Figure 8, inset 4) and showed regular striation in longitudinal section (not shown). These results suggested the presence of tubular-like structures (TLS) which were composed of multimers of the recombinant NP. Very similar structures were identified in the inclusions of MBGV-infected cells representing aggregated nucleocapsids (Figure 8). Thus, it was concluded that recombinant NP forms nucleocapsid-like structures (Kolesnikova *et al.*, 2000). Interestingly, the inclusions built of recombinant NP were in most cases closely associated with membranes of the rough ER.

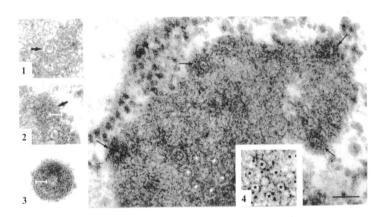

Figure 8: Comparative analysis of ultrathin sections of viral inclusions in MBGV-infected cells and aggregates formed by recombinant NP. Large photo shows a cross section of viral inclusion in MBGV-infected Vero cells at 43 h p.i. Preformed nucleocapsids are arranged in a hexagonal pattern (white asterisks) and embedded in electron-dense matrices. Some of the thin-walled TLS in the periphery of the inclusion are surrounded by more electron-dense matrices and represent mature nucleocapsids (arrows). Inset 1 shows thin-walled TLS surrounded by narrow electron-translucent regions (arrow). Inset 2 represents thin-walled TLS surrounded by a region of high electron density. Inset 3 shows a cross section of a MBGV particle. The thickness and electron density of the matrix surrounding the TLS within MBGV inclusion and within viral particle are almost identical. White bar: 25 nm. Inset 4 shows a cross sectioned TLS formed by recombinant NP. Photo of inset 4 is printed in the same magnification as the photo of MBGV inclusion. Black bar: 70 nm.

One difference between the inclusions in MBGV-infected cells and the NP-induced aggregates was their degree of electron density. Aggregates formed by recombinant NP were less electron dense in comparison with the inclusions in MBGV-infected cells (Figure 8, inset 4). This is caused by the accumulation of proteinaceous material surrounding the TLS in MBGV-infected cells resulting in thick walled electron dense TLS (Figure 8, inset

2). The electron dense TLS can also be detected outside the inclusions and in the mature virions and are therefore believed to represent the mature nucleocapsids (Figure 8, inset 3). The sequence of events finally leading to budding virirons is hypothesized as follows. The thin-walled TLS represent the precursors of the nucleocapsids which mature by recruiting additional viral proteins (VP35, VP30, L). These mature and electron dense nucleocapsids become transport-competent, are subsequently released from the inclusions, transported to the plasma membrane and enwrapped by a lipid bilayer in which the viral surface protein is incorporated.

The ability of nucleoproteins to form nucleocapsid-like structures is not unique to MBGV. It has also been shown for paramyxo- and rhabdoviruses that their respective nucleoproteins are able to self assemble into structures which closely resemble the nucleocapsids of the virions (Egelman *et al.*, 1989; Fooks *et al.*, 1993; Iseni *et al.*, 1998; Meric *et al.*, 1994; Schoehn *et al.*, 2001). Thus, it seems to be a general feature of the nucleoproteins of viruses of the order Mononegavirales to determine the structure of the nucleocapsid.

REFERENCES

Becker, S., and Mühlberger, E. (1999). Co- and posttranslational modifications and functions of Marburg virus proteins. *Curr Top Microbiol Immunol* **235**, 23-34.

Becker, S., Huppertz, S., Klenk, H. D., and Feldmann, H. (1994). The nucleoprotein of Marburg virus is phosphorylated. *J Gen Virol* **75**(Pt 4), 809-18.

Becker, S., Klenk, H.-D., and Mühlberger, E. (1996). Intracellular transport and processing of the Marburg virus surface protein in vertebrate and insect cells. *Virology* **225**(1), 145-55.

Becker, S., Rinne, C., Hofsäss, U., Klenk, H.-D., and Mühlberger, E. (1998). Interactions of Marburg virus nucleocapsid proteins. *Virology* **249**(2), 406-17.

Becker, S., Spiess, M., and Klenk, H. D. (1995). The asialoglycoprotein receptor is a potential liver-specific receptor for Marburg virus. *J Gen Virol* **76**(Pt 2), 393-9.

Bukreyev, A. A., Volchkov, V. E., Blinov, V. M., Dryga, S. A., and Netesov, S. V. (1995). The complete nucleotide sequence of the Popp (1967) strain of Marburg virus: a comparison with the Musoke (1980) strain. *Arch Virol* **140**(9), 1589-600.

Chan, S. Y., Empig, C. J., Welte, F. J., Speck, R. F., Schmaljohn, A., Kreisberg, J. F., and Goldsmith, M. A. (2001). Folate receptor-α is a cofactor for cellular entry by Marburg and Ebola virues. *Cell* **106**, 117–126.

Chenik, M., Chebli, K., Gaudin, Y., and Blondel, D. (1994). In vivo interaction of rabies virus phosphoprotein (P) and nucleoprotein (N): existence of two N-binding sites on P protein. *J Gen Virol* **75**(Pt 11), 2889-96.

Collins, P. L., Hill, M. G., Cristina, J., and Grosfeld, H. (1996). Transcription elongation factor of respiratory syncytial virus, a nonsegmented negative-strand RNA virus. *Proc Natl Acad Sci U S A* **93**(1), 81-5.

Conzelmann, K. K. (1996). Genetic manipulation of non-segmented negative-strand RNA viruses. *J Gen Virol* **77**(Pt 3), 381-9.

Davis, N. L., Arnheiter, H., and Wertz, G. W. (1986). Vesicular stomatitis virus N and NS proteins form multiple complexes. *J Virol* **59**(3), 751-4.

Davis, N. L., Arnheiter, H., and Wertz, G. W. (1986). Vesicular stomatitis virus N and NS proteins form multiple complexes. *J Virol* **59**(3), 751-4.

Davis, R. J. (1993). The mitogen-activated protein kinase signal transduction pathway. *J Biol Chem* **268**(20), 14553-6.

Egelman, E. H., Wu, S. S., Amrein, M., Portner, A., and Murti, G. (1989). The Sendai virus nucleocapsid exists in at least four different helical states. *J Virol* **63**(5), 2233-43.

Elliott, L. H., Kiley, M. P., and McCormick, J. B. (1985). Descriptive analysis of Ebola virus proteins. *Virology* **147**(1), 169-76.

Feldmann, H., Bugany, H., Mahner, F., Klenk, H. D., Drenckhahn, D., and Schnittler, H. J. (1996). Filovirus-induced endothelial leakage triggered by infected monocytes/macrophages. *J Virol* **70**(4), 2208-14.

Feldmann, H., Muhlberger, E., Randolf, A., Will, C., Kiley, M. P., Sanchez, A., and Klenk, H. D. (1992). Marburg virus, a filovirus: messenger RNAs, gene order, and regulatory elements of the replication cycle. *Virus Res* **24**(1), 1-19.

Feldmann, H., Will, C., Schikore, M., Slenczka, W., and Klenk, H. D. (1991). Glycosylation and oligomerization of the spike protein of Marburg virus. *Virology* **182**(1), 353-6.

Fisher-Hoch, S. P., and McCormick, J. B. (1999). Experimental filovirus infections. *Curr Top Microbiol Immunol* **235**, 117-43.

Fooks, A. R., Stephenson, J. R., Warnes, A., Dowsett, A. B., Rima, B. K., and Wilkinson, G. W. (1993). Measles virus nucleocapsid protein expressed in insect cells assembles into nucleocapsid-like structures. *J Gen Virol* **74**(Pt 7), 1439-44.

Garcia, J., Garcia-Barreno, B., Vivo, A., and Melero, J. A. (1993). Cytoplasmic inclusions of respiratory syncytial virus-infected cells: formation of inclusion bodies in transfected cells that coexpress the nucleoprotein, the phosphoprotein, and the 22K protein. *Virology* **195**(1), 243-7.

Garcia-Barreno, B., Delgado, T., and Melero, J. A. (1996). Identification of protein regions involved in the interaction of human respiratory syncytial virus phosphoprotein and nucleoprotein: significance for nucleocapsid assembly and formation of cytoplasmic inclusions. *J Virol* **70**(2), 801-8.

Gear, J. S., Cassel, G. A., Gear, A. J., Trappler, B., Clausen, L., Meyers, A. M., Kew, M. C., Bothwell, T. H., Sher, R., Miller, G. B., Schneider, J., Koornhof, H. J., Gomperts, E. D., Isaacson, M., and Gear, J. H. (1975). Outbreak of Marburg virus disease in Johannesburg. *Br Med J* **4**(5995), 489-93.

Geisbert, T. W., and Jaax, N. K. (1998). Marburg hemorrhagic fever: report of a case studied by immunohistochemistry and electron microscopy. *Ultrastruct Pathol* **22**(1), 3-17.

Geisbert, T. W., and Jahrling, P. B. (1995). Differentiation of filoviruses by electron microscopy. *Virus Res* **39**(2-3), 129-50.

Gombart, A. F., Hirano, A., and Wong, T. C. (1995). Nucleoprotein phosphorylated on both serine and threonine is preferentially assembled into the nucleocapsids of measles virus. *Virus res* **37**, 63-73.

Horikami, S. M., Curran, J., Kolakofsky, D., and Moyer, S. A. (1992). Complexes of Sendai virus NP-P and P-L proteins are required for defective interfering particle genome replication in vitro. *J Virol* **66**(8), 4901-8.

Hsu, C. H., and Kingsbury, D. W. (1982). Topography of phosphate residues in Sendai virus proteins. *Virology* **120**(1), 225-34.

Huber, M., Cattaneo, R., Spielhofer, P., Orvell, C., Norrby, E., Messerli, M., Perriard, J. C., and Billeter, M. A. (1991). Measles virus phosphoprotein retains the nucleocapsid protein in the cytoplasm. *Virology* **185**(1), 299-308.

Iseni, F., Barge, A., Baudin, F., Blondel, D., and Ruigrok, R. W. (1998). Characterization of rabies virus nucleocapsids and recombinant nucleocapsid-like structures. *J Gen Virol* **79**(Pt 12), 2909-19.

Kawai, A., Anzai, J., Honda, Y., Morimoto, K., Takeuchi, K., Kohno, T., Wakisaka, K., Goto, H., and Minamoto, N. (1997). Monoclonal antibody #5-2-26 recognizes the phosphatase-sensitive epitope of rabies virus nucleoprotein. *Microbiol Immunol* **41**(1), 33-42.

Kolesnikova, L., Mühlberger, E., Ryabchikova, E., and Becker, S. (2000). Ultrastructural organization of recombinant Marburg virus nucleoprotein: comparison with Marburg virus inclusions. *J Virol* **74**(8), 3899-904.

Liston, P., Batal, R., DiFlumeri, C., and Briedis, D. J. (1997). Protein interaction domains of the measles virus nucleocapsid protein (NP). *Arch Virol* **142**(2), 305-21.

Lötfering, B., E, M. h., Tamura, T., Klenk, H. D., and Becker, S. (1999). The nucleoprotein of Marburg virus is target for multiple cellular kinases. *Virology* **255**(1), 50-62.

Luytjes, W., Krystal, M., Enami, M., Pavin, J. D., and Palese, P. (1989). Amplification, expression, and packaging of foreign gene by influenza virus. *Cell* **59**(6), 1107-13.

Martini, G. A. (1971). Marburg virus disease. Clinical syndrome. *In* "Marburg virus disease" (G. A. Martini, and R. Siegert, Eds.), pp. 1-9. Springer, Berlin, Heidelberg, New York.

Martini, G. A. (1971). Marburg virus disease. Clinical syndrome. *In* "Marburg virus disease" (G. A. Martini, and R. Siegert, Eds.), pp. 1-9. Springer, Berlin, Heidelberg, New York.

Masters, P. S., and Banerjee, A. K. (1988). Resolution of multiple complexes of phosphoprotein NS with nucleocapsid protein N of vesicular stomatitis virus. *J Virol* **62**(8), 2651-7.

Meric, C., Spehner, D., and Mazarin, V. (1994). Respiratory syncytial virus nucleocapsid protein (N) expressed in insect cells forms nucleocapsid-like structures. *Virus Res* **31**(2), 187-201.

Mühlberger, E., Lotfering, B., Klenk, H.-D., and Becker, S. (1998). Three of the four nucleocapsid proteins of Marburg virus, NP, VP35, and L, are sufficient to mediate replication and transcription of Marburg virus-specific monocistronic minigenomes. *J Virol* **72**(11), 8756-64.

Mühlberger, E., Sanchez, A., Randolf, A., Will, C., Kiley, M. P., Klenk, H. D., and Feldmann, H. (1992). The nucleotide sequence of the L gene of Marburg virus, a filovirus: homologies with paramyxoviruses and rhabdoviruses. *Virology* **187**(2), 534-47.

Mühlberger, E., Trommer, S., Funke, C., Volchkov, V., Klenk, H.-D., and Becker, S. (1996). Termini of all mRNA species of Marburg virus: sequence and secondary structure. *Virology* **223**(2), 376-80.

Mühlberger, E., Weik, M., Volchkov, V. E., Klenk, H.-D., and Becker, S. (1999). Comparison of the transcription and replication strategies of Marburg virus and Ebola virus by using artificial replication systems. *J Virol* **73**(3), 2333-42.

Murphy, F. A., Simpson, D. I., Whitfield, S. G., Zlotnik, I., and Carter, G. B. (1971). Marburg virus infection in monkeys. Ultrastructural studies. *Lab Invest* **24**(4), 279-91.

Muyembe-Tamfum, J.-J., Borchert, M., Swanepoel, R., Bausch, D. G., Tshioko, F. K., Campbell, P., Roth, C., Sleurs, H., Olinda, L. A., Libande, M., Colebunders, R., Rodier, G., Leirs, H., Zeller, H., Van der Stuyft, P., and Rollin, P. E. (2000). *Symposium on Marburg and Ebola viruses, Marburg.*

Nishio, M., Tsurudome, M., Kawano, M., Watanabe, N., Ohgimoto, S., Ito, M., Komada, H., and Ito, Y. (1996). Interaction between nucleocapsid protein (NP) and phosphoprotein (P) of human parainfluenza virus type 2: one of the two NP binding sites on P is essential for granule formation. *J Gen Virol* **77**(Pt 10), 2457-63.

Pattnaik, A. K., Ball, L. A., LeGrone, A. W., and Wertz, G. W. (1992). Infectious defective interfering particles of VSV from transcripts of a cDNA clone. *Cell* **69**(6), 1011-20.

Peters, D., Müller, G., and Slenczka, W. (1971). Morphology, development, and classification of the Marburg virus. *In* "Marburg virus disease", pp. 68-83. Springer, Berlin, Heidelberg, New York.

Rippey, J. J., Schepers, N. J., and Gear, J. H. (1984). The pathology of Marburg virus disease. *S Afr Med J* **66**(2), 50-4.

Ryabchikova, E. I., Kolesnikova, L. V., and Netesov, S. V. (1999). Animal pathology of filoviral infections. *Curr Top Microbiol Immunol* **235**, 145-73.

Ryabchikova, E., Strelets, L., Kolesnikova, L., Pyankov, O., and Sergeev, A. (1996). Respiratory Marburg virus infection in guinea pigs. *Arch Virol* **141**(11), 2177-90.

Ryan, K. W., Murti, K. G., and Portner, A. (1990). Localization of P protein binding sites on the Sendai virus nucleocapsid. *J Gen Virol* **71**(Pt 4), 997-1000.

Sanchez, A., Kiley, M. P., Klenk, H. D., and Feldmann, H. (1992). Sequence analysis of the Marburg virus nucleoprotein gene: comparison to Ebola virus and other non-segmented negative-strand RNA viruses. *J Gen Virol* **73**(Pt 2), 347-57.

Sanchez, A., Kiley, M. P., Klenk, H. D., and Feldmann, H. (1992). Sequence analysis of the Marburg virus nucleoprotein gene: comparison to Ebola virus and other non-segmented negative-strand RNA viruses. *J Gen Virol* **73**(Pt 2), 347-57.

Schnittler, H. J., Mahner, F., Drenckhahn, D., Klenk, H. D., and Feldmann, H. (1993). Replication of Marburg virus in human endothelial cells. A possible mechanism for the development of viral hemorrhagic disease. *J Clin Invest* **91**(4), 1301-9.

Schoehn, G., Iseni, F., Mavrakis, M., Blondel, D., and Ruigrok, R. W. (2001). Structure of recombinant rabies virus nucleoprotein-RNA complex and identification of the phosphoprotein binding site [In Process Citation]. *J Virol* **75**(1), 490-8.

Schwemmle, M., Salvatore, M., Shi, L., Richt, J., Lee, C. H., and Lipkin, W. I. (1998). Interactions of the borna disease virus P, N, and X proteins and their functional implications. *J Biol Chem* **273**(15), 9007-12.

Slenczka, W. G. (1999). The Marburg virus outbreak of 1967 and subsequent episodes. *Curr Top Microbiol Immunol* **235**, 49-75.

Sokol, F., and Clark, H. F. (1973). Phosphoproteins, structural components of rhabdoviruses. *Virology* **52**(1), 246-63.

Spehner, D., Drillien, R., and Howley, P. M. (1997). The assembly of the measles virus nucleoprotein into nucleocapsid-like particles is modulated by the phosphoprotein. *Virology* **232**(2), 260-8.

Stille, W., and Böhle, E. (1971). Clinical course and prognosis of Marburg virus (green monkey) disease. *In* "Marburg virus disease" (G. A. Martini, and R. Siegert, Eds.), pp. 10-18. Springer, Berlin, Heidelberg, New York.

Sutter, G., Ohlmann, M., and Erfle, V. (1995). Non-replicating vaccinia vector efficiently expresses bacteriophage T7 RNA polymerase. *FEBS Lett* **371**(1), 9-12.

Swanepoel, R., Smit, S., Burt, F. J., Rollin, P. E., Ksiazek, T. G., Bowen, M. D., Trappier, S. G., McMullan, L., Bausch, D. G., Zaki, S. R., and Nichol, S. T. (2000). *Symposium on Marburg and Ebola viruses, Marburg.*

Will, C., Muhlberger, E., Linder, D., Slenczka, W., Klenk, H. D., and Feldmann, H. (1993). Marburg virus gene 4 encodes the virion membrane protein, a type I transmembrane glycoprotein. *J Virol* **67**(3), 1203-10.

Zaki, S. R., and Goldsmith, C. S. (1999). Pathologic features of filovirus infections in humans. *Curr Top Microbiol Immunol* **235**, 97-116.

3. Viral determinants for capsid formation and packaging

Chapter 3.1

Packaging DNA into Herpesvirus Capsids

Jay C. Brown[1], Michael A. McVoy[2] and Fred L. Homa[3]

[1] *Department of Microbiology and Cancer Center, University of Virginia Health System, Charlottesville, Virginia 22908, USA*
[2] *Department of Pediatrics, Medical College of Virginia campus of Virginia Commonwealth University, Richmond, Virginia 23298, USA*
[3] *Infectious Disease Research, Pharmacia Corp., Kalamazoo, Michigan 49001, USA*

1. INTRODUCTION

Injection of DNA into a pre-formed capsid is a central event in herpesvirus replication. Similar packaging of DNA into a pre-formed shell is also observed during replication of dsDNA bacteriophage such as T4 and λ; adenoviruses may encapsidate DNA in the same way. Herpesvirus DNA packaging takes place in the infected cell nucleus where capsid assembly and DNA replication also occur. The substrates for packaging are capsids plus the multi-genome, concatemeric DNA that is the product of virus DNA replication. During the encapsidation process, double strand cuts are made at specific sites (pac sites) in the DNA concatemer so that one complete genome is packaged into each capsid.

Genetic studies with herpes simplex virus 1 (HSV-1) have resulted in identification of seven virus genes whose protein products are specifically involved in DNA encapsidation. None is required, for example, for capsid formation or DNA replication. The seven proteins are expected to be involved in processes such as introduction of specific cuts in the DNA concatemer, formation of a portal through which virus DNA can enter the

Structure-Function Relationships of Human Pathogenic Viruses, Edited by
Holzenburg and Bogner, Kluwer Academic/Plenum Publishers, New York, 2002

capsid, provision of the energy required for DNA translocation into the capsid and sealing the capsid once it is filled.

Investigators are studying herpesvirus DNA packaging with the idea that it constitutes an attractive target for novel therapeutic agents directed against herpesvirus replication. Encapsidation is particularly appealing as a target because it is required for herpesvirus growth, and because most, if not all, of the proteins involved are virus-encoded.

Recent studies of herpesvirus DNA packaging have defined the basic nature of the process and provided information about the components involved in individual steps. Here we summarize recent progress with emphasis on areas such as pac site recognition and the function of the processing/packaging proteins where there has been the most interest. Discussion is focused on HSV-1 with other herpesviruses, particularly cytomegalovirus, mentioned when relevant studies have been done. The mechanism of DNA encapsidation as it occurs in dsDNA bacteriophage is described briefly because the same basic mechanism is expected to apply in herpesviruses and because, as studies with phage are generally more advanced, they have suggested productive lines of research with herpesviruses. We conclude with a brief description of small molecule packaging inhibitors that have the potential to be developed as anti-herpes therapeutics.

2. HERPESVIRUS REPLICATION

All herpesviruses consist of an icosahedral capsid surrounded by a membrane envelope. The viral DNA is contained inside the capsid, and a layer of protein called the tegument is found between the capsid and the membrane. Herpesvirus replication begins when the virus binds one or more receptors on the surface of a susceptible cell. Virus and cell membranes then fuse, and as a result the capsid is introduced into the cytoplasm. There it makes its way to the nucleus, docks at a nuclear pore, and injects its DNA as illustrated in Fig. 1. The parental capsid does not enter the nucleus. Once inside the nucleus, the virus DNA is replicated and viral genes are transcribed. Later, progeny capsids assemble in the nucleus and are packaged with DNA as described in detail below. Filled capsids then leave the nucleus by budding through the two nuclear membranes, acquire tegument and a membrane in the cytoplasm and exit the host cell (Roizman and Sears, 1996; Weller, 1995). The replication pathway described above applies broadly among herpesviruses undergoing lytic growth. The events of

latent infection are distinct and differ significantly among individual
members of the herpesvirus family.

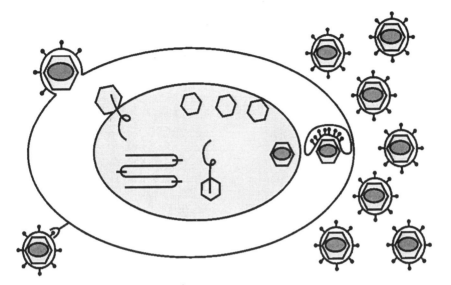

Herpesvirus Replication

Figure 1: Schematic representation of events in herpesvirus replication. Note that DNA
replication and capsid formation occur in the infected cell nucleus. DNA is packaged into a
pre-formed capsid, a process that also occurs in the nucleus.

3. HSV-1 CAPSID STRUCTURE AND ASSEMBLY

3.1 Capsid structure

The clearest view of herpesvirus capsid structure has come from the
results of studies involving electron cryomicroscopy and three dimensional
image reconstruction. Structures have now been determined in this way for
the capsids of HSV-1, equine herpesvirus 1, human cytomegalovirus and
Kaposi's sarcoma-associated herpesvirus (Baker *et al.*, 1990; Chen *et al.*,
1999; Nealon *et al.*, 2001; Trus *et al.*, 1999; Trus *et al.*, 1995; Zhou *et al.*,
2000). Most reconstructions are calculated at resolutions of ~20 Å, although

a structure for the HSV-1 capsid is now available at 8.5 Å resolution. In this structure one can identify elements of protein secondary structure such as alpha helices and regions of beta sheet (Zhou *et al.*, 2000). As all four capsid structures are similar in their basic features, the discussion below will focus on HSV-1.

The HSV-1 capsid is an icosahedral shell 125 nm in diameter and approximately 15 nm in thickness. Its major structural features are 162 capsomers that lie on a T=16 icosahedral lattice as shown in Fig. 2. The presence of a capsid with 162 capsomers is a defining property for members of the herpesvirus family. Each capsomer consists of a roughly cylindrical protruding domain that projects radially outward from a floor layer 3-4 nm thick. The protruding domain is approximately 10 nm in diameter, and it projects 10 nm-11 nm above the floor. Each capsomer has an axial channel approximately 4 nm in diameter. Capsomers are of two types, hexons and pentons. The 150 hexons (red in Fig. 2) form the edges and faces of the capsid icosahedron while one penton (orange in Fig. 2) is found at each of the 12 capsid vertices. VP5, the major HSV-1 capsid protein, is the structural subunit of both the hexons and pentons (see Table 1). Hexons are hexamers of VP5 while pentons are pentamers (Homa and Brown, 1997; Rixon, 1993; Steven and Spear, 1996)

Table 1: Major protein components of the herpes simplex virus capsid

Protein	Gene	Protein MW	Copies per Capsid	Position in Capsid
VP5	UL19	149,075	960	Capsomers
VP19C	UL38	50,260	320	Triplexes
VP23	UL18	34,268	640	Triplexes
VP26	UL35	12,095	900	Hexon Tips

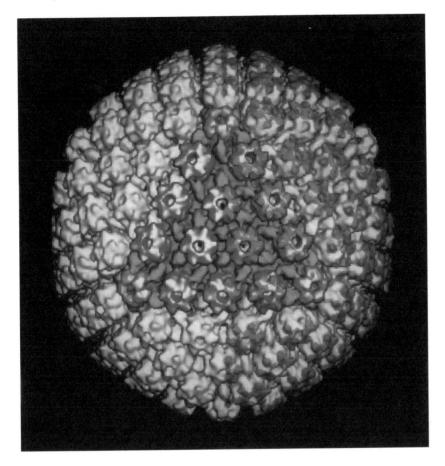

Figure 2: Structure of the HSV-1 capsid. The structure was determined by electron cryomicroscopy and three-dimensional image reconstruction. The capsomers (162 in all) are the major structural units observed in the reconstruction. In the color-coded region, hexons are red, pentons orange and triplexes green. VP6 is colored blue in half the structure.

The capsomers are connected in groups of three by the triplexes (green in Fig. 2), small, trivalent structures that lie above the capsid floor. There are a total of 320 triplexes, and in the mature capsid they are thought to reinforce the floor layer in holding the capsomers together while keeping them separated by an appropriate distance. In the procapsid, however, the floor layer is rudimentary, and the triplexes are the only links holding the capsomers together (Newcomb *et al.*, 2000; Newcomb *et al.*, 1996; Trus *et al.*, 1996). Most, if not all, triplexes are molecular heterotrimers consisting of two copies of VP23 and one of VP19C. Heterotrimers with the same composition are formed in solution from the purified proteins (Baxter and Gibson, 1997; Spencer *et al.*, 1998).

Capsids also contain 900 copies of VP26, a fourth abundant capsid protein, which is located at the tips of the hexons. One VP26 (blue in Fig. 2) is bound to each hexon-associated VP5 (Trus *et al.*, 1995; Zhou *et al.*, 1995). Pentons lack VP26 entirely. VP26 is not required for HSV-1 growth in cell culture, but it appears to enhance neurovirulence (Desai *et al.*, 1998).

In addition to the four major structural proteins mentioned above, HSV-1 capsids also contain ~150 copies of the maturational protease (VP24) plus trace amounts of four proteins involved in DNA encapsidation. Filled capsids in virions, for example, contain the products of the genes UL6, UL15 and UL25 while procapsids contain the same three plus the UL28 gene product (Salmon and Baines, 1998; Sheaffer *et al.*, 2001; Yu and Weller, 1998a).

3.2   Capsid assembly

Assembly of the HSV-1 capsid has been studied in HSV-infected cells, in insect cells infected with recombinant baculoviruses encoding HSV-1 capsid proteins, in extracts of insect cells containing HSV-1 proteins and in a mixture of purified capsid proteins (Newcomb *et al.*, 1999; Newcomb *et al.*, 1994; Newcomb *et al.*, 1996; Tatman *et al.*, 1994; Thomsen *et al.*, 1994). Results with the four systems are in substantial agreement regarding the mechanism by which capsids are formed. Assembly of morphologically normal capsids requires the major capsid protein (VP5), the two triplex proteins (VP19C and VP23) and one or both of two scaffolding proteins, pre-VP22a and VP21 (Newcomb *et al.*, 1994; Tatman *et al.*, 1994; Thomsen *et al.*, 1994). The scaffolding proteins are involved in capsid assembly, but later they are lost and are not found in the mature virion. Both scaffolding proteins are encoded in the viral UL26 gene and their amino acid sequences are overlapping. Pre-VP22a, the more abundant scaffolding protein in procapsids, consists of UL26 amino acids 307-635 while VP21 is amino acids 248-635 (Homa and Brown, 1997; Rixon, 1993).

Using an in vitro system in which capsids are formed in insect cell extracts containing HSV-1 proteins, Newcomb *et al.* (Newcomb *et al.*, 1996) employed electron microscopy to define the structure of intermediates in capsid assembly. Mature capsids were formed by way of partial procapsid and procapsid intermediates as shown diagrammatically in Fig. 3. Partial procapsids are angular wedges or domes in which a region of capsid shell partially surrounds a region of core. The shell contains VP5 and the triplex proteins while the core is composed entirely of scaffolding protein. Regions of shell and core grow until the shell closes to create the procapsid, a spherical structure with the same diameter, number of capsomers and icosahedral symmetry (i.e. T=16) as the mature capsid. The structure of the HSV-1 procapsid has been determined at a resolution of 18 Å by electron cryomicroscopy and three-dimensional image reconstruction (Newcomb et al, 2000; Trus *et al.*, 1996).

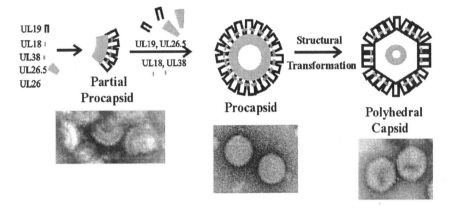

Figure 3: Assembly of the HSV-1 capsid. The pathway is shown schematically above with micrographs of intermediates below. Note that the mature, icosahedral capsid is formed by way of partial procapsid and procapsid intermediates.

Shortly after it is formed, the procapsid is transformed structurally into the mature icosahedral capsid morphology. The scaffolding protein exits the capsid during the morphological transformation, which is accomplished without further protein addition. Structural maturation of the procapsid is promoted by activity of the maturational protease (VP24) whose major substrate is the scaffolding protein. Both scaffolding proteins are cleaved. Cleavage results in release of the region (the C-terminal 25 amino acids) involved in scaffolding protein attachment to VP5 permitting the scaffold to exit the maturing capsid (Homa and Brown, 1997). Procapsid transformation is accompanied by a dramatic increase in structural stability. The procapsid, for instance, is disassembled by incubation at 2°C while the mature capsid is unaffected by the same treatment (Newcomb *et al.*, 1999).

Electron microscopic examination of in vitro capsid assembly reaction mixtures has demonstrated the presence of a continuous distribution of partial procapsids with sizes ranging from wedges of only a few tens of arc degrees to complete structures (Newcomb *et al.*, 1996). This continuous distribution suggests partial procapsids grow in small rather than large increments. Small complexes containing VP5 and the scaffolding protein are suggested as the assembly subunits because: (1) VP5 is a monomer in solution and does not self-associate in the absence of the scaffolding protein; (2) purified VP5 and scaffolding protein associate in solution to form small oligomers containing 1-2 VP5 and 2-6 scaffolding protein molecules; and (3) in the presence of triplexes, such small complexes have the capacity to assemble to form procapsids (Newcomb et al, 1999). Genetic and biochemical studies have resulted in identification of specific regions in VP5 (the N-terminal 85 amino acids) and in the scaffolding protein (the C-terminal 25 amino acids) that are required for productive interaction between the two proteins (Desai and Person, 1999; Hong *et al.*, 1996; Oien *et al.*, 1997; Warner *et al.*, 2000).

In its basic features, the pathway for assembly of the HSV-1 capsid resembles that observed in dsDNA bacteriophage such as P22 and T4 (Black *et al.*, 1994; Casjens and Hendrix, 1988; Murialdo and Becker, 1978; Prevelige and King, 1993). For instance, as in HSV-1 capsid formation, assembly of the icosahedral bacteriophage capsid proceeds by way of a spherical, mechanically fragile procapsid, a structure that undergoes a morphological transformation (prohead expansion) to create the mature capsid form. Assembly of the bacteriophage capsid involves participation of a scaffolding protein that, like the HSV-1 scaffolding proteins, binds the major capsid protein, but is not present in the mature virion. The similarity of HSV-1 capsid assembly to that of bacteriophage has suggested that DNA packaging may also conform to the phage pathway, a well-studied process whose fundamental features are understood as described below. The expected similarity of HSV-1 and bacteriophage DNA packaging has provided the framework of ideas that underlie current studies of herpesvirus DNA packaging.

4.   DNA PACKAGING IN dsDNA BACTERIOPHAGE

4.1   Packaging mechanism

The process of DNA encapsidation in dsDNA bacteriophage has been studied for more than thirty years with the result that its basic features have now been defined. Both genetic and biochemical approaches have been employed, and studies have been carried out with a wide range of phages including λ, T4, P22, T7, T3, φ29 and SPP1 (Black, 1989; Catalano *et al.*, 1995; Murialdo, 1991). Although there are differences in the packaging mechanism among the phages (e.g. in use of specific DNA pac sites as opposed to a headfull packaging mechanism), the fundamental events of packaging are broadly the same. The starting materials are: (1) newly replicated phage DNA, a multi-genome concatemer in which individual viral genomes are linked in a head-to-tail fashion; (2) the phage procapsid containing, at one site, a portal ring through which DNA enters the capsid; and (3) the phage-encoded terminase, an enzyme with multiple functions in the packaging process as described below.

Packaging begins when the terminase makes a double strand cut in the concatemer DNA. The terminase-DNA end complex then docks onto the procapsid by way of the portal complex as shown in Fig. 4, steps 1 and 2. The cut creates one end of the progeny virus genome, and cutting may occur at a specific nucleotide sequence (e.g. *cosN* in the case of phage λ). After cutting, DNA begins to enter the procapsid and continues until terminase makes a second cut in the concatemer DNA (Catalano, 2000; Leffers and Rao, 2000). The second cut may occur at a second pac site or after a headfull of DNA has been injected. As DNA is entering, the procapsid is transformed into its mature, icosahedral morphology. For example, in phage T4, procapsid transformation occurs after ≥8% of the DNA has been encapsidated (Jardine and Coombs, 1998). Once the last DNA end has entered the capsid, the portal is closed and the capsid stabilized by addition of head completion proteins (gp2 and gp4 in the case of phage T4). The force required for translocation of DNA into the capsid is thought to be delivered by turning of the portal ring in the procapsid, a process that requires ATP hydrolysis by the terminase (Simpson *et al.*, 2000). Measurements with the phage φ29 packaging system have demonstrated that one ATP is hydrolyzed for each ~2 DNA base pairs encapsidated (Guo *et al.*, 1987). Similar studies with phage T3 have demonstrated that DNA is packaged at a rate of 2×10^4 bp/min at 30°C (Shibata *et al.*, 1987).

DNA Packaging in dsDNA Bacteriophage

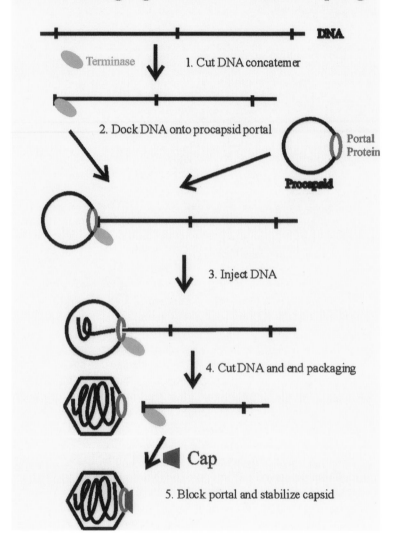

Figure 4: Steps in the pathway of DNA encapsidation as it occurs in dsDNA bacteriophage such as T4 and λ.

4.2   Structure of the packaged DNA mass

DNA is by far the predominant component inside the filled bacteriophage capsid (Black *et al.*, 1994; Earnshaw and Casjens, 1980). Trace amounts of proteins are found in the capsid cavity of some phage (e.g. T4 and T7), but even in these cases the proportion of protein is small. DNA is considered to be the only component present inside the mature, filled HSV-1 capsid (Booy *et al.*, 1991; Newcomb *et al.*, 1993). In filled capsids, all proteins are components of the capsid shell.

Electron cryomicroscopy has been employed to examine the arrangement of DNA in the capsids of HSV-1 and of dsDNA bacteriophage (Adrian *et al.*, 1984; Booy *et al.*, 1991; Lepault *et al.*, 1987). In such micrographs, individual DNA strands can be identified and their organization studied. Micrographs of phage and HSV-1 capsids show local regions where DNA strands are ordered in parallel arrays, the so-called "fingerprint" patterns. The DNA strand spacing in such regions is approximately 26Å suggesting the DNA is highly condensed in the capsid. Several local regions containing parallel DNA strands may be observed in a single capsid. There is little evidence, however, for longer-range organization of DNA. Further, the pattern of organized regions observed in different virions cannot be interpreted to define an organizational principle that applies even to all virions of the same virus species. An exception occurs in certain preparations of phage T7 in which the DNA is arranged in concentric shells inside the capsid (Cerritelli *et al.*, 1997).

5.   ELECTRON MICROSCOPY OF HSV-1 CAPSIDS PACKAGING DNA

Basic information about HSV-1 DNA packaging can be obtained from electron micrographs of capsids in the process of packaging DNA in vivo (Fig. 5). Such micrographs show capsids with a region of DNA apparently making its way into the interior. Arrows in Fig. 5 indicate two such capsids, and others can be seen in the same micrograph. Such images support the idea that DNA is packaged into a preformed capsid. Only closed capsids are found to be in the process of filling. From the same images, one can observe that many capsids in the process of filling are round rather than angular in cross section suggesting they are spherical rather than icosahedral in shape. A representative round capsid image is starred in Fig. 5. The spherical morphology suggests packaging begins with the procapsid as observed with dsDNA bacteriophage. Procapsid angularization may occur as DNA is encapsidated. DNA in the process of entering a capsid appears highly condensed and derived from a much larger pool of less condensed DNA that surrounds the filling capsids (Fig. 5).

6.   HSV-1 GENOME STRUCTURE AND DNA REPLICATION

6.1   Genome structure

The genome of HSV-1 is a 152-kb linear double stranded DNA molecule. The presence of repeated sequences and invertible regions renders the genome structure highly complex. Long (L) and short (S) components are delineated by two pairs of large inverted repeats (Fig. 6). The L component consists of a long unique region (U_L) flanked by inverted copies of the b sequence. Similarly, the S component consists of a short unique region (U_S) flanked by inverted copies of the c sequence. A single copy of a smaller repeated sequence called the a sequence is found at the S component terminus adjacent to the c sequence and one to several reiterated copies are found at the L component terminus adjacent to the b sequence. One to several a sequence copies are also found at the L/S junction. The terminal a sequences are in direct orientation to one another but in inverted orientation relative to the internal a sequences at the L/S junction. Consequently, the HSV-1 genome can be represented by a_n-b-U_L-b'-a'_m-c'-U_S-c-a, where n and m represent the number of a sequence reiterations and a', b', and c' represent inverted copies of the a, b, and c sequences, respectively. During replication, homologous recombination occurs between the inverted repeats, resulting in efficient inversion of both the L and S components. The consequence of these inversions is that DNA within HSV-1 virions is an equimolar mixture of four isomers in which the L, S, or L and S components are inverted relative to the prototype orientation (Roizman and Sears, 1996).

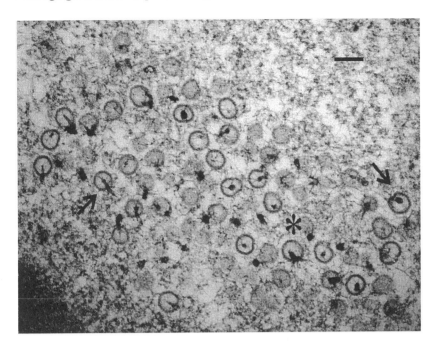

Figure 5: Electron micrograph showing HSV-1 capsids in the process of packaging DNA in the nucleus of an infected cell. The arrows indicate two such capsids, but others can be seen in the same field. Note that DNA is packaged into pre-formed capsids. Many capsids are round in profile (e.g. the capsid below the *) suggesting they are procapsids rather than mature, icosahedral capsids. Bar: 150 nm.

6.2   DNA replication

HSV-1 DNA replication is believed to begin with circularization of the linear genome shortly after infection (Garber *et al.*, 1993; Poffenberger and Roizman, 1985). This appears to be a direct ligation of the termini, although homologous recombination between terminal *a* sequences is also a possible mechanism (McVoy *et al.*, 1997; Mocarski and Roizman, 1982; Yao *et al.*, 1997; Yao and Elias, 2001). One consequence of circularization is that the junctions formed by fusion of L and S termini are indistinguishable from the internal L/S junctions (Mocarski and Roizman, 1982). DNA synthesis begins at three origins of replication and produces a large replicative intermediate composed of unit genomes linked in a head-to-tail (*i.e.*, L termini linked to S termini) concatemeric organization (Bataille and Epstein, 1994; Jacob *et al.*, 1979; Martinez *et al.*, 1996; Severini *et al.*, 1994; Zhang *et al.*, 1994). However, the replicative intermediate is clearly more complex than a simple long linear concatemer. Complete digestion with restriction enzymes that cut once per genome fails to render a large proportion of replicative intermediate DNA competent to migrate into pulsed-field gels, and this has been taken to indicate the presence of frequent branches (Bataille and Epstein, 1997; Martinez *et al.*, 1996). Such branching may arise from homologous recombination or strand invasion priming of DNA synthesis reminiscent of the replication strategy of bacteriophage T4 (Mosig, 1998). Consistent with this hypothesis, the observation that replicative intermediate DNA contains inverted L components clearly demonstrates that homologous recombination frequently occurs within replicative intermediate DNA (Bataille and Epstein, 1994; Bataille and Epstein, 1997;Martinez *et al.*, 1996; Severini *et al.*, 1994; Slobedman *et al.*, 1999;Zhang *et al.*, 1994).

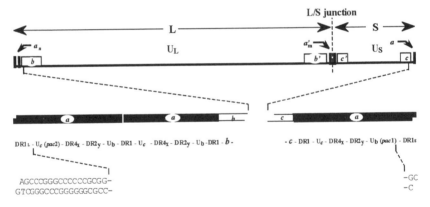

Figure 6: Structure of the HSV-1 genome. The L (long) and S (short) components are indicated above the genome, while details of the terminal *a* sequences are illustrated below. S termini contain only one *a* sequence, whereas L/S junctions and L termini contain one to several (two are illustrated). The terminal U_b and U_c sequences include (*pac*1) and (*pac*2) notations to indicate the presence of these elements near the respective termini. For simplicity such notations have been omitted from the more distal U_b and U_c sequences. Nucleotide sequences of the two partial DR1 repeats that are found at the L and S termini ($DR1_L$ and $DR1_S$, respectively) are shown at the bottom. Note that annealing and ligation of the termini reconstitutes a complete DR1 (figure modified from Mocarski and Roizman, 1982).

7.   *CIS* CLEAVAGE/PACKAGING SEQUENCES

7.1   *a* sequence structure and location of pac sites

As in the large dsDNA bacteriophage, converting herpesvirus concatemeric replicative intermediates into unit genomes that have distinct, specific terminal sequences requires the presence of certain sequences near the point of cleavage to direct the enzymatic machinery to the precise site to be cleaved. In HSV-1, these *cis*-acting cleavage signals are located within the terminal *a* sequence repeats. This was initially suggested by recombinant viruses in which *a* sequence-containing insertions were engineered within U$_L$. These "ectopic" *a* sequence insertions were able to direct packaging of novel HSV-1 genomes having termini formed by cleavage at the ectopic *a* sequences (Mocarski *et al.*, 1980; Mocarski and Roizman, 1981; Mocarski and Roizman, 1982; Varmuza and Smiley, 1985). Studies using HSV-1 defective genomes or artificially engineered HSV-1 "amplicons" served to further refine the role of the *a* sequence in DNA cleavage and packaging (Barnett *et al.*, 1983; Deiss *et al.*, 1986; Deiss and Frenkel, 1986; Frenkel, 1981; Nasseri and Mocarski, 1988; Spaete and Frenkel, 1982; Spaete and Frenkel, 1985; Stow *et al.*, 1983; Varmuza and Smiley, 1985; Vlazny and Frenkel, 1981; Vlazny *et al.*, 1982). Using this system, bacterial plasmids containing an HSV-1 origin of replication were replicated into concatemeric DNA within HSV-1 infected cells. The DNA was not packaged into viral capsids, however, unless an *a* sequence was present in *cis* within the plasmid (Stow *et al.*, 1983). Therefore, the *a* sequence is both necessary and sufficient for HSV-1 DNA cleavage and packaging.

Within the *a* sequence, functional elements were further inferred by comparing the terminal regions from the genomes of several different herpesviruses (Fig. 7). Although these sequences were generally non-homologous, two small islands of conservation were clearly evident approximately 30-35 bp from the genomic termini. Their conserved proximity to the termini, conservation among herpesviruses, and presence within the HSV-1 *a* sequence (an element known to contain packaging elements) prompted their designations as *pac*1 and *pac*2 following the bacteriophage convention for packaging signals. The distinguishing characteristics of *pac*1 motifs are a 3 to 7 bp A-or T-rich region flanked on each side by 5 to 7 Cs; *pac*2 motifs consist of a 5- to 10-base pair A-rich region that is often associated with a nearby CGCGGCG motif (Deiss *et al.*, 1986; Tamashiro *et al.*, 1984).

The structure of the HSV-1 a sequence is shown in Fig. 6. Bordering the a sequence are two copies of a 20 bp direct repeat called DR1. Just internal to the terminal DR1 repeats and at opposite ends of the a sequence lie $pac1$ and $pac2$ within non-repetitive 64- and 58-bp regions called U_b and U_c. The interior of the a sequence is composed of directly repeated arrays of a 12 bp G/C-rich sequence (DR2) and in some strains an additional directly repeated 37 bp sequence (DR4). Thus, the a sequence can be represented by DR1-$U_b(pac1)$-DR4$_m$-DR2$_n$-$U_c(pac2)$-DR1. The size of the a sequence varies with strain from 200 to 500 bp depending on the number of internal repeats (Davison and Wilkie, 1981; Mocarski et al., 1985; Mocarski and Roizman, 1981; Umene, 1991; Varmuza and Smiley, 1985). Each end of the HSV-1 genome has at least one copy of the a sequence such that $pac1$ lies 30-bp from the S terminus and $pac2$ lies 32 bp from the L terminus (Mocarski and Roizman, 1982).

7.2   Functional roles of the pac sites

The roles of $pac1$ and $pac2$ in the cleavage and packaging processes have been investigated by examining partial and complete cleavage sites that occur naturally in some viruses (Chowdhury et al., 1990; Davison, 1984; Hammerschmidt et al., 1988; McVoy et al., 1997) and by directed mutagenesis in defective genomes (Deiss et al., 1986; Deiss and Frenkel, 1986; Nasseri and Mocarski, 1988; Varmuza and Smiley, 1985; Zimmermann and Hammerschmidt, 1995) and recombinant viruses (McVoy et al., 1997; McVoy et al., 1998; Smiley et al., 1990; Varmuza and Smiley, 1985). Initial studies used the ability to serially propagate defective virus genomes as an assay for packaging and the effects of deleting certain regions from the a sequence were determined. Removal of the DR2 array had no effect, whereas deletion of U_c completely blocked packaging, indicating that $pac2$-containing U_c sequences are necessary for packaging. Curiously, serial passage occurred when U_b was deleted, but the defective genomes had acquired intact a sequences from the HSV-1 helper virus that was used to infect the cells (Deiss et al., 1986). Thus, restoration of $pac1$-containing U_b sequences was required for packaging. The fact that U_c deletions were not "rescued" by recombination in a similar manner to U_b sequences indicates that distinctions must exist in how $pac1$ and $pac2$ function.

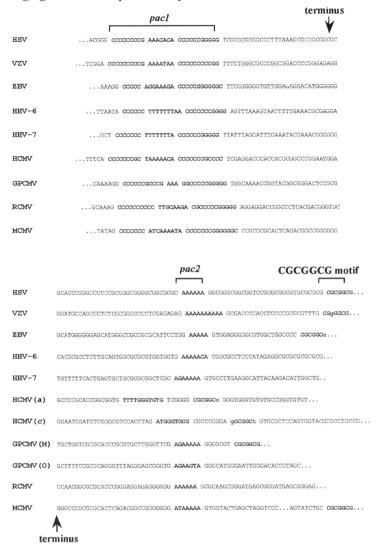

Figure 7: Terminal nucleotide sequences from selected herpesviruses. Conserved *pac*1 and *pac*2 elements are shown offset in bold. Note that two right termini can be formed by HCMV (a and c) and by GPCMV (M and O), and that HCMV(a), HCMV(c) and GPCMV(O) termini lack a poly(A) tract. In these cases presumptive cryptic *pac*2 sequences were assigned on the basis of sequence conservation between the two termini (McVoy *et al.*, 1997; McVoy *et al.*, 2000). Note that *pac*1 sequences are contiguous while there are intervening sequences between the *pac*2 A-rich and CGCCGCG motifs (for sequence references see McVoy *et al.*, 1998).

A sub-fragment of the *a* sequence that retained *pac*1 and *pac*2 but lacked the left DR1, part of U_b, and part of the right DR1 was also able to support serial propagation of defective viruses, indicating that DR1, which contains the actual point at which cleavage occurs (Mocarski and Roizman, 1982), is dispensable for cleavage and packaging (Varmuza and Smiley, 1985). When the same fragment was inserted into the thymidine kinase gene within the U_L region of the HSV-1 genome, novel termini were formed within the adjacent thymidine kinase coding sequences by cleavage on either side of the ectopically inserted *a* sequence sub-fragment (Varmuza and Smiley, 1985). The distances from these novel termini to *pac*1 or *pac*2 were unaltered despite the fact that the sequences at the point of cleavage were different. These results suggested that separate cleavage events are mediated by *pac*1 and *pac*2 and that these reactions resemble cleavage by bacteriophage lambda terminase (Higgins and Becker, 1994) and by type IIS restriction endonucleases in that cleavage site specificity is determined not by sequences at the point of cleavage but rather by *cis* sequences located a metered distance (in this case 30 and 32 bp) from the point of cleavage.

Nested deletions from each end of the *a* sequence were similarly tested for cleavage function in recombinant viruses by ectopic insertion in U_L. As before, deletions that removed DR1 and parts of U_b but left *pac*1 intact were cleaved within flanking thymidine kinase sequences. Further deletions that also removed *pac*1 were also cleaved; however, analysis of the resulting termini indicated that recombinational events had juxtaposed a wild type *a* sequence adjacent to the mutant *a* sequence to provide the missing *pac*1 element. Nested deletions on the U_c side of the *a* sequence followed a similar pattern but analysis of recombinational repair of the *pac*2 deletions was not done (Smiley *et al.*, 1990). Thus, both within defective genomes and within the context of the viral genome these studies supported the hypothesis that *pac*1 and *pac*2 constitute herpesvirus cleavage/packaging signals.

Two factors related to the complexity of the HSV-1 genome appear to underlie the recombinational events that frustrated the mutational analyses of the *a* sequence described above. The first is the presence of the internal inverted repeats that include copies of the *a* sequence, such that regardless of the orientation or location of the ectopic *a* sequence insertions, native *a* sequences are invariably present in inverted orientation elsewhere in the genome. The propensity for inverted copies of the *a* sequence to mediate segment inversion (Chou and Roizman, 1985) then creates considerable potential for homologous recombination between ectopic and native *a* sequences. The second factor stems from the fact that *a* sequences frequently become amplified into directly repeated *a* sequence arrays (Deiss and

Frenkel, 1986; Smiley *et al.*, 1990; Varmuza and Smiley, 1985). In the studies described above, both factors apparently conspired to create ectopic cleavage sites that frequently contained multiple *a* sequences composed of a mixture of wild type and mutant *a* sequences.

7.3   Studies with other herpesviruses

The difficulties described above prompted mutational analyses of *pac*1 and *pac*2 in the context of less complicated herpesviral genomes. The murine cytomegalovirus (MCMV) genome is comparatively quite simple. The genome lacks inverted repeats and invertible segments (Ebeling *et al.*, 1983; Mercer *et al.*, 1983), and has only one copy of a 30-bp terminal repeat at each terminus (Marks and Spector, 1988). A cleavage site consisting of a fusion of the MCMV termini inserted ectopically within the MCMV genome was shown to be cleaved with efficiency equal to the naturally occurring cleavage sites. This permitted evaluation of the efficiency of cleavage at ectopic cleavage sites bearing mutations. Alteration of just 4-bp within the poly(C) tracts that flank the central A-rich region of *pac*1 resulted in greater than 30-fold reductions in cleavage, whereas similar mutations within the *pac*1 A-rich region had only a modest 3-fold effect. Likewise, mutation of the *pac*2 A-rich region had a greater than 40-fold effect on cleavage efficiency but a mutation in the CGCGGCG motif associated with *pac*2 had an intermediate 20-fold effect (McVoy *et al.*, 1998; Nixon and McVoy, unpublished observations).

The studies described above clearly established the importance of conserved sequence elements within *pac*1 and *pac*2 in the cleavage and packaging process; however, assignment of *cis* cleavage/packaging signals on the basis of sequence homology remains unreliable. For example, studies with guinea pig cytomegalovirus (GPCMV) indicate the presence of a "cryptic *pac*2 element" that functions like a *pac*2 yet lacks overt *pac*2 sequence characteristics (McVoy *et al.*, 1997). Furthermore, human cytomegalovirus (HCMV) lacks any obvious *pac*2 element at the expected location within the HCMV *a* sequence (Broll *et al.*, 1999; Chee *et al.*, 1990; Kemble and Mocarski, 1989; Mocarski *et al.*, 1987); however, parallels between the genome structures of HCMV and GPCMV suggest that HCMV may also utilize cryptic *pac*2 elements (McVoy *et al.*, 2000). Sequences outside the immediate *pac*1-*pac*2 region may also be important for cleavage. The L/S junction within the varicella zoster virus genome includes both *pac*1 and *pac*2 yet cleavage at this site only occurs 5% of the time (Davison, 1984), and in Epstein-Barr virus, deletion of sequences adjacent to the *pac*1-

*pac*2 region resulted in a failure to package DNA in a plasmid-based assay (Zimmermann and Hammerschmidt, 1995). Clearly, more work is needed to fully define the *cis*-acting sequence elements that are important for cleavage and packaging of herpesvirus DNA.

8.   THE MECHANISM OF DNA CLEAVAGE AND PACKAGING

8.1   Processive packaging

In the large dsDNA phage, DNA packaging is initiated by docking of empty capsids at termini that exist on concatemeric DNA as described above (Fig. 4). The DNA is then translocated into the capsid until a length of one genome has entered and cleavage occurs when the appropriate *cis* sequences are encountered. Cleavage releases the newly packaged genome within the capsid while generating a new terminus on the concatemer that can serve for initiation of the next round of packaging. Thus, the process is processive in that it moves inward from concatemer ends one genome at a time. The process is also directional, since one of the two genomic termini predominates on concatemers and serves as the preferred substrate for initiation of packaging (Becker and Murialdo, 1990; Black, 1989 ;Chung *et al.*, 1990; Fujisawa *et al.*, 1990; Khan *et al.*, 1995; Serwer *et al.*, 1992; Son *et al.*, 1993). In phage lambda, directionality stems from the presence of a binding site for terminase proteins on one side of the point of cleavage such that after cleavage, terminase proteins remain associated with the newly formed concatemer terminus. This protein/DNA complex then serves to dock the concatemer end to an empty capsid initiating the next round of packaging (Catalano *et al.*, 1995).

Current experimental results are consistent with a similar mechanism for herpesvirus DNA packaging. Concatemeric DNA of HSV-1, HCMV, MCMV, GPCMV and equine herpesvirus 1, all exhibit a strong predominance of one genomic terminus over the other (Martinez *et al.*, 1996; McVoy and Adler, 1994; Severini *et al.*, 1994; Slobedman and Simmons, 1997; Zhang *et al.*, 1994) and in all but HCMV these ends contain *pac*2 elements (what constitutes the HCMV *pac*2 element remains uncertain [see above], but *pac*1-containing ends are absent while the ends that do not contain *pac*1 are present on HCMV concatemers (McVoy and Adler, 1994)). This observation has led to speculation that *pac*2 may define packaging directionality in herpesviruses by mediating initiation of concatemer packaging in a manner similar to that described for phage lambda (McVoy *et al.*, 2000). It should be noted, however, that while directional packaging is well supported in the bacteriophage by physical and biochemical data, similar data demonstrating that one end of herpesvirus genomes is preferentially inserted into capsids is lacking. Thus, more work on packaging directionality in the herpesviruses is warranted.

Genetic studies have defined seven HSV-1 proteins as having a probable role in DNA cleavage and packaging. One or more of these proteins would be predicted to bind specifically to viral *cis* cleavage/packaging sequences. Indeed, *in vitro* DNA binding assays have demonstrated that one of these proteins, HSV-1 pUL28, binds to the HSV-1 *pac*1 (Adelman *et al.*, 2001), while similar studies found that the pUL28 homolog in HCMV, pUL56, binds to the HCMV *pac*1 (Bogner *et al.*, 1998). The relevance to cleavage and packaging of pUL28 binding was further inferred by a strong correlation between the ability of mutations in the HSV-1 *pac*1 to prevent binding of pUL28 with the ability of analogous mutations in the MCMV *pac*1 to block cleavage and packaging *in vivo* (Adelman *et al.*, 2001; McVoy *et al.*, 1998). Thus, mutations analogous to those that failed to be cleaved in the context of the viral genome also failed to bind to pUL28 *in vitro*, whereas mutations that only modestly reduced cleavage also bound pUL28 with affinities similar to that of wild type *pac*1. Curiously, only oligonucleotides used in the binding assays that formed a higher order (apparently nonlinear) structure were bound by pUL28, suggesting the possibility that binding may be determined more by DNA structure than by specific sequence recognition (Adelman *et al.*, 2001).

A final question remains as to how the processing/packaging proteins function in concert to cleave and package DNA. Unfortunately, the process appears to be much more complex than a simple recognition of *cis* elements by the appropriate proteins and cleavage of the DNA. One complexity arises from the observation that HSV-1 genomes contain an internal cleavage site at the L/S junction, while HSV-1 defective genomes contain numerous internal cleavage sites. This has been taken as evidence that, like the large dsDNA bacteriophage, the cleavage reaction is restricted by a "head-full" requirement that prevents cleavage, even at authentic cleavage sites, unless a genomic-length of DNA has entered the capsid (Roizman and Sears, 1996). Also consistent with this hypothesis is the fact that DNA cleavage has never been observed in the absence of capsid formation or under circumstances in which the DNA was not at least transiently packaged into capsids (McNab *et al.* 1998). How the cleavage machinery is linked to a "head-full sensor" in the context of the capsid remains unknown.

8.2   *a* sequence amplification

A second complexity arises from the duplication of terminal repeats during cleavage. Early on, direct sequencing of HSV-1 genomic termini revealed that one end of the genome consists of an *a* sequence plus part of a DR1 and that the other end consists of an *a* sequence plus the remaining bases of the DR1. The two ends also had complementary single base overhangs (Fig. 6). Thus, a model was proposed in which cleavage involves a single base staggered cut within the DR1 that lies between two *a* sequences (note that reiterated *a* sequences are separated by a single DR1). Conversely, annealing and ligation of these two ends, as occurs during circularization of the genome after infection, would reconstruct the single DR1 between two *a* sequences (Mocarski and Roizman, 1982). This model predicts that the DNA substrate for the cleavage reaction is the *pac*1-DR1-*pac*2 region that forms at the junction of two *a* sequences. Indeed, fragments containing such a junction can be cleaved and can support HSV-1 defective genome propagation (Nasseri and Mocarski, 1988).

Although this model is attractive for its simplicity, that cleavage would only occur between two *a* sequences seems inconsistent with observations that single *a* sequence-containing L/S junctions are frequent within HSV-1 virion and concatemeric DNA (Locker and Frenkel, 1979; Wagner and Summers, 1978). It was further found that a single *a* sequence, when used to make defective genomes or when ectopically inserted into the viral genome, gives rise to reiterated *a* sequence arrays as well as to genomes with one or more *a* sequences at both ends. These findings prompted proposals that *a* sequences are duplicated during cleavage (Deiss and Frenkel, 1986; Varmuza and Smiley, 1985). Supporting data was gleaned from the ectopic cleavage site system in MCMV, where it was found that duplication of the 30 bp MCMV terminal repeat is dependent on cleavage. Sites in which small mutations prevented cleavage maintained the same number of terminal repeats (one or two) that were present in the plasmids used for virus construction. In contrast, sites that were cleaved even inefficiently evolved in the course of virus construction to contain a mixture of one and two copies of the repeat (McVoy *et al.*, 1998).

In bacteriophage T3 and T7, small direct repeats are present at each end of the phage genomes yet these sequences are not duplicated within concatemeric DNA (Dunn and Studier, 1983; Fujisawa and Sugimoto, 1983; Langman *et al.*, 1978). Because the substrate for cleavage (the concatemer) contains a single copy of the repeat and the reaction products (the two termini) each bear one copy of the repeat, every cleavage event must result in a duplication of the repeat. This was unclear for herpesviruses since in HSV-1 both the cleavage sites and the termini can have variable numbers of *a* sequence reiterations and in both the HSV-1 and MCMV studies the duplications might have occurred only rarely, but cumulatively, during the process of constructing and propagating these genomes. This situation prompted examination of the concatemeric DNA from a recombinant GPCMV that forms genomes having only one copy of a terminal repeat at each end. It was found that the vast majority of junctions between genomes within concatemeric DNA contained only one copy of the terminal repeat (Nixon and McVoy, unpublished observations). Thus, duplication of terminal repeats must occur with virtually every cleavage event.

How terminal repeat duplication occurs remains a matter of speculation for both bacteriophage and herpesviruses. Figure 8A shows a model for HSV-1 cleavage and *a* sequence duplication that draws heavily on previous models proposed for phage T7 (Chung *et al.*, 1990; White and Richardson, 1987). In this model, DNA packaging initiates at a *pac2* end on a concatemer and proceeds until a full genome has entered the capsid. A nick then occurs adjacent to the *a* sequence and is used to prime DNA synthesis. Passage of the replication fork through the *a* sequence creates a small branch containing a duplicate *a* sequence. Cleavage mediated by *pac1* releases an encapsidated genome having an *a* sequence at each end while a second *pac2*-mediated cleavage trims off a small fragment to restore a *pac2* end on the concatemer. Finally, resolution of the branch created by the replication fork produces a concatemeric substrate suitable for the next round of packaging.

When duplication at tandem *a* sequences is considered (Figure 8B), resolution produces small *a* sequence-containing fragments that are authentically cleaved at the *pac2* end while containing variable amounts of adjacent *b* sequences (depending on how far the replication fork travels before resolution occurs). DNA fragments with precisely these characteristics have been abundantly detected in HSV-1-infected cells (Umene, 1994). Thus, this model provides an explanation for the derivation of these fragments that were heretofore mysterious. It also incorporates independent *pac*1- and *pac*2-mediated cleavage events consistent with termini formed at ectopic *a* sequence insertions (Varmuza and Smiley, 1985). While several aspects of this model are compelling, the validity of this model awaits rigorous confirmation.

Finally, it should be noted that duplication of terminal sequences is neither universal nor essential for cleavage since many herpesviruses do not have terminal repeats. Sequence comparisons of termini and concatemer junctions from several of these viruses are consistent with the simple cleavage model (Chowdhury *et al.*, 1990; Davison, 1984; Hammerschmidt *et al.*, 1988; Marks and Spector, 1988; McVoy *et al.*, 1997; Mocarski *et al.*, 1987; Tamashiro and Spector, 1986). For GPCMV, the cleavage machinery appears to be able to mediate both simple and duplicative mechanisms. While the evidence described above suggests that GPCMV genomes having one terminal repeat at each end arise predominantly from duplicative cleavage, half of the genomes formed by wild type GPCMV lack a terminal repeat at the right end (Gao and Isom, 1984) and are therefore not produced by a duplicative mechanism. Thus, the GPCMV cleavage machinery appears to be able to mediate both simple and duplicative mechanisms. By extension, HSV-1 may also have the ability to utilize either mechanism as circumstances dictate. As shown in Fig. 8C, a failure to nick and initiate DNA synthesis at double *a* sequences would result in a simple cleavage between the two *a* sequences. Thus it could be envisioned that at tandem *a* sequences either a simple or a duplicative cleavage might occur, whereas upon encountering a single *a* sequence duplicative cleavage would be obligatory.

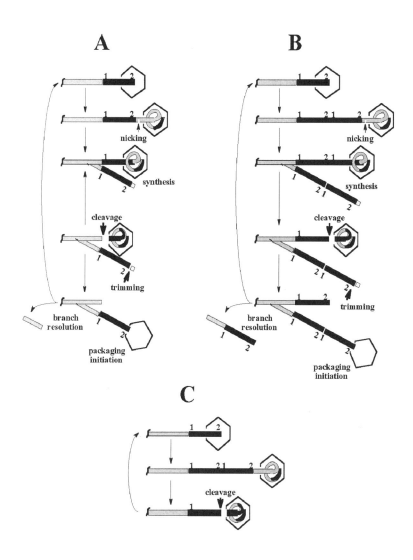

Figure 8: Suggested mechanism for *a* sequence duplication during herpesvirus DNA packaging. The mechanism is illustrated at junctions containing one *a* sequence (A) and multiple *a* sequences (B). Simple cleavage without *a* sequence duplication is illustrated in C.

9.   ESSENTIAL DNA CLEAVAGE AND PACKAGING PROTEINS

Cleavage and packaging of HSV-1 DNA are tightly linked processes and studies with viral mutants have identified seven genes that are essential for the overall process. The seven genes are *UL6, UL15, UL17, UL25, UL28, UL32*, and *UL33* (Addison *et al.*, 1990; Ali *et al.*, 1996; Al-Kobaisi *et al.*, 1991; Baines *et al.*, 1994; Cavalcoli *et al.*, 1993; Chang *et al.*, 1996; Lamberti and Weller, 1996; McNab *et al.*, 1998; Patel *et al.*, 1996; Poon and Roizman, 1993; Preston *et al.*, 1983; Reynolds *et al.*, 2000; Salmon *et al.*, 1998; Taus *et al.*, 1998; Tengelsen *et al.*, 1993; Yu *et al.*, 1997). Mutations that map to any of these genes (either temperature sensitive or null mutations) result in a virus that fails to produce infectious progeny virions. All such mutants retain the ability to replicate viral DNA to near wild type levels. DNA replication results in the formation of head-to-tail concatemers composed of tandem repeats of the viral genome. As described above, these are subsequently cleaved into unit-length monomers and packaged into preformed capsids.

DNA packaging takes place in the infected cell nucleus, but active DNA replication is not required at the time encapsidation takes place (Church *et al.*, 1998; Ward *et al.*, 1996). There does appear to be an energy requiring step in this process since depletion of ATP from infected cells blocks DNA packaging (Dasgupta and Wilson, 1999). Mutations in six of the processing/packaging genes (all but UL25) block cleavage of concatemeric DNA molecules, demonstrating that the proteins expressed from the *UL6, UL15, UL17, UL28, UL32*, and *UL33* genes play essential roles in the cleavage and subsequent packaging of viral DNA. Thin sections of cells infected with these mutants show that only mature, scaffold-containing capsids (B capsids) reside in the nucleus and no A (empty) or C (DNA containing) capsids are found.

Four of the seven-cleavage/packaging proteins (UL6, UL15, UL25 and UL28) are found to be associated with capsids (McNab *et al.*, 1998; Ogasawara *et al.*, 2001; Patel and MacLean, 1995; Salmon and Baines, 1998; Taus and Baines, 1998; Tengelsen *et al.*, 1993; Yu and Weller, 1998a). Interestingly, capsid association of the four is not dependent on the presence of the other three. This suggests that cleavage and packaging of viral DNA requires the four proteins but they do not appear to require each other in order to be incorporated into capsids.

Studies with a *UL25* null virus demonstrated that some combination of the UL6, UL15, and UL28 proteins make up the machinery that results in the cleavage of viral DNA (McNab *et al.*, 1998). The UL17, UL32 and UL33 proteins appear to serve more of an accessory function, by chaperoning the proteins involved in packaging (UL6, UL15, UL28) to the proper location in cells where DNA encapsidation takes place (Lamberti and Weller, 1998; Reynolds *et al.*, 2000; Taus, *et al.*, 1998). In the absence of the UL25 protein, cleavage of DNA is observed, but the DNA fails to be incorporated into capsids. Instead, large numbers of A (empty) capsids are found in cells infected with a *UL25* null virus, suggesting the function of the UL25 protein is to retain DNA in the capsid following the cleavage event. The findings with the *UL25* mutant demonstrate that all the components for interaction of viral DNA with capsids and its eventual cleavage are intact but the mechanism to retain DNA is lost. In the absence of the UL25 protein, DNA appears to enter the capsid and is cleaved by the cleavage/packaging machinery but then the DNA is lost generating a capsid (A capsid) that lacks both DNA and scaffold proteins.

Despite the identification of the seven genes required for encapsidation of viral DNA, the mechanism of packaging is not thoroughly understood. Although the role each protein plays in the cleavage/packaging process is not defined, some recent biochemical studies have revealed interesting properties for several of these proteins and may suggest their function. A summary of what is known about the protein products for each of the seven essential cleavage/packaging proteins is listed below.

UL6. The *UL6* gene was initially identified as a cleavage/packaging gene through the characterization of three ts mutants and more recently through the use of *UL6* null mutants (Lamberti and Weller, 1996; Patel *et al.*, 1996; Preston *et al.*, 1983; Sherman and Bachenheimer, 1988; Sherman and Bachenheimer, 1987). The product of the *UL6* gene is a 75-kDa protein (Patel and MacLean, 1995). Immunofluorescence studies have shown that the UL6 protein is found exclusively in the nucleus when expressed transiently in transfected cells in the absence of other HSV-1 proteins (Patel *et al.*, 1996). Homologs of the UL6 protein have been found in all herpesviruses whose genomes have been sequenced, indicating a highly conserved function for this protein (Lamberti and Weller, 1996; Patel *et al.*, 1996; Patel and MacLean, 1995). The UL6 protein is found associated with all three capsid forms, procapsids, filled capsids and empty capsids (Sheaffer *et al.*, 1995). Immuno-electron microscopy studies with intact capsids have recently demonstrated that the UL6 protein is located specifically at one of the twelve-capsid vertices. In the same study, the purified UL6 protein was also observed to form rings (Fig. 9) composed of twelve UL6 monomers (Newcomb *et al.*, 2001). The unique location of the UL6 protein on the capsid and the fact that it forms a ring structure are analogous to what is found for bacteriophage T4 and P22 portal proteins (Bazinet and King, 1985; Valpuesta and Carrascosa, 1994). The results support the view that the UL6 protein forms the portal through which DNA enters the capsid as it is packaged.

UL15. *UL15* is the most extensively studied of the seven cleavage/packaging genes. Several ts and null mutants have been isolated and used to demonstrate that the UL15 protein is required for DNA encapsidation (Baines *et al.*, 1994; Baines *et al.*, 1997; Bogner *et al.*, 1998; Dolan *et al.*, 1991; Giesen *et al.*, 2000a; Giesen *et al.*, 2000b; Koslowski *et al.*, 1999; Koslowski *et al.*, 1997; Poon and Roizman, 1993; Salmon and Baines, 1998; Salmon *et al.*, 1999; Yu and Weller, 1998a; Yu and Weller, 1998b; Yu *et al.*, 1997). *UL15* is the most highly conserved gene in the herpesvirus family. Based on homology with the ATP binding terminase (gp17) of bacteriophage T4, it has been proposed that *UL15* codes for the herpes terminase. Direct evidence awaits the purification of the UL15 protein and demonstration that the purified protein has endonuclease activity. The *UL15* mRNA is expressed from two exons, but HSV-1 replicates to high titers when *UL15* is expressed as single cDNA copy (Ward *et al.*, 1996).

Two proteins are expressed from the *UL15* open reading frame (Salmon and Baines, 1998; Salmon *et al.*, 1999; Yu and Weller, 1998), a full-length 81- to 83-kDa form and a 30-kDa protein translated from a second reading frame (in-frame to the full-length protein) located in the 3' end of the *UL15* mRNA. Only the full-length protein is detected in capsids. Truncated forms of the full-length protein have been reported (Salmon and Baines, 1998; Salmon *et al.*, 1999). An 80-kDa and a 79-kDa form of UL15, which may be generated by proteolytic cleavage, are found associated with mature capsids lacking DNA (B capsids). Association of the truncated proteins with capsids requires the intact DNA cleavage machinery since capsids isolated from viruses lacking the *UL6*, *UL17*, *UL28*, *UL32* or *UL33* genes contain only the full-length form of the UL15 protein. These results are supported by the fact that all three forms of the UL15 protein are found in capsids isolated from cells infected with a *UL25* null virus that cleaves but does not package DNA. In a separate report it was found that the amount of the full-length UL15 protein is reduced in DNA containing C capsids (Yu and Weller, 1998b). A larger 87-kDa protein was detected with UL15 antiserum, and this form was detected primarily in filled capsids (C capsids). The importance of the different forms of the UL15 protein remains to be resolved.

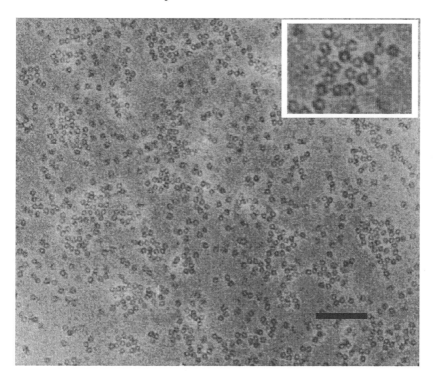

Figure 9: Electron micrograph showing the purified protein encoded by the HSV-1 *UL6* gene. Note that the protein forms small rings 16.5 nm in diameter with a 5 nm diameter axial channel. One ring (consisting of 12 UL6 monomers) is present at a unique vertex in the capsid where it forms the channel through which DNA enters. Bar: 200 nm.

Several lines of evidence indicate an interaction between UL15 and another cleavage/packaging protein, UL28 (Abbotts *et al.*, 2000; Koslowski *et al.*, 1999; Koslowski *et al.*, 1997). The UL15 and UL28 proteins have been shown to co-purify from infected cells and may interact to localize to the nucleus. Deletion studies have shown that truncated forms of the UL28 proteins retain the ability to interact with the UL15 protein.

UL17. Polyclonal rabbit antiserum raised against a bacterially expressed UL17 protein reacts with an abundant 77-kDa protein along with a weakly staining 72-kDa protein in HSV-1 infected cell lysates (Salmon *et al.*, 1998; Tengelsen *et al.*, 1993). UL17 protein is found in the tegument and is not associated with capsids. The UL17 protein appears to be required for the correct targeting of capsids and capsid proteins to intra-nuclear sites where viral DNA is processed and packaged.

UL25. As described above, HSV-1 capsid assembly and DNA packaging is similar to that of dsDNA bacteriophage. Therefore, it is likely that the over-abundance of empty capsids found in cells infected with a *UL25* null mutant results from an abortive packaging event (McNab *et al.*, 1998). A similar phenomenon has been observed in mutants defective in the gene products gp4, gp10, and gp26 of phage P22. In cells infected with such mutants, filled capsids are unstable and lose mature DNA within the cells, resulting in the accumulation of empty capsids (Black, 1988; Prevelige and King, 1993). It has been proposed that these phage gene products function to close the channel at the unique vertex where DNA is added to the capsid. The UL25 gene expresses a 60-kDa protein that is found to associate with HSV-1 A, B and C capsids. Recent immuno-electron microscopic studies have shown that there are approximately 42 copies of the UL25 protein per B capsid and it is present on both hexons and pentons (Ogasawara *et al.*, 2001). The amount of UL25 protein is reduced in procapsids compared to B capsids, while there is approximately 2-3 fold more UL25 in C capsids versus B capsids (Ogasawara *et al.*, 2001; Sheaffer *et al.*, 2001). The increased UL25 in DNA containing capsids supports the idea that it plays a role in retaining DNA inside capsids following DNA cleavage. The UL25 protein associates with VP5 and VP19C of virus capsids and shows a preference for binding to penton structures. Gel mobility shift analysis demonstrated that UL25 protein has the potential to bind DNA, which would be expected if its function is to hold DNA in the capsids once its cleaved (Ogasawara *et al.*, 2001).

UL28. Viruses containing mutations in *UL28* have been studied in HSV-1 and pseudorabies virus (PRV). In both, UL28 is an essential gene and *UL28* mutants are unable to form filled capsids (Cavalcoli *et al.*, 1993; Mettenleiter *et al.*, 1993; Pederson and Enquist, 1991Tengelsen *et al.*, 1993). The pseudorabies virus UL28 protein remains in the cytoplasm when transiently expressed in cells. Co-expression with the PRV UL15 protein enables UL28 to enter nuclei (Koslowski *et al.*, 1997). The interaction of the HSV-1 UL28 protein with HSV-1 UL15 protein has also been demonstrated (Abbotts *et al.*, 2000;Koslowski *et al.*, 1999). The human cytomegalovirus UL56 protein (homolog of HSV-1 UL28) has been shown to specifically bind to the HCMV packaging elements, *pac*1 and *pac*2 (Bogner *et al.*, 1998; Giesen *et al.*, 2000a; Giesen *et al.*, 2000b). In addition, a specific endonuclease activity was found to be associated with baculovirus-expressed UL56 using circular plasmid DNA containing HCMV cleavage/packaging sequences (Bogner *et al.*, 1998).

The HSV-1 *UL28* gene encodes an 85-kDa protein, which has been expressed in bacteria and the protein purified (Adelman *et al.*, 2001). The purified protein has been shown to recognize conserved HSV-1 DNA packaging sequences. Specifically, it was found to bind to the *pac*1 site but only when the DNA was heat denatured. The *pac*1 site appears to adopt a novel DNA conformation when denatured and UL28 binds to only one strand of the *pac*1 motif. The novel DNA binding activity of the UL28 protein and its interaction with a putative nuclease, UL15 protein, suggest a complex process for recognizing and cleaving DNA at the proper cleavage site in the viral genome.

UL32 and UL33. The *UL32* and *UL33* genes of HSV-1 encode proteins of 65- and 19-kDa, respectively (Lamberti and Weller, 1998). The UL32 protein is a cysteine rich, zinc-binding protein that accumulates in both the cytoplasm and nucleus of infected cells (Chang *et al.*, 1996; Lamberti and Weller, 1998). The UL33 protein accumulates predominately in the nucleus of infected cells and is found in DNA replication compartments (Reynolds *et al.*, 2000). Both genes are required for DNA packaging but neither is found associated with capsids (Al-Kobaisi *et al.*, 1991;Chang *et al.*, 1996; Lamberti and Weller, 1998; Reynolds *et al.*, 2000). Little is known about these two genes but the absence of UL32 and UL33 from capsids suggests that they may not play a direct role in the cleavage/packaging event. The two proteins may function to ensure that the other proteins (UL6, UL15, UL25, UL28) assemble into a functional cleavage/packaging complex. For instance, they may act as chaperones or as transport (nuclear localization) proteins for assembly of the cleavage/packaging machinery.

10. ANTI-HERPESVIRUS COMPOUNDS THAT TARGET DNA PROCESSING AND PACKAGING

Efforts to identify small molecule inhibitors of herpesvirus replication have led to the discovery of three types of compounds, benzimidazole ribosides, substituted thioureas and BAY 38-4766, that act by inhibiting steps in DNA processing and encapsidation. The benzimidazole ribosides are specific for cytomegalovirus although a related compound, 1263W94, inhibits Epstein-Barr virus (EBV) replication. The substituted thioureas are most active against HSV-1 with lesser inhibitory effects for HSV-2, HCMV and varicella-zoster virus (VZV). BAY 38-4766 inhibits replication of HCMV and MCMV.

10.1   Benzimidazole ribosides

Studies   with   BDCRB   (2-bromo-5,6,-dichloro-1-β-D-ribofuranosyl benzimidazole; Fig. 10), the best-studied of the benzimidazole ribosides, demonstrated that it is a potent inhibitor of HCMV replication although it does not affect virus DNA replication, mRNA synthesis or protein synthesis (Underwood *et al.*, 1998). Formation of genome ends, however, was found to be inhibited in a dose-dependent fashion suggesting the drug acts by inhibiting the step in which virus genomes are cut from the DNA concatemer (i.e. step 1 in Fig. 4). In support of the same view, BDCRB-resistant mutants were selected and found to have lesions in UL89, the HCMV homolog of the HSV-1 presumptive terminase subunit UL15. In studies with a related benzimidazole riboside, TCRB, it was observed that resistant mutants could have lesions in the genes encoding either terminase subunit, UL89 or UL56, the homologs of HSV-1 genes UL15 and UL28, respectively (Krosky *et al.*, 1998).

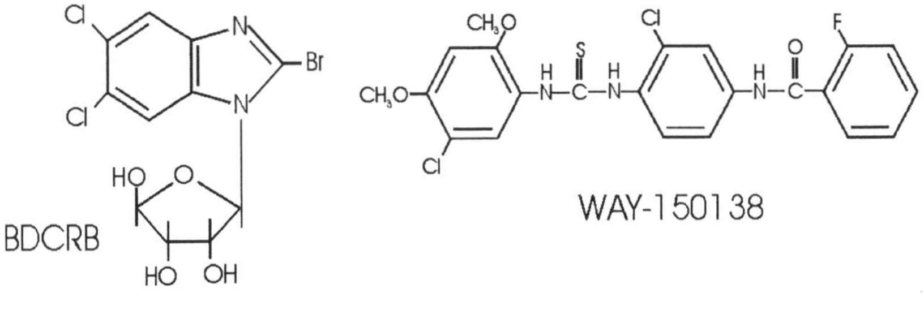

Figure 10: Molecular structures of BDCRB and WAY-150138, two inhibitors of HSV-1 growth that act by inhibiting DNA encapsidation.

Zacny and colleagues observed that 1263W94, a benzimidazole riboside with an L rather than a D ribose linkage, is a potent inhibitor of EBV replication. Experiments measured lytic EBV growth after induction of a Burkitt's lymphoma cell line latently infected with EBV (Zacny *et al.*, 1999). 50% inhibitory concentrations of 1263W94 were observed in the range of 0.15-1.1 µM. Maturation of DNA ends was found to be inhibited in a dose-dependent manner suggesting 1263W94 may act like BDCRB by blocking cleavage of the DNA concatemer by terminase.

10.2   Other inhibitors

WAY-150138 (Fig. 10) is the most potent of a new class of substituted thioureas identified as inhibitors of HSV-1 replication in a screen carried out at Wyeth-Ayerst. Like BDCRB, it was found to inhibit maturation of DNA ends and appearance of filled capsids (van Zeijl *et al.*, 2000). Resistant mutants map to the HSV-1 *UL6* gene, the gene encoding the portal protein, suggesting WAY-150138 may affect portal function, portal interaction with other processing/packaging proteins or perhaps incorporation of the portal into capsids. In order to exert a significant inhibitory effect on HSV-1 replication, WAY-150138 needs to be present during the first four hours after cultured cells are infected. This observation is particularly intriguing in light of the fact that DNA cleavage and packaging are late events in HSV-1 growth occurring for the most part well after four hours post infection. It suggests that some early event of HSV-1 infection is able to prevent WAY-150138 from inhibiting virus growth.

BAY 38-4766 is a member of a new class of non-nucleoside compounds found to inhibit replication of HCMV and MCMV in cultured cells. MCMV mutants resistant to the drug have been isolated and the mutagenic lesions located in the genes, M89 and M56, encoding the two presumptive terminase subunits (Buerger *et al.*, 2001). Resistance in these genes suggests BAY 38-4766 acts by inhibiting DNA processing or encapsidation, possibly by affecting cleavage of the DNA concatemer.

ACKNOWLEDGMENTS

Work in the authors' laboratories was supported by NIH grants AI41644 (JCB), AI46668 MAM) and AI43527 (MAM), by NSF grant MCB 9904879 (JCB), and by an award from the Virginia Commonwealth Health Research Board (MAM and JCB).

REFERENCES

Abbotts, A. P., Preston, V. G., Hughes, M., Patel, A. H., and Stow, N. D., 2000, Interaction of the herpes simplex virus type 1 packaging protein UL15 with full-length and deleted forms of the UL28 protein. *J. Gen. Virol.* **81 Pt 12:** 2999-3009.

Addison, C., Rixon, F. J., and Preston, V. G., 1990, Herpes simplex virus type 1 UL28 gene product is important for the formation of mature capsids. *J. Gen. Virol.* **71:** 2377-2384.

Adelman, K., Salmon, B., and Baines, J. D., 2001, Herpes simplex virus DNA packaging sequences adopt novel structures that are specifically recognized by a component of the cleavage and packaging machinery. *Proc. Natl. Acad. Sci. U.S.A* **98:** 3086-3091.

Adrian, M., Dubochet, J., Lepault, J., and McDowall, A. W., 1984, Cryo-electron microscopy of viruses. *Nature* **308:** 32-36.

Al-Kobaisi, M. F., Rixon, F. J., McDougall, I., and Preston, V. G., 1991, The herpes simplex virus UL33 gene product is required for the assembly of full capsids. *Virology* **180:** 380-388.

Ali, M. A., Forghani, B., and Cantin, E. M., 1996, Characterization of an essential HSV-1 protein encoded by the UL25 gene reported to be involved in virus penetration and capsid assembly. *Virology* **216:** 278-283.

Baines, J. D., Cunningham, C., Nalwanga, D., and Davison, A., 1997, The UL15 gene of herpes simplex virus type 1 contains within its second exon a novel open reading frame that is translated in frame with the UL15 gene product. *J. Virol.* **71:** 2666-2673.

Baines, J. D., Poon, A. P., Rovnak, J., and Roizman, B., 1994, The herpes simplex virus 1 UL15 gene encodes two proteins and is required for cleavage of genomic viral DNA. *J. Virol.* **68:** 8118-8124.

Baines, J. D. and Roizman, B., 1992, The cDNA of UL15, a highly conserved herpes simplex virus 1 gene, effectively replaces the two exons of the wild-type virus. *J. Virol.* **66:** 5621-5626.

Baker, T. S., Newcomb, W. W., Booy, F. P., Brown, J. C., and Steven, A. C., 1990, Three-dimensional structures of maturable and abortive capsids of equine herpesvirus 1 from cryoelectron microscopy. *J. Virol.* **64:** 563-573.

Barnett, J. W., Eppstein, D. A., and Chan, H. W., 1983, Class I defective herpes simplex virus DNA as a molecular cloning vehicle in eucaryotic cells. *J. Virol.* **48:** 384-395.

Bataille, D. and Epstein, A., 1994, Herpes simplex virus replicative concatemers contain L components in inverted orientation. *Virology* **203:** 384-388.

Bataille, D. and Epstein, A. L., 1997, Equimolar generation of the four possible arrangements of adjacent L components in herpes simplex virus type 1 replicative intermediates. *J. Virol.* **71:** 7736-7743.

Baxter, M. K. and Gibson, W. The putative cytomegalovirus triplex proteins minor capsid protein (mCP) and mCP-binding protein (MCP-BP) form a heterotrimeric complex that localizes to the cell nucleus in the absence of other viral proteins. 22nd International Herpesvirus Workshop . 1997.

Bazinet, C. and King, J., 1985, The DNA translocating vertex of dsDNA bacteriophage. *Annu. Rev. Microbiol.* **39:** 109-129.

Becker, A. and Murialdo, H., 1990, Bacteriophage lambda DNA: the beginning of the end. *J. Bacteriol.* **172:** 2819-2824.

Black, L. W., 1988, DNA packaging in dsDNA bacteriophages. *In* "The Bacteriophages Vol. 2" (R. Calendar, Ed.), pp. 321-373. Plenum Press, New York.

Black, L. W., 1989, DNA packaging in dsDNA bacteriophages. *Annu. Rev. Microbiol.* **43:** 267-292.

Black, L. W., Showe, M. K., and Steven, A. C., 1994, Morphogenesis of the T4 head. *In* "Bacteriophage T4" (J. D. Karam, Ed.), pp. 218-258. ASM Press, Washington, D.C.

Bogner, E., Radsak, K., and Stinski, M. F., 1998, The gene product of human cytomegalovirus open reading frame UL56 binds the pac motif and has specific nuclease activity. *J. Virol.* **72:** 2259-2264.

Booy, F. P., Newcomb, W. W., Trus, B. L., Brown, J. C., Baker, T. S., and Steven, A. C., 1991, Liquid-crystalline, phage-like packing of encapsidated DNA in herpes simplex virus. *Cell* **64:** 1007-1015.

Broll, H., Buhk, H. J., Zimmermann, W., and Goltz, M., 1999, Structure and function of the prDNA and the genomic termini of the gamma2-herpesvirus bovine herpesvirus type 4. *J. Gen. Virol.* **80 (Pt 4):** 979-986.

Buerger, I., Reefschlaeger, J., Bender, W., Eckenberg, P., Klenk, H. D., Ruebsamen-Waigmann, H., and Hallenberger, S. Mechanism of antiviral action of BAY 38-4766--resistance to a novel non-nucleoside inhibitor of cytomegalovirus replication. 2001 Cytomegalovirus Workshop , 72. 2001.

Casjens, S. and Hendrix, R., 1988, Control mechanisms in dsDNA bacteriophage assembly. *In* "The Bacteriophages" (R. Calendar, Ed.), Vol. 1, pp. 15-91. Plenum Press, New York.

Catalano, C. E., 2000, The terminase enzyme from bacteriophage lambda: a DNA-packaging machine. *Cell Mol. Life Sci.* **57:** 128-148.

Catalano, C. E., Cue, D., and Feiss, M., 1995, Virus DNA packaging: the strategy used by phage lambda. *Molecular Microbiology* **16:** 1075-1086.

Cavalcoli, J. D., Baghian, A., Homa, F. L., and Kousoulas, K. G., 1993, Resolution of genotypic and phenotypic properties of herpes simplex virus type 1 temperature-sensitive mutant (KOS) tsZ47: evidence for allelic complementation in the UL28 gene. *Virology* **197:** 23-34.

Cerritelli, M. E., Cheng, N., Rosenberg, A. H., McPherson, C. E., Booy, F. P., and Steven, A. C., 1997, Encapsidated conformation of bacteriophage T7 DNA. *Cell* **91:** 271-280.

Chang, Y. E., Poon, A. P., and Roizman, B., 1996, Properties of the protein encoded by the UL32 open reading frame of herpes simplex virus type 1. *J. Virol.* **70:** 3938-3946.

Chee, M. S., Bankier, A. T., Beck, S., Sohni, R., Brown, C. M., Cerny, R., Horsnell, T., Hutchinson, C. A., Kouzarides, T., Marignetti, J. A., Preddie, E., Satchwell, S. C., Tomlinson, P., Weston, K., and Barrell, B. G., 1990, Analysis of the protein-coding content of the sequence of human cytomegalovirus strain AD169. *In* "Cytomegaloviruses" (J. K. McDougall, Ed.), Vol. 154, pp. 125-169. Springer-Verlag, New York.

Chen, D. H., Jiang, H., Lee, M., Liu, F., and Zhou, Z. H., 1999, Three-dimensional visualization of tegument/capsid interactions in the intact human cytomegalovirus. *Virology* **260:** 10-16.

Chou, J. and Roizman, B., 1985, Isomerization of herpes simplex virus 1 genome: identification of the cis-acting and recombination sites within the domain of the a sequence. *Cell* **41:** 803-811.

Chowdhury, S. I., Buhk, H. J., Ludwig, H., and Hammerschmidt, W., 1990, Genomic termini of equine herpesvirus 1. *J. Virol.* **64:** 873-880.

Chung, Y. B., Nardone, C., and Hinkle, D. C., 1990, Bacteriophage T7 DNA packaging. III. A "hairpin" end formed on T7 concatemers may be an intermediate in the processing reaction. *J. Mol. Biol.* **216:** 939-948.

Church, G. A., Dasgupta, A., and Wilson, D. W., 1998, Herpes simplex viurus DNA packaging without measurable DNA synthesis. *J. Virol.* **72:** 2745-2751.

Dasgupta, A. and Wilson, D. W., 1999, ATP depletion blocks herpes simplex virus DNA packaging and capsid maturation. *J. Virol.* **73:** 2006-2015.

Davison, A. J., 1984, Structure of the genome termini of varicella-zoster virus. *J. Gen. Virol.* **65 (Pt 11):** 1969-1977.

Davison, A. J. and Wilkie, N. M., 1981, Nucleotide sequences of the joint between the L and S segments of herpes simplex virus types 1 and 2. *J. Gen. Virol.* **55:** 315-331.

Deiss, L. P., Chou, J., and Frenkel, N., 1986, Functional domains within the a sequence involved in the cleavage-packaging of herpes simplex virus DNA. *J. Virol.* **59:** 605-618.

Deiss, L. P. and Frenkel, N., 1986, Herpes simplex virus amplicon: cleavage of concatameric DNA is linked to packaging and involves amplification of the terminally reiterated a sequence. *J. Virol.* **57:** 933-941.

Desai, P., DeLuca, N. A., and Person, S., 1998, Herpes simplex virus type 1 VP26 is not essential for replication in cell culture but influences production of infectious virus in the nervous system of infected mice. *Virology* **247:** 115-124.

Desai, P. and Person, S., 1999, Second site mutations in the N-terminus of the major capsid protein (VP5) overcome a block at the maturation cleavage site of the capsid scaffold proteins of herpes simplex virus type 1. *Virology* **261:** 357-366.

Dolan, A., Arbuckle, M., and McGeoch, D. J., 1991, Sequence analysis of the splice junction in the transcript of herpes simplex virus type 1 gene UL15. *Virus Research* **20:** 97-104.

Dunn, J. J. and Studier, F. W., 1983, Complete nucleotide sequence of bacteriophage T7 DNA and the locations of T7 genetic elements. *J. Mol. Biol.* **166:** 477-535.

Earnshaw, W. C. and Casjens, S. R., 1980, DNA packaging by the double-stranded DNA bacteriophages. *Cell* **21:** 319-331.

Ebeling, A., Keil, G. M., Knust, E., and Koszinowski, U. H., 1983, Molecular cloning and physical mapping of murine cytomegalovirus DNA. *J. Virol.* **47:** 421-433.

Frenkel, N., 1981, Defective interfering herpesviruses. *In* "The Human Herpesviruses--An Interdisciplinary Perspective" (A. H. Hahmias, W. R. Dowdle, and R. S. Schinazy, Eds.), pp. 91-120. Elsevier Science Publishing, Inc., New York.

Fujisawa, H., Kimura, M., and Hashimoto, C., 1990, In vitro cleavage of the concatemer joint of bacteriophage T3 DNA. *Virology* **174:** 26-34.

Fujisawa, H. and Sugimoto, K., 1983, On the terminally redundant sequences of bacteriophage T3 DNA. *Virology* **124:** 251-258.

Gao, M. and Isom, H. C., 1984, Characterization of the guinea pig cytomegalovirus genome by molecular cloning and physical mapping. *J. Virol.* **52:** 436-447.

Garber, D. A., Beverley, S. M., and Coen, D. M., 1993, Demonstration of circularization of herpes simplex virus DNA following infection using pulsed field gel electrophoresis. *Virology* **197:** 459-462.

Giesen, K., Radsak, K., and Bogner, E., 2000b, Targeting of the gene product encoded by ORF UL56 of human cytomegalovirus into viral replication centers. *FEBS Lett.* **471:** 215-218.

Giesen, K., Radsak, K., and Bogner, E., 2000a, The potential terminase subunit of human cytomegalovirus, pUL56, is translocated into the nucleus by its own nuclear localization signal and interacts with importin alpha [In Process Citation]. *J. Gen. Virol.* **81 Pt 9:** 2231-2244.

Guo, P., Peterson, C., and Anderson, D., 1987, Prohead and DNA-gp3-dependent ATPase activity of the DNA packaging protein gp16 of bacteriophage phi 29. *J. Mol. Biol.* **197:** 229-236.

Hammerschmidt, W., Ludwig, H., and Buhk, H. J., 1988, Specificity of cleavage in replicative-form DNA of bovine herpesvirus 1. *J. Virol.* **62:** 1355-1363.

Higgins, R. R. and Becker, A., 1994, The lambda terminase enzyme measures the point of its endonucleolytic attack 47 +/- 2 bp away from its site of specific DNA binding, the R site. *EMBO Journal* **13:** 6162-6171.

Homa, F. L. and Brown, J. C., 1997, Capsid assembly and DNA packaging in herpes simplex virus. *Reviews in Medical Virology* **7:** 107-122.

Hong, Z., Beaudet-Miller, M., Durkin, J., Zhang, R., and Kwong, A. D., 1996, Identification of a minimal hydrophobic domain in the herpes simplex virus type 1 scaffolding protein which is required for interaction with the major capsid protein. *J. Virol.* **70:** 533-540.

Jacob, R. J., Morse, L. S., and Roizman, B., 1979, Anatomy of herpes simplex virus DNA. XII. Accumulation of head-to-tail concatemers in nuclei of infected cells and their role in the generation of the four isomeric arrangements of viral DNA. *J. Virol.* **29:** 448-457.

Jardine, P. J. and Coombs, D. H., 1998, Capsid expansion follows the initiation of DNA packaging in bacteriophage T4. *J. Mol. Biol.* **284:** 661-672.

Kemble, G. W. and Mocarski, E. S., 1989, A host cell protein binds to a highly conserved sequence element (pac- 2) within the cytomegalovirus a sequence. *J. Virol.* **63:** 4715-4728.

Khan, S. A., Hayes, S. J., Watson, R. H., and Serwer, P., 1995, Specific, nonproductive cleavage of packaged bacteriophage T7 DNA in vivo. *Virology* **210:** 409-420.

Koslowski, K. M., Shaver, P. R., Casey II, J. T., Wilson, T., Yamanaka, G., Sheaffer, A. K., Tenney, D. J., and Pederson, N. E., 1999, Physical and functional interactions between the herpes simplex virus UL15 and UL28 DNA cleavage and packaging proteins. *J. Virol.* **73:** 1704-1707.

Koslowski, K. M., Shaver, P. R., Wang, X. Y., Tenney, D. J., and Pederson, N. E., 1997, The pseudorabies virus UL28 protein enters the nucleus after coexpression with the herpes simplex virus UL15 protein. *J. Virol.* **71:** 9118-9123.

Kronsky, P. M., Underwood, M. R., Turk, S. R., Feng, K. W., Jain, R. K., Ptak, R. G., Westerman, A. C., Biron, K. K., Townsend, L. B., and Drach, J. C., 1998, Resistance of human cytomegalovirus to benzimidazole ribonucleosides maps to two open reading frames: UL89 and UL56. *J. Virol.* **72:** 4721-4728.

Lamberti, C. and Weller, S. K., 1996, The herpes simplex virus type 1 UL6 protein is essential for cleavage and packaging but not for genomic inversion. *Virology* **226:** 403-407.

Lamberti, C. and Weller, S. K., 1998, The herpes simples virus type 1 cleavage/packaging protein, UL32, is involved in efficient localization of capsids to replication compartments. *J. Virol.* **72:** 2463-2473.

Langman, L., Paetkau, V., Scraba, D., Miller, R. C., Jr., Roeder, G. S., and Sadowski, P. D., 1978, The structure and maturation of intermediates in bacteriophage T7 DNA replication. *Can. J. Biochem.* **56:** 508-516.

Leffers, G. and Rao, V. B., 2000, Biochemical characterization of an ATPase activity associated with the large packaging subunit gp17 from bacteriophage T4. *J. Biol. Chem.* **275:** 37127-37136.

Lepault, J., Dubochet, J., Baschong, W., and Kellenberger, E., 1987, Organization of double-stranded DNA in bacteriophages: a study by cryo- electron microscopy of vitrified samples. *EMBO Journal* **6:** 1507-1512.

Locker, H. and Frenkel, N., 1979, BamI, KpnI, and SalI restriction enzyme maps of the DNAs of herpes simplex virus strains Justin and F: occurrence of heterogeneities in defined regions of the viral DNA. *J. Virol.* **32:** 429-441.

Marks, J. R. and Spector, D. H., 1988, Replication of the murine cytomegalovirus genome: structure and role of the termini in the generation and cleavage of concatenates. *Virology* **162:** 98-107.

Martinez, R., Sarisky, R. T., Weber, P. C., and Weller, S. K., 1996, Herpes simplex virus type 1 alkaline nuclease is required for efficient processing of viral DNA replication intermediates. *J. Virol.* **70**: 2075-2085.

McNab, A. R., Desai, P., Person, S., Roof, L. L., Thomsen, D. R., Newcomb, W. W., Brown, J. C., and Homa, F. L., 1998, The product of the herpes simplex virus type 1 UL25 gene is required for encapsidation but not for cleavage of replicated viral DNA. *J. Virol.* **72**: 1060-1070.

McVoy, M. A. and Adler, S. P., 1994, Human cytomegalovirus DNA replicates after early circularization by concatemer formation, and inversion occurs within the concatemer. *J. Virol.* **68**: 1040-1051.

McVoy, M. A., Nixon, D. E., and Adler, S. P., 1997, Circularization and cleavage of guinea pig cytomegalovirus genomes. *J. Virol.* **71**: 4209-4217.

McVoy, M. A., Nixon, D. E., Adler, S. P., and Mocarski, E. S., 1998, Sequences within the herpesvirus-conserved pac1 and pac2 motifs are required for cleavage and packaging of the murine cytomegalovirus genome. *J. Virol.* **72**: 48-56.

McVoy, M. A., Nixon, D. E., Jur, J. K., and Adler, S. P., 2000, The ends on herpesvirus DNA replicative concatemers contain pac2 cis cleavage/packaging elements and their formation is controlled by terminal cis sequences. *J. Virol.* **74**: 1587-1592.

Mercer, J. A., Marks, J. R., and Spector, D. H., 1983, Molecular cloning and restriction endonuclease mapping of the murine cytomegalovirus genome (Smith Strain). *Virology* **129**: 94-106.

Mettenleiter, T. C., Saalmuller, A., and Weiland, F., 1993, Pseudorabies virus protein homologous to herpes simplex virus type 1 ICP18.5 is necessary for capsid maturation. *J. Virol.* **67**: 1236-1245.

Mocarski, E. S., Deiss, L. P., and Frenkel, N., 1985, Nucleotide sequence and structural features of a novel US-a junction present in a defective herpes simplex virus genome. *J. Virol.* **55**: 140-146.

Mocarski, E. S., Liu, A. C., and Spaete, R. R., 1987, Structure and variability of the a sequence in the genome of human cytomegalovirus (Towne strain). *J. Gen. Virol.* **68** (Pt 8): 2223-2230.

Mocarski, E. S., Post, L. E., and Roizman, B., 1980, Molecular engineering of the herpes simplex virus genome: insertion of a second L-S junction into the genome causes additional genome inversions. *Cell* **22**: 243-255.

Mocarski, E. S. and Roizman, B., 1981, Site-specific inversion sequence of the herpes simplex virus genome: domain and structural features. *Proc. Natl. Acad. Sci .U.S.A* **78**: 7047-7051.

Mocarski, E. S. and Roizman, B., 1982, Structure and role of the herpes simplex virus DNA termini in inversion, circularization and generation of virion DNA. *Cell* **31**: 89-97.

Mosig, G., 1998, Recombination and recombination-dependent DNA replication in bacteriophage T4. *Annu. Rev. Genetics* **32**: 379-413.

Murialdo, H., 1991, Bacteriophage lambda DNA maturation and packaging. *Annu. Rev. Biochem.* **60**: 125-153.

Murialdo, H. and Becker, A., 1978, Head morphogenesis of complex double-stranded deoxyribonucleic acid bacteriophages. *Microbiological Reviews* **42**: 529-576.

Nasseri, M. and Mocarski, E. S., 1988, The cleavage recognition signal is contained within sequences surrounding an a-a junction in herpes simplex virus DNA. *Virology* **167**: 25-30.

Nealon, K., Newcomb, W. W., Pray, T. R., Craik, C. S., Brown, J. C., and Kedes, D. H., 2001, Lytic replication of Kaposi's sarcoma-associated herpesvirus results in the formation of multiple capsid species: isolation and molecular characterization of A, B, and C capsids from a gammaherpesvirus. *J. Virol.* **75**: 2866-2878.

Newcomb, W. W., Homa, F. L., Thomsen, D. R., Booy, F. P., Trus, B. L., Steven, A. C., Spencer, J. V., and Brown, J. C., 1996, Assembly of the herpes simplex virus capsid: characterization of intermediates observed during cell-free capsid assembly. *J .Mol. Biol.* **263**: 432-446.

Newcomb, W. W., Homa, F. L., Thomsen, D. R., Trus, B. L., Cheng, N., Steven, A. C., Booy, F. P., and Brown, J. C., 1999, Assembly of the herpes simplex virus procapsid from purified components and identification of small complexes containing the major capsid and scaffolding proteins. *J. Virol* **73**: 4239-4250.

Newcomb, W. W., Homa, F. L., Thomsen, D. R., Ye, Z., and Brown, J. C., 1994, Cell-free assembly of the herpes simplex virus capsid. *J. Virol.* **68**: 6059-6063.

Newcomb, W. W., Juhas, R. M., Thomsen, D. R., Homa, F. L., Burch, A. D., Weller, S. K., and Brown, J. C., 2001, The UL6 gene product forms the portal for entry of DNA into the herpes simplex virus capsid. *J. Virol.* In press.

Newcomb, W. W., Trus, B. L., Booy, F. P., Steven, A. C., Wall, J. S., and Brown, J. C., 1993, Structure of the herpes simplex virus capsid: molecular composition of the pentons and the triplexes. *J. Mol. Biol.* **232**: 499-511.

Newcomb, W. W., Trus, B. L., Cheng, N., Steven, A. C., Sheaffer, A. K., Tenney, D. J., Weller, S. K., and Brown, J. C., 2000, Isolation of herpes simplex virus procapsids from cells infected with a protease-deficient mutant virus. *J. Virol.* **74**: 1663-1673.

Ogasawara, M., Suzutani, T., Yoshida, I., and Azuma, M., 2001, Role of the UL25 Gene Product in Packaging DNA into the Herpes Simplex Virus Capsid: Location of UL25 Product in the Capsid and Demonstration that It Binds DNA. *J. Virol.* **75**: 1427-1436.

Oien, N. L., Thomsen, D. R., Wathen, M. W., Newcomb, W. W., Brown, J. C., and Homa, F. L., 1997, Assembly of herpes simplex virus capsids using the human cytomegalovirus scaffold protein: critical role of the C-terminus. *J. Virol.* **71**: 1281-1291.

Patel, A. H. and MacLean, J. B., 1995, The product of the UL6 gene of herpes simplex virus type 1 is associated with virus capsids. *Virology* **206**: 465-478.

Patel, A. H., Rixon, F. J., Cunningham, C., and Davison, A. J., 1996, Isolation and characterization of a herpes simplex virus type-1 mutant defective in the UL6 gene. *Virology* **217**: 111-123.

Pederson, N. E. and Enquist, L. W., 1991, Overexpression in bacteria and identification in infected cells of the pseudorabies virus protein homologous to herpes simplex virus type 1 ICP18.5. *J. Virol.* **65**: 3746-3758.

Poffenberger, K. L. and Roizman, B., 1985, A noninverting genome of a viable herpes simplex virus 1: presence of head-to-tail linkages in packaged genomes and requirements for circularization after infection. *J. Virol.* **53**: 587-595.

Poon, A. P. W. and Roizman, B., 1993, Characterization of a temperature-sensitive mutant of the UL15 open reading frame of herpes simplex virus 1. *J. Virol.* **67**: 4497-4503.

Preston, V. G., Coates, J. A. V., and Rixon, F. J., 1983, Identification and characterization of a herpes simplex virus gene product required for encapsidation of virus DNA. *J. Virol.* **45**: 1056-1064.

Prevelige, P. E. and King, J., 1993, Assembly of bacteriophage P22: A model for ds-DNA virus assembly. *Progress in Medical Virology* **40**: 206-221.

Reynolds, A. E., Fan, Y., and Baines, J. D., 2000, Characterization of the U(L)33 gene product of herpes simplex virus 1. *Virology* **266**: 310-318.

Rixon, F. J., 1993, Structure and assembly of herpesviruses. *Seminars in Virology* **4**: 135-144.

Rixon, F. J. and McNab, D., 1999, Packaging-competent capsids of a herpes simplex virus temperature-sensitive mutant have properties similar to those of in vitro-assembled procapsids. *J. Virol.* **73**: 5714-5721.

Roizman, B. and Sears, A. E., 1996, Herpes simplex viruses and their replication. *In* "Fields Virology" (B. N. Fields, D. M. Knipe, P. M. Howley, R. M. Chanock, J. L. Melnick, T. P. Monath, B. Roizman, and S. E. Straus, Eds.), Vol. 2, pp. 2231-2295. Lippincott-Raven, Philadelphia.

Salmon, B. and Baines, J. D., 1998, Herpes simplex virus DNA cleavage and packaging: association of multiple forms of UL15-encoded proteins with B capsids requires at least the UL6, UL17 and UL28 genes. *J. Virol.* **72:** 3045-3050.

Salmon, B., Cunningham, C., Davison, A. J., Harris, W. J., and Baines, J. D., 1998, The herpes simplex virus type 1 UL17 gene encodes virion tegument proteins that are required for cleavage and packaging of viral DNA. *J. Virol.* **72:** 3779-3788.

Salmon, B., Nalwanga, D., Fan, Y., and Baines, J. D., 1999, Proteolytic cleavage of the amino terminus of the UL15 gene product of herpes simplex virus type 1 is coupled with maturation of viral DNA into unit-length genomes. *J. Virol.* **73:** 8338-8348.

Serwer, P., Watson, R. H., and Hayes, S. J., 1992, Formation of the right before the left mature DNA end during packaging- cleavage of bacteriophage T7 DNA concatemers. *J. Mol. Biol.* **226:** 311-317.

Severini, A., Morgan, A. R., Tovell, D. R., and Tyrrell, D. L., 1994, Study of the structure of replicative intermediates of HSV-1 DNA by pulsed-field gel electrophoresis. *Virology* **200:** 428-435.

Sheaffer, A. K., Newcomb, W. W., Gao, M., Yu, D., Weller, S. K., Brown, J. C., and Tenney, D. J., 2001, Herpes simplex virus DNA cleavage and packaging proteins associate with the procapsid prior to its maturation. *J. Virol.* **75:** 687-698.

Sherman, G. and Bachenheimer, S., 1987, DNA processing in temperature-sensitive morphogenic mutants of HSV-1. *Virology* **158:** 427-430.

Sherman, G. and Bachenheimer, S. L., 1988, Characterization of intranuclear capsids made by ts morphogenetic mutants of HSV-1. *Virology* **163:** 471-480.

Shibata, H., Fujisawa, H., and Minagawa, T., 1987, Characterization of the bacteriophage T3 DNA packaging reaction in vitro in a defined system. *J. Mol. Biol.* **196:** 845-851.

Simpson, A. A., Tao, Y., Leiman, P. G., Badasso, M. O., He, Y., Jardine, P. J., Olson, N. H., Morais, M. C., Grimes, S., Anderson, D. L., Baker, T. S., and Rossmann, M. G., 2000, Structure of the bacteriophage phi29 DNA packaging motor. *Nature* **408:** 745-750.

Slobedman, B. and Simmons, A., 1997, Concatemeric intermediates of equine herpesvirus type 1 DNA replication contain frequent inversions of adjacent long segments of the viral genome. *Virology* **229:** 415-420.

Slobedman, B., Zhang, X., and Simmons, A., 1999, Herpes simplex virus genome isomerization: origins of adjacent long segments in concatemeric viral DNA. *J. Virol.* **73:** 810-813.

Smiley, J. R., Duncan, J., and Howes, M., 1990, Sequence requirements for DNA rearrangements induced by the terminal repeat of herpes simplex virus type 1 KOS DNA. *J. Virol.* **64:** 5036-5050.

Son, M., Watson, R. H., and Serwer, P., 1993, The direction and rate of bacteriophage T7 DNA packaging in vitro. *Virology* **196:** 282-289.

Spaete, R. R. and Frenkel, N., 1982, The herpes simplex virus amplicon: a new eucaryotic defective-virus cloning-amplifying vector. *Cell* **30:** 295-304.

Spaete, R. R. and Frenkel, N., 1985, The herpes simplex virus amplicon: analyses of cis-acting replication functions. *Proc. Natl. Acad. Sci. U.S.A* **82:** 694-698.

Spencer, J. V., Newcomb, W. W., Thomsen, D. R., Homa, F. L., and Brown, J. C., 1998, Assembly of the herpes simplex virus capsid: pre-formed triplexes bind to the nascent capsid. *J. Virol.* **72:** 3944-3951.

Steven, A. C. and Spear, P. G., 1996, Herpesvirus capsid assembly and envelopment. *In* "Structural Biology of Viruses" (R. Burnett, W. Chiu, and R. Garcea, Eds.), pp. 312-351. Oxford University Press, New York.

Stow, N. D., McMonagle, E. C., and Davison, A. J., 1983, Fragments from both termini of the herpes simplex virus type 1 genome contain signals required for the encapsidation of viral DNA. *Nucleic Acids Res.* **11:** 8205-8220.

Tamashiro, J. C., Filpula, D., Friedmann, T., and Spector, D. H., 1984, Structure of the heterogeneous L-S junction region of human cytomegalovirus strain AD169 DNA. *J. Virol.* **52:** 541-548.

Tamashiro, J. C. and Spector, D. H., 1986, Terminal structure and heterogeneity in human cytomegalovirus strain AD169. *J. Virol.* **59:** 591-604.

Tatman, J. D., Preston, V. G., Nicholson, P., Elliott, R. M., and Rixon, F. J., 1994, Assembly of herpes simplex virus type 1 capsids using a panel of recombinant baculoviruses. *J. Gen.l Virol.* **75:** 1101-1113.

Taus, N. S. and Baines, J. D., 1998, Herpes simplex virus 1 DNA cleavage/packaging: the UL28 gene encodes a minor component of B capsids. *Virology* **252:** 443-449.

Taus, N. S., Salmon, B., and Baines, J. D., 1998, The herpes simplex virus 1 UL17 gene is required for localization of capsids and major and minor capsid proteins to intranuclear sites where viral DNA is cleaved and packaged. *Virology* **252:** 115-125.

Tengelsen, L. A., Pederson, N. E., Shaver, P. R., Wathen, M. W., and Homa, F. L., 1993, Herpes simplex virus type 1 DNA cleavage and encapsidation require the product of the UL28 gene: isolation and characterization of two UL28 deletion mutants. *J. Virol.* **67:** 3470-3480.

Thomsen, D. R., Roof, L. L., and Homa, F. L., 1994, Assembly of herpes simplex virus (HSV) intermediate capsids in insect cells infected with recombinant baculoviruses expressing HSV capsid proteins. *J. Virol.* **68:** 2442-2457.

Trus, B. L., Booy, F. P., Newcomb, W. W., Brown, J. C., Homa, F. L., Thomsen, D. R., and Steven, A. C., 1996, The herpes simplex virus procapsid: structure, comformational changes upon maturation, and roles of the triplex proteins VP19C and VP23 in assembly. *J. Mol. Biol.* **263:** 447-462.

Trus, B. L., Gibson, W., Cheng, N., and Steven, A. C., 1999, Capsid structure of simian cytomegalovirus from cryoelectron microscopy: evidence for tegument attachment sites. *J. Virol.* **73:** 2181-2192.

Trus, B. L., Homa, F. L., Booy, F. P., Newcomb, W. W., Thomsen, D. R., Cheng, N., Brown, J. C., and Steven, A. C., 1995, Herpes simplex virus capsids assembled in insect cells infected with recombinant baculoviruses: Structural authenticity and localization of VP26. *J. Virol.* **69** 7362-7366.

Umene, K., 1991, Recombination of the internal direct repeat element DR2 responsible for the fluidity of the a sequence of herpes simplex virus type 1. *J. Virol.* **65:** 5410-5416.

Umene, K., 1994, Excision of DNA fragments corresponding to the unit-length a sequence of herpes simplex virus type 1 and terminus variation predominate on one side of the excised fragment. *J. Virol.* **68:** 4377-4383.

Underwood, M. R., Harvey, R. J., Stanat, S. C., Hemphill, M. L., Miller, T., Drach, J. C., Townsend, L. B., and Biron, K. K., 1998, Inhibition of human cytomegalovirus DNA maturation by a benzimidazole ribonucleoside is mediated through the UL89 gene product. *J. Virol.* **72:** 717-725.

Valpuesta, J. M. and Carrascosa, J. L., 1994, Structure of viral connectors and their function in bacteriophage assembly and DNA packaging. *Q. Rev. Biophys.* **27:** 107-155.

van Zeijl, M., Fairhurst, J., Jones, T. R., Vernon, S. K., Morin, J., LaRocque, J., Feld, B., O'Hara, B., Bloom, J. D., and Johann, S. V., 2000, Novel class of thiourea compounds that inhibit herpes simplex virus type 1 DNA cleavage and encapsidation: resistance maps to the UL6 gene [In Process Citation]. *J. Virol.* **74:** 9054-9061.

Varmuza, S. L. and Smiley, J. R., 1985, Signals for site-specific cleavage of HSV DNA: maturation involves two separate cleavage events at sites distal to the recognition sequences. *Cell* **41:** 793-802.

Vlazny, D. A. and Frenkel, N., 1981, Replication of herpes simplex virus DNA: localization of replication recognition signals within defective virus genomes. *Proc. Natl. Acad. Sci. U.S.A* **78:** 742-746.

Vlazny, D. A., Kwong, A., and Frenkel, N., 1982, Site-specific cleavage/packaging of herpes simplex virus DNA and the selective maturation of nucleocapsids containing full-length viral DNA. *Proc. Natl. Acad. Sci. U.S.A* **79:** 1423-1427.

Wagner, M. J. and Summers, W. C., 1978, Structure of the joint region and the termini of the DNA of herpes simplex virus type 1. *J. Virol.* **27:** 374-387.

Ward, P. L., Ogle, W. O., and Roizman, B., 1996, Assemblons: Nuclear structures defined by aggregation of immature capsids and some tegument proteins on herpes simplex virus 1. *J. Virol.* **70:** 4623-4631.

Warner, S. C., Desai, P., and Person, S., 2000, Second-site mutations encoding residues 34 and 78 of the major capsid protein (VP5) of herpes simplex virus type 1 are important for overcoming a blocked maturation cleavage site of the capsid scaffold proteins. *Virology* **278:** 217-226.

Weller, S. K., 1995, Herpes simplex virus DNA replication and genome maturation. *In* "The DNA Provirus: Howard Temin's Scientific Legacy" (G. M. Cooper, R. G. Temin, and B. Sugden, Eds.), pp. 189-213. American Society for Microbiology, Washington, D.C.

White, J. H. and Richardson, C. C., 1987, Processing of concatemers of bacteriophage T7 DNA in vitro. *J. Biol. Chem.* **262:** 8851-8860.

Yao, X. D. and Elias, P., 2001, Recombination during early herpes simplex virus type 1 infection is mediated by cellular proteins. *J. Biol. Chem.* **276:** 2905-2913.

Yao, X. D., Matecic, M., and Elias, P., 1997, Direct repeats of the herpes simplex virus a sequence promote nonconservative homologous recombination that is not dependent on XPF/ERCC4. *J. Virol.* **71:** 6842-6849.

Yu, D., Sheaffer, A. K., Tenney, D. J., and Weller, S. K., 1997, Characterization of ICP6::lacZ insertion mutants of the UL15 gene of herpes simplex virus type 1 reveals the translation of two proteins. *J. Virol.* **71:** 2656-2665.

Yu, D. and Weller, S. K., 1998b, Genetic analysis of the UL15 gene locus for the putative terminase of herpes simplex virus type 1. *Virology* **243:** 32-44.

Yu, D. and Weller, S. K., 1998a, Herpes simplex virus type 1 cleavage and packaging proteins UL15 and UL28 are associated with B but not C capsids during packaging. *J. Virol.* **72:** 7428-7439.

Zacny, V. L., Gershburg, E., Davis, M. G., Biron, K. K., and Pagano, J. S., 1999, Inhibition of Epstein-Barr virus replication by a benzimidazole L- riboside: novel antiviral mechanism of 5, 6-dichloro-2-(isopropylamino)- 1-beta-L-ribofuranosyl-1H-benzimidazole. *J. Virol.* **73:** 7271-7277.

Zhang, X., Efstathiou, S., and Simmons, A., 1994, Identification of novel herpes simplex virus replicative intermediates by field inversion gel electrophoresis: implications for viral DNA amplification strategies. *Virology* **202:** 530-539.

Zhou, Z. H., Dougherty, M., Jakana, J., He, J., Rixon, F. J., and Chiu, W., 2000, Seeing the herpesvirus capsid at 8.5 A. *Science* **288:** 877-880.

Zhou, Z. H., He, J., Jakana, J., Tatman, J. D., Rixon, F. J., and Chiu, W., 1995, Assembly of VP26 in herpes simplex virus-1 inferred from structures of wild-type and recombinant capsids. *Nature Struct. Biol.* **2:** 1026-1030.

Zimmermann, J. and Hammerschmidt, W., 1995, Structure and role of the terminal repeats of Epstein-Barr virus in processing and packaging of virion DNA. *J. Virol.* **69:** 3147-3155.

Chapter 3.2

From Concatemeric DNA into Unit-length Genomes – a Miracle or Clever Genes?

ANDREAS HOLZENBURG[*] and ELKE BOGNER[#]
*Texas A&M University, Electron Microscopy Center, Dept. Biology and Dept. Biochemsitry & Biophysics, College Station, TX 77843-2257, USA; #Institut für Molekulare und Klinische Virologie, Universität Erlangen-Nürnberg, Schloßgarten 4, D-91054 Erlangen, Germany

1. INTRODUCTION

Herpesviruses have large (130-235 kb) double-stranded linear DNA genomes that circularise shortly after infection (Garber et al., 1993; McVoy and Adler, 1994; Roizman and Sears, 1996). During infection, viral DNA is replicated in nuclei of infected cells and accumulates as head-to-tail concatemers (Ben-Porat, 1983; Jacob et al., 1979; Martinez et al., 1996; McVoy and Adler, 1994; Roizman and Sears, 1996; Severini et al., 1994; Zhang et al., 1994). Unit-length genomes are cleaved from intranuclear concatemeric viral DNA and packaged into preassembled capsids. This process is the first essential step in the production of infectious particles, i.e. virions. Considering that current anti-herpesviral drugs like acyclovir, ganciclovir, cidofovir and foscarnet have limited effects and often exhibit dose-dependent toxicity, new antiviral therapeutics ought to be considered. This, in turn, would necessitate the characterisation of new targets supporting different modes of action. To this end, an active interference with the cleavage and packaging mechanisms of the viral DNA may offer viable alternatives. This chapter is exclusively concerned with the packaging of the viral genome into the procapsid. For a review of the assembly mechanisms of capsids and procapsids from their proteinaceous building blocks, the reader is referred to chapter 3.1.

DNA-packaging in herpesviruses is a very complex biological process, and while it is commonly accepted that ATP-hydrolysis is the driving force behind it, the molecular mechanisms of DNA-translocation and genome packaging remain a mystery. In general, the following five steps are involved:

(i)   The recognition of concatemeric DNA by a specific protein (-complex) able to

(ii)   bind and cut the DNA at specific sequence motifs (packaging signals, e.g. *pac1* and *pac2*),

(iii)   translocation of the DNA-protein complex into the procapsid,

(iv)   packaging of one unit-length genome worth of DNA, and

(v)   completion of the packaging process by cutting off excess DNA at the portal region.

Shortly after infection, the viral DNA is released into the nucleus and immediately circularised. Following on from this, rolling circle replication leads to concatemeric (or intermediate) DNA, which has the ability to undergo recombination (McVoy and Adler, 1994). The herpesvirus genome has a unique long (U_L) and short segment (U_S), separated by two inverted repeats *b* and *c* (Stinski, 1991; Sheldrick and Berthelot, 1975). Furthermore, short but repetitive *a* sequences are found within the internal (IR) and terminal repeats (TR) that flank the U_L and U_S segments (Fig.3.2.1). Cleavage of DNA occurs within the *a* sequence, resulting in an U_L terminus with a variable number of *a* sequences and an U_S terminus with a single or no *a* sequence (Chou and Roizman, 1985; Deiss et al., 1986; Deiss and Frenkel, 1986; Frenkel, 1981; Nasseri and Mocarski, 1988; Spaete and Frenkel, 1982, Spaete and Frenkel, 1985; Spaete and Mocarsci, 1985; Tamashiro and Spector, 1986; Vlazny and Frenkel, 1981; Vlazny et al., 1982). A specific characteristic of herpesvirus DNA replication is the free inversion of the covalently linked segments, U_L and U_S relative to each other, giving rise to four isomeric forms in equimolar amounts (Stinski, 1991). These genomic inversions are brought about by recombination of concatemeric DNA after rolling circle replication (see above) or, alternatively, be linked to the early rounds of replication (Zhang et al., 1994). However, the exact inversion mechanism remains to be established. It is interesting to note that concatemeric DNA is not found in mammalian DNA replication.

Since only concatemers with accessible U_S termini have been reported, *cis*-acting packaging and cleaving sequences must exist at the termini. The two *cis*-acting motifs *pac1* and *pac2* are located in the *a* sequence, and are necessary for packaging of unit-length genomes (Stow et al., 1983). Only *pac2* is located near the ends of concatemers suggesting that these elements drive the directional packaging by binding proteins that initiate the translocation of concatemeric DNA into procapsids (McVoy et al., 2000).

Initiation as well as termination of DNA-packaging requires endonuclease-induced dsDNA cuts. It is speculated that initiation of DNA-

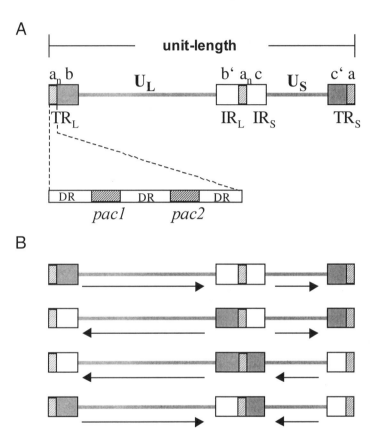

Figure 1. Genome of human cytomegalovirus. A) The unique long (U_L) and the unique short (U_S) segments are flanked by terminal (TR) and internal (IR) repeats. The *a* sequence are present in one or several copies at the end of the TR_L and at the L-S junction of the IR, whereas only one copy is located in the TR_S region. The *pac1* and *pac2* sequences were flanked by direct repeats (DR). B) The four equimolar isomers of the HCMV genome.

packaging is mediated by the attachment of a capsid-associated protein (-complex) to *pac* sites. The resulting DNA molecule is translocated into procapsids and encapsidated prior to a second DNA cleavage step. This mechanism is analog to the one in the dsDNA bacteriophages (Becker and Murialdo, 1990; Black, 1988; Higgins and Becker, 1994; Smith and Feiss,

1993). The formation of unit length-genomes during DNA-packaging in bacteriophage T4 is catalysed by a class of proteins known as terminases, which catalyse the ATP-dependent translocation of genomic DNA into the bacteriophage procapsids; they also bind and cleave concatenated DNA (Black, 1988; Feiss & Becker, 1983). Most bacteriophage terminases are hetero-oligomers with each subunit carrying a different function (Black et al., 1994; Casjens and Hendrix, 1988; Eppler et al., 1991; Fujisawa and Hendrix, 1994). The small terminase subunit gp16 is required for DNA recognition and the large subunit gp17 displays ATP-hydrolysing activity (Bhattacharyya and Rao, 1994; Kuebler and Rao, 1998). Mutations in any of the encoding genes lead to an accumulation of empty procapsids (proheads) and DNA concatemers (Black, 1995).

A common principle of viral DNA replication is the formation of "endless" DNA concatemers from which unit-length genomes are cut off prior to packaging. This chapter deals with the characterisation of two highly conserved proteins which are thought to play an important role in packaging of HCMV genomes.

2.   ANALYSIS OF TWO SUBUNITS OF A PUTATIVE HCMV TERMINASE

2.1   Characterization of two HCMV proteins: pUL56 and pUL89

A common feature of all herpesviruses is the organisation of genes encoding structural proteins or proteins involved in DNA replication in seven conserved sequence or gene blocks (Chee et al., 1990). While the arrangements of these blocks vary in all herpesviruses the sequences themselves are conserved. A prominent gene block (A) is the one containing the viral DNA polymerase (UL54), the major envelope protein gB (UL55), the single stranded binding protein (UL57) and the gene of ORF UL56. Human cytomegalovirus (HCMV) pUL56 (p130) is the homolog of the HSV-1 ICP 18.5 (pUL28) and conserved throughout all herpesviruses. The amino acid sequence of this early protein contains motifs suggesting its involvement in (i) DNA-binding (a Zinc finger motif towards the amino-terminus), (ii) protein-protein interaction (a leucin zipper), and (iii) nuclear localisation (stretch of basic amino acids at the carboxy-terminus resembling a nuclear localisation signal (NLS); see Fig.2). pUL56 is predominantly found in virions and therefore constitutes a structural protein. An analysis of subviral fractions revealed the presence of pUL56 in the capsid- but not in the tegument-fraction leading to the assumption that pUL56 is a capsid-associated rather than a tegument protein (Bogner et al., 1993).

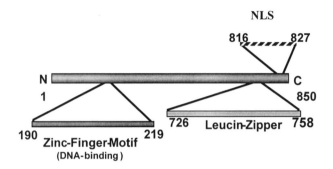

Figure 2. Structural features of HCMV pUL56. The Zinc-Finger motif towards the N-terminus could constitute the DNA-binding domain. The carboxy-terminus comprises a potential Leucin-Zipper and a cluster of basic amino acids representing the NLS.

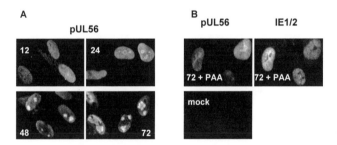

Figure 3. Intranuclear distribution of HCMV pUL56.. A) Permissive cells (HFF) were mock-infected or infected with HCMV for 12, 24, 48 and 72 h and subjected to immunofluorescence with antibody pabUL56 against pUL56. B) HFF cells infected with HCMV in the presence of PAA were double stained with antibodies against pUL56 and against IE1/2 at 72 h p.i., respectively

After cytoplasmic translation, pUL56 is translocated into the nucleus. The transition from the cytoplasm into the nucleus is a very fast one and takes only minutes to complete (Giesen, 1996). Studies by Giesen et al. (2000b) analysing the nuclear localisation of pUL56 using indirect immunofluorescence, show that the distribution and intensity of fluorescence changed during the course of infection: the nuclear staining increased with time until the late stages of infection when additional bright nuclear patches are detected (Fig.3).

Early after infection this protein is also detected in nucleolar regions (Fig.3) for reasons as yet unknown. The observation of staining in the presence of an inhibitor of viral DNA replication (e.g. phosphono acidic acid [PAA]), demonstrated that pUL56 is an early protein (Fig.3; Giesen et al., 2000a). Since a nuclear localisation is also observed after transfection of permissive (U373) and nonpermissive (COS-7) cells, the protein is imported by its own nuclear localisation signal (NLS). Deletion mutants demonstrated that the NLS is located between aa 816-827 (RRVRATRKRPRR) and could even translocate a cytoplasmic reporter protein into the nucleus. Alanin-scans showed that aa 822 and 823 (Arginine and Lysine) are essential for nuclear import of pUL56 (Giesen et al., 2000b).

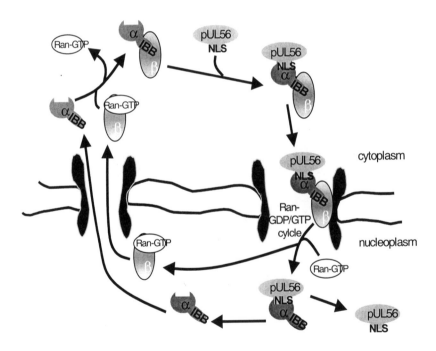

Figure 4. Nuclear import of pUL56. In the cytoplasm the heterodimeneric importin complex is formed. Translocation of pUL56 into the nucleus is mediated by the importin-dependent pathway after interaction of the NLS with importin α. In the nucleus pUL56 dissociates from importin α. This diagram is modified from D.Görlich, Curr.Op.Cell.Biol., 1997.

It is known that nuclear transport exhibits energy- and signal-dependence and is carrier-mediated (Dingwall & Laskey, 1991; Newmeyer *et al.*, 1986; Weis et al., 1996a; Zasloff, 1983). Active transport of proteins across the nuclear pore complex is achieved by the importin-system. Several cellular proteins have been identified that are involved in this process (Görlich et al., 1995; Görlich et al., 1996; Görlich, 1997). Interestingly, it was demonstrated that HCMV pUL56 can be recognized via its NLS by importin α (hSRPα1; Weis et al., 1995; Weis et al., 1996b), the adapter protein of the heterodimeric importin complex, and is subsequently transported into the nucleus using the importin-dependent pathway (Giesen et al., 2000b; Fig 4.).

First electron microscopic data of purified pUL56 revealed that the protein has a ring-shaped structure constituting a new class of toroidal proteins involved in DNA metabolism (Hingorani and O'Donnell, 1998). Upon interaction with DNA, pUL56 was found to undergo large-scale structural changes and could toggle between at least two structural states (Scheffczik et al., 2002). Such structural changes require some intrinsic structural flexibility. As pUL56 consists of several discrete domains, this architectural feature in itself could enable flexibility by a e.g. re-arrangement of the domains relative to each other.

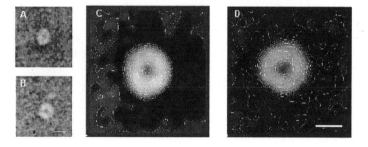

Figure 5: Electron microscopic analysis of purified pUL56 after negative staining. A) and B) Representative molecules belong to the two different projections (view 1 and 2, respectively). Averaged projections after digital image analysis are depicted in C (view 1) and D (view 2). The scale bars correspond to 10 (A,B) and 5 nm (C,D).

The apex of the pentagonal outline in Fig. 5, for instance, coincides with the location of the cleft, again suggesting that conformational flexibility could well be within the remit of, if not a prerequisite for pUL56's biological activity. It is therefore possible that the different structural states observed correspond to different functional states. It has been speculated that the cleft, together with the central hole, are intimately involved in the biological function of pUL56 by facilitating protein-DNA interaction and increasing the surface at the protein-DNA interface (Scheffczik et al., 2002).

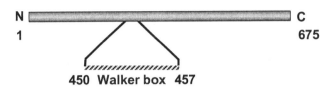

Figure 6. Structural features of HCMV pUL89. The walker box motif comprises a potential ATP-binding site.

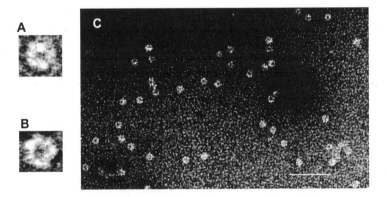

Figure 7: Electron microscopic analysis of purified pUL89 after negative staining. A) and B) Representative molecules. An overview is shown in C. The scale bar corresponds to 50 nm

Another highly conserved HCMV protein is the gene product of ORF UL89, the homolog of HSV pUL15. This protein belongs to the gene block F and is known as the homolog of the terminase subunit gp17 of bacteriophage T4 (Davison, 1992). It is an 675 aa protein and the *in vitro* translation product as well as the recombinant protein (rpUL89) possesses a molecular mass of approximately 70–75kDa. Interestingly, the protein sequence contains, as does its T4 homolog, a conserved ATP-binding motif (*walker box*, AYDYFGKT aa 450-457), implying an involvement in translocating genomes into procapsids. In analogy to its counterpart in bacteriophage T4 it is suggested that pUL89 is in contrast to pUL56 a nonstructural protein. In addition, pUL89 interacts specifically with the carboxy-terminus of pUL56, demonstrating that these proteins normally carry out complementary functions. Furthermore, EM analysis of purified pUL89 demonstrated that this monomer is also a toroidal DNA-metabolizing protein. The ring-shaped structure has in comparison with pUL56, however, a more irregular outline. The rings have a diameter of approximately 8 nm and the central protein deficit is approximately 2 nm across (Scheffczik et al, 2002, Fig.7). Nuclease activity together with EM analysis demonstrated that pUL89 is necessary for the cleavage of viral DNA. It is therefore possible that HCMV pUL56 and pUL89 are analogous to the T4 gp16 and gp17 subunit, respectively.

2.2   Interaction with DNA

In recent years, DNA-packaging and -cleavage have become a major focus point in virus research. First information about this process was obtained from virus mutants of the pUL56-homologous proteins of herpes simplex virus 1 (HSV-1) ICP18.5 (pUL28) and pseudorabies virus (Addison et al., 1990; Tengelson et al., 1993; Mettenleiter et al., 1993). These mutants showed a nuclear accumulation of naked nucleocapsids and uncleaved concatemeric DNA. Experiments were undertaken to verify whether HCMV pUL56 has functions similar to the ICP18.5 homologues of HSV-1 and pseudorabies virus. Using electrophoretic mobility shift assays, it was demonstrated that baculovirus-UL56-infected recombinant cell extracts containing pUL56 bind specifically to the *cis*-acting packaging motifs *pac1* and *pac2* of the viral DNA *a* sequence (Bogner et al., 1998). These studies showed that HCMV pUL56 is a DNA-binding protein, which forms specific DNA-protein complexes with HCMV DNA-packaging motifs. Recently, it has been reported that HSV-1 pUL28 also interacts with the packaging motif *pac1* (Adelman et al., 2001). The *pac1* DNA-binding, however, was only obtained after heat treatment. The authors concluded that during viral packaging the interaction of proteins with DNA depends on the structure of the viral DNA.

Since a final cut into unit-length genomes is required during packaging, experiments concerning a specific endonuclease activity of pUL56 and pUL89 were conducted. Interestingly, both proteins were able to convert supercoiled plasmid DNA containing the *a* sequence into open circular as well as linear molecules (Bogner et al., 1998; Bogner, 1999). These observations are in line with the structure of the proteins, where a toroidal shape has been shown to be conducive for DNA metabolism activities (Hingorani & O'Donnell, 1998). While either protein is able to cut DNA *in vitro*, the full *in vivo* activity is only achieved when both proteins form a complex, i.e. the terminase complex.

When incubating pUL56 with linear DNA containing an *a* sequence, specificDNA-pUL56 complexes are formed, which can be visualized (Fig.8a, b). Subsequent addition of pUL89 leads to the DNA being cut (Fig.8c, d) while a simultaneous incubation of DNA with pUL56 and pUL89 results in the DNA being densely covered with pUL89 but not cut.

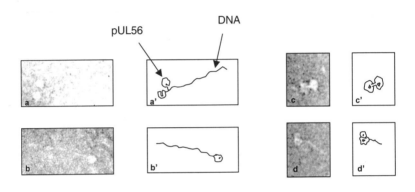

Figure 8. . Electron micrographs showing proteins incubated with linear DNA containing an *a* sequence. DNA incubated with pUL56 alone remains uncleaved (a,b). Cutting of DNA occurs after subsequent incubation with pUL89 (c,d).

2.3   Targeting into viral replication centers

It is known that herpesvirus transcription and packaging occur in defined intranuclear globular structures called replication centers (de Bruyn and Knipe, 1994; de Bruyn et al., 1998). More recently, Ahn et al. (1999) reported the formation of viral DNA replication compartments in HCMV-infected cells is initiated within granular structures that bud from the periphery of some of the promyelocytic leukemia protein oncogenic domains (PODs) and subsequently merge into larger structures that are flanked by PODs. The replication centers are dense chromatin structures and are surrounded by areas where capsids are formed (Fig.9).

In vitro observations showing that pUL56 and pUL89 have the ability to interact with specific DNA packaging motifs and cleave DNA bearing these motifs, led to the suggestion that both proteins are localized in viral replication centers. In infected cells pUL56 and the gene products of pUL112-113, the proteins which form prereplication centers and are required for DNA replication, partially co-localize in viral replication centers. However, in the absence of DNA replication both proteins show a different intracellular distribution (Giesen et al., 2000a). The UL112-113 gene products are found in the prereplication structures and redistribute in viral replication centers only during DNA replication (Anders and McCue, 1996). In the case of pUL44, a protein of the replication fork where it increases DNA-binding specificity of the viral DNA (Gibson, 1981; Plachter et al., 1992; Weiland et al., 1994), pUL56 and pUL44 are co-localize. Using co-immunoprecipitations it was demonstrated that both proteins directly interact with each other (Giesen et al., 2000a). This observation was later confirmed in infected cells by immunostaining of thin sections (unpublished observations; Fig.9).

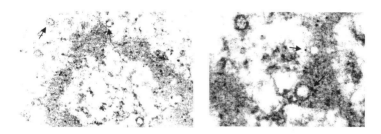

Figure 9. Detection of pUL56 in viral replication centers. Immunostaining of thin sections was performed by using UL56-specific antibody and a secondary colloidal gold-labeled antibody (5nm). The arrows indicate the position of procapsids to which pUL56 has bound.

The molecular mechanisms of DNA translocation into procapsids remain unclear, but it is commonly accepted that ATP hydrolysis is the driving force behind it (Higgins & Becker, 1994). Most bacteriophage terminases are hetero-oligomers with each subunit carrying a different function (Black, 1989; Casjens and Hendrix, 1988; Eppler et al., 1991, Fujisawa and Hearin, 1994). The large subunit catalyzes the ATP-dependent translocation of unit-length DNA into the bacteriophage procapsids, the small unit binds and cleave concatenated DNA (Bahattacharyya and Rao, 1994; Kuebler and Rao, 1998). Leffers and Rao (2000) reported that in the case of bacteriophage T4 the ATPase activity is associated with large terminase subunit gp17. In HCMV, this enzymatic activity is exclusively associated with pUL56 (Hwang and Bogner, 2002). It is interesting to note that, although not endowed with its own ATPase activity, pUL89 can enhance the pUL56-

associated ATP hydrolysis by 30 percent. Comparable observations were made with bacteriophage T4 (Leffers and Rao, 2000).

These findings suggest a concerted function of pUL56 and pUL89 during HCMV-packaging, indicating that these proteins could be defined as the two subunits making up a HCMV-specific terminase.

3.   CONCLUSIONS

During the last couple of years the list of genes which might be involved in DNA-packaging increased up to seven genes. In this chapter we have presented details of two essential proteins that are involved in this process, and suggest for the first time a model of pUL56´s interaction with DNA (fig.10). Based on recent observations it is reasonable to assume that initiation of DNA-packaging is mediated by attachment of pUL56 to pac sites on the concatemeric DNA and specific cleavage requiring the interaction of pUL89 with pUL56.

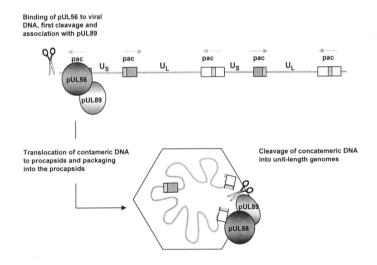

Figure 10. Model of pUL56 and pUL89 action In the first step pUL56 binds to the pac site and a first cleavage occurs. After interaction with pUL89 the DNA is translocated to procapsids. In the next step pUL56 binds to the procapsid and the DNA is packaged. A second cleavage into unit-length genomes finishes the process.

In HCMV, it is postulated that pUL56 is the translocating ATPase. The capsid association of pUL56 implies that it could function as a portal protein and that the attachment to procapsids may also be catalyzed by pUL56. In the light of the current data, the recent studies by Giesen et al. (2000a) in conjuction with immunostaining on thin sections of infected cells showing that pUL56 is associated with viral replication centers seem to suggest that DNA packaging may even be coupled with replication. As a mechanism that is centrally important to the life cycle of HCMV and other viruses, more light should perhaps be shed on the possibilities of utilizing DNA packaging as target for the development of new drugs. The proteins driving this process are highly specific, conserved, and do not have a counterpart in humans.

ACKNOWLEDGMENTS

We would like to thank all present and past members of the AH and EB labs and colleagues who have helped in many invaluable ways in the production of this chapter. The research has been supported by the Deutsche Forschungsgemeinschaft (Bo 1214/4-1) and by the Johannes and Frieda Marohn-Stiftung.

REFERENCES

Addison, C., Rixon, F.J., and Preston, V.G., 1990, Herpes simplex virus type 1 UL28 gene product is important for the formation of mature capsids. *J.Gen. Virol.* **71:** 2377-2384

Adelman, K., Salmon, B., and Baines, J.D., 2001, Herpes simplex virus DNA packaging sequences adopt novel structures that are specifically recognized by a component of the cleavage and packaging machinery. *Proc.Natl.Acad.Sci.USA* **98:** 3086-3091

Ahn, J.H., Jang, W.J., and Hayward, G.S., 1999, The Human Cytomegalovirus IE2 and UL112-113 Proteins Accumulate in Viral DNA Replication Compartments That Initiate from the Periphery of Promyelocytic Leukemia Protein-Associated Nuclear Bodies (PODs or ND10) *J. Virol.* **73:** 10458-10471

Anders, D.G., and McCuew, L.A., 1996, The human cytomegalovirus genes and proteins required for DNA synthesis. *Intervirology* **39:** 378-388

Becker, A., and Murialdo, H., 1990, Bacteriophage lambda DNA: the beginning of the end. *J. Bacteriol.* **172:** 2819-2824

Ben-Porat, T, 1983, Replication of herpesvirus DNA. In *The herpesviruses* (R.Roizman, ed.), Plenum Press, New York, pp. 81-106.

Bhattacharyya, S.P., and Rao, V.B., 1994, Structural analysis of DNA cleaved *in vivo* by bacteriophage T4 terminase. *Gene* **146:** 67-72

Black, L.W., 1988, DNA packaging in dsDNA bacteriophages. In *The bacteriophages* (R.Calender, ed.), Plenum Press, New York, pp. 321-373

Black, L.W., 1989, DNA packaging in dsDNA bacteriophages. *Annu.Rev.Microbiol.* **43:** 267-292

Black , L.W., 1995, DNA packaging and cutting by phage terminases: control in phage T4 by a synaptic mechanism. *BioEssay* **17:** 1025-1030

Black, L.W., Showe, M.K., and Steven, A.C., 1994, Morphogenesis of the T4 head. In *Molecular biology of bacteriophage T4* (Karam, J., Drake, J.W., Kreuzer, K.N., Mosig, G., Hall, D.H., Eiserling, F.A., Black, L.W., Spicer, E.K., Kutter, E., Carlson, K. and Miller, E.S., eds), American Society for Microbiology, Wahington, DC, pp218-258.

Bogner, E., 1999, Human cytomegalovirus HCMV) nuclease: implications for new strategies in gene therapy. *Gene Ther.Mol.Biol.* **3:** 75-78

Bogner, E., Radsak, K., and Stinski, M.F., 1998, The gene product of human cytomegalovirus open reading frame UL56 binds the pac motif and has specific nuclease activity. *J. Virol.* **72:** 2259-2264

Bogner E., Reis, B., Reschke, M., Mochenhaupt, T., and Radsak, K., 1993, Identification of the gene product encoded by ORF UL56 of human cytomegalovirus genome. *Virology* **196:** 290-293

Casjens, S., and Hendrix, R., 1988, Control mechanisms in ds DNA bacteriophage assembly. In *The bacteriophages* (R.Calender, ed.), Plenum Press, New York, pp15-91

Chee MS, Bankier AT, Beck S, Bohni R, Brown CM, Cerny R, Horsnell T, Hutchison CA 3rd, Kouzarides T, Martignetti JA, et al., 1990, Analysis of the protein-coding content of the sequence of human cytomegalovirus strain AD169. *Curr Top Microbiol Immunol.* **154:**125-69

Chou, J., and Roizman, B., 1985, The isomerization of the herpes simplex virus 1 genome: identification of the *cis*-acting and recombination sites within the domain of the *a* sequence. *Cell* **41:** 803-811

Davison, A.J., 1992, channel catfish virus: a new type of herpesvirus, *Virology*, **59:** 605-618

de Bruyn, A., and Knipe, D.M., 1994, Preexisting nuclear architecture defines the intranuclear location of herpesvirus DNA replication structures *J. Virol.* **68:** 3512-3526

de Bruyn, a., Uprichard, S.L., Chen, M., and Knipe, D.M., 1998, Comparison of the Intranuclear Distributions of Herpes Simplex Virus Proteins Involved in Various Viral Functions *Virology* **252:** 162-178

Deiss, L.P., Chou, J., and Frenkel, N., 1986, Functional domains within the a sequence involved in the cleavage-packaging of herpes simplex virus DNA.*J.Virol.* **59:** 605-618

Deiss, L.P., and Frenkel, N., 1986, Herpes simplex virus amplicon: cleavage of concatameric DNA is linked to packaging and involves amplification of the terminally reiterated a sequence. *J. Virol.***57:**933-941

Dingwall, C. & Laskey, R. A. (1991). Nuclear targeting sequences – a consensus? Trends Biochem. Sci. **16:** 478-481.

Eppler, K., Wyckoff, E., Goates, J., Parr, R., and Casjens, S., 1991, Nucleotide sequence of the bacteriophage P22 genes required for DNA packaging. *Virology* **183:** 519-538

Feiss, M, and Becker, A., 1983, DNA packaging and cutting. In *DNA packaging and cutting* (R.W. Hendrix, J.W. Roberts, F.W. Stahl and R.A. Weisberg, eds.), Lambda II, Cold Spring Harbor, New York, pp. 305-330

Frenkel, N., 1981, Defective interfering herpesviruses. In *The human herpesviruses – An interdisciplinary perspective.* (Hahmias, A.H., Dowdle, W.R., and Schinazy, R.S., eds.), Elsevier Science Publishing, Inc. New York, pp91-120

Fujisawa, H., and Hearin, P., 1994, Structure, function and specificity of the DNA packaging signals in double-stranded DNA viruses. *Semin.Virol.* **5:** 5-13

Garber, D., Beverly, S., and Coen, D., 1993, Demonstration of circularization of herpes simplex virus DNA following infection using pulsed field gel electrophoresis. *Virology* **197:** 459-462

Gibson, W., 1981, Structural and nonstructural proteins of strain Colburn cytomegalovirus. *Virology* **111:** 251-262

Giesen, K., Radsak, K., and Bogner, E., 2000a, Targeting of the gene product encoded by ORF UL56 of human cytomegalovirus into viral replication centers.*FEBS letters* **471**: 215-218

Giesen, K., Radsak, K., and Bogner, E, 2000b, The potential terminase subunit pUL56 of HCMV is translocated into the nucleus by its own NLS and interacts with importin α. *J.Gen.Virol.* **81**: 2231-2244

Görlich, D., Kostka, S., Kraft, R., Dingwall, D., Laskey, R.A., Hartmann, E., and Prehn, S., 1995, Two different subunits of importin co-operate to recognize nuclear localization signals and bind them to the nuclear envelope. *Curr. Biol.* **5**: 383-392.

Görlich, D., Panté, N., Kutay, U., Aebi, U., and Bischoff, R. R., 1996, Identification of different roles for RanGDP and RanGTP in nuclear protein import. *EMBO J.* **15**: 5584-5594

Görlich, D., 1997, Nuclear protein import. *Curr. Op. Cell. Biol.* **9**: 412-419.

Higgins R.R., and Becker, A., 1994, Chromosome end formation in phage lambda, catalyzed by terminases, is controlled by two DNA elements of cos, cosN and R3, and by ATP. *EMBO J.* **13**: 61526161

Hwang, J.S., and Bogner, E., 2002, ATPase activity of the terminase subunit pUL56 of human cytomegalovirus, *J. Biol. Chem.* In press

Jacob, R.J., Morse, L.S., and Roizman -, R., 1979, Anatomy of herpes simplex virus DNA. XII. Accumulation of head-to-tail concatemers in nuclei of infected cells and their role in the generation of the four isomeric arrangements of viral DNA. *J.Virol.* **29**: 448-457

Kuebler, D., and Rao, V.B., 1998, Functional analysis of the DNA-packaging/terminase protein gp17 from bacteriophage T4. *J.Mol.Biol.* **281**: 803-814

Leffers, G., and Rao V.B., 2000, Biochemical characterization of an ATPase activity associated with the large packaging subunit gp17 from bacteriophage T4. *J.Biol.Chem.* **275**: 37127-37136

Martinez, R., Sarisky, R., Webber, P., and Weller, S.K., 1996, Herpes simplex virus type 1 alkaline nuclease is required for efficient processing of viral DNA replication intermediates. *J. Virol.* **70**: 2075-2085

McVoy, M.A., and Adler, S.P., 1994, Human cytomegalovirus DNA replicates after early ciruclarization by concatemer formation, and inversion occurs within the concatemers. *J. Virol.* **68**: 1040-1051

McVoy, M.A., Nixon, D.E., Hur, J.K., and Adler, S.P., 2000, The ends on herpesvirus DNA replicative concatemers contain *pac2 cis* cleavage/packaging elements and their formation is controlled by terminal *cis* sequences. *J. Virol.* **74**: 1587-1592

Mettenleiter, T.C., Saalmüller, A., and Weiland, F., 1993, Pseudorabies virus protein homologous to herpes simplex virus type 1 ICP 18.5 is necessary for capsid maturation. *J.Virol.* **67**: 1236-1245

Nasseri, m., and Mocarski, E.S., 1988, The cleavage recognition signal is contained within sequences surrounding an a-a junction in herpes simplex virus DNA. *Virology* **167**: 25-30

Newmeyer, D. D., Lucocq, J. M., Bürglin, T. R. & De Robertis, E. M. (1986). Assembly in vitro of nuclei active in nuclear transport: ATP is required for nucleoplasmin accumulation. EMBO J. 5, 501-510.

Poffenberger, K.L., and Roizman, R., 1985, Studies on non-iverting genome of a viable herpes virus 1. Presence of head to tail linkages in packaged genomes and requirements for cirularization after infection. *J Virol.* **53**: 589-595

Roizman, R., and Sears, A.E., 1996, Herpes simplex viruses and their replication. In: *The human herpesviruses* (B.N. Fields, D.M. Knipe, P.M. Howlex, R.M. Chanock, J.L. McUnick, T.P. Monath, and C. Lopez, eds.), Raven Press, New York, pp. 1048-1066

Scheffczik, H., Savva, C.W., Holzenburg, A., Kolesnikova, L., and Bogner E., 2002, The terminase subunits pUL56 and pUL89 are DNA-metabolizing proteins with toridal structure. *Nucl. Acid Res*. In press

Sheldrick, P., and N. Berthelot, 1975, Inverted repetititons in the chromosome of herpes simplex virus. *Cold Spring HarborSymp.Quant.Biol.* **39**: 667-678

Severini, A., Morgan, A.R., Tovell, D.R., and Tyrell, D.L., 1994, Study of the structure of replicative intermediates of HSV-1 DNA by pulsed-field gel electrophoresis. *Virology* **200**: 428-435

Smith, M.P., and Feiss, M., 1993, Sites and gene products involved in lambdoid phage DNA packaging. *J. Bacteriol.* **175**: 2393-2399

Spaete, R.R., and Frenkel, N., 1982, The herpes simplex virus amplicon: a new eucaryotic defective-virus cloning-amplifying vector. *Cell* **30**:295-304

Spaete, R.R., and Frenkel, N., 1982, The herpes simplex virus amplicon: analyses of cis-acting replication functions. *Proc.Natl.Acad.Sci.U.S.A.* **82**:694-698

Spaete, R.R., and Mocarski, E.S., 1985, The *a* sequence of the cytomegalovirus genome functions as a cleavage/packaging signal of herpes simplex virus defective genomes. *J.Virol.* **54**: 817-824

Stinski, M., 1991, Cytomegalovirus and its replication. In: *Fundamental virology* (B.N. Fields, and D.M. Knipe, eds.), Raven Press, New York, pp. 929-950

Stow, N.D., McMonagle, E.C., and Davison, A.J., 1983, Fragments from both termini of the herpes simplex virus type 1 genome contains signals required for the encapsidation of viral DNA. *NucleicAcidsRes.* **11**: 8205-8220

Tamashiro, J.C., and Spector, D.H., 1986, Terminal structure and heterogeneity in human cytomegalovirus strain AD169. *J. Virol.* **59**: 591-604

Tengelsen, L. A., Pederson, N. E., Shaver, P. R., Wathen, M. W. & Homa, F. L. (1993). Herpes simplex virus type 1 DNA cleavage and encapsidation require the product of the UL28 gene: isolation and characterization of two UL28 deletion mutants. J. Virol. 67, 3470-3480.

Vlazny, D.A., and Frenkel, N., 1981, Replication of herpes simplex virus DNA: localization of replication recognition signals within defective virus genomes. *Proc.Natl.Acad.Sci.U.S.A.* **78**: 742-746

Vlazny, D.A., and Frenkel, N., 1982, Site-specific cleavage/packaging of herpes simplex virus DNA and the selective maturation of nucleocapsids containing full-length viral DNA. *Proc.Natl.Acad.Sci.U.S.A.* **79**: 1423-1427

Weis, K., Mattaj, I. W., and Lamond, A. I., 1995, Identification of hSRP1α as a functional receptor for nuclear localization sequences. *Science* **268**: 1049-1053.

Weis, K., Dingwall, C., and Lamond, A. I., 1996a, Characterization of the nuclear protein import mechanism using Ran mutants with altered nucleotide binding specificities. *EMBO J.* **15**: 7120-7128.

Weis, K., Ryder, U., and Lamond, A. I., 1996b, The conserved amino terminal domain of hSRP1α is essential for nuclear protein import. *EMBO J.* **15**: 1818-1825.

Zasloff, M., 1983, TRNA transport from the nucleus in a eucaryotic cell: carrier-mediated translocation process. *Proc. Natl. Acad. Sci.* **80** 6436-6440.

Zhang, X., Efstathiou, S., and Simmns, A., 1994, Identification of novel herpes simplex virus replicative intermediates by field inversion gel electrophoresis: implication for viral DNA amplification strategies. *Virology* **202**: 530-539

4. Determinants for viral maturation

Chapter 4.1

Role of Envelope Proteins in Measles Virus Assembly

ANDREA MAISNER
Institute of Virology, Philipps University Marburg, Robert-Koch-Str. 17, D-35037 Marburg, Germany

1. INTRODUCTION

Over the centuries, the measles virus (MV) has been responsible for severe epidemics throughout the world with a substantial degree of morbidity and a significant mortality (Robbins, 1962). Despite the availability of an efficient and safe live vaccine for over 30 years, measles remains a frequent and fatal disease of young children in developing countries (Clements and Cutts, 1995). MV is spread by the respiratory route and rapidly establishes viremia that allows systemic spread to several organs. Despite the development of an effective immune response leading to virus clearance from blood and tissues, immunological abnormalities arise. A severe immunosuppression that lasts for weeks or even months contributes to the susceptibility to secondary infections that accounts for most of the mortality associated with measles (Miller, 1964; Beckford et al., 1985)

After isolation of MV in 1954 (Enders and Peebles, 1954), the first live attenuated measles vaccine, the Edmonston B virus strain, was developed in 1962 (Enders, 1962). Many of today's vaccine strains are derived from this severely attenuated virus. In 1995, the generation of recombinant MV from cloned Edmonston B virus cDNA was achieved in the laboratory of M. A. Billeter (Radecke *et al.,* 1995). Recently, a similar rescue system based on the MV wildtype strain, Ichinose B, has been described (Takeda *et al.,* 2000). Nowadays the application of the reverse genetics nowadays allows the introduction of targeted mutations into the virus genome facilitating the study of MV replication and pathogenesis. This chapter focuses on one

Structure-Function Relationships of Human Pathogenic Viruses, Edited by
Holzenburg and Bogner, Kluwer Academic/Plenum Publishers, New York, 2002   173

aspect of the complex virus biology: the role of the three viral envelope proteins M, H and F during MV assembly in infected cells.

2.   VIRION STRUCTURE AND VIRAL LIFE CYCLE

MV is a member of the Morbillivirus genus in the order *Mononegavirales* of the *Paramyxovirus* family. It is an enveloped virus containing a single-stranded 50S RNA genome of negative polarity (Baczko *et al.*, 1983). The genome encodes for only six structural proteins in the gene order, 3′ leader-N-P-M-F-H-L-trailer 5′ (Fig. 1). The viral envelope is derived from the host cell plasma membrane and surrounds the viral genome which is composed of a helical nucleocapsid containing the RNA (15894 nucleotides), tightly encapsidated by the viral nucleoprotein (N). Also associated with the nucleocapsid is the large protein (L) as well as the phosphoprotein (P) which make up the RNA polymerase complex. Protruding from the viral envelope are two glycoproteins, the hemagglutinin (H) and the fusion protein (F). The sixth structural MV protein is the matrix protein (M) that is located at the inner surface of the lipid bilayer tethering the ribonucleocapsid (RNP) to the envelope, probably by interaction with both the glycoproteins and the RNP (Fig. 2).

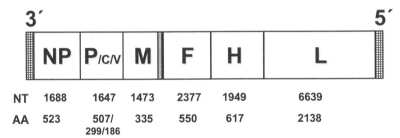

Figure 1. Genome organization of measles virus. The coding region is flanked at its 3′ and 5′ ends by leader and trailer sequences that contain promotor binding sites and signals for encapsidation. The reading frames of the genes are separated by intergenic boundaries at which the polymerase enzyme complex detaches and reinitiates transcription. Consequently, mRNAs are synthesized with decreasing efficiency as the distance to the 3′ promotor increases. From each gene a monocistronic mRNA is transcribed, except for the P gene that encodes two additional nonstructural proteins, C and V. C is translated from an overlapping reading frame whereas V is encoded by an edited P mRNA. Between M and F the noncoding region is especially long and contains about 1,000 nucleotides. The numbers beneath each gene represents its contents of nucleotides (NT) and amino acids (AA).

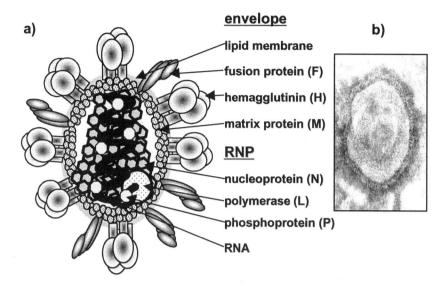

Figure 2. (a) Schematic drawing of the structural components of measles virus. (b) Electron micrograph of measles virus Edmonston strain purified from infected Vero cell cultures.

Like all paramyxoviruses, MV replicates in the cytoplasm of the infected cell. After binding of the virion to the cell surface via a specific host cell receptor, fusion of the virus envelope and the plasma membrane takes place. Thereby, the viral RNP is liberated into the cytoplasm. The next step in the viral replication cycle is the transcription of the negative-sense RNA genome into mRNAs by the viral polymerase complex to produce virus-specific proteins. Since the polymerase detaches from the template at each intergenic junction and reinitiates at the 3´end of the next gene, each gene is transcribed into a monocystronic mRNA. The amount of mRNA and the amount of the corresponding viral protein depend on the distance of the gene from the 3-terminal promotor. The mRNAs for the viral surface glycoproteins are translated at ribosomes associated with the endoplasmatic reticulum (ER). The glycoproteins which are inserted into the ER membrane are then transported through the secretory pathway via the Golgi apparatus and the trans-Golgi network to the plasma membrane. All other MV proteins, N, P and L are synthesized at free polyribosomes in the cytoplasm, and are transported to the inner leaflet of the plasma membrane. When the L protein is overproduced, the polymerase complex switches from transcription (mRNA synthesis) to replication (synthesis of full-length antigenomes and genomes). The newly synthesized viral RNA genomes are immediately encapsidated by the N protein and RNPs are rapidly formed. The final assembly is believed to occur at the plasma membrane where the subsequent budding of new virions from the surface of infected cells takes place.

3.   VIRAL ENVELOPE PROTEINS

3.1   Hemagglutinin

3.1.1   Structure

The H protein is an integral type II glycoprotein. There is no signal peptide at the amino terminus, but presumably the transmembrane region fulfills this function. H consists of 617 amino acids and forms disulfide-linked homodimers (Hardwick and Bussell, 1978). The few experimental data available on the oligomeric state of fully processed H indicate that it resides on the surfaces of virions and infected cells as a homotetramer (Ogura et al., 1991; Malvoisin and Wild, 1993). H possesses five potential glycosylation sites but only four are used (Alkhatib and Briedis, 1986, Hu et al., 1994). The protein can be structurally divided into four regions (Fig. 3).

i) Head: Beyond amino acid 181 lies the carboxy-terminal globular head which includes the receptor binding site(s) (Bartz et al., 1996; Buchholz et al., 1997; Hsu et al., 1998). It contains 13 strongly conserved cysteine residues influencing either the conformational structure and/or dimerization (Hu and Norrby, 1994).

ii) Stalk: Amino acid 59-181 are proposed to form a slender stalk forming a protease-sensitive hinge region within the molecule (Sato et al., 1995; Langedijk et al., 1997). Cysteines at positions 139 and 154 have been shown to be critical in mediating the formation of covalently linked homodimers (Plemper et al., 2000). Amino acid 98 is required for fusion helper function (Hummel and Bellini, 1995).

iii) Transmembrane domain: Residues 35-58 span the lipid bilayer (Alkhatib and Briedis, 1986).

iv) Tail: At the amino terminus, H possesses a 34 amino acid-long cytoplasmic tail which is responsible for polarized transport and endocytosis of the protein. Basolateral targeting and rapid endocytosis of H critically depend on the only tyrosine in the cytoplasmic tail (Maisner et al., 1998; Moll et al., 2001).

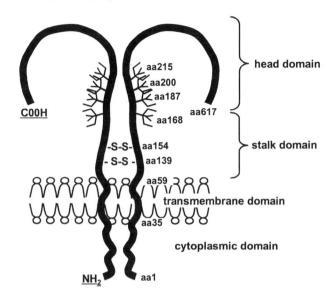

Figure 3. Diagram of the measles virus hemagglutinin.

3.1.2 Function

The name hemagglutinin stems from its ability to bind and agglutinate primate erythrocytes. Thus, the primary function of H is the attachment to receptor(s) on host cells. Two cellular surface molecules are currently known to function as MV receptor. One is the membrane cofactor protein (MCP; CD46), a complement-regulatory protein present on all nucleated human cells (Naniche et al., 1993; Dorig et al., 1993). The second MV receptor is a signaling lymphocytic activation molecule (SLAM; CD150) that is only expressed on human B and T cells (Tatsuo et al., 2000). Both, CD46 and SLAM have been shown to confer binding and replication to previously nonsusceptible cells. Whereas CD46 is supposed to be the main receptor for MV vaccine strains, SLAM is presumably one of the receptors that are required for the entry of MV wildtype strains (Bartz et al., 1998; Tatsuo et al., 2000). Agglutination of African Green Monkey erythrocytes is the consequence of H binding to CD46 which is absent on human but present on primate red blood cells (Liszewski and Atkinson, 1992).

Besides its function as receptor binding protein, H is necessary for (i) the promotion of F-mediated fusion and (ii), together with the F protein, the immunosuppressive properties of MV. The glycoprotein complex is able to suppress lymphoproliferative responses in uninfected cells even in the absence of other viral proteins (Schlender et al., 1996).

3.2   Fusion Protein

3.2.1   Structure

In contrast to the H protein, F is a type I membrane protein of 550 amino acids containing a putative 28-residue signal sequence at the amino terminus. After synthesis and cotranslational glycosylation, F0, a 60 kD precursor protein, oligomerizes into trimers (R. Buckland, personal communication) F0 trimers are then transported through the Golgi to the trans-Golgi network where F0 is cleaved into two disulfide-linked subunits, F1 and F2. Cleavage is mediated by a subtilisin-like endoprotease, furin, which recognizes the multibasic cleavage site R-H-K-R, at positions 109-112 of the F protein (Watanabe et al., 1995; Bolt et al., 1998). Fusion activity and with it the infectivity of MV essentially depends on the proteolytic activation of the F protein (Maisner et al., 2000). The F1 cleavage product (41 kD), derived from the carboxy terminus of F0, contains several functional domains:

i) At the carboxy terminus, the 33-residue cytoplasmic tail is located. It is responsible for polarized targeting and endocytosis of the F protein. Like the H protein, F possesses a tyrosine-dependent sorting and internalization signal (Moll et al., 2001).

ii) The cytoplasmic tail is followed by the transmembrane domain which is responsible for anchoring the protein in lipid membranes.

iii) Ten amino acids from the transmembrane domain towards the amino terminal, a leucine zipper motif is located. This region is known to be important for the formation of the fusion pore (Buckland et al., 1992).

iv) The very amino terminus of F1 is formed by a stretch of hydrophobic amino acids, called the fusion peptide. The fusion peptide is strongly conserved among the *Paramyxoviruses* and is indispensable for the biological activity of viral fusion proteins. Its role is to insert into the target membrane, thereby, initiating the fusion process (Gething et al., 1978). Fusion peptide analogues are able to inhibit both virus penetration and cell-to-cell fusion (Richardson et al., 1980).

The second cleavage product, F2 (18 kD), derived from the amino terminus of F0, contains all three potential glycosylation sites. All of them have to be used to allow efficient surface transport of the F protein (Alkathib et al., 1994). The tetrabasic cleavage site forms the carboxy terminal end of the F2 subunit.

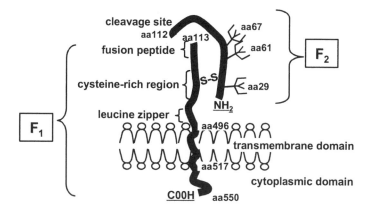

Figure 4. Diagram of the measles virus F protein.

3.2.2   Function

The main function of F is the mediation of fusion. Like the F protein of other *Paramyxoviruses*, MV F allows fusion at neutral pH. Two fusion processes are important for a successful MV infection. Fusion of the viral envelope with the target cell membrane is necessary for virus entry into host cells. Later on in the infectious cycle, fusion of infected cells with noninfected neighboring cells is required for virus spread from cell to cell. Cell-to-cell fusion results in the formation of large syncytia that can be observed in infected cell cultures (*in vitro*) and *in vivo* in infected human tissues (Enders and Peebles, 1954). The fusion activity of MV F depends on several preconditions. First, F must be cleaved into F1/F2 in order to expose the fusion peptide (Alkhatib et al., 1994a; Maisner et al., 2000). Secondly, F must be *N* glycosylated at Asn 29 (Alkhatib et al., 1994b). Palmitoylation at cysteines in the transmembrane domain has been shown to influence the fusion efficiency (Caballero and Celma, 1998). Most importantly, F-mediated fusion depends on the simultaneous coexpression of the H protein. Interaction of H and F critically depends on a 45-residue segment of the cysteine-rich region in the stalk domain of F (amino acid 337-381) (Wild et al., 1994). In the fusion-active F/H complex both proteins are tightly associated and the glycoproteins can only be exchanged for H and F proteins of the closely related canine distemper virus (Von Messling et al., 2001).

Besides its function as a fusion molecule, MV F plays an important role during virus assembly. The cytoplasmic tail is supposed to mediate the contact between the H/F complex and the RNP by interaction with the viral M protein (Spielhofer et al., 1998; Cathomen et al., 1998b).

F takes also part in the immunosuppressive glycoprotein complex. In order to suppress lymphoproliterative responses in target cells, F has to be proteolytically cleaved although fusion of H/F expressing cells with the target cells is not required (Weidmann et al., 2000).

3.3   Matrix Protein

3.3.1   Structure

The M protein is a positively charged (basic) protein of 335 amino acids, 17% of which are arginine and lysine residues. In contrast to H and F, M is synthesized at ribosomes in the cytoplasm as a nonglycosylated 37 kD protein. M possesses no hydrophobic sequence long enough to function as transmembrane domain but several hydrophobic regions. The sequence between amino acid 235 and 279 begins and ends with α–helices which sandwiching a ß sheet containing region. The combination of nonpolar hydrophobic and ß-sheet characteristics suggests a possible interaction with lipid membranes (Bellini et al., 1986).

3.3.2   Function

While no functional domains have been located as yet, M must possess several (different) regions responsible for the following interactions:

i) *M interaction with itself.* Self-association results in the formation of salt-resistant M aggregates. Varsanyi et al. (1984) proposed the model in which 6 molecules form a disc of 8 nm in diameter with a central hole.

ii) *M binding to lipid membranes.* Like the M proteins of several other viruses (Caldwell and Lyles, 1980; Chong and Rose, 1993; Gregoriades, 1980), a substantial portion of MV M has been shown to associate with lipid membranes (Hirano et al., 1992). Binding of M to lipids must be due to hydrophobic or to electrostatic interactions of M with membranes since it has been shown to be independent from interactions with the viral glycoproteins H and F (Riedl et al., 2001). M attachment to intracellular membranes may serve to increase the local M concentration, so as to prime self association.

iii) *M association with the cytosceleton.* It has been known for a long time that destruction of actin filaments prevents virus budding (Stallcup et al., 1983). Furthermore, actin is found as one structural component of MV (Stallcup et al., 1979). It is suggested that the M protein is directly associated with actin filaments (Guiffre et al., 1982; Bohn et al., 1986).

vi) *M interaction with RNPs*. The positively charged carboxy terminus of M is supposed to interact with the acidic, negatively charged carboxy terminus of the N protein in RNP complexes (Stallcup et al., 1979; Hirano et al., 1992).

v) *M interaction with viral glycoproteins*. The existence of a physical contact between the cytoplasmic portions of the glycoproteins and M had been deduced from the observation that M, H and F protein cocap at surface membranes after incubation with glycoprotein-specific antibodies (Tyrell and Ernst, 1979).

All of the listed binding activities are presumably required for the main function of M, that is the organization of MV assembly. While the M protein is undoubtedly involved in the orchestration of the different components to the budding site, the exact mechanism remains an enigma.

Besides its function during MV assembly, M is responsible for the inhibition of transcription from the moment it binds to the RNPs. This function often is defective in cells persistently infected with MV (Suryanarayana et al., 1994). Furthermore, M protein regulates glycoprotein-mediated fusion in infected cells (Cathomen et al., 1998b) and inhibits endocytosis of H and F (Moll et al., 2001). In the absence of a M, fusion is greatly enhanced and both glycoproteins are rapidly internalized from the cell surface via clathrin-coated pits.

4. VIRUS ASSEMBLY IN INFECTED CELLS

4.1 Interactions of Viral Envelope Proteins

4.1.1 General remarks

Virus assembly is necessary for the formation of new virions to be released from the infected cells. Correct virus assembly is the prerequisite for the spread of infection by cell-free viruses in body fluids or cell culture supernatants. Budding of MV is known to occur at cell surface membranes, but, depending on the cell culture system, more than 90% of the infectivity remains cell-associated (Udem, 1984). Therefore, in order to recover maximal amounts of infectious virus, infected culture cells have to be disrupted by several freezing and thawing cycles. Because of the inefficient release of cell-free measles virions from infected cells it is assumed that MV is mainly spread *in vivo* by direct contact between infected and uninfected cells. Efficient virus assembly may not be required for a systemic virus distribution within one individual. Nevertheless, for virus transmission from host to host, correct virus assembly at plasma membranes with subsequent release of infectious virions is probably indispensable.

4.1.2   Assembly models

There are two principal models for the assembly of *Mononegavirales* at plasma membranes. The first is described for the vesicular stomatitis virus (VSV). VSV has only one surface glycoprotein, the G protein. G and the VSV matrix protein can reach the cell surface independently (Bergmann and Fusco, 1988). Although the VSV M mediates the contact between the RNP and the G protein, virion formation also occurs in the absence of M. This is explained by an M-independent transport of RNPs to the plasma membrane and an intrinsic vesiculation or "pull activity" of the G protein (Mebatsion et al., 1996). In contrast to VSV, budding of Sendai virus crucially depends on the presence of a functional M protein (Mottet et al., 1999). M interacts with RNPs in the cytoplasm of infected cells (Stricker et al., 1994). Transport of the M appears to depend on a simultaneous transport of the Sendai virus glycoproteins HN and F (Sanderson et al., 1993; Ali et al., 2000). Thus, M-RNP complexes are cotransported with the viral glycoproteins. Interaction of M with the cytoplasmic tails is thought to occur at intracellular membranes, in early Golgi compartments or even in the ER. For interaction with the M protein a SYWST motif in the HN cytoplasmic tail was defined (Takimoto et al., 1998). Although the binding site in the F cytoplasmic tail is not yet identified, it is clear that only M-F interaction but not M-HN interaction is essential for virus assembly and budding (Fouillot-Coriou and Roux, 2000).

4.1.3   Importance of MV M protein for virus assembly

For MV not that much is known, but it appears that MV assembly does neither fit the model of VSV nor that of Sendai virus assembly. Similar to Sendai M protein, MV M is thought to play the organizer role during MV assembly. Recombinant MV generated from cloned cDNA lacking the M protein (MVΔM) are completely defective in virus assembly although virus spread by cell-to-cell fusion is not only maintained but even improved (Cathomen et al., 1998a). In the absence of M, viral RNPs cannot interact with the viral glycoproteins. Thus, virion formation is impossible in the absence of functional M protein.

4.1.4 Glycoprotein distribution on the plasma membrane

The MV glycoproteins are anchored by their transmembrane domain in ER membrane and reach the plasma membrane following the constitutive secretory pathway. Since MV glycoproteins, in contrast to VSV-G, have no "pull-activity" they are believed to move freely in the plane of the membrane until they come into contact with an array of M proteins complexed with RNPs. H and F then gather in patches from which cellular proteins are excluded and bud from the plasma membrane to yield a newly formed virus.

4.1.5 Function of the glycoprotein tails

It is currently not known which portion of the M protein is necessary for the interaction with H and F. However, it is clear that the cytoplasmic tails of the glycoproteins are involved in M binding whereby the F tail appears to be more important. Several findings indicate that M directly interacts with the cytoplasmic tail of the F protein. Only chimeric glycoproteins containing the F cytoplasmic tail are efficiently incorporated into the envelope of recombinant MV (Spielhofer et al., 1998). Furthermore, recombinant MV with a deleted F cytoplasmic tail shows an altered protein composition. The amount of M protein in the virions is drastically reduced and a large amount of cellular proteins are incorporated (Cathomen et al., 1998b). In this regard, the requirements for MV assembly agree in principle with those for Sendai virus. As with Sendai virus, M binding to only one of the glycoprotein tails appears to be sufficient to allow MV budding. However, in contrast to Sendai virus, MV particles are even formed and released when both glycoprotein tails are truncated (Cathomen et al., 1998b).

4.1.6 Intracellular transport of M-RNP complexes

While the intracellular location where M-RNP complexes encounter viral glycoproteins is currently unknown, it has been shown that retention of the MV glycoproteins in the Golgi does not result in the simultaneous accumulation of M in this compartment (Riedl et al., 2001). Thus, in contrast to Sendai M protein, MV M is not cotransported with H and F to the plasma membrane where budding is initiated. Since M protein colocalizes with the glycoproteins at the plasma membrane of infected cells, M seems to reach the cell surface independently from the glycoproteins, just as VSV-M. A mechanism how internal viral components can reach the cell surface has already been proposed several years ago. In this model, progeny RNPs attach themselves to growing actin filaments, possibly by their association with nucleocapsid-bound M protein. Subsequently, the vectorial growth of the actin filament is used to transport RNPs from the cytoplasm to the plasma membrane (Giuffre et al., 1982; Stallcup et al., 1983; Bohn et al. 1986). Here, the internal components of MV encounter the glycoproteins that are already inserted in the plasma membrane.

4.1.7   MV assembly and rafts

Whether it is indeed the plasma membrane where all MV components assemble has been questioned recently. Manie et al. (2000) propose another attractive model: Membrane rafts as platforms for MV assembly. Rafts are described as glycosphingolipids and cholesterol-rich microdomains that are formed in the trans-Golgi network and then migrate to the plasma membrane. Rafts can incorporate specific integral and peripheral membrane proteins among which are many glycophosphatidylinositol (GPI)-anchored proteins (for a review see Simons and Ikonen, 1997). In at least two cell lines, a substantial amount of all MV proteins have been found to be associated with rafts (Manie et al., 2000; Vincent et al., 2000). Despite these interesting data, further work is required to prove that raft microdomains represent indeed the intracellular location for interaction of MV envelope proteins and thus for MV assembly.

4.1.8   Further consequences of envelope protein interactions

Interaction of M with the cytosolic portions of MV glycoproteins does not only result in the formation of new virions, but also affects surface expression and fusion activity of H and F.

Influence on surface expression: The tyrosine-dependent signals in the cytoplasmic tail mediating rapid endocytosis of both glycoproteins are not functional in the presence of M (Moll et al., 2001). In infected cells expressing functional M protein, internalization of both viral glycoproteins is completely inhibited, whereas both MV glycoproteins are efficiently endocytosed in cells infected with MVΔM, a recombinant MV lacking the M protein (unpublished observations). This indicates that interaction of the M protein with the cytoplasmic tails of H and/or F hinders the recognition of the internalization signals by the cellular endocytosis machinery. This suggests that MV has developed a system that removes those H and F proteins from the cell surface that are not associated with the M protein and are therefore not destined for incorporation into virions. Similar mechanisms to regulate the expression of viral envelope proteins by interaction with viral core proteins have been proposed for Sendai virus as well as the human immunodeficiency virus (Roux et al., 1985; Egan et al., 1996).

Influence on fusion activity: In comparison to cells infected with standard MV, cells infected with recombinant viruses which express H and/or F proteins with truncations in the cytoplasmic tails rapidly form syncytia (Cathomen et al., 1998b). Similarly, recombinant MV lacking the M protein also spreads very rapidly through the cell monolayer by cell-to-cell fusion (Cathomen et al., 1998a). The enhanced fusion competence of the glycoproteins is probably due to the impaired H/F tail-M interaction. It is reasonable to speculate that in the absence of this interaction the lateral mobility of the glycoproteins is enhanced thereby facilitating the formation of fusion pore complexes.

4.2   Persistent Infections

In vitro and *in vivo*, MV can establish a persistent infection of cells. In cell cultures, MV normally causes rapid cell death. Cells surviving the lytic infection can establish persistent infections that usually are accompanied by a marked decrease of virus-induced cytopathic effects and a limited virus production (Barry et al., 1976). *In vivo*, a persistent infection can cause a subacute sclerosing panencephalitis (SSPE), a rare complication occuring 5 to 10 years after acute MV infection leading to death after a period of a few months to several years (ter Meulen et al., 1983). The disease develops on the basis of a persistent infection of the central nervous system. In the brains of patients who died from SSPE, high amounts of inclusion bodies containing viral core particles were found in cells of both, the gray and white matter. Over the years, MV apparently propagates without budding, only by cell-to-cell fusion spreading through the brain by synaptic transmission of RNPs (Paula-Barbosa and Cruz, 1981). Persistent SSPE infection is mainly characterized by a defective virus assembly allowing the virus to persist in the brain, while being inaccessible to host immune reactions. Molecular analysis of the MV gene expression in infected brains demonstrated that overall MV-RNA transcription is downregulated leading to a drop in genomic RNA and mRNA synthesis with a steep transcriptional gradient from the 3′ to the 5′mRNAs. As a consequence, only low amounts of envelope proteins are expressed (Haase et al., 1985). Cloning of MV genes directly from diseased human brains revealed that SSPE associated MV strains not only downregulate their transcription but also have multiple genetic changes in the envelope proteins (Cattaneo et al., 1988). H proteins often show mutations that affect the glycosylation and consequently folding, antigenicity and transport to the cell surface (Cattaneo and Rose, 1993; Hu et al., 1994). Truncations, mutations, and deletions in the cytoplasmic domain of F which do not interfere with the fusion activity are almost universal in SSPE strains. In most of the persistent viruses the M protein is either completely eliminated or functionally inactivated. This is achieved by a massive hypermutation in the M gene (Cattaneo et al., 1988/1989). It must be assumed that the lack of virus production in persistently infected cells, whether in SSPE brains or in persistently infected cell cultures, is due to defective virus assembly. Such a defective assembly is the result of the complete or functional elimination of the M protein or the mutation of the glycoprotein domains required for the interaction with M.

4.3   Infection of Polarized Epithelial Cells

Epithelial cells line all body cavities and surfaces and often represent the primary barrier for a virus infection. Respiratory epithelial cells are one of the primary target cells for aerosol-transmitted MV that is transmitted by aerosols. Therefore, efficient replication in epithelia may be the prerequisite for the establishment of a productive MV infection. Polarized epithelia differ from normal cells in that their plasma membrane is divided by tight junctions into an apical and a basolateral domain that differ both in composition and function (Fig. 5). The apical membrane faces the external environment and the basolateral membrane mediates the contact with internal tissues and blood vessels. A specialized sorting apparatus in the trans-Golgi network ensures that newly synthesized proteins and lipids are segregated and transported to their respective domains (Rodriguez-Boulan and Nelson, 1989).

It has been proposed that restriction of virus entry due to a polarized expression of the viral receptor or polarized virus release has significant implications for the pathogenesis *in vivo*. (Tucker and Compans, 1993) Viruses that are released from the apical side normally establish only local infections restricted to site of virus entry. In contrast to apical release, budding from the basolateral surface is suggested to facilitate virus spread from epithelia to underlying tissues. Fitting into this hypothesis, viruses that cause systemic infections such as VSV or HIV are released from the basolateral membrane of polarized epithelial cells. Viruses that cause restricted infections such as influenza virus, parainfluenza viruses, Sendai virus or respiratory syncytial virus are selectively released from the apical side of epithelia (Tucker and Compans, 1993).

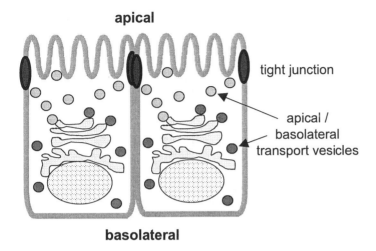

Figure 5. Schematic drawing of a typical polarized epithelial cell.

4.3.1 Measles virus entry into polarized epithelia

Since MV efficiently infects the respiratory tract, it must be assumed that MV binds to a receptor that is present on the apical side of bronchial epithelial cells. At present time, it is still unclear whether one of the known receptors, CD46 or SLAM (Naniche et al., 1993; Tatsuo et al., 2000) is responsible for the entry of MV into polarized respiratory epithelia. But none of both is a very probable candidate. SLAM is only expressed on lymphocytes but not on the surfaces of epithelial cells. In contrast to SLAM, CD46 is abundantly expressed on epithelia. But it possess a basolateral targeting signal (Maisner et al., 1996/1997), therefore, only small amounts of the protein should be present on the surface of human respiratory epithelial cells. The future will show if there is an additional MV receptor that mediates virus entry into polarized epithelia from the apical side.

4.3.2 Measles virus assembly and release from polarized epithelia

MV, as most other viruses that infect the respiratory tract, buds from the apical surface of polarized epithelia (Blau and Compans, 1995; Maisner et al., 1998). This explains the efficient transmission from host-to-host by aerosols. But whereas most other respiratory viruses only cause a local infection restricted to site of virus entry, MV establishes a systemic infection that, in general, is only caused by viruses which bud from the basolateral surface of epithelial cells. This clearly indicates that MV must have developed a maturation strategy that allows systemic virus spread without direct virus release from infected bronchial epithelia to underlying blood vessels. The current model of MV assembly and budding in polarized cells can be summarized as follows (Fig. 6):

In contrast to what is known about viruses such as influenza virus, VSV, HIV and many others, the MV glycoproteins H and F do not determine the site of virus budding. Despite abundant expression of both glycoproteins on the basolateral surface of infected epithelia, no virus particles are formed (Maisner et al., 1998). The selective assembly of new virions at the apical surface of infected cells is determined by the viral matrix protein that almost exclusively accumulates at apical cell membranes (Naim et al., 2000; Riedl et al., 2001). Due to the lack of M protein, budding at basolateral membranes is prevented. Although basolateral expression of the glycoproteins does not result in the formation of new virions it may be of biological importance. In the absence of basolateral H and F expression, fusion of polarized cells does not occur (Moll et al., 2001) suggesting that *in vivo* basolateral expression of both glycoproteins is required for MV spread from primarily infected respiratory epithelia to underlying tissues and blood by direct cell-to-cell fusion.

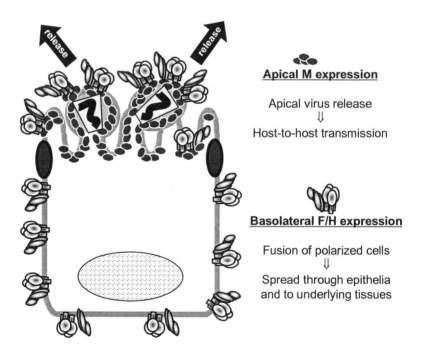

Figure 6. Distribution of MV envelope proteins in infected epithelial cells.

ACKNOWLEDGMENTS

I would like to thank all colleagues who have helped in many ways in the production of this chapter, in particular Markus Moll for critical reading the manuscript.

REFERENCES

Alkathib, G., Roder, J., Richardson, C. D., Briedis, D., Weinberg, R., Smith, D., Taylor, J., Paoletti, E., and Shen, S.-H., 1994a, Characterization of a cleavage mutant of measles virus fusion protein defective in syncytium formation. *J. Virol.* **68**:6770-6774.

Alkhatib, G., and Briedis, D., 1986, The predicted primary structure of measles virus hemagglutinin. *Virology* **150**:479-490.

Alkhatib, G., Shen, S. H., Briedis, D., Richardson, C., Massie, B., Weinberg, R., Smith. D., Taylor, J., Paoletti, E., and Roder, J., 1994b, Functional analysis of N-linked glycosylation mutants of the measles virus fusion protein synthesized by recombinant vaccinia virus vectors. *J. Virol.* **68**:1522-1531.

Baczko, K., Billeter, M. A., and ter Meulen, V., 1983, Purification and molecular weight determination of measles virus genomic RNA. *J. Gen. Virol.* **64**:1409-1413.

Barry, D. W., Sullivan, J. L., Lucas, S. L., Dunlap, R. C., and Albrecht, P., 1976, Acute and chronic infection of human lymphoblastoid cell lines with measles virus. *J. Immunol.* **116**:89-98.

Bartz, R., Brinckmann, U., Dunster, L. M., Rima, B., ter Meulen, V., and Schneider-Schaulies, J., 1996, Mapping amino acids of the measles virus hemagglutinin responsible for receptor (CD46) downregulation. *Virology* **224**:334-337.

Bartz, R., Firsching, R., Rima, B., ter Meulen, V., and Schneider-Schaulies, J., 1998, Differential receptor usage by measles virus strains. *J. Gen. Virol.* **79**:1015-25.

Bellini, W. J., Englund, G., Richardson, C. D., Rozenblatt, S., and Lazzarini, R. A., 1986, Matrix genes of measles virus and canine distemper virus: cloning, nucleotide sequences, and deduced amino acid sequences. *J. Virol.* **58**:406-416.

Bergmann, J. E., and Fusco, P. J., 1988, The M protein of Vesicular stomatitis virus associates specifically with the basolateral membrane of polarized epithelial cells independently of the G protein. *J. Cell Biol.* **107**, 1707-1715.

Blau, D. M., and Compans, R. W., 1995, Entry and release of measles virus are polarized in epithelial cells. *Virology* **210**:91-99.

Bohn, W., Rutter, G., Hohenberg, H., Mannweiler, K., and Nobis, P., 1986, Involvement of actin filaments in budding of measles virus: studies on cytoskeletons of infected cells. *Virology* **149**:91-106.

Breckford, A. P., Kaschula, R. O. C., and Stephen, C., 1985, Factors associated with fatal cases of measles: a retrospective autopsy study. *S. Afr. Med. J.* **68**:858-863.

Buchholz, C., Koller, D. Devaux, P., Mumenthaler, C., Schneider-Schaulies, J., Braun, W., Gerlier, D., and Cattaneo, R., 1997, Mapping of the primary binding site of measles virus to its receptor CD46. *J. Biol. Chem.* **272**:22072-22079.

Buckland, R., Malvoisin, E., Beauverger, P., and Wild, F., 1992, A leucine zipper structure present in the measles virus fusion protein is not required for its tetramerization but is essential for fusion. *J. Gen. Virol.* **73**:1703-1707.

Caballero, M., Carabana, J., Ortego, J., Fernandez-Munoz, R., and Celma, M.L., 1998, Measles virus fusion protein is palmitoylated on transmembrane-intracytoplasmic cysteine residues which participate in cell fusion. *J. Virol.* **72**:8198-8204.

Cathomen, T., Mrkic, B., Spehner, B., Drillien, R., Naef, R., Pavlovic, J., Aguzzi, A., Billeter, M. A., and Cattaneo, R., 1998a, A matrix-less measles virus is infectious and elicits extensive cell fusion: consequence for propagation in the brain. *EMBO J.* **17**:3899-3908.

Cathomen, T., Naim, H. Y., and R. Cattaneo, 1998b, Measles virus with altered envelope protein cytoplasmic tails gain cell fusion competence. *J.Virol.* **72**:1224-1234.

Cattaneo, R., and Rose, J., 1993, Cell fusion by the envelope glycoproteins of persistent measles viruses that cause lethal human brain disease. *J. Virol.* **67**:1497-1502.

Cattaneo, R., Schmid, A., Eschle, D., Baczko, K., ter Meulen, V., and Billeter, M. A., 1988, Biased hypermutation and other genetic changes in defective measles viruses in human brain infections. *Cell* **55**:255-265.

Cattaneo, R., Schmid, A., Speilhofer, P., Kaelin, K., Baczko, K., ter Meulen, V., Pardowitz, J., Flanagan, S., Rima, B., Udem, S. A., and Billeter, M. A., 1989, Mutated and hypermutated genes of persistent measles viruses which caused lethal human brain diseases. *Virology* **173**:415-425.

Clements, C. J., and Cutts, F. T.; 1995; The epidemiology of measles: Thirty years of vaccination. In *Measles virus* (V. ter Meulen and M. A. Billeter, eds) Springer-Verlag, Berlin Heidelberg, pp. 13-30.

Dorig R., E., Marcil, A., Chopra, A., and Richardson C., D., 1993, The human CD46 molecule is a receptor for measles virus (Edmonston strain). *Cell* **75**:295-305.

Egan, M. A., Carruth, L. M., Rowell, J. F., Yu, X., and Siliciano, R. F., 1996, Human immunodeficiency virus type I envelope protein endocytosis mediated by a highly conserved intrinsic internalization signal in the cytoplasmic tail of gp41 is suppressed in the presence of the Pr55gag precursor protein. *J. Virol.* **70**:6547-6556.

Enders, J. F., and Peebles, T.C., 1954, Propagation in tissue cultures of cytopathogenic agents from patients with measles. *Proc. Soc. Exp. Biol. Med.* **86**:277-286

Enders, J.F., 1962, Measles virus: historical review, isolation and behaviour in various systems. *Am. J. Dis. Child.* **103**:282-287.

Fouillot-Coriou N, and Roux, L., 2000, Structure-function analysis of the Sendai virus F and HN cytoplasmic domain: different role for the two proteins in the production of virus particle. *Virology* **270**:464-75.

Gething, M. J., White, J. M., and Waterfield, M.D., 1978, Purification of Sendai virus: analysis of the NH2-terminal sequences generated during precursor activation. *Proc. Natl. Acad. Sci. USA* **75**:2737-2740.

Giuffre, R. M., Tovell, D. R., Kay, C. M., and Tyrell, L. D. (1982). Evidence for an interaction between the membrane protein of a paramyxovirus and actin. *J. Virol.* **42**:963-968.

Haase, A. T., Gantz, D., Eble, B. Walker, D., Stowring, L., Ventura, P., Blum, H., Wietgrefe, S., Zupanic, M., Tourtellotte, W., Gibbs, C. J., Norrby, E., and Rozenblatt, S., 1985, Natural history of restricted synthesis and expression of measles virus genes in subacute sclerosing panencephalitis. *Proc. Natl. Acad. Sci. USA* **82**:3020-3024.

Hardwick, J. M., and Bussell, R. H., 1978, Glycoproteins of measles virus under reducing and non-reducing conditions. *J. Virol.* **25**:687-692.

Hirano, A., Wang, A. H., Gombart, A. F., and Wong, T. C., 1992, The matrix protein of neurovirulent subacute scerosing panencephalitis virus and its acute measles progenitor are functionally different. *Proc. Natl. Acad. Sci. U.S.A.* **89**:8745-8749.

Hsu, E. C., Sarangi, F., Iorio, C., Sidhu, M. S., Udem, S. A., Dillehay, D. L., Xu, W., Rota, P. A., Bellini, W. J., Richardson, C. D., 1998, A single amino acid change in the hemagglutinin protein of measles virus determines its ability to bind CD46 and reveals another receptor on marmoset B cells. *J. Virol.* **72**:2905-2916.

Hu, A., Cattaneo, R., Schwarz, S., and Norrby, E., 1994, Role of N-linked oligosaccharide chains in the processing and antigenicity of measles virus haemagglutinin. *J. Gen. Virol.* **74**: 1043-1053.

Hu, A., and Norrby, E., 1994, Role of individual cysteine residues in the processing and antigenicity of the measles virus haemagglutinin protein. *J. Gen. Virol.* **75**:2173-2181.

Hummel, K. B., and Bellini, W.J., 1995, Localization of monoclonal antibody epitopes and functional domains in the hemagglutinin protein of measles virus. *J. Virol.* **69**:1913-1916.

Langedijk, J. P., Daus, F. J., and van Oirschot, J. T., 1997, Sequence and structure alignment of Paramyxoviridae attachment proteins and discovery of enzymatic activity for a morbillivirus hemagglutinin. *J. Virol.* **71**:6155-6167.

Liszewski, K., and Atkinson, J. P., 1992, Membrane cofactor protein. *Curr. Top. Microbiol. Immunol.* **178**:7-60.

Maisner, A., Klenk, H.-D. and Herrler, G., 1998, Polarized budding of measles virus is not determined by viral glycoproteins. *J.Virol.* **72**:5276-5278.

Maisner, A., Liszewski, M. K., Atkinson, J.. P., Schwartz-Albiez, R. and Herrler, G., 1996, Two different cytoplasmic tails direct isoforms of the membrane cofactor protein (CD46) to the basolateral surface of Madin-Darby canine kidney cells. *J. Biol. Chem.* **271**:18853-18858.

Maisner, A, Mrkic, B., Herrler, G., Moll, M., Billeter, M. A., Cattaneo, R., and Klenk, H.-D., 2000, Recombinant measles virus requiring an exogenous protease for activation of infectivity. *J. Gen. Virol.* **81**:441-449.

Maisner, A., Zimmer, G., Liszewski, M. K., Lublin, M. D., Atkinson, J. P. and Herrler, G., 1997. "Membrane cofactor protein (MCP; CD46) is a basolateral protein that is not endocytosed: Importance of the tetrapeptide F-T-S-L at the carboxy terminus". *J. Biol. Chem.* **272**:20793-20799.

Malvoisin, E., and Wild, F., 1993, Measles virus glycoproteins: studies on the structure and interaction of the haemagglutinin and fusion proteins. *J. Gen. Virol.* **74**: 2365-2372.

Manie. S. N., Debreyene, D. S., Vincent, S., and Gerlier, D., 2000, Measles virus structural components are enriched into raft microdomains: Potential cellular location for virus assembly. *J. Virol.* **74**, 305-311.

Mebatsion, T., Konig, M.., and Conzelmann, K. K., 1996, Budding of rabies virus particles in the absence of the spike glycoprotein. *Cell* **84**:941-951.

Miller, D. L., 1964, Frequency of complications of measles. *Br. Med. J.* **2**;75-78.

Moll, M., Klenk, H.-D., Herrler, G., and Maisner, A., 2001, A single amino acid change in the cytoplasmic domains of measles virus glycoproteins H and F alters targeting, endocytosis and fusion in polarized Madin-Darby canine kidney cells. *J. Biol. Chem..* in press.

Mottet, G., Müller, V., and Roux, L., 1999, Characterization of Sendai virus M protein mutants that can partially interfere with virus particle production. *J. Gen. Virol.* **80**:2977-2986.

Naim, H. Y., Ehler, E. and Billeter, M. A., 2000, Measles virus matrix protein specifie apical virus release and glycoprotein sorting in epithelial cells. *EMBO J.* **19**:3576-3585.

Naniche, D., Varior-Krishnan, G., Cervoni, F., Wild T. F., Rossi, B., Rabourdin-Combe, C., and Gerlier, D., 1993, Human membrane cofactor protein (CD46) acts as a cellular receptor for measles virus. *J. Virol.* **67**:6025-6032.

Ogura, H., Sato, H., Kamiya, S., and Nakamura, S., 1991, Glycosylation of measles virus haemagglutinin protein in infected cells. *J. Gen. Virol.* **72**:2679-2684.

Paula-Barbosa, M. M., and Cruz, C., 1981, Nerve cell fusion in a case of subacute sclerosing panencephalitis. *Ann. Neurol.* **9**:400-403.

Peebles, 1991, Paramyxovirus M proteins. Pulling it all together and taking it on the road, In *The Paramyxoviruses* (Kingsbury, D.W., ed.) Plenum Press, New York, pp 427-456.

Plemper, R. K., Hammond, A. L., and Cattaneo, R., 2000, Characterization of a region of the measles virus hemagglutinin sufficient for its dimerization. J. Virol. 74:6485-6493.

Richardson, C. D., Scheid, A., and Choppin, P.W., 1980, Specific inhibition of paramyxovirus and myxovirus replication by oligopeptides with amino acid sequences similar to those at the N-termini of F1 or HA2 viral polypeptides. *Virology* 105:205-222.

Riedl, P., Moll, M., Klenk, H.-D., and Maisner, A., 2001, Measles virus matrix protein is not cotransported with the viral glycoproteins but requires virus infection for efficient surface targeting. *Virus Res.*, in press.

Robbins, F.C., 1962, Measles clinical features. Pathogenesis, pathology, and complications. *Am. J. Dis. Child.* 102:266-273.

Rodriguez-Boulan, E. and Nelson, W. J., 1989, Morphogenesis of the polarized epithelial cell phenotype. *Science* 245:718-725.

Roux, L., Beffy, P., and Portner, A., 1985, Three variations in the cell surface expression of the haemagglutinin-neuraminidase glycoprotein of Sendai virus. *J. Gen. Virol.* 66:987-1000.

Sanderson, C. M., McQueen, N. L., and Nayak, D. P., 1993, Sendai virus assembly: M protein binds to viral glycoprotein in transit through the secretory pathway. *J. Virol.* 67, 651-663.

Sato, T.A., Enami, M., and Kohama, T., 1995, Isolation of the measles virus hemagglutinin protein in a soluble form by protease digestion. *J. Virol.* 69:513-516.

Scheiffele, P., Roth, M. G., and Simons, K., 1997, Interaction of influenza haemagglutinin with sphinogilipid-cholesterol membrane domains via its transmembrane domain. *EMBO J.* 16, 5501-5508

Schlender, J., Schnorr, J. J., Spielhofer, P., Cathomen, T., Cattaneo, R., Billeter, M., ter Meulen, V., and Schneider-Schaulies, S., 1996, Interaction of measles virus glycoproteins with the surface of uninfected peripheral blood lymphocytes induces immunosuppression in vitro. *Proc. Natl. Acad. Sci. USA* 93:13194-13199.

Spielhofer, P., Bächi, T., Fehr, T., Christiansen, G., Cattaneo, R., Kaelin, K., Billeter, M. A., and Naim, H. Y., 1998, Chimeric measles virus with a foreign envelope. *J. Virol.* 72:2150-2159.

Stallcup, K. C., Raine, C. S., and Fields, B. N., 1983, Cytocholasin B inhibits the maturation of measles virus. *Virology* 124:59-74.

Stricker, R., Mottet, G., and Roux, L., 1994, The Sendai virus matrix protein appears to be recruited in the cytoplasm by the viral nucleocapsid to function in viral assembly and budding. *J. Gen. Virol.* 75:1031-1042.

Suryanarayana, K., Baczko, K., ter Meulen, V., and Wagner, R. R., 1994, Transcription inhibition and other properties of matrix proteins expressed by M genes cloned from measles virus and diseased human brain tissue. *J. Virol.* 68:1532-1543.

Takeda, M., Takeuchi, K., Miyajima, N., Kobune, F., Ami, Y., Nagata, N., Suzari, Y., Nagai, Y., and Tashiro, M., 2000, Recovery of pahogenic measles virus from cloned cDNA. *J. Virol.* 74:6643-6647.

Tatsuo, H., Ono, N., Tanaka, K., Yanagi Y., 2000, SLAM (CDw150) is a cellular receptor for measles virus. *Nature* 406:893-897.

Ter Meulen, V., Stephenson, J. R., and Kreth, H. W., 1983, Subacute sclerosing panencephalitis. *Compr. Virol.* 18:105-185.

Tucker, S. P., and Compans, R. W., 1993, Virus infection of polarized epithelial cells. *Adv. Virus Res.* 42:187-247.

Tyrell, D. L. J., and Ernst, A., 1979, Transmembrane communication in cells chronically infected with measles virus. *J. Cell. Biol.* 81:396-402.

Udem, S.A., 1984, Measles virus:conditions for the propagation and purification of infectious virus in high yield. *J. Virol. Methods* **8**:123-136.

Varsanyi, T. M., Utter, G., and Norrby, E., 1984, Purification, morphology, and antigenic characterization of measles virus envelope proteins. *J. Gen. Virol.* **65**:335-366.

Vincent, S., Gerlier, D., and Manié, S. N., 2000, Measles virus assembly within membrane rafts. *J. Virol.* **74**:9911-9915.

Von Messling, V., Zimmer, G., Herrler, G., Haas, L., and Cattaneo, R., 2001, The hemagglutinin of canine distemper virus determines tropism and cytopathogenicity. *J. Virol.* **75**:6418-6427.

Weidmann, A., Maisner, A., Garten, W., Seufert, M., ter Meulen, V., and Schneider-Schaulies, S., 2000. Proteolytic cleavage of the fusion protein but not membrane fusion is required for measles virus-induced immunosuppression in vitro. *J. Virol.* **74**:1985-1993.

Wild T. F., Malvoisin, E., and Buckland, R., 1991, Measles virus: both the haemagglutinin and fusion glycoproteins are required for fusion.. *J. Gen. Virol.* **72**:439-442.

Wild, T. F, Fayolle, J., Beauverger, P., and Buckland, R., 1994,. Measles virus fusion: role of the cysteine-rich region of the fusion glycoprotein. *J Virol.* **68**:7546-7548.

Chapter 4.2

The Predominant Varicella-zoster Virus gE and gI Glycoprotein Complex

CHARLES GROSE
Departments of Microbiology and Pediatrics, University of Iowa, Iowa City IA, 52242, USA

1. INTRODUCTION

Varicella-zoster virus (VZV) is classified as a member of the family Herpesviridae (human herpesvirus 3) and further subdivided into the Alphaherpesviridae. Alphaherpesviruses are defined by their ability to replicate and spread efficiently, to destroy their host cells during infection and to establish latent infection in the sensory ganglia. Within the Alphaherpesviridae, VZV has been reclassified as the prototype of the genus Varicellovirus while herpes simplex virus type 1 is the prototype of the genus Simplexvirus. VZV causes two common diseases; these include chickenpox (varicella) during childhood and shingles (herpes zoster) during late adulthood. Chickenpox is the primary infection, after which the neurotropic virus travels to the sensory ganglia along the spinal column, where it remains in a quiescent or latent state for decades. During the sixth decade of life and thereafter the virus occasionally reactivates, whence it traverses the same sensory nerve and emerges onto the skin in a dermatomal distribution known as herpes zoster (Hope-Simpson, 1965). Reactivation is accompanied by a pronounced anamnestic VZV specific immune response (Abramson, 1946; Weigle and Grose, 1984)

VZV was first isolated in cell culture by the Nobel laureate Weller (1953). The VZV genome consists of a linear double stranded DNA molecule 125 kbp in length, which was sequenced in its entirety in 1986 (Davison and Scott, 1986). An often overlooked fact is that the VZV genome is the smallest among the human herpesviruses. The genome consists of a

Structure-Function Relationships of Human Pathogenic Viruses, Edited by
Holzenburg and Bogner, Kluwer Academic/Plenum Publishers, New York, 2002

unique long region flanked by both internal and terminal repeated regions, and a unique short region flanked by repeated regions. Two isomeric forms predominate by virtue of inversion of the short unique DNA segment and its flanking repeated elements. The viral genome encodes at least 69 open reading frames (ORFs); according to Davison and Scott (1986), the ORFs are designated numerically beginning in the unique long segment. Based on the HSV model, the VZV genome is replicated by a rolling circle mechanism. One complete copy of the viral genome is inserted into each nascent capsid during assembly in the nucleus of the infected cell. The capsid exits the nucleus, transits the perinuclear membranes and enters the cytoplasm, where it acquires an envelope containing many of the viral glycoproteins (Gershon et al., 1973; Grose et al, 1983; Harson and Grose, 1995). The process includes several intermediary steps which have been called envelopment—deenvelopment—reenvelopment (Jones and Grose, 1988; Zhu et al., 1995). Finally the enveloped viral particles within cytoplasmic vacuoles emerge onto the outer cell membrane by exocytosis. Unlike most other herpesviruses, however, the VZV particles remain attached to the plasma membranes, often in long rows called viral highways (Harson and Grose, 1996).

The VZV genome encodes several glycoproteins, which are now designated according to the name given to the HSV homologous protein. Since VZV has a smaller genome than other herpesviruses, it has fewer glycoproteins but does code for gE, gI, gB, gC, gH and gL. Among these ORFs, only gI and gE are found in the unique short segment, where they are designated as genes 67 and 68, respectively. The genes in the short segment are of particular interest, because this segment is considered the most recent addition in evolutionary time and therefore may represent that portion of the genome which determines the particular niche of a herpesvirus, i.e., the gene products in the unique short segment provide attributes which give VZV its distinguishing properties. Like HSV-1 gE, the VZV gE product is usually found in a complex with gI. Unlike HSV-1 gE, however, VZV gE is the predominant glycoprotein in the infected cell and not a minor component (Weigle and Grose, 1983). As will be described in the chapter, the two VZV glycoproteins gE and gI form an omnipresent pluripotential complex which is involved in numerous aspects of the VZV like cycle such as cell-to-cell spread (Dingwell et al., 1994; Santos et al., 2000). In addition, the two VZV glycoproteins contain several antigenic sites which define the immune response to the virus (Grose et al., 1981;Grose and Friedrichs, 1982; Weigle and Grose, 1984; Ito et al., 1985; Arvin et al., 1986; Bergen et al., 1991). The latter finding is very important because the first naturally occurring VZV mutant virus to be recovered from an infected child contains a mutated gE glycoprotein.

2.   VZV GLYCOPROTEIN E (ORF 68)

2.1   VZV gE sequence

Based on the sequence and structure of gE, it is designated a typical type 1 transmembrane glycoprotein which is 623 amino acids in length with a 24 amino acid cleavable signal sequence (Davison and Scott, 1986). The glycoprotein consists of three major regions: a 544 amino acid hydrophilic extracellular region, a 17 amino acid hydrophobic transmembrane region and a 62 amino acid charged cytoplasmic tail (Grose, 1990). Examination of the predicted amino acid sequence of gE ectodomain reveals two cysteine-rich regions, three N-linked glycosylation signals and a juxtamembrane domain favoring O-linked glycans (Figure 1).

2.2   VZV gE N-linked and O-linked glycosylation

Like other type 1 glycoproteins, the gE molecule is formed through a series of intermediary products as the protein traffics from the endoplasmic reticulum through the Golgi apparatus en route to the plasma membrane (Montalvo et al, 1985). The molecular mass of the polypeptide backbone has been calculated under conditions of both in vitro coupled transcription/translation as well as infection in cultured cells. In the latter case, the Mr was estimated to be 73 kdal. However, in the former experiment, the Mr was estimated to be 64 kdal. Formation of mature gE results in a protein of Mr 98 kdal, hence the original designation of gE as gp98 (Grose, 1991). The presence of both high-mannose and complex type glycans on gE, as well as O-linked glycans, has been well documented (Edson et al., 1985; Grose, 1990). Thus, the potential N-linked Ser-Xxx-Asp sites and the potential O-linked Ser-Thr-Pro sites on the ectodomain are utilized. Furthermore, VZV gE is heavily sialylated and sulfated (Edson, 1993).

2.3   VZV gE palmitylation and myristylation

Other VZV gE modifications include the attachment of fatty acids. For this biochemical investigation, VZV-infected cells were radiolabeled with either tritiated palmitic acid or tritiated myristic acid (Namazue et al., 1989; Harper and Kangro, 1990). Analysis of the immunoprecipitates by

electrophoresis and fluorography showed that gE contained both palmitic
acid and myristic acid.

```
  1 MGTVNKPVVG VLMGFGIITG TLRITNPVRA SVLRYDDFHT DEDKLDTNSV
    -------------------------------

 51 YEPYYHSDHA ESSWVNRGES SRKAYDHNSP YIWPRNDYDG FLENAHEHHG

101 VYNQGRGIDS GERLMQPTQM SAQEDLGDDT GIHVIPTLNG DDRHKIVNVN
                                                        ~
151 QRQYGDVFKG DLNPKPQGQR LIEVSVEENH PFTLRAPIQR IYGVRYTETW
    ```````````

201 SFLPSLTCTG DAAPAIQHIC LKHTTCFQDV VVDVDCAENT KEDQLAEISY

                  >
251 RFQGKKEADQ PWIVVNTSTL FDELELDPPE IEPGVLKVLR TEKQYLGVYI

301 WNMRGSDGTS TYATFLVTWK GDEKTRNPTP AVTPQPRGAE FHMWNYHSHV

351 FSVGDTFSLA MHLQYKIHEA PFDLLLEWLY VPIDPTCQPM RLYSTCLYHP
                                                     >
401 NAPQCLSHMN SGCTFTSPHL AQRVASTVYQ NCEHADNYTA YCLGISHMEP

451 SFGLILHDGG TTLKFVDTPE SLSGLYVFVV YFNGHVEAVA YTVVSTVDHF
                                  >
501 VNAIEERGFP PTAGQPPATT KPKEITPVNP GTSPLLRYAA WTGGLAAVVL
               ~   ~~   ~   ~~   ======
551 LCLVIFLICT AKRMRVKAYR VDKSPYNQSM YYAGLPVDDF EDSESTDTEE
    =============   **** | || |
                                                + +
601 EFGNAIGGSH GGSSYTVYID KTR
```

```
--- Signal sequence    >  N-linked glycan signal
=== Transmembrane sequence   ~  O-linked glycan signal
 *  Endocytosis motif   +  Casein kinase I site
    MAb 3B3 epitope      |  Casein kinase II site
```

Figure 1. Amino acid sequence of VZV gE. The gE gene of the VZV-MSP strain was
sequenced and compared with the sequence of the prototypic VZV-Dumas sequence.
Domains discussed in the text are annotated.
```

The precise sites were not determined, but palmitic acid attaches to cysteine residues while myristic acid is usually present only on an N-terminal glycine residue.

2.4   VZV gE serine/threonine phosphorylation

The gE endodomain is also highly modified. Among all the herpesviral glycoproteins, the VZV gE product was the first to be studied because of its phosphorylation status (Montalvo and Grose, 1986; Yao et al., 1993; Yao et al., 1992). Phosphorylation of gE was first suspected because gE bears homology to many mammalian surface receptors which in turn are phosphorylated on their cytoplasmic tails (Kishimoto et al., 1987). The VZV glycoprotein was shown to be phosphorylated by an ubiquitous cellular protein kinase. Casein kinase II is a serine/threonine protein kinase with properties shared by few other kinases. First, casein kinase II is capable of utilizing GTP almost as well as ATP as a phosphate donor; secondly, casein kinase II activity is potently inhibited by heparin. These two characteristics were utilized to investigate gE phosphorylation in a recombinant gE-baculovirus expression system (Olson et al, 1997). The protein kinase assay was initiated by the addition of either 32P-ATP and 32P-GTP to immunoprecipitated gE protein. With 32P-ATP, gE was phosphorylated in a similar manner to that demonstrated in cell culture, as shown by detection of a prominent phosphorylated Mr 98 kdal monomer. These results showed that the cellular kinase coprecipitated with its substrate, namely gE; in turn, the same kinase phosphorylated gE in the in vitro protein kinase assay. When heparin was added to the protein kinase assay, phosphorylation of gE was greatly diminished. Thus, the phosphorylation event was inhibited by heparin. When a gE precipitate was heat treated, the associated kinase activity was inactivated and unable to phosphorylate gE. However, phosphorylation of gE was restored to the heat inactivated precipitate when fresh Hela cell lysate was added to the assay; phosphorylation of gE was restored also by addition of fresh lysate of insect cells. These results indicate that casein kinase II present in either human cells or insect cells can phosphorylate VZV gE.

VZV gE contains a sequence within its cytoplasmic tail with a consensus motif for casein kinase II phosphorylation (Grose et al, 1989). This amino acid sequence from residues 590-602 contains the following: Phe-Glu-Asp-Ser-Glu-Ser-Thr-Asp-Thr-Glu-Glu-Glu-Phe. The notable phosphorylatable serine and threonine residues are surrounded by acidic residues. To determine whether this sequence was the authentic phosphorylation site for casein kinase II, the gE sequence was mutated and the phosphorylation of the gE mutant protein was analysed. The two serine residues at 593 and 595 were changed to alanine and phenylalanine respectively and the two threonine residues at 595 and 598 were both changed to alanines (Olson et al, 1997). The gE construct with 4 altered amino acids was designated gE-AFAA. When analysed by laser scanning confocal microscopy, the gE-AFAA protein expressed by a recombinant baculovirus was detected both in the cytoplasm and on the plasma membrane; thus, the mutant gE protein was distributed in a similar pattern as wild type gE although the precise trafficking pattern of each construct needs further study. Secondly, the biosynthesis of the mutant construct was measured by immunoprecipitation and found to be similar to wild type gE. Thirdly, the phosphorylation of the mutant protein was compared with that of the wild type gE. Instantimager analysis of the two phosphorylated proteins showed a 70% reduction on the mutant gE as compared with wild type gE.

Based on these dramatic results, the assumption has been made that gE is phosphorylated primarily by casein kinase II at the canonical phosphorylation consensus sequence in its cytoplasmic tail. The VZV encoded ORF 47 protein kinase resembles casein kinase II and may also phosphorylate gE at the same consensus site in the endodomain (Ng and Grose, 1992; Ng et al., 1994). There is also a casein kinase I consensus site in this region but it has not been further investigated (Grose et al, 1989).

2.5   VZV gE tyrosine phosphorylation

In the studies described above, the phosphorylated gene product mainly represented the VZV gE monomer. In further studies with gE, a phosphorylated dimeric form was also observed. Of interest the phosphorylation of the gE dimer was abolished when GTP was the radiolabel, a result which indicated that the dimer was not primarily phosphorylated by casein kinase II which can utilize GTP nearly as well as ATP as the phosphate source. Again, examination of the gE sequence provided clues as to the nature of the second phosphate acceptor, namely, six tyrosine residues were present in the cytoplasmic tail. To investigate whether tyrosine phosphorylation was involved, insect cells were infected with recombinant baculovirus containing the wild type gE insert (Olson et al, 1997). The precipitated gE protein was transferred to a nitrocellulose membrane and probed with a biotinylated P-tyr-1 antibody, which specifically recognizes phosphotyrosine linkages but not phosphoserine sites. After visualization by chemiluminescense, the anti-phosphotyrosine antibody detected a single protein band at Mr 130 kdal, which was previously identified as the dimeric form of the gE in insect cells.

Because of the importance of the above observation, the experiment was repeated in mammalian cells. Since the transfected mammalian cells did not produce as much gE protein as the baculovirus infected insect cells, the gE transfected cell lysates were concentrated several fold to obtain sufficient gE protein for immunoprecipitaion and immunoblot analysis. When the nitrocellulose membrane was probed with the same P-tyr-1 antibody, the phosphotyrosine antibody again detected a protein at Mr 130 which corresponded to the gE protein recognized in the insect cells. Thus, this result documented the presence of phosphotyrosine linkages in gE dimers present in mammalian cells. These dimeric gE forms are present in limited quantities and are not easily detectable in the absence of a highly radioactive 32P-label.

2.6 VZV gE homology to cell surface receptors

Knowledge of the VZV gE sequence led to computer assisted analyses of its similarity to that of mammalian cell surface receptors, including one of the human Fc receptors and the low density lipoprotein receptor (Litwin et al, 1990; Olson et al, 1997). A computer assisted Bestfit homology comparison between gE and an Fc receptor showed 44% amino acid similarity and 23% amino acid identity between the two ORFs (Litwin and Grose, 1992). As a comparison, VZV gE showed 47% amino acid similarity and 27% amino acid identity with its alpha herpesviral homolog HSV-1 gE. One feature of cell surface receptors is a propensity to exist as dimers on the cell surface in order to bind their ligand and to become activated. Of note, some dimer interactions have been determined to be among the strongest protein-to-protein interactions in nature. These dimer interactions may be relevant to functions of the tyrosine phosphorylated VZV gE dimer described previously.

2.7 VZV gE endocytosis assay by confocal microscopy

Another common feature of cell surface receptors is endocytosis. Endocytosis is the process by which proteins are continually internalised within the cell through specific interactions at the plasma membrane (Letourneur and Klausner, 1992; Trowbridge et al., 1993). Receptors typically undergo endocytosis from the cell membrane through interactions with clathrin coated pits and are often recycled through the endosomes back to the cell membrane. Since VZV gE contains tyrosine based internalisation motifs in its cytoplasmic tail, gE was analysed to determine if it was endocytosed like other receptors, such as the human Fc receptor (Alconada et al., 1996; Olson and Grose, 1997).

In one of the investigations of gE internalisation, a new assay was developed which exploited an unrealised potential of the laser scanning confocal microscope (Olson and Grose, 1997). To this end, Hela cells were transfected with an expression plasmid containing the gE insert. After 16 hr, the cells were incubated with a primary murine monoclonal antibody (MAb) to gE at 4° for 30 min. Then individual monolayers were returned to 37° for increasing intervals of 0, 10, 15, 30, 30, 45, and 60 min to allow internalisation of gE. Thereafter, secondary anti-mouse antibody conjugated to a dye was incubated with the transfected cells to identify the localisation of gE. Subsequently the cells were observed by confocal microscopy with digital tomography in one micron increments from the surface of the cell to its attachment plane on the plastic surface; the zeta series of gE confocal images were stored in a computer. When the images were subsequently analysed, the following scenario was evident (Figure 2). At the initial timepoint at 4°, VZV gE was easily detected on the cell surface. At the 5 min timepoint, gE remained localised mainly in a clumping pattern on the membrane although some gE was observed within the cell. By 10-15 min at 37°, however, more gE was detected in the cytoplasm, often within small vesicles. After 30-60 min at 37°, large amounts of gE were detected in the cytoplasm in a typical pattern consisting of clusters of small vesicles.

To verify that gE internalisation was clathrin mediated, a second set of experiments was performed in which gE transfected cells were double stained with an IgG antibody to gE and an IgM antibody to clathrin (Olson and Grose, 1997). The confocal microscope has the capacity to store duplicate images of each time point. When the images of gE and clathrin were merged and observed, they documented that gE was clustering in clathrin coated pits forming the clathrin coated vesicles at the cell membrane. These confocal microscopic analyses demonstrated conclusively that VZV gE is internalised into the cytoplasm from the cell membrane in a pattern consistent with endocytosis of other cell surface receptors.

The transferrin receptor has been long recognised as a cell surface protein which undergoes receptor mediated endocytosis in clathrin coated vesicles (Collawn et al., 1990). The receptor is subsequently transported to sorting and recycling endosomes which facilitate the transport of the transferrin receptor back to the plasma membrane. To determine if VZV gE was following the same endocytosis pathway, gE trafficking was compared with that of the transferrin receptor. For this experiment, Hela cells were transfected with the gE expression plasmid and the cells were subsequently probed with rabbit polyclonal monospecific gE antibody and murine transferrin receptor MAb (Olson and Grose, 1997). At all timepoints, images were analysed by double staining confocal microscopy. When the images were merged and observed, there was a marked degree of colocalization at each time point; in other words, gE was internalised from the cell surface in a trafficking pattern very similar or identical to that of the transferrin receptor. These data provide additional support for the conclusion that VZV gE is internalised and recycled in a clathrin mediated endocytosis pathway. Like the transferrin receptor, gE also recycles to the cell surface (Olson and Grose, 1997).

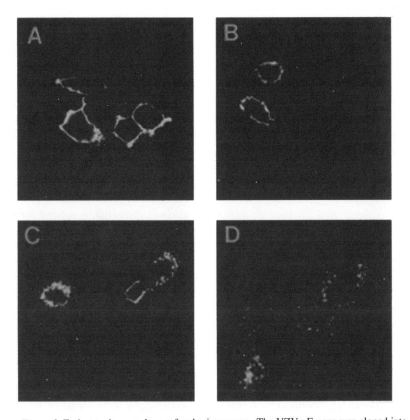

Figure 2. Endocytosis assay by confocal microscopy. The VZV gE gene was cloned into a plasmid expression vector. After transient transfection of 4 monolayers, the internalisation of gE from the cell surface was documented at 4 timepoints by confocal microscopy. Panel A: at the initial 0 timepoint, gE formed a broad fluorescent rim on the surface of each transfected cell. Panels B-D: at 15 minute intervals thereafter, internalisation of gE in clathrin coated pits led to an increasingly large number of endocytic vesicles in the cytoplasm of each transfected cell, especially noticeable at the 60 minute timepoint (Panel D).

2.8   VZV gE tyrosine based endocytosis motif

Receptor endocytosis has been shown to be dependent on specific cytoplasmic tail motifs. Deletional analysis studies of the endodomains from several receptors including the low density lipoprotein receptor, the Fc receptor, and the transferrin receptor, have shown less efficient internalisation of a tailless receptor as compared with the nonmutated receptor. To determine if VZV gE endocytosis was likewise dependent on a specific internalisation motif, a tailless gE protein was first constructed by deletion of the entire endodomain and trafficking of this mutant was analysed by a confocal microscopic endocytosis assay. The tailless mutant molecule reached the cell surface, but remained on the surface in subsequent timepoints when the wild type molecule was largely internalised. Thus, this experiment confirmed that like other receptors, VZV gE required the cytoplasmic tail sequence for efficient internalisation.

The VZV gE endodomain contains 6 tyrosine residue but one tyrosine in particular is part of a defined internalisation motif, namely, the so-called YXXL sequence (Olson and Grose, 1997). As part of a detailed endocytosis analysis, this tyrosine residue was placed by a glycine residue. As in precious experiments, HeLa cells were transfected with the gene encoding the gE-Y582G product and the monolayer was analysed by confocal microscopy. When the gE-Y582G product was first observed at timepoint 0, the protein was located at the cell surface. When other cultures were observed at the 15 and 30 min timepoints, the mutant protein remained on the cell surface and was not internalised. Therefore, the concluding result of these experiments was that the tyrosine in the gE endodomain YAGL motif was required for efficient internalisation of the gE molecule. In all aspects, therefore, VZV gE endocytosis mimics that of other cell surface receptors.

2.9   VZV gE targeting to the trans-Golgi network

Additional targeting signals have been discovered in the cytoplasmic tail of gE. One of these sequences, namely AYRV, was sufficient to cause expressed gE protein to colocalize with marker proteins for the trans-Golgi network (Zhu et al., 1996). This targeting was abrogated when the tyrosine residue in the AYRV sequence was replaced with glycine or lysine, when arginine was replaced with glutamic acid, when valine was substituted with lysine or when alanine was replaced with aspartic acid. In contrast, tyrosine could be replaced by phenylalanine and valine could be substituted with leucine. In addition to the AYRV sequence, the acidic amino acid region associated with the previously described casein kinase II phosphorylation consensus sequence also facilitated gE targeting to the trans-Golgi network (Zhu et al, 1996).

3.   VZV GLYCOPROTEIN I (ORF 67)

3.1   VZV gI sequence

Based on the sequence and structure of VZV gI, it is designated a typical type 1 transmembrane glycoprotein comprised of a total of 354 amino acids (Davison and Scott, 1986). The ectodomain includes 279 amino acids, the transmembrane domain 17 amino acids and the ectodomain 58 amino acids (Grose, 1990). The current terminology for ORF 67 is based on its homology to the HSV-1 gI glycoprotein (Figure 3). Like the other VZV glycoproteins, VZV gI had historical designations such as gp64 and gpIV. In particular, some confusion has arisen in the literature between the new designation gI (capitol I) and the old designation for gE, which was gpI (Roman numeral one).

```
                              >                        >
  1 MFLIQCLISA VIFYIQVTNA LIFKGDHVSL QVNSSLTSIL IPMQNDNYTE
    --------------------------
                              >
 51 IKGQLVFIGE QLPTGTNYSG TLELLYADTV AFCFRSVQVI RYDGCPRIRT
                              >
101 SAFISCRYKH SWHYGNSTDR ISTEPDAGVM LKITKPGIND AGVYVLLVRL

151 DHSRSTDGFI LGVNVYTAGS HHNIHGVIYT SPSLQNGYST RALFQQARLC

201 DLPATPKGSG TSLFQHMLDL RAGKSLEDNP WLHEDVVTTE TKSVVKEGIE
                                                 ~~    ~ ~
251 NHVYPTDMST LPEKSLNDPP ENLLIIIPIV ASVMILTAMV IVIVISVKRR
    ~    ~~    ~                     ==================
301 RIKKHPIYRP NTKTRRGIQN ATPESDVMLE AAIAQLATIR EESPPHSVVN
                         **                      +++
351 PFVK

--- Signal sequence
=== Transmembrane sequence
  * Endocytosis motif
  > N-linked glycan signal
  ~ O-linked glycan signal
  + Cylin-dependent kinase site
```

Figure 3. Amino acid sequence of VZV gI. The gI gene of the VZV-MSP strain was sequenced and compared with the sequence of the prototypic VZV-Dumas strain. Domains discussed in the text are annotated.

3.2   VZV gI glycosylation

The VZV gI ectodomain contains four potential N-linked glycosylation sites as well as a juxtamembrane region compatible with O-linkages. Several studies with radiolabeled glycans and inhibitors of glycosylation have confirmed the presence of both complex type N-linked glycans and O-linked glycans on VZV gI. In early studies of gE, the fact that gI coprecipitated as a gE/gI complex was not recognised (see below); however, these papers can be reanalysed and reinterpreted to define the gI specific modifications (Grose, 1980; Montalvo et al., 1985; Cohen and Nguyen, 1997; Yao et al., 1993).

3.3   VZV gI serine phosphorylation

Like VZV gE, gI is phosphorylated following virus infection in cell culture. Initial studies in a transient transfection system documented that gI expressed in the absence of any other VZV protein was phosphorylated on its cytoplasmic tail by a serine protein kinase present in cultured cells (Yao and Grose, 1994). Yet, examination of the gI sequence quickly established that no casein kinase II phosphorylation consensus motif was present in the endodomain. Therefore, others serine residues were the phosphoacceptors. One sequence on the gI endodomain Ser-Pro-Pro was of special interest, because of its semblance to the phosphorylation consensus motif for cyclin dependent kinases (Ye et al, 1999). To investigate this possibility, the three residues in the gI endodomain were individually mutated. Three expression plasmids containing the following mutations were studied in detail: gI-S343A, gI-P344A and gI-P345A. After 32P-radiolabeling, phosphorylation of each gI mutant protein was compared with wild type protein by phosphorimaging analysis. The phosphorylation signals of gI-S343A and gI-P344A were about 10% of that seen with wild type gI while that of gI-P345A was about 30% of with type gI. These results indicated that both serine residue 343 and proline residue 344 were essential for gI phosphorylation. In contrast, proline 345 modulated but did not abrogate the phosphorylation of gI. In other words, the principal gI phosphorylation site is a serine-proline sequence in the cytoplasmic tail.

Since the above sequence is usually associated with phosphorylation by a cyclin dependent kinase (CDK), an in vitro kinase assay was performed with gI as the substrate and CDK1 and CDK2 as the protein kinases. Of note, the gI product was phosphorylated by both kinases. When the gI phosphorylation signals of CDK1 and CDK2 were compared, a 4-fold higher phosphorylation signal accompanied CDK1 in the protein kinase assay. As a specificity control, CDK6 was included in the kinase assay but no phosphorylation of gI was observed. The above results strongly suggest that the gI endodomain is phosphorylated by CDK1.

3.4   VZV gI dileucine endocytosis motif

VZV gI expressed in a transient transfection system was investigated in a similar manner as previously described for gE. To this end, gI transfected cells were incubated with antibody to gI and analysed for endocytosis by laser scanning confocal microscopy and digital tomography at increasing time intervals (Olson and Grose, 1998). As seen with gE, gI was detectable on the surface at timepoint 0, while some gI was found within the cytoplasm within 15 min. By 30-60 min, large amounts of gI were observed within the cytoplasm in small vesicle-filled clusters. These results clearly indicated that VZV gI underwent endocytosis from the cell membrane in a similar pattern to gE under the same transfection conditions.

Endocytosis of cell surface receptors is known to be dependent on specific amino acid sequence within the cytoplasmic tail. As described above, the gE endodomain contains a YAGL internalisation motif. Upon examination of the gI cytoplasmic tail, no tyrosine containing internalisation motif was identified. However, the gI endodomain did contain another potential endocytosis motif, namely, a variation on the dileucine motif (Bremnes et al., 1994). Several receptors, such as the invariant chain associated with the major histocompatibility complex class II molecule as well as the CD4 molecule, contain a dileucine motif which effects efficient internalisation of the receptor from the plasma membrane. The cytoplasmic tail of gI contains a Met-Leu sequence at residues 328-329, which corresponds to a dileucine motif. In order to determine if the Met-Leu sequence was responsible for gI endocytosis, the sequence was changed to Ala-Ala by site directed mutagenesis. Subsequently, HeLa cells were transfected with the wild type gI gene or the mutant gI gene designated gI-AA. When gI-AA was expressed in cells and analysed by confocal microscopy, the mutant protein was present on the cell surface at timepoint 0. After 30 min, gI-AA was still localised mainly to the cell membrane. In contrast, wild type gI was internalised within the cytoplasm after 30 min. These results showed that the dileucine motif in the VZV gI endodomain was functioning as a signal for endocytosis.

3.5   VZV gI chaperone functions

Increasing evidence suggests that VZV gI behaves as a chaperone protein when coexpressed with gE (Mallory et al., 1997; Olson and Grose, 1998). Complex formation between gE and gI will be discussed in detail in the next section. This paragraph will concentrate on the chaperone functions of gI. Both gE and gI, when expressed individually by transient transfection, can travel from the endoplasmic reticulum through the Golgi en route to the plasma membrane, after which the glycoprotein is internalised (Yao et al. 1993). When gE and gI are coexpressed, the gE/gI complex is endocytosed more efficiently than gE alone (Olson and Grose, 1998). Presumably the two signals for clathrin mediated endocytosis in the same gE/gI complex enhance the efficiency of internalisation (Figure 4).

Further, the Y582G mutation to the gE endocyctosis signal results in markedly diminished gE endocytosis, as described previously. This impairment in gE-Y582G endocytosis was overcome by cotransfection with gI containing its intact dileucine internalisation motif. (Table 1). The percent internalisation of radiolabeled VZV Y582G-gE/gI complex was the same as the percent internalisation of the nonmutated gE/gI complex. However, the reverse situation was not true, ie, an endocytosis gI mutant was not efficiently endocytosed even when associated with wildtype gE. Thus, VZV gI exhibits a chaperone function which facilitates the trafficking and internalisation of the gE protein.

Table 1. Quantitative internalisation assay for VZV gI with mutant gE

VZV Gene	counts	time 0	time 15	time 30	time 60
Y582G-gE	cpm	0	1949	3706	3588
Alone	percent	0	4.1	6.5	6.9
Y582G-gE	cpm	0	10565	11026	11112
With gI	percent	0	52.7	54.3	55.3

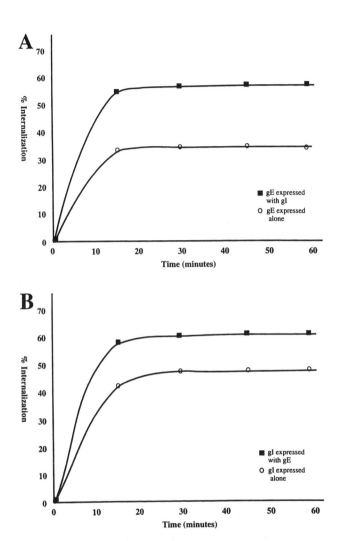

Figure 4. Endocytosis of the VZV gE/gI complex. Hela cells were transfected with either the gE gene or the gI gene or, alternatively, cotransfected with both genes. After a 16 hour incubation, anti-gE or anti-gI antibody was added at 4° for 30 minutes. The cells were then returned to 37° for 0, 15, 30, 45, and 60 minutes. Thereafter the cells were treated with trypsin to remove surface proteins. The cells were lysed and the VZV proteins were precipitated. The precipitated proteins were separated by electrophoresis and the gels analysed by an Instantimager. The percent internalised VZV protein was derived from a formula comparing the total protein labeled on the surface at 0 minutes with the amount of protein internalised at the subsequent timepoints.

4.   COMPLEX FORMATION

4.1   Ectodomain interactions

VZV glycoproteins gE and gI form a noncovently linked complex (Vafai et al, 1988; Litwin et al, 1992; Yao et al, 1993). In order to investigate criteria for complex formation, the ectodomains of both glycoprotein genes were cloned into baculovirus, using either native or insect derived signal peptides (Kimura et al, 1997). Each recombinant virus yielded soluble VZV protein in culture medium. When insect cells were simultaneously infected with both recombinant VZV gE and gI viruses, a soluble heterodimeric gE/gI complex was detected by a combination of immunoprecipitation and immunoblot analyses. Site specific mutations in the amino terminus of gI abrogated its ability to bind with the gE protein. In short, these findings indicate that the mature gI N-terminus is required for heterodimeric complex formation by the ectodomains of VZV gE and gI.

4.2   Endodomain interactions

Based on the results described in the previous sections of this chapter, it is possible to construct a figure that includes a network of phosphorylation interactions between the two components of the VZV gE/gI complex. The cytoplasmic tail of the gE constituent of this complex is phosphorylated by casein kinase II at threonine and seine residues between amino acids 593 and 598. In addition, casein kinase II binds to and coprecipitates with gE. Casein kinase II is normally active in mammalian cells; further, both the alpha and beta subunits of this enzyme are phosphorylated in turn by the cyclin dependent kinase called CDK1 (Litchfield et al., 1995; Bosc et al., 1995). Therefore, the VZV model depicts potential interactions between CDK1, casein kinase II and the gE endodomain, as well as interactions between CDK1 and the gI endodomain. (Figure 5). This series of phosphorylation and dephosphorylation reactions may provide signals for internalisation and trafficking of the VZV glycoproteins; a similar strategy has been proposed for the cytomegalovirus glycoprotein gB (Tugizov et al, 1998).

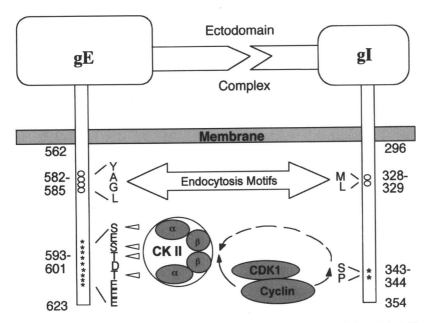

Figure 5. Potential mechanisms for cross talk between the components of the VZV gE/gI complex. The locations of the endocytosis motifs and phosphorylation consensus sites within the cytoplasmic tails are designated with their respective amino acid sequences. Potential interactions between the protein kinases modifying the gE and gI endodomains are demarcated by arrows.

5.   GLYCOPROTEIN E MUTANT VIRUS

5.1   Initial characterisation of VZV-MSP

As mentioned earlier, the VZV genome of a Dutch strain called Dumas has been completely sequenced (Davison and Scott, 1986). At the time, the prevailing opinion was that the VZV genome was so stable that the sequence was applicable to virtually all VZV laboratory strains and wild type isolates. However, in the late 1990s, a wild type virus was isolated in Minnesota, USA, which had a missence mutation in the gE ectodomain. In turn, this mutation led to loss of a B cell epitope. This mutant strain was designated VZV-MSP (Santos et al., 1998; Santos et al., 2000).

VZV-MSP was initially isolated in commercial human fibroblast monolayers. The cytopathic effect was compatible with VZV, but the isolate was poorly reactive with antibodies in a commercial VZV diagnostic kit. Because no isolate had been discovered over the prior 15 years which failed to react in this kit, the virus isolate was subjected to extensive further analysis. Initially, virus infected monolayers were prepared for examination by laser scanning confocal microscopy. The low-passage VZV-32 strain was included in separate dishes as a control virus. When cytopathic effects covered ~70% of each monolayer, the infected moolayers were probed with MAb 3B3 against gE and with MAb 6B5 against gI and examined by confocal microscopy. These two antibodies do not cross-react with other viral or cellular proteins (Grose, 1990). As expected from numerous published experiments, MAb 3B3 and MAb 6B5 reacted with the laboratory strain VZV-32. In marked contrast, the anti-gE MAb 3B3 did not attach to cells infected with the VZV-MSP strain, even though MAb 6B5 did bind the infected cells strongly. As an additional control, the anti-gH MAb 206 was added to cultures individually infected with both VZV strains; all VZV-infected cultures were positive in this assay.

5.2   Sequence analysis of VZV-MSP gE

To further investigate the nature of the VZV gE mutation, polymerasechain reaction (PCR) amplification techniques were employed to first determine that a full-length gE gene (VZV ORF 68) was present in the mutant strain.   Thereafter, other primers were used that amplified overlapping portions of the gE gene, each ~300 bases in size. Each fragment was subjected to DNA sequencing, and each sequence was compared with the published sequence of the Dumas strain (GenBank accession no. X04370).   After analysis of all 623 codons of VZV-MSP gE ORF, the most important base change occurred in codon 150; the substitution involved a replacement of a guanine by an adenine. Of great interest, this point mutation led to a change in amino acid from aspartic acid to asparagine (Figure 6). Because this alteration in VZV gE occurred in the vicinity of the previously deduced 3B3 epitope, codons 151-161, the sequence data strongly suggested that amino acid 150 was an important contributor to the B cell epitope (Santos et al., 1998).

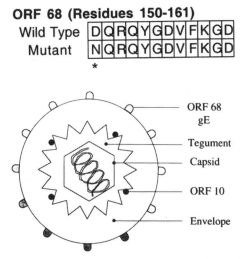

Figure 6. Missence mutation in the VZV gE ectodomain. The wild type sequence is VZV-Dumas; the mutant sequence is VZV-MSP. The 12 amino acids in the wild type sequence comprise the B cell epitope defined by MAb 3B3. The lower diagram illustrates the location of gE (ORF 68) on the virion envelope; the 3B3 epitope is exposed on the exterior of the virion. ORF 10 is a tegument protein.

5.3   Epitope mapping of VZV-MSP gE

In a preliminary experiment, the 3B3 epitope in the VZV gE ectodomain was mapped by inserting a 11-amino-acid sequence into the unrelated VZV ORF 60, namely, the gL glycoprotein (Duus and Grose, 1996; Hatfield, et al, 1997). The epitope tag within gL was recognised by MAb 3B3 when observed by laser scanning confocal microscopy. To evaluate the contribution of the aspartic acid residue to formation of the epitope, the gL epitope mapping and tagging experiment was repeated with the addition of the aspartic acid residue in its correct location at the N-terminus of the 3B3 epitope. To obtain the proper parameters for the mutagenesis primers, one additional codon was inserted along with an aspartic acid. The pTM-1 expression plasmids, including gL-3B3.11 and gL-3B3.13, were transfected into HeLa cells and observed by confocal microscopy after labelling with MAb 3B3. Cells transfected with the gL-3B3.11 plasmid were weakly positive in a restricted cytoplasmic pattern, as previously described by Duus and Grose (1996). Not only were cells transfected with the gL-3B3.13 plasmid more intensely stained, but also the pattern was more widely distributed through the cytoplasm (Santos et al., 1998). Thus, aspartic acid was shown to be an important component of the B cell epitope recognised by MAb 3B3. The above experiments are critical because they provide verification that VZV-MSP is an authentic gE ectodomain mutant virus.

6.   GENETIC POLYMORPHISMS IN E AND I

6.1   Overview of the VZV genome

The recent discovery of a genetically variant virus called VZV-MSP led to a reconsideration of the concept of VZV immutability (Grose, 1999). Subsequent genetic analyses of the ORFs coding for surface proteins were undertaken with the presumption that these genes would be subject to change (Faga et al., 2001). This postulate is based on the fact that genes within the VZV genome can be modified by the two processes of selection and random genetic drift. Selection is the means by which fitter viral variants increase in frequency in the population and less fit viral variants disappear from the population. Selection can be heavily influenced by the immune response of the host; thus the VZV glycoprotein genes were among the first to be examined. On the other hand, random genetic drift describes the appearance of variant progeny viruses which may be neither more nor less fit than the parent virus.

In the case of VZV-MSP, the VZV ORF 68 variant virus was considered to be a VZV gE escape mutant virus which was the likely end result of antibody selection in the infected human host (Grose, 1999). In other words, the virus has lost an epitope considered to be present in most wildtype strains. In 1998, however, there was limited VZV sequence data available other than that of the Dumas strain and the Oka vaccine strain. Because of the above observations, a more extensive genetic analysis of the major VZV structural genes was undertaken. The protocol for the VZV genetic studies was patterned on the studies being performed as part of the Human Genome Project (Collins et al., 1997; Wang et al., 1998; Brookes, 1999). In this regard, the goal was to identify patterns of single nucleotide polymorphisms (SNPs) in these VZV coding regions, which could distinguish individual VZV strains. As with the human genome, the discovery of SNPs would allow further clustering of VZV strains into groups with common ancestry and perhaps groups with differences in virulence or attenuation. By definition, a SNP must be found in at least 1% of a population under study.

Ten viruses were examined (Takahashi et al., 1975; Faga et al., 2001). VZV-MSP was isolated in Minnesota in late 1995 (GenBank accession no. AY005330-5). VZV-32 was isolated in Texas in the 1970s (no. AF314215-21). VZV-Oka vaccine strain was isolated in Japan and attenuated in the 1970s (no. AY016462-67). VZV-VSD was a wildtype virus collected in South Dakota in the 1980s (no. AY013747-52). VZV-VIA was isolated in Iowa from a child with chickenpox in the 1990s (no. AF325436-41). VZV-Ellen was originally isolated in Georgia from a child with chickenpox in the 1960s and obtained from the American Type Culture Collection (no. AY010902-07). VZV 80-2 was originally isolated in Pennsylvania from an adult with herpes zoster in the 1980s (no. AY016457-61). Two isolates collected in the Los Angeles, California, area in 1993 were called LAX1 and LAX2 (no. AY016445-50 and AY16451-56). VZV-Iceland was collected in Iceland (no. AF322634-40)

PCR amplification was performed with primers flanking the VZV region of interest (Table 2). The Expand High Fidelity PCR System provided by Roche Molecular Biochemicals was used in the PCR amplification procedure. This system includes Taq DNA and Pwo DNA polymerases, with the 3'-5' proofreading activity of the Pwo DNA polymerase to increase fidelity. VZV DNA was subsequently analysed with an Applied Biosystems Model 373A stretch fluorescent automated sequencer at the University of Iowa DNA Facility. Sequences were further analysed using the program DNASIS V2.0 from the Hitachi Software Engineering Company. Any region of a VZV genome which differed in sequence from that of the prototypic VZV-Dumas was re-amplified in a second PCR step and subjected to a second sequencing analysis.

6.2   Polymorphisms in the VZV gE gene

The VZV gE gene was of greatest interest because of the discovery of the gE mutant strain VZV-MSP (Santos et al., 1998). Surprisingly, six gE polymorphisms were found among the 10 tested strains and isolates, four of which cause amino acid substitutions. To be called a SNP, the polymorphism must be detected in at least 2 of 10 strains, ie, even if 98 as yet unsequenced VZV strains lack the polymorphism, it would still be present in 1% of strains and therefore an authentic SNP (Brookes, 1999). VZV-Ellen, VZV-Iceland, and VZV 80-2 had three identical polymorphisms (Figure 7).   One was a

Table 2. Primers for VZV gE and gI genes

Gene Primer and nucleotide numbers

gE   Ip2 (A)   2287 to 2308   Sequence/Amplification
     CGCTCTAGAACTAGTGGATCCCCCGGGGAATTTGTCACAGGCTTTT
gE   Nco gpl (S) 1 to 11   Amplification
     CGACCCGGGGAGCTCCCATGGGGACAGTTAATAAACC
gE   Sp 1 (S)   1452 to 1470   Sequence
     GCATGTTGAAGCCGTAGCA
gE   Sp 3 (S)   199 to 218   Sequence
     ATGCGCGGCTCCGATGGTAC
gE   Sp 4 (A)   486 to 505   Sequence
     GGCCTTGGGGTTTTGGATTA
gE   Sp 6 (S)   -70 to –48   Sequence
     GTCCATGGTTTTAGACCTCGGG
gE   Sp 7 (S)   543 to 561   Sequence
     GTTTACTTTACGCGCACCG
gE   Sp 8 (S)   823 to 840   Sequence
     GAATTAGACCCCCCCGAG
gE   Sp 9 (S)   1127 to 1146   Sequence
     TAGAGTGGTTGTATGTCCCC
gE   Sp 10 (S)   1601 to 1620   Sequence
     CACTTCTACGATATGCCGCA
gE   Sp 12 (A)   827 to 844   Sequence
     CAATCTCGGGGGGGGTCTA
gE   Sp 13 (A)   1772 to 1790   Sequence
     TCCGTAGATTCCGAGTCCT

gI   ScP1 (S)   -60 to –33   Sequence/Amplification
     CGGCTCACAGAGCTGCTCTTCGGTGTAG
gI   ScP2 (A)   1137 to 1154   Sequence/Amplification
     TAATCCTTCCCCTCATATCACAACGCGT
gI   Sp 1 (A)   978 to 995   Sequence
     GCGGCCTCCAACATCACA
gI   Sp 2 (A)   365 to 383   Sequence
     CCAGCATCCGGCTCTGTTG
gI   Sp 4 (S)   314 to 337   Sequence
     CGTGTAGGTACAAACATTCGTGGC

---

synonymous mutation within codon 220. Two non-synonymous mutations in these strains caused amino acid substitutions within codons 40 (T → I) and 536 (L → I). The vaccine strain VZV-Oka also contained the mutation within codon 40, but lacked the other two mutations found within VZV-Ellen, VZV-Iceland, and VZV 80-2. VZV-VSD was the only strain tested which contained a polymorphism within the cytoplasmic domain of gE. Interestingly, this change within codon 603 (G → D) inserted an additional acidic amino acid adjacent to the acidic casein kinase II phosphorylation consensus site of gE. VZV-32 and VZV-VIA were the only strains tested that did not contain gE substitutions when compared to the Dumas strain. Since the D150N mutation previously found in VZV-MSP gE ectodomain was not discovered in any other strain, VZV-MSP gE retained a unique genotype among all currently tested strains and isolates.

bp	119	448	660	1606	1808
aa	T > I	D > N	silent	L > I	G > D
Dumas	CACACCGAT	GTGGACCAA	ATATGTTTA	CTTCTACGA	TTTGGTAAC
LAX2	CACACCGAT	GTGGACCAA	ATATGTTTA	CTTCTACGA	TTTGGTAAC
MSP	CACACCGAT	GTGAACCAA	ATATGTTTA	CTTCTACGA	TTTGGTAAC
Ellen	CACATCGAT	GTGGACCAA	ATATGCTTA	CTTATACGA	TTTGGTAAC
Iceland	CACATCGAT	GTGGACCAA	ATATGCTTA	CTTATACGA	TTTGGTAAC
80-2	CACATCGAT	GTGGACCAA	ATATGCTTA	CTTATACGA	TTTGGTAAC
Oka	CACATCGAT	GTGGACCAA	ATATGTTTA	CTTCTACGA	TTTGGTAAC
LAX1	CACATCGAT	GTGGACCAA	ATATGTTTA	CTTCTACGA	TTTGGTAAC
VSD	CACACCGAT	GTGGACCAA	ATATGTTTA	CTTCTACGA	TTTGATAAC
32	CACACCGAT	GTGGACCAA	ATATGTTTA	CTTCTACGA	TTTGGTAAC
VIA	CACACCGAT	GTGGACCAA	ATATGTTTA	CTTCTACGA	TTTGGTAAC

Figure 7. Polymorphisms in the VZV gE gene. The gE genes from 10 different VZV strains were sequenced. All sequences were compared with the prototypic VZV-Dumas sequence. Polymorphisms are underlined.

6.3   Polymorphisms in the VZV gI gene

The discovery of several polymorphisms in the VZV gE gene of the 10 strains was unexpected because of the presumption that the VZV genome was inherently stable and therefore not subject to nonsynonymous mutation. Since VZV gE and gI proteins are commonly found in a complex in the infected cell culture, the gI gene was the next obvious candidate for further genetic analysis. Sequencing of ORF 67 led to the discovery of two changes from the published Dumas sequence (Figure 8). VZV-32 and VZV-VIA had an A to C substitution at bp 15 that resulted in a glutamine to histidine substitution; this mutation occurred in the signal sequence. VZV-Oka also had a silent change at bp 546. In short, VZV ORF 67 had fewer polymorphisms than ORF 68, possibly reflecting the role of gI as a chaperone protein.

bp	15	546
aa	Q > H	silent
Dumas	ATCCAATGT	TCTCCGTCT
LAX2	ATCCAATGT	TCTCCGTCT
MSP	ATCCAATGT	TCTCCGTCT
Ellen	ATCCAATGT	TCTCCGTCT
Iceland	ATCCAATGT	TCTCCGTCT
80-2	ATCCAATGT	TCTCCGTCT
Oka	ATCCAATGT	TCTCCATCT
LAX1	ATCCAATGT	TCTCCATCT
VSD	ATCCAATGT	TCTCCGTCT
32	ATCCACTGT	TCTCCGTCT
VIA	ATCCACTGT	TCTCCGTCT

Figure 8. Polymorphisms in the VZV gI gene. The gI genes from 10 different strains were sequenced. All sequences were compared to the prototypic VZV-Dumas sequence. Polymorphisms are underlined.

6.4   Summation

In the previous sections of the chapter, the diverse functions of the VZV gE/gI complex have been examined. The endocytosis property was given particular emphasis because of the unusual juxtaposition of a tyrosine based internalisation signal in gE and a dileucine motif in gI. Interactions between the two components of the gE/gI complex can occur between endodomains or ectodomains. Phosphorylation reactions with casein kinase II and a cyclin dependent kinase may provide cross talk within the VZV gE/gI complex. The discovery of the first VZV gE escape mutant virus was unexpected, as was the subsequent observation that the mutant virus exhibited an accelerated cell-to-cell spread phenotype (Santos et al., 2000). Whether the mutant VZV-MSP is also more virulent is an intriguing question requiring further study. Finally, the genetic analyses to define nucleotide polymorphisms in gE and gI suggest that the gE gene, in particular, is not as stable as previously thought (Faga et al., 2001). In short, the pluripotential VZV gE/gI complex plays a central role in the life cycle of this herpesvirus, and any mutation of the complex may dramatically alter the virus phenotype.

ACKNOWLEDGMENTS

I would like to thank all the undergraduate and graduate students, postdoctoral fellows, research associates and collaborating scientists who have generously participated in the virology research and have been included in the cited publications from my laboratory over the past years. The research has been supported primarily by grants from the National Institutes of Health (Bethesda). Additional support has been received from the VZV Research Foundation (New York City).

REFERENCES

Abramson, A.W., 1944, Varicella and herpes zoster: an experiment. *Br. Med. J.* 1:812-13

Alconada, A., Bauer, U., and Hoflack, B., 1996, A tyrosine-based motif and a casein kinase II phosphorylation site regulate the intracellular trafficking of the varicella-zoster virus glycoprotein I, a protein localized in the trans-Golgi network. *Embo J.* 15:6096-6110

Arvin, A.M., Kinney-Thomas, E., Shriver, K., Grose, C., Koropchak, C.M., Scranton, E., Wittek, A.E., and Diaz, P.S., 1986, Immunity to varicella-zoster viral glycoproteins, gpI (gp90/58) and gpIII (gp118), and to nonglycosylated protein. *J. Immunol.* 137:1346-1351

Bergen, R.E., Sharp, M., Sanchez, A., Judd, A.K., and Arvin, A.M., 1991, Human T cells recognize multiple epitopes on an immediate early/tegument protein (IE62) and glycoprotein I of varicella zoster virus. *Viral Immunol.* 4:151-166

Bosc, D.G., Slominski, E., Sichler, C., and Litchfield, D.W., 1995, Phosphorylation of casein kinase II by p34 cdc2. *J. Biol. Chem.* **270**:25872-25878

Bremnes, B., Madsen, T., Gedde-Dahl, M., and Bakke, O., 1994, An LI and ML motif in the cytoplasmic tail of the MHC-associated invariant chain mediates rapid internalization. *J. Cell Scienc.* **107**:2021-2032

Brookes, A.J., 1999, The essence of SNPs. *Gene.* **234**:177-186

Cohen, J. I., and Nguyen, H., 1997, Varicella-zoster virus glycoprotein I is essential for growth of virus in Vero cells. *J. Virol.* **71**:6913-6920

Collawn, J.F., Stangel, M., Kuhn, L.A., Esekogwu, V., Jing, S., Trowbridge, I.S., and Trainer, J.A., 1990, Transferrin receptor internalization sequence YXRF implicates a tight turn as the structural recognition motif for endocytosis. *Cell* **63**:1061-1072

Collins, F.S., Guyer, M.S., and Chakravarti, A., 1997, Variations on a theme: Cataloguing human DNA sequence variation. *Science* **278**:1580-1581

Davison, A.J., and Scott, J.E., 1986, The complete DNA sequence of varicella-zoster virus. *J. Gen. Virol.* **67**:1759-1816

Dingwell, K.S., Brunetti, C.R., Hendriks, R.L., Tang, Q., Tang, M., Rainbow, A.J., and Johnson, D.C., 1994, Herpes simplex virus glycoproteins E and I facilitate cell-to-cell spread in vivo and across junctions of cultured cells. *J. Virol.* **68**:834-845

Duus, K., and Grose, C., 1996, Multiple regulatory effects of varicella-zoster virus gL on trafficking patterns and fusogenic properties of VZV and gH. *J. Virol.* **70**:8961-8971

Edson, C.M., 1993, Tyrosine sulfation of varicella-zoster virus envelope glycoprotein gpI. *Virology* **197**:159-165

Edson, C.M., Hosler, B.A., Poodry, C.A., Schooley, R.T., Waters, D.J., and Thorley-Lawson, D.A., 1985, Varicella-zoster virus envelope glycoproteins: biochemical characterization and identification in clinical material. *Virology* **145**:62-71

Faga, B., Maury, W., Bruckner, D.A., and Grose, C., 2001, Identification and mapping of single nucleotide polymorphisms in the varicella-zoster virus genome. *Virology* **280**:1-6

Gershon, A., Cosio, L., and Brunell, P.A., 1973, Observations on the growth of varicella-zoster virus in human diploid cells. *J. Gen. Virol.* **18**:21-31

Grose, C., 1980, The synthesis of glycoproteins in human melanoma cells infected with varicella-zoster virus. *Virology* **101**:1-9

Grose, C., 1990, Glycoproteins encoded by varicella-zoster virus: biosynthesis, phosphorylation, and intracellular trafficking. *Ann. Rev. Microbiol.* **44**:59-80

Grose, C., 1991, Glycoproteins of varicella-zoster virus and their herpes simplex virus homologs. *Rev. Infect. Dis.* **13**:S960-963

Grose, C., 1999, Varicella-zoster virus: Less immutable than once thought. *Pediatrics* **103**:1027-1028

Grose,C., Edmond, B.J., and Friedrichs, W.E., 1981, Immunogenic glycoproteins of laboratory and vaccine strains of varicella-zoster virus. *Infect. Immun.* **31**:1044-1053

Grose, C., and Friedrichs, W.E., 1982, Immunoprecipitable polypeptides specified by varicella-zoster virus. *Virology* **118**:86-95

Grose, C., Friedrichs, W.E., and Smith, G.C., 1983, Purification and molecular anatomy of the varicella-zoster virion. *Biken J.* **26**:1-15

Grose, C., Jackson, W., and Traugh, J.A., 1989, Phosphorylation of varicella-zoster virus glycoprotein gpI by mammalian casein kinase II and casien kinase I. *J. Virol.* **63**:3912-3918

Harper, D.R., and Kangro, H.O., 1990, Lipoproteins of varicella-zoster virus. *J. Gen. Virol.* **71**:459-463

Harson, R., and Grose, C., 1995, Egress of varicella-zoster virus from the melanoma cell: a tropsim for the melanocyte. *J. Virol.* **69**:4994-5010

Hatfield, C., Duus, K.M., Jones, D.J., and Grose, C., 1997, Epitope mapping and tagging by recombination PCR mutagenesis.*BioTechniques* **22**:332-337

Hope-Simpson, R.E., 1965, The nature of herpes zoster: a long-term study and new hypothesis. *Proc. Roy. Soc. Med. (Lond.).* **58**:9-20

Ito, M., Ihara, T., Grose, C., and Starr, S., 1985, Human leukocytes kill varicella-zoster virus infected fibroblasts in the presence of murine monoclonal antibodies to virus-specific glycoproteins. *J.Virol.* **54**:98-102

Jones, F., and Grose, C., 1988, Role of cytoplasmic vacuoles in varicella-zoster virus glycoprotein trafficking and virion envelopment. *J. Virol.* **62**:2701-2711

Kimura, H., Straus, S.E., and Williams, R.K., 1997, Varicella-zoster virus glycoproteins E and I expressed in insect cells form a heterodimer that requires the N-terminal domain of glycoprotein I. *Virology* **233**:382-391

Kishimoto, A., Brown, M.S., Slaughter, C.A., and Goldstein, J.L., 1987, Phosphorylation of serine 833 in cytoplasmic domain of low intensity lipoprotein receptor by a high molecular weight enzyme resembling casein kinase II. *J. Biol. Chem.* **262**:1344-1351

Letourneur, F., and Klausner, R.D., 1992, A novel di-leucine motif and a tyrosine-based motif independently mediate lysosomal targeting and endocytosis of CD3 chains. *Cell* **69**:1143-1157

Litchfield, D.W., Bosc, D.G., and Slominski, E., 1995, The protein kinase from mitotic human cells that phosphorylates Ser-209 on the casein kinase II β-subunits is p34 cdc2. *Biochim. Biophys. Acta* **1269**:69-78

Ltiwin, V., and Grose, C., 1992, Herpesviral Fc receptors and their relationship to the human Fc receptors. *Immunologic Research* **11**:226-38

Litwin, V., Jackson, W., and Grose, C., 1992, Receptor properties of two varicella-zoster virus glycoproteins gpI and gpIV homologous to herpes simples virus gE and gI. *J. Virol.* **66**:3643-51

Litwin, V., Sandor, M., and Grose, C., 1990, Cell surface expression of the varicella-zoster virus glycoproteins and Fc receptor. *Virology* **178**:263-272

Mallory, S., Sommer, M., and Arvin. A.M., 1997, Mutational analysis of the role of glycoprotein I in varicella-zoster virus replication and its effects on glycoprotein E conformation and trafficking. *J. Virol.* **71**:8279-8288

Montalvo, E.A., and Grose, C., 1986, Varicella zoster virus glycoprotein gpI is selectively phosphorylated by a virus-induced protein kinase.*Proc. Nat. Acad. Sci. USA.* **83**:8967-8971

Montalvo, E.A., Parmley, R.T., and Grose, C., 1985, Structural analysis of the varicella-zoster virus gp98-gp62 complex: Posttranslational addition of N-linked and O-linked oligosaccharide moieties. *J. Virol.* **53**:761-770

Namazue, J., Kato, T., Okuno, T., Shiraki, K,. and Yamanishi, K., 1989, Evidence for attachment of fatty acid to varicella-zoster virus glycoproteins and effect of cerulenin on the maturation of varicella-zoster virus glycoproteins. *Intervirology.* **30**:268-277

Ng, T., and Grose, C., 1992, Serine protein kinase associated with varicella-zoster virus ORF 47. *Virology* **191**:9-18

Ng, T., Keenan, L., Kinchington, P.R., and Grose, C., 1994, Phosphorylation of varicella-zoster virus open reading frame (ORF) 62 regulatory product by viral ORF 47 associated protein kinase. *J. Virol.* **68**:1350-1359

Olson, J., Bishop, G., and Grose, C., 1997, Varicella-zoster virus Fc receptor gE glycoprotein: Serine/threonine and tyrosine phosphorylation of monomeric and dimeric forms. *J. Virol.* **71**:110-119

Olson, J.K., and Grose, C., 1997, Endocytosis and recycling of varicella-zoster virus Fc receptor glycoprotein gE: internalization mediated by YXXL motif in the cytoplasmic tail.

J. Virol. **71**:4042-4054

Olson, J.K., and Grose, C., 1998, Complex formation facilitates endocytosis of the varicella-zoster virus gE:gI Fc receptor. *J Virol.* **72**:1542-1551

Santos, R.A., Hatfield, C.C., Cole, N.L., Padilla, J.A., Moffat, J.F., Arvin, A.M., Ruyechan, W.T., Hay, J., and Grose, C., 2000, Varicella-zoster virus gE escape mutant VZV-MSP exhibits an accelerated cell-to-cell spread phenotype in both infected cell cultures and SCID-hu mice. *Virology* **275**:306-317

Santos, R.A., Padilla, J.A., Hatfield, C.C., and Grose, C., 1998, Antigenic variation of caricella zoster virus Fc receptor gE: loss of a major B cell epitope in the ectodomain. *Virology* **249**:21-31

Takahashi, M., Okuno, Y., Otsuka, T., Osame, J., Takamizawa, A., Sasada, T., and Kubo, T., 1975, Development of a live attenuated varicella vaccine. *Biken J.* **18**:25-33

Trowbridge, I.S., Collawn, J.F., and Hopkins, C.R., 1993, Signal-dependent trafficking in the endocytic pathway. *Ann. Rev. Cell Biol.* **9**:129-161

Tugizov, S., Maidji, E., Xiao, J., Zheng, Z., and Pereira, L., 1998, Human cytomegalovirus glycoprotein B contains autonomous determinants for vectorial targeting to apical membranes of polarized epithelial cells. *J. Virol.* **72**:7374-7386

Vafai, A., Wroblewska, Z., Mahalingham, R, Cabirac, G., Wellish, M., Cisco, M., Gilden D. Recognition of similar epitopes on varicella-zoster virus gpI and gpIV by monoclonal antibodies. J. Virol. 62: 2544-2551.

Wang, D.G., Fan, J.B., Siao, C.J., Berno, A., Young, P., Sapolsky, R., Ghandour, G., Perkins, N., Winchester, E., Spencer, J., Kruglyak, L., Stein, L., Hsie, L., Topaloglou, T., Hubbel, E., Robinson, E., Mittmann, M., Morrris, M.S, Shen, N., Kilburn, D., Rioux, J., Nusbaum, C., Rozen, S., Hudson, T.J., Liptshutz, R., Chee, M., and Lander, E.S., 1998, Large-scale identification, mapping, and genotyping of single-nucleotide polymorphisms in the human genome. *Science* **280**:1077-1082

Weigle, K.A., and Grose, C., 1983, Common expression of varicella-zoster virus glycoprotein antigens in vitro and in chickenpox and zoster vesicles. *J. Infect. Dis.* **148**:630-638

Weigle, K.A., and Grose, C., 1984, Molecular dissection of the humoral immune response to individual varicella-zoster virus proteins during chickenpox, quiescence, reexposure, and reactivation. *J. Infect. Dis.* **149**:741-749

Weller, T.H., 1953, Serial propagation in vitro of agents producing inclusion bodies derived from varicella and herpes zoster. *Proc. Soc. Exp. Biol. Med.* **83**:340-346

Yao, Z., and Grose, C., 1994, Unusual phosphorylation sequence on the gpIV (gI) component of the varicella-zoster virus (VZV) gpI-gpIVglycoprotein complex (VZV gE-gI complex). *J. Virol* **68**:4202-4211

Yao, Z., Jackson, W., Forghani, B., and Grose, C., 1993, Varicella-zoster virus glycoprotein gpI/gpIV receptor: Expression, complex formation, and antigenicity within the vaccinia virus –T7 RNA polymerase transfection system. *J. Virol.* **67**:305-314

Yao, Z., Jackson, W., and Grose, C., 1993, Identification of the phosphorylation sequence in the cytoplasmic tail of the varicella-zoster virus Fc receptor glycoprotein gpI. *J. Virol.* **67**:4464-4473

Yao, Z., Jones, D.H., and Grose, C., 1992, Site-directed mutagenesis of herpesvirus glycoprotein phosphorylation sites by recombination polymerase chain reaction. *PCR Methods Appl.* **1**:205-207

Ye, M., Duus, K.M., Peng, J., Price, D.H., and Grose, C., 1999, Varicella-zoster virus Fc receptor component gI is phosphorylated in its endodomain by a cyclin dependent kinase. *J. Virol.* **73**:1320-1330

Zhu, Z., Gershon, M.D., Hao, Y., Ambron, R.T., Gabel, C.A., and Gershon, A.A., 1995, Envelopment of varicella-zoster virus: targeting of viral glycoproteins to the trans-Golgi network. *J. Virol.* **69**:7951-7959

Zhu, Z., Hao, Y., Gershon, M.D., Ambron, R.T., and Gershon, A.A., 1996, Targeting of glycoprotein I (gE) of varicella-zoster virus to the trans-Golgi network by an AYRV sequence and an acidic amino-acid rich patch in the cytosolic domain of the molecule. *J. Virol.* **70**:6563-6575

Chapter 4.3

Expression Strategy and Functions of the Filoviral Glycoproteins

VIKTOR E. VOLCHKOV[1] and HEINZ FELDMANN[2]

[1] *Biologie des Filovirus, Claude Bernard University Lyon-1, 21, Avenue Tony Garnier, 69007, France* [2] *Canadian Science Centre for Human and Animal Health, 1015 Arlington Street, Winnipeg, Manitoba R3E 3R2, Canada*

1. INTRODUCTION

Infections with filoviruses cause a fulminate hemorrhagic disease in human and non-human primates. Among all viral hemorrhagic fevers, Marburg and Ebola infections are characterised as the most severe forms with case-fatality rates ranging from 22%-90%. Infection results in a rapidly progressive, severe illness characterised by bleeding, capillary leakage, multisystem dysfunction, and a shock-like state culminating in death (Bwaka et al., 1999; Peters and LeDuc, 1999). Molecular mechanisms that induced devastating pathophysiological changes in infected patients however are not well understood as yet. Morphological studies on experimentally infected animals (Geisbert et al., 1992; Ryabchikova et al., 1996 and 1999) showed that monocytes/macrophages and fibroblasts may be the preferred sites of virus replication in early stages of infection, whereas other cell types become also involved as the diseases progresses. Studies performed in cell cultures confirmed theses observations and showed that filoviruses are pantropic, exhibited a very broad host range, and infect a variety of different cell lines from multiple species

Structure-Function Relationships of Human Pathogenic Viruses, Edited by
Holzenburg and Bogner, Kluwer Academic/Plenum Publishers, New York, 2002

and tissue types (van der Groen et al., 1978). Filovirus infections cause a moderate cytopathogenic effect in target cells leading to cell lysis. This suggests that cytopathic effects induced by virus replication may be an important pathophysiological parameter. Clinical and biochemical findings support the anatomical observations of extensive liver involvement, renal damage, changes in vascular permeability including endothelial dysfunction, fluid distribution problems and activation of the clotting cascade (Murphy et al., 1978; Fisher-Hoch et al., 1985). Although the studies on infected non-human primates did not identify endothelial cells as sites of massive virus replication, post-mortem observations of human cases and in vitro studies demonstrated that endothelial cells of human origin are suitable targets for virus replication (Schnittler et al., 1999). Besides evidences for direct vascular damage caused by virus replication, active mediator molecules released from infected monocytes/macrophages seems also play an important role in the pathogenesis of the filovirus infection. It has been demonstrated that supernatants of filovirus-infected monocytes/ macrophages cultures are capable of increasing paraendothelial permeability in an vitro model (Feldmann et al., 1996). Significant increase in the level of secreted TNF-α, the prototype cytokine of macrophages, and MCP-1 have been observed in supernatants of monocytes\macrophages infected with filoviruses (Schnittler and Feldmann, 1999, Gupta et al., 2001a). Evidences for the involvement of inflammatory mediators in the pathogenesis of Ebola virus infection have been also obtained in the study of both symptomatic and asymptomatic infected individuals who have been found to have increased levels of circulating TNF-α, IL-1β, IL-6, MIP-1α, and MCP-1 (Leroy et al., 2000; Villinger et al., 1999). The combination of the virus-induced cytokine release from infected monocytes/macrophages and viral replication in fibroblasts and endothelial cells provides an explanation for a distinct proinflammatory endothelial phenotype that could trigger the coagulation cascade and later induce fluid distributions problems. All data available to date strongly support the concept of a mediator-induced vascular instability and thus increased permeability as a key mechanism for the development of the shock syndrome seen in severe and fatal cases. A model summarising these pathophysiological events is illustrated in Figure 1.

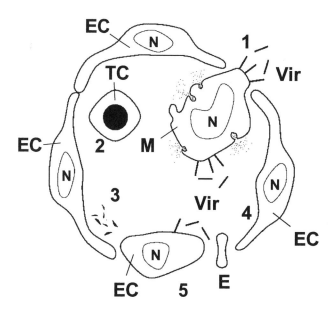

Figure 1: Schematic drawing illustrating the possible role of macrophages and endothelial cells in the development of hemorrhagic fever caused by filoviruses. EC, endothelial cell; M, macrophage; Vir, virus particles; E, erythrocyte; TC, T cells; N, nucleus. 1, monocytes/macrophages are the preferred sites of virus replication in early stages of infection, whereas other cell types may become involved as the disease progresses. Macrophages play an important role in spread of the virus through the organism; 2, massive apoptotic death of blood leucocytes that, most likely, is induced by uncontrolled release of different cytokines from virus-infected antigen presenting cells; 3, infection of the mononuclear phagocytes and cells of fibroblastic reticular system could results in widespread fibrin deposition in the blood vessels; 4, vascular instability and increased permeability is a key mechanism for the development of hemorrhages and shock syndrome seen in severe and fatal cases; 5, strong cytotoxic effect in endothelial cells induced by viral replication may promote the coagulation cascade and can also explain the hemorrhagic manifestation characteristic of filovirus infection. Taken from Feldmann and Klenk, 1996 with modifications.

Filoviruses are enveloped, nonsegmented negative-stranded RNA viruses and constitute a separate family within the order Mononegavirales. The family consists of the genera *Marburg virus* (MBGV) and *Ebola virus* (EBOV). The genus *Ebola virus* is further subdivided into three distinct African species, Cote d'Ivoire, Sudan and Zaire, and the Asian species Reston (van Regenmortel et al. 2000). Filoviruses consist of a single negative-stranded linear RNA genome which is non-infectious and does not contain a poly(A) tail. Upon entry into the cytoplasm of host cells is transcribed to generate polyadenylated subgenomic mRNA species. The genome shows the following characteristic gene order: 3' leader - nucleoprotein (NP) - virion protein (VP) 35 - VP40 - glycoprotein (GP) - VP30 - VP24 - polymerase (L) protein   - 5' trailer. Transcription and translation lead to the synthesis of seven structural polypeptides with presumed identical functions for the different filoviruses. Four proteins are associated with the viral genomic RNA in the ribonucleoprotein complex: NP, VP30, VP35, and the L protein. $GP_{1,2}$ is a type I transmembrane protein, that forms surface spikes. The two remaining structural proteins VP24 and VP40 are membrane-associated and are probably located at the inner side of the membrane. A non-structural, secreted glycoprotein (sGP) is expressed by EBOV but not MBGV (Feldmann & Kiley, 1999; Volchkov, 1999).

Despite the fact that all viral components may contribute to viral pathogenesis, the glycoproteins of filoviruses are considered to be the key players.

2.   FILOVIRAL GLYCOPROTEINS

2.1   Expression Strategies of the Glycoprotein Genes

Filoviral glycoproteins are encoded by gene four (GP gene) of the nonsegmented negative-strand RNA genome (Fig. 2). MBGV gene four encodes for a single open reading frame that translates into the transmembrane glycoprotein GP (Feldmann et al., 1992; Will et al., 1993; Bukreyev et al., 1995, Sanchez et al., 1998a). In contrast, the expression strategy of gene four of all EBOVs involves transcriptional RNA-editing and gives rise to different glycosylated proteins (Volchkov et al., 1995 and 1999; Sanchez et al., 1996). The primary structure of the editing site is a run of seven uridine residues on the genomic sequence. Here, transcriptional editing is performed by the viral RNA-dependent RNA polymerase (L protein). Unedited GP gene mRNA species (~ 80%) encode for a non-structural secreted glycoprotein (sGP). The transmembrane glycoprotein (GP) is translated from edited GP gene-specific mRNA species, which are the result of the addition of a single adenosine residue at the editing site during transcription (Fig. 2). This event, which shifts the open reading frame into -1, seems to occur in about 20% of the GP gene-specific transcripts (Volchkov et al., 1995; Sanchez et al., 1996). RNA editing has been described for a variety of paramyxoviruses, like simian virus 5 (Thomas et al., 1988), Sendai virus (Vidal et al., 1990a and 1990b), measles virus (Cattaneo et al., 1989), mumps virus (Paterson and Lamb, 1990; Takeuchi et al., 1990), and parainfluenza virus types 2 and 4 (Southern et al., 1990; Ohgimoto et al., 1990; Kondo et al., 1990). With each of these viruses, editing has been shown to occur by the insertion of an additional G residue at the specific sequence 3'-UU^U/$_C$UCCC-5' of the P gene. The P gene of paramyxoviruses is reported to be edited exclusively by viral RNA-dependent RNA polymerases (Horikami et al., 1991; Pelet et al.,1991; Matsuoka et al.,1991; Vidal et al., 1990a and b). In contrast, the editing site of EBOV GP is also recognised by DNA dependent RNA polymerases. The mechanism by which editing of the GP gene occurs, appears to be similar with RNA- and DNA- dependent RNA polymerases, since both types of enzymes insert the same non-template nucleotide (A, mRNA sense) in exactly the same region. The editing site in EBOV genomic RNA (7Us, genomic sense) is different from the editing site found in the P gene of paramyxoviruses, rather resembling the polyadenylation site (transcription stop signal) of EBOV mRNAs (3'-UAAUUCUUUUUU, genomic sense). The EBOV editing site is also similar to the transcription stop signal of the vaccinia virus polymerase (3'-UUUUUNU) (Moss, 1990). Presumably, temporary pausing of the viral RNA polymerase and both investigated DNA-dependent RNA-polymerases at the editing site of EBOV GP gene enables the transcription complex to slip backward or forward on the vRNA template before the next nucleotide is incorporated. This mechanism was described for editing of P gene of paramyxoviruses (Vidal et al., 1990b) and obviously could be used in editing of EBOV GP. In this case the similarity with polyadenylation sites may at least partly explain the relatively broad spectrum of RNA polymerases recognising the editing site of EBOV GP.

Variation at the genomic level of the number of uridine residues at the editing site (8 instead of 7 residues) has been independently described by two research groups (Sanchez et al., 1993; Volchkov et al., 2000a). These variants of EBOV were selected during characterisation of the individual viral clones after

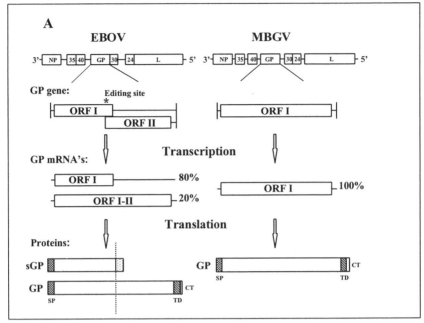

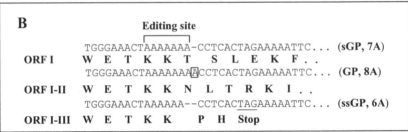

Figure 2: Expression strategies of the glycoprotein (GP) gene of filoviruses. (A) Unlike MBGV, the EBOV transmembrane glycoprotein GP$_{1,2}$ can only be expressed through transcriptional RNA editing. The primary product of the EBOV GP gene is the secreted glycoprotein sGP that is expressed from the unedited transcripts (80%). The second mRNA species encodes transmembrane GP and comprises only about 20% of total GP mRNAs. The corresponding open reading frames (ORF) are shown as boxes. The amino terminal 295 amino acids of sGP are identical with GP, but the last 69 a.a are different. (B) Nucleotide and deduced amino acid sequences of different EBOV GP-specific mRNA's at the editing site. The third mRNA species encoding ssGP had been observed as an editing product of only EBOV-8U variant or when recombinant GP was expressed Abbreviations: SP, signal peptide; TD, transmembrane domain; CT, cytoplasmic tail.

plaque purification. This observation suggests that the editing site is a hot spot for insertion of nucleotides not only at the level of transcription but also of genome replication. Insertion of the uridine residue drastically alters the expression strategy of the GP gene. With either EBOV-8U variants synthesis of the virion $GP_{1,2}$ is facilitated without the need of RNA editing. Whereas sGP is the dominant expression product with wild type virus and GP is synthesised in smaller amounts, the proportions are inverted with EBOV-8U variant (Volchkov et al., 2000a). Furthermore, the insertion of the additional uridine residue substantially changed the efficacy and accuracy of editing compared with wild-type virus. Several GP gene specific proteins not observed with wild type virus were synthesised with EBOV-8U. (Volchkov et al., 1995; Volchkova et al., 1998). Deletion of one or insertion of two adenosine residues allows a switch into a third open reading frame (-2) which terminates 2 amino acids downstream of the editing site and generates a third non-structural secreted glycoproteins (ssGP) (Volchkova et al., 1998; Volchkov et al., 2000a) that was previously observed only in recombinant expression systems. The second secreted glycoprotein (ssGP) of EBOV resembles a natural carboxy-terminal truncated variant of sGP. Due to the lack of three carboxy-terminal cysteine residues including the one involved in oligomerisation of sGP (position 306), ssGP is secreted in a monomeric form (Volchkov et al., 1995; Volchkova et al., 1998).

Recently, a reverse genetics system allowing the generation of infectious Ebola virus from cloned cDNA has been established (Volchkov et al., 2001). This system permits the study of the underlying principles that give rise to the high pathogenicity of Ebola virus. Using the system, all viral genes or regions can now be modified and the impact on virus-cells and/or virus-host interactions can be investigated. In order to understand biological significance of the RNA editing in EBOV replication, a recombinant virus in which the editing site of the GP gene was modified by several mutations has been constructed and rescued (Fig. 3). These substitutions have two major effects: synthesis of sGP is totally blocked and only one type of mRNA's are synthesized from the GP gene encoding only surface GP. Due to the changes in GP gene expression strategy significant increase in synthesis of surface GP was observed that resulted in enhanced virus cytotoxicity and early detachment of infected cells. These results indicate that expression of GP gene of Ebola virus is tightly controlled by RNA editing mechanism which allowed expression of surface GP from only small part of the GP-specific transcripts. It is therefore very likely that editing of the GP gene of EBOV is evolutionary linked with the need to down-regulate cytotoxicity caused by expression of surface GP. It appears also that sGP may

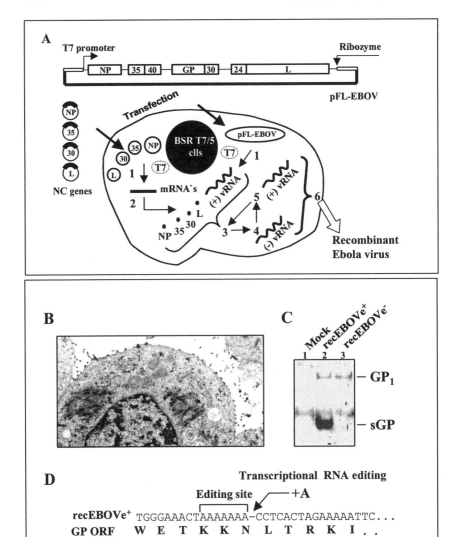

Figure 3: (A) Schematic representation of the rescue of recombinant EBOV. The plasmid construct containing full-length EBOV antigenome sequence (the Gene Bank accession number AF086833) pFL-EBOV has the following features: T7 promoter is adjacent to the viral leader

region, and the viral trailer region is constructed to be adjacent to a ribozyme sequence followed by tandem terminators of T7 transcription. In this case, the correct 3'end of the transcribed EBOV antigenome, free of additional nucleotides, is generated by self-cleavage of the ribozyme. To rescue recEBOV cells were transfected with plasmids encoding EBOV antigenome and nucleocapsid proteins NP, VP35, VP30, and L (NC genes). 1, synthesis of the EBOV specific mRNA's and full-length EBOV antigenomic RNA by using T7 polymerase stably expressed in BSR T7/5 cells; 2, synthesis of the EBOV nucleocapside proteins NP, VP35, VP30 and L polymerase; 3, encapsidation of the antigenomic EBOV (+)vRNA by nucleocapside proteins; 4, synthesis of the full-length genomic (-) vRNA ; 5, synthesis of the viral antigenome (+) vRNA and virus specific mRNA's; 6, assembly and budding of the recombinant EBOV.

(B) Electron micrograph of Vero cells infected with recombinant EBOV. (C) Synthesis of the sGP is totally blocked with recEBOVe⁻. (D) The sequences at the editing site of recEBOVe⁺ and recEBOVe⁻ are indicated. The editing site has been changed so that two adenosines were substituted by guanosines to interrupt the poly-adenosine sequence and an additional adenosine was inserted to link the two overlapping reading frames encoding GP.

not be essential for the replication of EBOV in cell culture but may play a role as a biologically active protein in infection in humans or in the yet unknown natural host. Confirmation of these hypotheses must however await further studies. In this respect, it is very interesting to investigate the pathogenicity of the "no editing-no sGP" EBOV mutant in animal models.

2.2   Biosynthesis, Processing and Maturation of Glycoproteins

2.2.1.   Transmembrane Glycoprotein (GP). The open reading frames for

the transmembrane glycoproteins (GP) of MBGV, strains Musoke and Popp, and EBOV, strain Mayinga, encode for 681 and 676 amino acids in length, respectively. They are type I transmembrane proteins and can be subdivided into a large ectodomain, a lipid membrane spanning domain of approximately 30 amino acids, and a short cytoplasmic tail of 4 (EBOV) and 8 amino acids (MBGV). GP undergoes a complex sequence of processing events in the endoplasmic reticulum. This includes the removal of the signal peptide (Will et al., 1993; Sanchez et al., 1998b), N-glycosylation (Feldmann et al., 1991 and 1994; Volchkov et al., 1995 and 1998; Becker et al., 1996; Sanchez et al., 1998), and oligomerization. ER processing is followed by acylation in a pre-Golgi compartment (Funke et al., 1995; Ito et al., 2001), and by O-glycosylation and maturation of N-glycans in the Golgi apparatus (Feldmann et al., 1991, 1994;

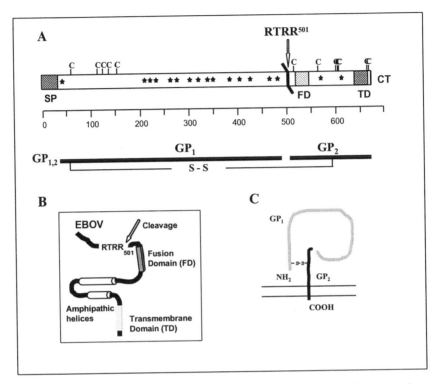

Figure 4: (A) The transmembrane glycoprotein GP of EBOV. The type I transmembrane glycoprotein GP carries three hydrophobic domains (gray boxes): a signal peptide (SP) at the amino-terminal end, a fusion domain (FD), and a transmembrane domain (TD) at the carboxy-terminal end. The glycoprotein precursor is proteolytically cleaved into the subunits GP_1 and GP_2. Arrows indicate the cleavage sites. Both subunits are disulfide-linked in the mature molecule. Potential N-linked carbohydrate sites are indicated by asterisks; 'C', cysteine residue; S-S - disulfide bridge. (B) Proposed structure of GP_2. The ectodomain of GP_2 contains the fusion peptide, followed by an amino-terminal helix, a peptide loop, and a carboxy-terminal helix. Helices were proposed by the GARNIER program of PC/GENE (IntelliGenetics Inc.) (Volchkov et al., 1993; Weissenhorn et al., 1998a and b). (C) A model for the structure of the mature GP monomer.

Geyer et al., 1992; Will et al., 1993; Becker et al., 1996). Depending on the host cell, there are wide variations in the amount of neuraminic acid present on MBGV GP (Feldmann et al., 1994). Finally, GP gets proteolytically cleaved into a larger amino-terminal (GP_1) and smaller carboxy-terminal (GP_2) subunit in the trans-Golgi network by the subtilisin-like proprotein convertases (Volchkov et al., 1998a and 2000b) (Fig. 4).

The mature envelope glycoprotein ($GP_{1,2}$) forms spikes on the surface of virions and virus-infected cells. It is anchored in the membrane by a carboxy-terminal hydrophobic domain of GP_2 (Volchkov et al., 1998b) (Fig. 4). The middle region of $GP_{1,2}$ is variable, extremely hydrophilic, and carries the bulk of N- and O-glycans that account for more than one third of the molecular weight of the mature protein (Geyer et al., 1992; Will et al., 1993; Feldmann et al., 1991; Volchkov et al., 1995; Becker et al., 1996). Oligosaccharide side chains differ in their terminal sialylation patterns, which seem to be isolate- as well as cell line-dependent (Feldmann et al., 1994). Comparison of the GP sequences of MBGV and EBOV shows conservation at the amino-terminal and carboxy-terminal ends. Two carboxy-terminal cysteine residues are acylated (Funke et al., 1995, Ito et al., 2001). GP_2 contains a sequence of several uncharged and hydrophobic amino acids at a distance of 22 (EBOV) or 91 (MBGV) amino acids from the cleavage site which bears some structural similarity to the fusion peptides of retroviruses (Volchkov et al., 1992 and 2000b; Gallaher, 1996) (Fig. 4).

The special arrangement of the cysteine residues in the molecule allows an intramolecular disulfide bond formation between the two cleavage products which suggests a stem region consisting of GP_1 and GP_2 and a crown-like domain on the top formed by GP_1 carrying the mass of the carbohydrate side chains (Fig. 4). Mature $GP_{1,2}$ is a trimer consisting of disulfide-bonded $GP_{1,2}$ molecules (Feldmann et al., 1991, Sanchez et al., 1998b). X-ray crystallography demonstrated that the central structural feature of the GP_2 ectodomain is a long triple-stranded coiled coil followed by a disulfide-bonded loop which reverses the chain direction and connects to a helix packed antiparallel to the core helices (Fig. 4B) (Weissenhorn et al., 1998a, b, Malashkevich et al., 1999). Although, direct experimental evidences are not available yet, it can be assumed that cysteine residue 53 is also critical in maintaining the structure of $GP_{1,2}$ as has been shown for sGP (Volchkova et al., 1998).

Significant amount of GP is released into the culture medium in non-virion forms. Part of the released GP represents intact $GP_{1,2}$ complexes incorporated into virosome-like particles (Volchkov et al., 1998). The spikes on these vesicles are morphologically indistinguishable from virion spikes indicating that assembly of spikes is independent of other viral proteins. Further studies have to clarify if GP expression triggers a specific mechanism which is involved in virus budding or that the release of virosomes is resulted from a cytopathic effect caused by EBOV. Other part of the non-virion GP in the medium is soluble GP_1 subunit that is shed after release of its disulfide linkage to the transmembrane subunit GP_2. (Fig. 5). The mechanism of the release of GP_1, which seem to be cell type-dependent is not well understood. One possible explanation of this phenomenon is ineffective bonding between cysteine residues of the two GP subunits in the endoplasmic reticulum. This allows GP_1 to be released from the transmembrane GP_2 during GP transport through the late Golgi where proteolytic processing takes place (Volchkov et al., 1998b).

In addition, a significant amount of truncated $GP_{1,2}$ ($\Delta_{Tm}GP_{1,2}$) is shed into the culture medium from the surface of infected cells as soluble trimeric molecules (O. Dolnik , unpublished data). This shedding is mediated by an endoprotease (secretase) cleavage of surface GP close to the transmembrane spanning domain of GP_2 subunit (Fig. 5). Structural similarity between virion $GP_{1,2}$ and shed $\Delta_{Tm}GP_{1,2}$ molecules suggests that this form of GP might have a decoy function during infection in humans or the still unknown natural host.

2.2.2. **Non-structural Secreted Glycoprotein (sGP).** Precursor of the non-structural glycoprotein sGP of EBOV, has a length of 364 amino acids (EBOV, species Zaire) and shares the amino-terminal 295 amino acids with the transmembrane glycoprotein (GP) (Fig. 2A). The different carboxy- terminus (69 amino acids) contains several charged residues as well as three conserved cysteine residues (Fig. 6). As with GP, sGP undergoes several co- and post-translational processing events, such as signal peptide cleavage, glycosylation, oligomerization and proteolytic cleavage (Volchkova et al., 1998 & 1999). The limiting step during maturation and transport seems to be oligomerisation in the ER (Volchkova et al., 1998). Once oligomerization has taken place, sGP is transported into the Golgi compartment where glycosylation is completed and posttranslational cleavage into the mature form (sGP) and a small peptide, called Δ-peptide, occurs (Volchkova et al., 1999). Cleavage is mediated by cellular subtilisine-like endoproteases, which are also responsible for cleavage of the transmembrane GP (Volchkov et al., 1998a) (Fig. 4). Due to a lack of a transmembrane anchor mature sGP is efficiently secreted from infected cells, a process which is independent on proteolytic cleavage (Volchkova et al., 1999). sGP appears as a disulfide-linked homodimer which shows an antiparallel

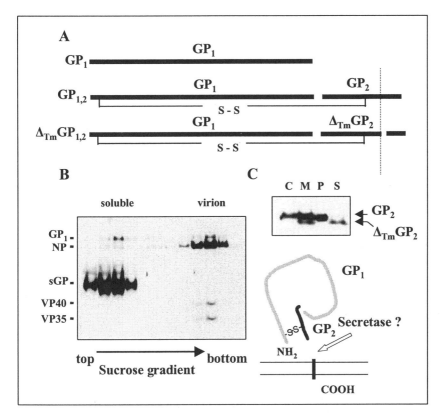

Figure 5: Schematic representation of different forms of GP released into the culture medium from EBOV-infected cells. GP_1 subunit is released in a soluble form as a monomer. $GP_{1,2}$ trimeric complexes are released with virosome-like particles containing spikes indistinguishable from the spikes on EBOV virions. Soluble trimeric GP shed into the culture medium from the surface of infected cells after proteolytic cleavage of GP_2 by unknown cellular endoproteases. (B) Release of EBOV glycoproteins from virus-infected cells. Vero E6 cells were infected with EBOV at 0.01 p.f.u. per cell. The culture medium was harvested at 5 days post-infection, clarified by low-speed centrifugation and subjected to sucrose equilibrium gradient analysis. Proteins in gradient fractions were analysed by SDS–PAGE followed by immunoblotting using anti-EBOV immunoglobulins. In addition to sGP, significant amount of soluble GP is released into the medium. (C) Soluble GP in the medium represents G1,2 complexes with a truncated GP2 subunit. Abbreviations: $\Delta_{Tm}GP_{1,2}$, $GP_{1,2}$ complexes containing truncated GP_2 subunits; C, cells; M, medium; P, pellet after sedimentation of membrane-associated GP by ultracentrifugation; S, supernatant after ultracentrifugation, contains soluble $GP_{1,2}$ complexes with truncated GP2 subunit.

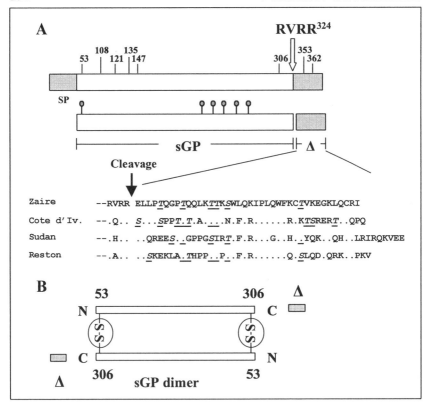

Figure 6: (A) Schematic representation of EBOV sGP precursors and proteolytically processed products, and amino acid sequence alignment of the Δ-peptides of four subtypes of EBOV. Signal peptide and the carboxy-terminal Δ-peptide are indicated by gray boxes. Threonine and serine residues, putative sites for O-glycosylation are underlined. Cysteine residues involved in sGP oligomerization are shown as S-S. Potential N-linked glycosylation sites are indicated. The cleavage site at position 324 is indicated by an arrow. (B) Schematic structure of an sGP dimer, which comprise antiparallel-orientated monomers stabilized by intermolecular disulfide bonds between cysteine residues at positions 53 and 306.

orientation of monomers (Fig. 6B) (Volchkova et al., 1998; Sanchez et al., 1998b). Dimerisation is due to an intermolecular disulfide linkage between the amino and carboxy-terminal cysteine residues at positions 53 and 306, respectively. The remaining four highly conserved cysteine residues at the amino-terminus seem to be involved in intramolecular folding of monomers

(Volchkova et al., 1998). Δ-peptide, the small cleavage product of the Golgi sGP precursor pre-sGP, varies in length between 40 and 48 amino acids for the different EBOVs (Volchkova et al., 1999) (Fig. 6). Its molecular mass of approximately 10 to 14 kDa is significantly larger than the one predicted from the amino acid sequence (~ 4.7 kDa). The difference is due to the attachment of several O-glycans that carry terminal sialic acids. In this respect it differs from sGP which seems to mainly carry N-linked carbohydrates. Δ-peptide is partially secreted from cells and this process seems to be not as effective as secretion of mature sGP (Volchkova et al., 1999). In particularly, this allows to speculate that while being secreted Δ-peptide may play any intracellular role. However the exact functions of the Δ-peptide as well as its significance with regards to sGP cleavage remain to be established. Analyses of EBOV variants carrying mutation in the sGP cleavage site and thus expressing noncleaved sGP are under way.

2.3. Potential Role of Glycoproteins in Filovirus Pathogenesis

2.3.1. Cell tropism, entry and virus spread. Transmembrane glycoprotein GP of filoviruses mediates receptor binding and subsequent fusion with susceptible cells. It was shown by several groups using vesicular stomatitis virus and retrovirus pseudotypes carrying EBOV or MBGV surface glycoprotein (Takada et al., 1997; Wool-Levis & Bates, 1998; Yang et al., 1998; Chan et al., 2000b). There is evidence that MBGV uses the asialoglycoprotein receptor to infect hepatocytes (Becker et al., 1995). Recently it was suggested that integrins, especially the β1 group, might interact with the EBOV glycoprotein GP intracellularly and perhaps be involved in EBOV entry into cells (Takada et al., 2000). More recent studies indicate that the folate receptor-α (FR-α) serves as a cofactor for cellular entry of MBGV and EBOV (Chan et al., 2001). It has also been demonstrated that soluble FR-α, anti-FR-α antisera, and even folic acid can inhibit entry of filoviruses. These findings provide a possible basis for establishing treatment procedures by preventing virus-cell interaction via FR-α.

It is interesting to note that early post-infection filovirus particles are associated with coated pits along the plasma membrane suggesting endocytosis as a possible mechanism for virus entry (Geisbert & Jahrling, 1995). This is also supported by studies employing lysosomotropic agents such as ammonium chloride or chloroquine indicating that infection with either filoviruses or pseudotypes carrying EBOV GP is very sensitive to pre-treatment of target cells with weak bases (Mariyankova et al., 1993; Wool-Lewis and Bates, 1998; Chan et al., 2000b). These data suggest that entry of filoviruses is a pH dependent process. However, using standard methodology such as syncitia formation fusion activity has never been demonstrated experimentally for filoviruses.

There is extensive structural homology between transmembrane proteins of filoviruses and number of oncoretroviruses, especially for the carboxy-terminal 181 amino acids (Volchkov et al., 1992). This region is highly conservative in retroviral glycoproteins and contain several functional domains critical for viral entry and cell fusion such as a central CX_6CC motif, the potential coiled coil, and a fusion peptide (Freed et al., 1990; Weissenhorn et al., 1998a and 1998b). A potential fusion peptide for EBOV has been postulated 22 amino acids down from the GP cleavage site (amino acids 524 to 539) (Gallaher 1996). Recently, it has been shown that the same synthetic peptide induces membrane destabilisation and fusion with liposomes *in vitro* (Ruiz-Arguello et al., 1998). The later observation together with mutational analysis of the putative fusion domain (Ito et al., 1999; Watanabe et al., 2000) strongly support a fusion peptide role for this conserved hydrophobic region in the EBOV transmembrane glycoprotein GP (Figs. 2 and 4). With MBGV, a similar domain can be found at a distance of 91 amino acids from the cleavage site (Volchkov et al., 2000b).

An important control mechanism of fusion activity of viral surface proteins is the processing by protein convertases (Klenk & Garten 1994a, b). Proteolytic cleavage, which often occurs next to a protein domain involved in fusion, is the first step in the activation of these fusion proteins and is followed by a conformational change resulting in the exposure of the fusion domain (Bullough et al., 1994: Chan et al., 1997; Weissenhorn et al., 1997). The conformational change may be triggered by low pH, such as in endosomes (Skehel et al., 1982), or by the interaction with a secondary receptor protein at the cell surface (Feng et al., 1996). The central structural features of the EBOV GP_2 ectodomain (Weissenhorn et al., 1998a and 1998b) suggests that the fusion peptide and membrane anchor domain are located at one end of the rod-like trimer. Such structures have been observed with the transmembrane subunits HA_2 of the influenza virus hemagglutinin (Bullough et al., 1994), the gp41 of the HIV env protein (Chan et al., 1997; Weissenhorn et al., 1997), and the F1 of the paramyxovirus fusion protein (Joshi et al., 1998). All of these proteins require cleavage and conformational changes to activate their fusogenic potency. Therefore, glycoprotein cleavage by furin and other host-cell proteases is absolutely necessary for the infectivity of these viruses. The structural similarities and the difference in the folding of uncleaved and cleaved GP, as judged by SDS-PAGE mobility (Volchkov et al., 1998b), strongly suggests that the fusion process of filoviruses occurs in a similar fashion.

Studies using pseudotype viruses, however, have demonstrated that proteolytic cleavage of the transmembrane glycoprotein is not essential for the replication of EBOV - at least in cell cultures (Wool-Levis & Bates, 1999; Ito et al., 2001). This observation was supported by the data obtained from the analysis of several recombinant EBOVs containing mutations at the GP cleavage site (Volchkov et al., submitted, Neumann et al., submitted). Although introduced mutations almost blocked the cleavage of GP, only a minor decrease in infectivity was noticed in cell culture experiments. To date mechanism of the filovirus entry is still a mystery with many open questions. The filamentous morphology of theses viruses which have a uniform diameter of 80nm up to a length of 14 µm, assumes that only a limited number of surface GP molecules are sufficient for the virus attachment and subsequent membrane fusion. Virus particles containing two forms of GP, cleaved and uncleaved, could therefore be still very infectious, especially in cell culture experiments. The cleavage sites of the filoviral glycoproteins are highly exposed as shown by experiments in which the site was replaced by a consensus chymotrypsin protease recognition motif that allowed cleavage of EBOV GP (Wool-Lewis and Bates, 1999). It is also possible that cleavage of GP is facilitated after arrival of viruses at the cell surface by yet unidentified endosomal endoproteases. It is interesting to note that in all filoviral GPs basic di- or tri-peptides (R-R, R-K, K-R, R-K-R or R-R-K) are either part of the furin-like cleavage site or located in the close proximity (Table 1). While the role of these basic paired residues in cleavage of GP has not been determined as yet, data on other viral glycoproteins, such as influenza HA and the F protein of paramyxoviruses, indicate that cleavage can occur at single basic residues (R or K) present in the cleavage sites (Kawaoka et al., 1990; Klenk & Rott, 1988; Klenk & Garten, 1994b). With filoviruses Wool-Lewis and Bates (1999) showed that showed cleavage of GP is occurred, with only one arginine residue remaining at the cleavage site.

The conservation of the cleavage site among all filoviruses point towards an essential function in the infected host or an unknown reservoir. Animal studies will have to verify whether proteolytic cleavage has a role in establishing filovirus infection *in vivo*. Here cleavage by furin may be a key factor in virus activation and an important event for the pantropism of infection. Furin is a processing enzyme of the constitutive secretory pathway and is expressed in most mammalian cells. It is localised predominantly in the trans-Golgi network (Molloy et al, 1994; Schäfer et al., 1995) and also secreted from cells in a truncated form (Wise et al., 1990; Vey et al., 1995). The enzyme belongs to the proprotein convertases, a family of subtilisin-like eukaryotic endoproteases that includes also PC1/PC3, PC2, PC4, PACE4, PC5/PC6 and LPC/PC7 (Seidah et al., 1996). These enzymes are differentially expressed in cells and tissues and display similar but not identical specificity for basic motifs, such as R-X-K/R-R, at the cleavage sites of their substrates. Variation in the cleavage site of the glycoprotein GP may account for differences in the pathogenicity of EBOV (Table 1) (Volchkov et al., 1998a). MBGV and all highly pathogenic EBOV strains display the canonical furin motif R-X-K/R-R at the cleavage site and are highly susceptible to cleavage. Only the glycoproteins of the EBOV Reston strains, which appear to be less pathogenic for humans and only moderately pathogenic for at least some monkey species (Fisher-Hoch et al., 1992), show a reduced cleavability because of the suboptimal cleavage site sequence K-Q-K-R (Volchkov et al., 1998a) (Table 1). Expression studies on the Reston virus glycoprotein demonstrated a lower cleavability of this protein that could be increased by a single amino acid change (Volchkov et al., 1998a). While on one hand highly pathogenic variants may emerge from Reston-like strains by mutations in the cleavage site, inhibition of furin cleavage may on the other hand be a valuable concept for treatment strategies of acute infections with filoviruses. Such an inhibition can, for instance, be achieved with peptidyl chloromethylketones (Anderson et al., 1993).

Table 1: Proteolytic cleavage sites of filovirus surface glycoproteins. The amino acid sequences of the cleavage sites are presented in positions -1 to -4. Proteolytic cleavage occurs at the carboxy-terminal of the arginine residue at position -1. The relative pathogenicity in human and non-human primates is indicated.

Virus species	-4 -3 -2 -1	Pathogenicity	
		Human	Monkey
EBOV-Zaire	-R-T-R-R-	+++	+++
EBOV-Sudan	-R-S-R-R-	+++	++
EBOV-Cote d'Iv.	-R-K-R-R-	+ (++)	+ (++)
*EBOV-Reston	-K-Q-K-R-	- (?)	+ / ++
MBGV	-R-R-K-R-	+++	+++

Filovirus infections lead to a moderate cytopathogenic effect in target cells. The mechanism causing cell destruction is, however, unknown. It is possible that a massive production and accumulation of vial proteins or intensive budding of viral particles at the plasma membrane are involved in this process. Alternatively, one of the viral proteins may have specific cytotoxic potential. Recently published studies demonstrated cell destruction upon expression of EBOV GP in 293 T cells (Chan et al., 2000a; Takada et al., 2000; Yang et al., 2000). In one study it was reported that a serine-threonine-rich mucin-like domain located on GP_1 mediates cytotoxicity. This could be in a part confirmed in vessel explants by infections with recombinant adenovirus vectors expressing EBOV GP (Yang et al., 2000). A second study demonstrated cell detachment of 293 T cells following expression of EBOV but not MBGV GP. Cell detachment in this case occurred in the absence of cell death. It was largely attributed to a domain within in the extracellular region of GP_2 and seemed to involve a phosphorylation-dependent signal cascade (Chan et al., 2000a). Another group published that the ectodomain of the glycoprotein and its anchorage to the membrane are required for GP-induced morphological changes of cells (Takada et al., 2000). Using a reverse genetic system it has been recently shown that the cytotoxicity of EBOV depends on the level of GP expression. Overexpression of GP leads to cytotoxicity and n early detachment of infected cells (Volchkov et al., 2001).

3.   INTERFERENCE WITH THE HOST DEFENSE SYSTEM

Immunosuppression seems to be an important factor in the pathogenesis of filoviral hemorrhagic fever. The exact mechanism leading to the immunosuppressed status of the host is currently being investigated. For EBOV, it has been reported that sGP interacts with the immune system by binding to neutrophils through CD16b, the neutrophil-specific form of the Fc receptor III thereby inhibiting early activation of these cells (Yang et al. 1998; Kindzelskii et al., 2000). This concept, however, has been challenged by a report from Maruyama and colleagues (1998).

Relatively high amounts of glycoprotein GP are released into the medium of filovirus-infected cells, and it has been suggested that this soluble form as well as sGP may effectively bind neutralising antibodies (Sanchez et al., 1996; Volchkov et al., 1998b).

In addition, filovirus transmembrane glycoprotein molecules possess a sequence close to the carboxy-terminus resembling a presumptive immunosuppressive domain found in retrovirus glycoproteins (Volchkov et al., 1992; Will et al. 1993; Bukreyev et al., 1995). Peptides synthesised according to this 26 amino acid long region inhibited the blastogenesis of lymphocytes in response to mitogens as well as the production of cytokines, and decreased proliferation of mononuclear cells in vitro (Ignatiev, 1999). It is not yet known whether the immunosuppressive domain on the GP is functional on mature molecules.

Disturbance of the blood tissue barrier, primarily controlled by endothelial cells, is another important factor in pathogenesis. The endothelium seems to be affected by two ways, directly by virus infection leading to activation and eventual cytopathogenic replication, and indirectly by a mediator-induced inflammatory response. Those mediators originate from virus-activated cells of the mononuclear phagocytic system (MPS), especially macrophages, which are the primary target cells (Schnittler & Feldmann, 1999). Current data indicate that the activation of endothelial and mononuclear phagocytotic cells could either be triggered by virus infection or binding of soluble viral or cellular factors produced during infection. Several soluble glycoproteins expressed and secreted or released upon infection with filoviruses might be directly involved in this process.

Recently neutralising anti-GP antibodies have be generated from several species, including human, which were immunised or infected with Ebola virus. These neutralising antibodies showed protective and therapeutic properties in animal models (Maruyama et al., 1999, Wilson et al., 2000; Gupta et al., 2001b). Protective properties, most likely due to neutralising antibodies, were also associated with convalescent sera (Mupapa et al., 1999). The successful use of the transmembrane glycoprotein in different immunisation approaches has clearly demonstrated the immunogenic and protective properties of this protein in small animal models and non-human primates (Hevey et al., 1998; Vanderzanden et al., 1998; Xu et al., 1998; Pushko et al., 2000; Sullivan et al., 2000). Thus, a step in the right direction towards a human vaccine is done, but there is still a long way to go (Burton & Parren, 2000; Klenk, 2000).

ACKNOWLEDGMENTS

V.E.V. and H.F. hold several grants on filoviruses provided by the Deutsche

Forschungsgemeinschaft (SFB 286, Fe 286/4-1), INSERM, Claude Bernard University Lyon-1 (BQR 2001-2002); Foundation Pour La Recherche Medicale, DGA (Convention N° 01.34.027.00.470.75.01), and the European Community (INCO-grant ERBIC 18 CT9803832). The authors are thankful for the contributions of several graduate students supported by these grants.

REFERENCES

Anderson, E.D., Thomas, L., Hayflick, J.S., and Thomas, G., 1993, Inhibition of HIV-1 gp160-dependent membrane fusion by a furin-directed alpha 1-antitrypsin variant. *J. Biol. Chem.* **268:** 24887-24891

Becker, S., Klenk, H.-D., and Mühlberger, E., 1996, Intracellular transport and processing of the Marburg virus surface protein in vertebrate and insect cells. *Virology* **225:** 145-155

Becker, S., Spiess, M., and Klenk, H.-D., 1995, The asialoglycoprotein receptor is a potential liver-specific receptor for Marburg virus. *J. Gen. Virol.* **76:** 393-399

Bukreyev A.A., Volchkov, V.E., Blinov, V.M., Dryga, S.A., and Netesov, S.V., 1995, The complete nucleotide sequence of the Popp (1967) strain of Marburg virus: A comparison with the Musoke (1980) strain. *Arch. Virol.* **140:** 1589-1600

Bullough, P.A., Hughson, F.M., Skehel, J.J. and Wiley, D.C., 1994, Structure of influenza haemagglutinin at the pH of membrane fusion. *Nature* **371:** 37-43

Burton, D.R., and Parren P.W.H.I., 2000, Fighting the Ebola virus. *Nature* **408:** 527-528

Bwaka, M.A., Bonnet, M.J., Calain, P., Colebunders, R., De Roo, A., Guimard, Y., Katwiki, K.R., Kibadi, K., Kipasa, M.A., Kuvula, K.J., Mapanda, B.B., Massamba, M., Mupapa, K.D., Muyembe-Tamfum, J.J., Ndaberey, E., Peters, C.J., Rollin, P.E., Van den Enden, E., and Van den Enden E., 1999, Ebola hemorrhagic fever in Kikwit, Democratic Republic of the Congo: clinical observations in 103 patients. *J Infect Dis* **179** Suppl 1: 1-7

Cattaneo, R., Kaelin, K., Baczko K., and Billeter, M A., 1989, Measles virus editing provides an additional cysteine-rich protein. *Cell* **56:** 759-764

Chan, D.C., Fass, D., Berger, J.M., and Kim, P.S., 1997, Core structure of gp41 from the HIV enevelope glycoprotein. *Cell* **89:** 263-273

Chan, S.Y., Empig, C.J., Welte, F.J., Speck, R.F., Schmaljohn, A., Kreisberg, J.F., and Goldsmith, M.A., 2001, Folate receptor-alpha is a cofactor for cellular entry by Marburg and Ebola viruses. *Cell* **106:** 117-126

Chan, S.Y., Ma, M.C., and Goldsmith, M.A., 2000a,. Differential induction of cellular detachment by envelope glycoproteins of Marburg and Ebola (Zaire) viruses. *J. Gen. Virol.* **81:** 2155-2159

Chan, S.Y., Speck, R.F., Ma, M.C., and Goldsmith, M.A., 2000b, Distinct mechanisms of entry by envelope glycoproteins of Marburg and Ebola (Zaire) viruses. *J. Virol.* **74:** 4933-4037

Feldmann H., Mühlberger, E., Randolf, A., Will, C., Kiley, M.P., Sanchez, A. and Klenk, H.-D., 1992, Marburg virus, a filovirus: messenger RNAs, gene order, and regulatory elements of the replication cycle. *Virus Res.* **24:** 1-19

Feldmann, H. and Kiley, M.P., 1999, Classification, structure, and replication of filoviruses. In: Marburg and Ebola viruses, Klenk, H.D., ed.; Springer Verlag; *Curr. Top. Microbiol. Immunol.*

235: 1-21

Feldmann, H., Bugany, H., Mahner, F., Klenk, H.-D., Drenckhahn, D. and Schnittler, H.-J., 1996, Filovirus-induced endothelial leakage triggered by infected monocytes/macrophages. *J Virol,* **70:** 2208-2214

Feldmann, H., Nichol, S.T., Klenk, H.-D., Peters, C.J., and Sanchez, A., 1994, Characterization of filoviruses based on differences in structure and antigenicity of the virion glycoprotein. *Virology* **199:** 469-473

Feldmann, H., Will, C., Schikore, M., Slenczka, W., and Klenk, H.-D., 1991, Glycosylation and oligomerization of the spike protein of Marburg virus. *Virology* **182:** 353-356

Feng, Y., Broder, C.C., Kennedy, P.E., and Berger, E.A., 1996,. HIV-1 entry cofactor: functional cDNA cloning of a seven-transmembrane, G protein-coupled receptor. *Science,* **272:** 872-877

Fisher-Hoch S.P, and Brammer L, 1992, Pathogenic potential of filoviruses: role of geographic origin of primate host and virus strain, *J. Infect. Dis.,* **166:** 753-63

Fisher-Hoch, S.P., Platt, G.S., Neild, G.H., Southee, T., Baskerville, A., Raymond, R.T., Lloyd, G., and Simpson, D.I.H., 1985, Pathophysiology of shock and hemorrhage in a fulminating viral infection (Ebola). *J. Infect. Dis.,* **152:** 887-894

Freed, E., Myers, D., and Risser, R., 1990, Characterization of the fusion domain of the human immunodeficiency virus type 1 envelope glycoprotein gp41. *Proc. Natl. Acad. Sci. USA* **87:** 4650-4654

Funke, C., Becker, B., Dartsch, H., Klenk, H.-D., and Mühlberger, E., 1995, Acylation of the Marburg virus glycoprotein. *Virology* **208:** 289-297

Gallaher, W.R., 1996, Similar structural models of the transmembrane proteins of Ebola and avian sarcoma viruses. *Cell (Letter)* **85:** 477-478

Geisbert TW, Jahrling PB, Hanes MA, and Zack PM, 1992, Association of Ebola-related Reston virus particles and antigen with tissue lesions of monkeys imported to the United States. *J Comp Path* **106:** 137-152

Geisbert, T.W., and Jahrling, P.B., 1995, Differentiation of filoviruses by electron microscopy. *Virus Res.* **39:** 129-150

Geyer, H., Will, C., Feldmann, H., Klenk, H.-D., and Geyer, R., 1992, Carbohydrate structure of Marburg virus glycoprotein. *Glycobiology* **2:** 299-312

Gupta, M., Mahanty, S., Ahmed, R., and Rollin, P.E., 2001a, Monocyte-derived human macrophages and peripheral blood mononuclear cells infected with ebola virus secrete MIP-1alpha and TNF-alpha and inhibit poly-IC-induced IFN-alpha in vitro. *Virology* **284:** 20-25

Gupta, M., Mahanty, S., Bray, M., Ahmed, R., and Rollin, P.E., 2001b, Passive transfer of antibodies protect immunocompetent and immunodeficient mice against lethal Ebola virus infection without complete inhibition of viral replication. *J. Virol.* **75:** 4649-4654

Hevey, M., Negley, D., Geisbert, J., Jahrling, P., and Schmaljohn, A., 1998, Antigenicity and vaccine potential of Marburg virus glycoprotein expressed by baculovirus recombinants. *Virology* **239:** 206-216

Horikami, S.M., and Moyer, S.A., 1991, Synthesis of leader RNA and editing of the P mRNA during transcription by purified measles virus. *J Virol* **65:** 5342-5347

Ignatyev G.M., 1999, Immune response to filovirus infections. In: Marburg and Ebola viruses, Klenk, H.D., ed.; Springer Verlag; *Curr. Top. Microbiol. Immunol.* **235:** 205-217

Ito, H., Watanabe, S., Sanchez, A., Whitt, M.A. and Kawaoka, Y., 1999, Mutational analysis of the putative fusion domain of Ebola virus gly,coprotein. *J. Virol.* **73:** 8907-8912

Ito, H., Watanabe, S., Takada, A., and Kawaoka, Y., 2001, Ebola virus glycoprotein: proteolytic processing, acylation, cell tropism, and detection of neutralizing antibodies. *J. Virol.,* in press.

Joshi, S.B., Dutch, R.E., and Lamb, R.A., 1998, A core trimer of the paramyxovirus fusion protein: parallels to influenza virus hemagglutinin and HIV-1 gp41. *Virology* **248**: 20-34

Kawaoka, Y., Yamnikova, S., Chambers, T.M., Lvov, D.K., and Webster, R.G., 1990, Molecular characterization of a new hemagglutinin, subtype H14, of influenza A virus. *Virology* **179**: 759-767

Kindzelskii, A.L., Yang, Z., Nabel, G.J., Todd III, R.F., and Petty, H.R., 2000, Ebola virus secretory glycoprotein (sGP) diminishes FcRIIIB-to-CR3 proximity on neutrophils. *J. Immunol.* **164**: 953-958

Klenk, H.-D., 2000, Will we have and why do we need an Ebola vaccine? *Nature Med.* **6**: 1322-1323

Klenk, H.-D., and Garten, W., 1994a, Activation cleavage of viral spike proteins by host proteases. In *Cellular Receptors for Animal Viruses E.* (Wimmer ed.), Cold Spring Harbor Laboratory Press, pp241-280

Klenk, H.-D., and Garten, W., 1994b, Host cell prateases controlling virus pathogenicicty. *Trends in Microbiol.* **2**: 39-43

Klenk, H.-D., and Rott, R., 1988, The molecular biology of Influenza virus pathogenicity. *Adv. Virus Res.* **34**: 247-281

Kondo, K., Bando, H., Tsurudome, M., Kawano, M., Nishio, M., and Ito, Y., 1990, Sequence analysis of the phosphoprotein (P) genes of human parainfluenza type 4A and 4B viruses and RNA editing at transcript of the P genes: the number of G residues added is imprecise. *Virology* **178**: 321-326

Leroy, E.M., Baize, S., Volchkov, V.E., Fisher-Hoch, S.P., Georges-Courbot, M.C., Lansoud-Soukate, J., Capron, M., Debre, P., McCormick, J.B., and Georges, A. J., 2000, Human asymptomatic Ebola infection and strong inflammatory response. *Lancet* **355**: 2210-2215

Malashkevich, V.N., Schneider, B.J., McNally, M.L., Milhollen, M.A., Pang, J.X,. and Kim, P.S., 1999, Core structure of the envelope glycoprotein GP2 from Ebola virus at 1.9-A resolution. *Proc. Natl. Acad. Sci. USA* **96**: 2662-2667

Mariyankova, R.F., Giushakowa, S.E., Pyzhik, E.V., and Lukashevich, I.S., 1993, Marburg virus penetration into eukaryotic cells. *Vopr Virusol* **2**: 74-76

Maruyama, T., Buchmeier, M.J., Parren, P.W.H.I., and Burton, D.R., 1998, Ebola virus, neutrophils and antibody specificity. *Science* **282**: 845a

Maruyama, T., Rodriguez, L.L., Jahrling, P.B., Sanchez, A., Khan, A.S., Nichol, S.T., Peters, C.J., Parren, P.W., and Burton, D.R., 1999, Ebola virus can be effectively neutralized by antibody produced in natural human infection. *J. Virol.* **73**: 6024-6030

Matsuoka, Y., Curran, J., Pelet, T., Kolakofsky, D., Ray, R., and Compans, R.W., 1991, The P gene of human parainfluenza virus type 1 encodes P and C proteins but not a cysteine rich V protein. *J Virol* **65**: 3406-3410

Molloy, S.S., Thomas, L., van Slyke, J.K., Stenberg, P.E., and Thomas, G., 1994, Intracellular trafficking and activation of the furin proprotein convertase: localization to the TGN and recycling from the cell surface. *EMBO J.* **13**: 18-33

Moss, B., 1990. Regulation of vaccinia virus transcription. *Annu.Rev.Biochem.,* **59**: 661-688

Mupapa, K.D., Massamba, M., Kibadi, K., Kuvula, K., Bwaka, A., Kipasa, M., Colebunders, R., and Muyembe-Tamfum, J.J., 1999, Treatment of Ebola hemorrhagic fever with blood transfusions from convalescent patients. *J. Infect. Dis.* (Suppl. 1) **179:** S18-S23

Murphy, F.A., van der Groen, G., Whitfield, S.G., and Lange, J.V., 1978, Ebola and Marburg virus morphology and taxonomy. In: *Ebola Virus Haemorrhagic Fever* (Pattyn, S.R., ed), Elsevier/North-Holland, Amsterdam, pp61-82

Ohgimoto, S., Bando, H., Kawano, M., Okamoto, K., Kondo, K., Tsurudome, M., Nishio, M., and Ito, Y., 1990, Sequence analysis of P gene of human parainfluenza type 2 virus: P and cysteine-rich proteins are translated by two mRNAs that differ by two nontemplated G residues. *Virology* **177:** 116-123

Paterson RG., and Lamb R.A., 1990, RNA editing by G-nucleotide insertion in mumps virus P-gene mRNA transcripts. *J Virol* **64:** 4137-4145

Pelet T., Curran, J., and Kolakofsky, D., 1991, The P gene of bovine parainfluenza virus 3 expresses all three reading frames from a single mRNA editing site. *EMBO J* **10:** 443-448

Peters, C.J., Sanchez, A., Rollin, P.E., Ksiazek, T.G., and Murphy, F.A., 1996,. Filoviridae: Marburg and Ebola viruses, In *Virology* (B. N. Fields, D. M. Knipe, et al., eds.), Raven Press, Philadelphia, 3rd edn., pp1161-1176

Peters, C.J. and LeDuc, J.W., 1999, Ebola: The virus and the disease. *J. Infect. Dis.* (Suppl. 1) **179:** 1-288

Pushko, P., Bray, M., Ludwih, G.V., Parker, M., Schmaljohn, A., Sanchez, A., Jahrling, P.B., and Smith, J.F., 2000, Recombinant RNA replicons derived from attenuated Venezuelan equine encephalitis virus protect guinea pigs and mice from Ebola hemorrhagic fever virus. *Vaccine* **19:** 142-153

Ruiz-Agüello, M.B., Goni, F.M., Pereira, F.B. and Nieva, J.L., 1998, Phosphatidylinositol-dependent membrane fusion induced by a put,ative fusogenic sequence of Ebola virus. *J. Virol.* **72:** 1775-1781

Ryabchikova, E., Kolesnikova, L.V., and Luchko, S.V., 1999, An analysis of features of pathogenesis in two animal models of Ebolka virus infection. *J. Infect. Dis.* (Suppl. 1) **179:** 199-202

Sanchez, A., Kiley, M.P., Holloway, B.P., and Auperin, D.D., 1993, Sequence analysis of the Ebola virus genome: organization, genetic elements, and comparison with the genome of Marburg virus. *Virus Res.* **29** : 215-240

Sanchez, A., Trappier, S.G., Mahy, B.W.J., Peters, C.J., and Nichol, S.T., 1996, The virion glycoprotein of Ebola viruses are encoded in two reading frames and are expressed through transcriptional editing. *Proc. Natl. Acad. Sci. USA* **93:** 602-3607

Sanchez, A., Trappier, S.G., Ströher, U., Nichol, S.T., Bowen, M.D., and Feldmann H., 1998a, Variation in the glycoprotein and VP35 genes of Marburg virus strains. *Virology* **240:** 138-146

Sanchez, A., Yang, Z.Y., Xu, L., Nabel, G.J., Crews, T., and Peters, C.J., 1998b, Biochemical analysis of the secreted and virion glycoproteins of Ebola virus. *J. Virol.* **72:** 6442-6447

Schäfer, W., Stroh, A., Berghöfer, S., Seiler, J., Vey, M., Kruse, M.L., Kern, H.F., Klenk, H.-D., and Garten, W., 1995, Two independent targeting signals in the cytoplasmic domain determine trans-Golgi network localization and endosomal trafficking of the proprotein convertase furin. *EMBO J.* **14:** 2424-2435

Schnittler, H.J., and Feldmann, H., 1999, Molecular pathogenesis of filovirus infections: role of macrophages and endothelial cells. In: Marburg and Ebola viruses, Klenk, H.D., ed.; Springer Verlag; *Curr. Top. Microbiol. Immunol.* **235:** 175-204

Seidah, N.G., Hamelin, J., Mamarbachi, M., Dong, W., Tadro, H., Mbikay, M., Chretien, M., and

Day, R., 1996, cDNA structure, tissue distribution, and chromosomal localization of rat PC7, a novel mammalian proprotein convertase closest to yeast kexin-like proteinases. *Proc. Natl. Acad. Sci. USA* **93**: 3388-3393

Skehel, J.J., Bayley, P.M., Brown, E.B., Martin, S.R., Waterfield, M.D., White, J.M., Wilson, I.A., and Wiley, D.C., 1982, Changes in the conformation of influenza virus hemagglutinin at the pH optimum of virus-mediated membrane fusion. *Proc. Natl. Acad. Sci. USA* **79**: 968-972

Southern, J.A., Precious, B., and Randall, R.E., 1990, Two nontemplated nucleotide additions are required to generate the P mRNA of parainfluenza virus type 2 since the genome encodes protein V. *Virology* **177**: 388-390

Sullivan N.J., Sanchez, A., Rollin, P.E., Yang, Z., and Nabel G.J., 2000, Development of a preventive vaccine for Ebola virus infection in primates. *Nature* **408**: 605-609

Takada, A., Robison, C., Goto, H., Sanchez, A., Murti, K G., Whitt, M.A., and Kawaoka, Y., 1997, A system for functional analysis of Ebola virus glycoprotein. *Proc. Natl. Acad. Sci. USA* **94**: 14764-14769

Takada, A., Watanabe, S., Ito, H., Okazaki, K., Kida, H., and Kawaoka, Y., 2000, Downregulation of beta1 integrins by Ebola virus glycoprotein: implication for virus entry. *Virology* **278**: 20-26

Takeuchi, K., Tanabayashi, K., Hishiyama, M., Yamada, Y. K., Yamada, A., and Sugiura, A,. 1990, Detection and characterization of mumps virus V protein. *Virology,* **178**: 247-253

Thomas, S.M., Lamb, R.A., and Paterson, R.G., 1988, Two mRNA's that differ by two nontemplated nucleotides encode the amino coterminal proteins P and V of the paramyxovirus SV5. *Cell* **54**: 891-902

van der Groen, G., Webb, P., Johnson, K., Lange, J., Linsday, L., and Elliott, L., 1978, Growth of Lassa and Ebola viruses in different cell lines. In *Ebola virus haemorrhagic fever* (Pattyn, S.R., ed), Elsevier/North-Holland Biomedical Press, Amsterdam, The Netherlands, pp225-260

van Regenmortel, M.H.V., Fauquet, C.M., Bishop, D.H.L., Carstens, E.B., Estes, M.K., Lemon, S.M., Maniloff, J., Mayo, M.A., McGeoch, D.J., Pringle, C.R., and Wickner, R.B., 2000, Virus Taxonomy: The Classification and Nomenclature of Viruses. The Seventh Report of the International Committee on Taxonomy of Viruses (book). Virus Taxonomy, VIIth report of the ICTV. Academic Press, San Diego, pp1167

Vanderzanden, L., Bray, M., Fuller, D., Roberts, T., Custer, D., Spik, K., Jahrling, P., Huggins, J., Schmaljohn, A., and Schmaljohn, C., 1998, DNA vaccines expressing either the GP or NP genes of Ebola virus protect mice from lethal challenge. *Virology* **246**: 134-144

Vey, M., Schäfer, W., Reis, B., Ohuchi, R., Britt, W., Garten, W., Klenk, H.-D., and Radsak, K., 1995, Proteolytic processing of human cytomegalovirus glycoprotein B (gpUL55) is mediated by the human endoprotease furin. *Virology* **206**: 746-749

Vidal, S., Curran, J., and Kolakofsky, D., 1990a, Editing of the Sendai virus P/C mRNA by G insertion occurs during mRNA synthesis via a virus encoded activity. *J Virol* **64**: 239-246

Vidal, S., Curran, J., and Kolakofsky, D., 1990b, A stuttering model for paramyxovirus P mRNA editing. *The EMBO J* **9**: 2017-2022

Villinger, F., Rollin, P.E., Brar, S.S., Chikkala, N.F., Winter, J., Sundstrom, J.B., Zaki, S.R., Swanepoel, R., Ansari, A.A., and Peters, C.J., 1999, Markedly elevated levels of interferon (IFN)-gamma, IFN-alpha, interleukin (IL)-2, IL-10, and tumor necrosis factor-alpha associated with fatal Ebola virus infection. *J Infect Dis.* (Suppl 1) **179**: 188-191

Volchkov, V.E., 1999, Processing of the Ebola virus glycoprotein. In *Marburg and Ebola viruses* (Klenk, H.-D., ed.), **235**, Springer-Verlag Berlin Heidelberg, pp35-47

Volchkov, V.E., Becker, S., Volchkova, V.A., Ternovoj, V.A., Kotov, A.N., Netesov, S.V., and Klenk, H.-D., 1995, GP mRNA of Ebola virus is edited by the Ebola virus polymerase and by T7 and vaccinia virus polymerases. *Virology* **214:** 421-430

Volchkov, V.E., Blinov, V.M., and Netesov, S.V. 1992, The envelope glycoprotein of bola virus contains an immunosuppressive-like domain s,imilar to oncogenic etroviruses. *FEBS Lett.* **305:**181-184

Volchkov, V.E., Chepurnov, A.A., Volchkove, V.A., Ternovoj, V.A,. and Klenk, H.-D., 2000a, Molecular characterization of guinea-pig-adapted variants of Ebola virus. *Virology* **277:** 147-155

Volchkov, V.E., Feldmann, H., Volchkova, V.A., and Klenk, H.-D., 1998a, Processing of the Ebola virus glycoprotein by the proprotein convertase furin. *Proc. Natl. Acad. Sci. USA* **95:** 5762-5767

Volchkov, V.E., Volchkova, V.A., Mühlberger, E., Kolesnikova, L.V., Weik, M., Dolnik, O., and Klenk, H.-D., 2001, Recovery of infectious Ebola virus from cDNA: transcriptional RNA editing of the GP gene controls viral cytotoxicity. *Science* **291:** 1965-1969

Volchkov, V.E., Volchkova, V.A., Slenczka, W., Klenk, H.-D. and Feldmann, H., 1998b, Release of viral glycoproteins during Ebola virus infection. *Virolog,y* **245:** 110-119

Volchkov, V.E., Volchkova, V.A., Strher, U., Cieplik, M., Becker, S., Dolnik, O., Garten, W., Klenk, H. D,. and Feldmann, H., 2000b, Proteolytic processing of Marburg virus glycoprotein. *Virology* **268:** 1-6

Volchkova, V.A., Feldmann, H., Klenk, H.-D., and Volchkov, V.E., 1998, The nonstructural small glycoprotein of Ebola virus is secreted as an antiparallel-orientated homodimer. *Virology* **250:** 408-414

Volchkova, V., Klenk, H.-D., and Volchkov, V., 1999, Delta-peptide is the carboxy-terminal cleavage fragment of the nonstructural small glycoprotein sGP of Ebola virus. *Virology* **265:** 64-171

Watanabe, S., Takada, A., Watanabe, T., Ito, H., Kida, H., and Kawaoka, Y., 2000, Functional importance of the coiled-coil of the Ebola virus glycoprotein. *J. Virol.* **74:** 10194-10201

Weissenhorn, W., Calder, L.J., Wharton, S.A., Skehel, J.J., and Wiley, D.C. 1998a, The central structural feature of the membrane fusion protein subunit from the Ebola virus glycoprotein is a long triple-stranded coiled coil. *Proc. Natl. Acad. Sci. USA* **95:** 6032-6036

Weissenhorn, W., Carfi, A., Lee, K.H., Skehel, J.J., and Wiley, D.C., 1998b, Crystal structure of the Ebola virus membrane fusion subunit, GP2, from the envelope glycoprotein ectodomain. *Mol. Cell* **2:** 605-616

Weissenhorn, W., Dessen, A., Harrison, S.C., Skehel, J.J., and Wiley, D.C., 1997, Atomic structure of the ectodomain from HIV-1 gp41. *Nature* **387:** 426-430

Will, C., Mühlberger, E., Linder, D., Slenczka, W., Klenk, H.-D., and Feldmann, H., 1993, Marburg virus gene 4 encodes the virion membrane protein, a type I transmembrane glycoprotein. *J. Virol.* **67:** 1203-1210

Wilson, J.A., Hevey, M., Bakken, R., Guest, S., Bray, M., Schmaljohn, A.L., and Hart, M.K., 2000, Epitopes involved in antibody-mediated protection from Ebola virus. *Science* **287:** 1664-1666

Wise, R.J., Barr, P.J., Wong, P.A., Kiefer, M., Brake, A.J., and Kaufman, R.J., 1990, Expression of a human proprotein processing enzyme: correct cleavage of the von Willebrand factor precursor at a paired basic amino acid site. *Proc. Natl. Acad. Sci. USA* **87:** 9378-9382

Wool-Levis, R.J., and Bates, P., 1998, Characterization of Ebola virus entry by using pseudotyped viruses: identification of receptor-deficient cell lines. *J. Virol.* **72:** 3155-3160

Wool-Levis, R.J. and Bates, P., 1999, Endoproteolytic processing of the Ebola virus envelope glycoprotein: cleavage is not required for function. *J. Virol.* **73:** 1419-1426

Xu, L., Sanchez, A., Yang, Z.Y., Zaki, S.R., Nabel, E.G., Nichol, S.T., and Nabel, G.J,. 1998, Immunization for Ebola virus infection. *Nature Med.* **4:** 37-42

Yang, Z., Delgado, R., Xu, L., Todd, R.F., Nabel, E.G., Sanchez, A., and Nabel, G.J., 1998, Distinct cellular interactions of secreted and transmembrane Ebola virus glycoproteins. *Science* **279:** 1034-1036

Yang, Z., Duckers, H. J., Sullivan, N., Sanchez, A., Nabel, E.G., and Nabel, G.J., 2000, Identification of the Ebola virus glycoprotein as the main viral determinant of vascular cell cytotoxicity and injury. *Nature Med.* **6:** 86-889

5. Pathogenesis

Chapter 5.1

Prions
When Proteins Take a Wrong Turn

ANNE BELLON AND MARTIN VEY
Aventis Behring GmbH, Postfach 1230, 35002 Marburg, Germany

1.   INTRODUCTION

Ever since the link between bovine spongiform encephalopathy (BSE) and a new form of Creutzfeldt-Jakob disease, vCJD, had been found in the mid 1990ies (Will *et al.* 1996), it was obvious that many people in Great Britain and elsewhere had been exposed to deadly human pathogens by contaminated beef products. The risk of acquiring such an usually rare but fatal disease had all of a sudden changed from remote into probable for hundreds of thousands of people not only in Great Britain but also in other BSE-affected European countries. BSE and vCJD belong to a family of incurable, brain wasting diseases which are caused by so-called prions (Prusiner, 1982), a novel class of pathogens completely different from viruses or bacteria: Prions are believed to be infectious, pathogenic particles solely composed of a protein which is called PrP^{Sc} (for prion protein scrapie). This abnormal protein not only seems to be the carrier of infectivity but it also seems to be involved in killing of brain cells. Many aspects of prion diseases are reminiscent of virus infections but, until today, no viral genome or other viral structures have ever been found in or isolated from afflicted humans or animals. When in the early 1980ies Stanley Prusiner introduced his prion hypothesis (Prusiner 1982), his views were ridiculed and heresy rather than scientific evidence was used to describe his contribution to the research of these fatal illnesses. Prusiner based his hypothesis on the discovery of the proteinaceous nature of the scrapie agent and its resistance against nucleic acid inactivation procedures. At the same time, these ridiculed prions were on the march to almost ruin Europe's beef industry and to elicit one of the most worrisome epidemics of modern society. As of today, almost twenty years later, the prion hypothesis has

Structure-Function Relationships of Human Pathogenic Viruses, Edited by
Holzenburg and Bogner, Kluwer Academic/Plenum Publishers, New York, 2002

been widely accepted and Prusiner's groundbreaking discoveries have set a firm foundation for the study and understanding of these novel, enigmatic pathogens and the terrible diseases they elicit (Prusiner 1998). The discovery of PrP^{Sc} has not only paved the way to modern molecular biological approaches for the study of prion diseases, it most importantly has provided the target for specific diagnostic tests to combat spread of today's creeping BSE and vCJD epidemics. In addition, it also has allowed rational drug design for therapeutic interventions. Some of the lead compounds have already been tested in animal systems of prion disease with some success. It is the aim of this chapter to illustrate the progress that has been made since Prusiner's seminal work on prions. Nevertheless, it is also the intention of this review to address some of the open questions to demonstrate the complexity and limitations of current prion research.

2.   PRION DISEASES

2.1   Characteristics of prion diseases.

Several different human prion diseases are known to date (Table 1). Most of them appear in the sixth, seventh, and eighth decade of life. The recently described vCJD seems to be an exception because it has been diagnosed mainly in young patients even in teenagers (Verity et al., 2000). Prion diseases of humans can have different origins but they all lead into a common pathologic pathway with the production of infectious prions. Prion replication then leads to destruction of the brain of the afflicted individuals. For most of the human prion diseases occurring today, initiation of prion formation is not clear and they appear in a sporadic pattern.

Almost 90% of human prion disease cases fall into the sporadic CJD category (Will et al., 1999). About ten percent of cases are caused by dominant germ-line mutations and are therefore inherited genetic diseases. The mutation leads to the production of prions and fatal disease in the brain of persons who carry the mutation (Gambetti et al., 1999). Less than 1% of all cases are caused by transmission either through prion-contaminated foodstuffs, therapeutics, surgical instruments or grafts which were in close contact with or derived from prion-infected human brains (Will et al., 1999). The incidence of prion disease seems to be constant worldwide with a rate of one new case per one million people per year.

Most of the naturally occurring cases of prion diseases in animals, such as scrapie in sheep or Chronic Wasting Disease in deer, seem to be caused by horizontal transmission but the routes of infection have not been clarified (Laplanche et al., 1999).

Prion Diseases of Humans and Animals

Human prion diseases

Sporadic
Somatic mutation or spontaneous conversion
of PrP^C into PrP^{Sc}

sCJD (sporadic Creutzfeldt-Jakob Disease)
FSI (Fatal Sporadic Insomnia)

Genetic
Germ-line mutations in PrP gene

FFI (Fatal Familial Insomnia)
GSS (Gerstmann-Sträussler-Scheinker disease)
fCJD (familial Creutzfeldt-Jakob Disease)

Transmitted

Infection by ingestion of prions

vCJD (variant Creutzfeldt-Jakob Disease)
Kuru

Infection by contaminated grafts, surgical
instruments, or therapeutics (hGH etc.),

iCJD (iatrogenic Creutzfeldt-Jakob Disease)

Animal prion diseases

Sheep, goats	Scrapie
Cattle	BSE (Bovine Spongiform Encephalopathy)
Mink	TME (Transmissible Mink Encephalopathy)
Mule, deer, elk	CWD (Chronic Wasting Disease)
Cat	FSE (Feline Spongiform Encephalopathy)

Table 1. Prion diseases of humans and animals.

Prion diseases are characterized by long incubation periods: prions replicate in the infected organism for a long time, years to decades for CJD, without causing signs of disease. After appearance of the first, mainly neurological or psychiatric symptoms, the deterioration of the brain functions progresses rapidly to dementia and other clinical syndromes. Finally, individuals die after a relatively short duration of the disease, for CJD it lasts 6 to 12 months (Will *et al.*, 1999, Gambetti *et al.*, 1999).

Although the different prion diseases of humans and animals (Table 1) may present with different clinical symptoms, they share some characteristic, pathologic changes mainly of the brain (DeArmond and Ironside, 1999): Microscopic analysis often reveals holes or vacuoles causing a sponge-like appearance of severely affected brain regions. Because of this spongiform degeneration in prion-infected brain, the term transmissible, spongiform encephalopathies (TSEs) is widely used as a synonym for prion diseases. Furthermore, activated glial cells can be detected in affected areas. This astrogliosis seems to be the only response of the individual to the prion infection: Neither specific immune responses nor inflammatory reactions are elicited by prions. Body defense mechanisms are blind to the destructive invasion by prions. Another typical finding for most of the prion diseases is the intracellular and extracellular deposition of PrP^{Sc}. Detection of PrP^{Sc} in brains with immunohistochemical methods or by Western blotting is a true diagnostic marker for prion disease. Due to the complete absence of immune responses in infected individuals, the lack of a pathogen genome, and the fact that only few organs produce detectable PrP^{Sc} amounts, an unambiguous diagnosis of the prion disease can be obtained only by post mortem neuropathologic examination.

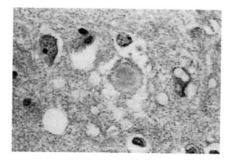

Figure 1. Histochemical staining of vCJD-infected brain. Vacuoles cause a sponge-like appearance indicative of a prion disease. The florid plaque in the center is typical for vCJD: extracellular PrP^{Sc} deposits are surrounded by vacuolized tissue. (Reprinted, with permission from Kopp *et al.,* 1996 [copyrights The Lancet])

Human prion diseases not only differ from each other with respect to their causes of initial prion formation, i.e. sporadic, familial, and iatrogenic, they also present with different clinical symptoms, different brain areas affected, differences in the forms of extracellular PrP^{Sc} deposits and even with differences in the physico-chemical properties of PrP^{Sc} itself (Gambetti *et al.,* 1999; Will *et al.*, 1999). For example, as first signs of vCJD, a transmitted prion disease, psychiatric symptoms appear which are followed by dementia. Duration of clinical phase is prolonged (12-24 months). PrP^{Sc} deposits in vCJD brains are widely disseminated and often appear in the form of so-called florid plaques, extracellular deposits containing PrP^{Sc} surrounded by vacuolized tissue (DeArmond and Ironside, 1999). During the course of Fatal Familial Insomnia or FFI, on the other hand, insomnia rather than dementia is a predominant clinical feature. Vacuoles and PrP^{Sc} deposits are mainly found in the thalamus region of the brain. No plaques can be found (Gambetti *et al.*, 1999). Similarly, prion diseases of animals differ from each other and from human prion diseases in many of the above-mentioned aspects (Laplanche *et al.*, 1999).

2.2   Route of infection

First experimental transmissions of Kuru and CJD to chimpanzees proved that these human spongiform encephalopathies, like scrapie of sheep and goats, were caused by infectious agents (Gajdusek *et al.*, 1966; Gibbs *et al.*, 1968). They further demonstrated that intracerebral inoculation was an efficient route of infection for primates. Transmission of prion disease to humans in most of the iatrogenic cases is caused by inoculation of prion-contaminated material which derived from brain or tissues which were in close contact with the central nervous system (CNS) of prion-infected individuals (Will *et al.*, 1999). Infection via dura mater or corneal grafts as well as infection through contaminated EEG electrodes allowed direct access of prions to neurons in the CNS of the recipients. The intracerebral route is also the most efficient mode of transmission in experimental prion transmissions (Prusiner *et al.*, 1999). Prions can be transmitted by peripheral inoculation as it was the case in young CJD patients which had received prion-contaminated growth hormones derived from cadaveric human pituitary glands (Billette de Villemeur, 1996). Intravenous infection requires ca. ten times more infectious prions than intracerebral infection (Brown *et al.*, 1999b). The infection via the gastrointestinal tract seems to be least efficient (Prusiner *et al.*, 1999): ca. 10^9 times more prions are needed to infect hamsters via the oral route. Nevertheless, oral consumption of BSE-contaminated food seems to be the most common mode of transmission for BSE and vCJD prions. Similar transmissions were the cause of Kuru epidemics among some Fore tribes in New Guinea in the middle of the 20th century who practised cannibalistic rituals. There has never been a documented case of prion transmission to humans through blood transfusion or blood products (Will *et al.*, 1999). Attempts to transmit CJD through blood to animals seemed to indicate positive transmission but could not be reproduced as summarized by Baron *et al.*, 1999. Vertical transmission of infectious prions does not seem to be an efficient route of prion spread in humans.

3.   NATURE OF INFECTIOUS AGENTS

3.1   The unusual properties of prions

Before prions were discovered, several different terms have been created for the agents causing TSEs such as "slow viruses"(Sigurdsson, 1954) based on the length of the incubation times or "unconventional viruses" based on the unusual physico-chemical properties of prions (Gajdusek, 1977). Prions are extremely resistant to conventional procedures used for inactivation of bacteria and viruses (Taylor, 2000): neither formaldehyde treatment, nor autoclaving at 121°C, UV irradiation (Alper et al., 1967), nor protease treatment (McKinley et al., 1983) are sufficient to completely destroy prion infectivity. Preferably procedures employing strong denaturants such as SDS, NaOH, chaotropic salts such as guanidinium (Prusiner et al., 1980), or free chlorine, significantly reduce prion infectivity. For decontamination and inactivation of prions, treatment with 1- 2% chlorine, extended autoclaving above 132°C or NaOH denaturation with additional boiling or autoclaving are efficient procedures (Taylor, 2000). Irradiation studies in 1967 gave a first hint at the non-viral nature of the scrapie agent (Alper et al., 1967) and many different biochemical substances that would withstand the irradiation treatment were proposed as candidate carriers of infectivity (Griffith, 1967). Nucleic acids seemed to be protected or non-existent in the agents since treatment with nucleases did not destroy scrapie infectivity. Scrapie research arrived at a turning point, when it was shown that the scrapie agent contains a hydrophobic protein (Prusiner et al., 1981a). Soon thereafter, the prion hypothesis was coined when it could be demonstrated that purified scrapie agents resisted procedures which inactivate nucleic acids and that they were composed of protein (Prusiner, 1982).

3.2   PrP27-30, PrP^{Sc}, and PrP^{C}

The prion protein was first identified as a protease-resistant protein with a molecular weight of 27-30kD (Bolton et al., 1982). Prion titers were proportional to the amount of PrP27-30 in the inoculum indicating that this protein was closely linked to infectivity: after denaturation, PrP27-30 became protease-sensitive and lost its infectivity (Prusiner et al., 1981b; McKinley et al., 1983). PrP27-30 is a glycoprotein (Bolton et al., 1985) which can contain up to several hundred different carbohydrate structures due to different modifications of N-linked side chains and chemical modifications of the GPI-anchor (Endo et al., 1989). Despite the use of sensitive analytical methods for the detection of nucleic acids in prion preparations, nucleic acids exceeding 50 nucleotides in length could not be found: this further demonstrated that infectivity does not depend on a viral or viroid genome (Kellings *et al.*, 1992). Determination of the N-terminal sequence of PrP27-30 (Prusiner *et al.*, 1984) enabled nucleotide screening for the genes encoding the prion protein (Chesebro *et al.*, 1985; Oesch *et al.*, 1985). Sequence analysis of PrP cDNA made clear that PrP27-30 represents only a fragment of a larger protein and that the gene which codes for PrP is a cellular gene (Basler *et al.*, 1986). It is not only expressed in brains of prion-infected organisms but to the same extent in non-infected brains (Chesebro *et al.*, 1985). The gene is highly conserved (Westaway and Prusiner, 1986) and it is expressed in almost all tissues (Bendheim *et al.*, 1992) with highest levels in brain and spleen. The protein is present on the surface of cells because a hydrophobic leader sequence directs translocation of the polypeptide into the lumen of the endoplasmic reticulum. It carries two sites for N-linked glycosylation and two cysteines form a disulfide bridge (Figure 2). At the COOH-terminus, PrP is tethered to the cellular membrane by a post-translationally attached glycosyl phosphatidyl inositol (GPI) anchor (Stahl *et al.*, 1987). This non-infectious, full length prion protein is designated PrP^C (Oesch *et al.*, 1985; Meyer *et al.*, 1986). It has a molecular weight of 33-35kDa and shows completely different biophysical properties compared to PrP27-30: it is protease-sensitive (Figure 3), it is soluble in detergents, and it can be released from cells by enzymes which cleave the GPI anchor from the COOH-terminus (Stahl *et al.*, 1987, 1990). PrP^C seems to be the precursor molecule for infectious prion proteins since a full length prion protein with the abnormal properties of prions is found during prion infection (Bolton *et al.*, 1987; Hope *et al.*, 1986). This full length, infectious prion protein is called PrP^{Sc}. PrP^{Sc} is hydrophobic and insoluble in the non-denaturing detergents just as PrP27-30. By protease treatment, ca. 67 amino acids are removed from the NH2-terminus of PrP^{Sc} and PrP27-30, the protease-resistant core of PrP^{Sc} is produced (Bolton *et al.*, 1982, 1987; Prusiner *et al.*, 1982; Figure 2, 3). UV-irradiation of purified prions does not reduce infectivity (Bellinger-Kawahara *et al.*, 1987) whereas γ-irradiation revealed a target size of an infectious unit of 55kD (Bellinger-Kawahara *et al.*, 1988) confirming the original observation on crude scrapie preparations made by Alper and colleagues in 1967, which first indicated that prions might propagate without replication of a nucleic acid (Alper *et al.*, 1967).

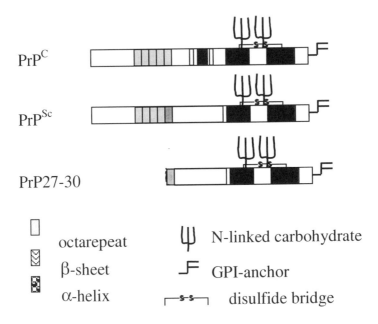

Figure 2. PrP^C, PrP^{Sc}, and PrP27-30. Full length PrP^C and PrP^{Sc} have the same primary structure spanning amino acids 23-231 (Hamster PrP nomenclature). The NH2-terminal hydrophobic leader peptide is cleaved and the COOH-terminus is covalently linked to a GPI-anchor tethering the molecules to the membrane. Secondary structures differ markedly: PrP^{Sc} and PrP27-30 are highly enriched for β-sheets. After refolding, the molecule is very stable and protease-resistant. The exact size of PrP27-30, the protease-resistant core of PrP^{Sc} can vary and depends on the conformation of the respective prion strain. The octarepeat region in PrP^C binds Cu^{2+} and is partially refolded into β-sheets in PrP^{Sc}. The α-helices at the COOH-terminus seem to remain untouched during refolding and are stabilized by a disulfide bridge. N-linked glycosylation at the two sites is not equally efficient producing different molecules, some of which contain only one or no N-linked carbohydrate side chain at all.

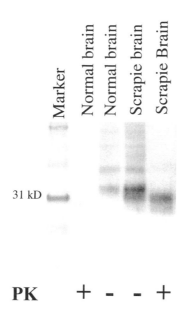

Figure 3. Prion proteins in normal and scrapie-infected hamster brains. Western blot of brain homogenates using the PrP-specific monoclonal antibody 3F4 (Kascsak *et al.*, 1987). Equal amounts of proteins were loaded. Non-infected hamster brains (Lanes: Normal brain) contain PrP^C which has a molecular weight of 33-35kD. Proteinase K treatment eliminates PrP^C. Syrian hamsters infected with the scrapie strain Sc237 (Lanes: Scrapie brain) show more intense staining of the 33-35kD bands due to the accumulation of full length PrP^{Sc}, in addition to PrP^C. After protease treatment, the protease-resistant core protein PrP27-30 can be detected. PrP molecules appear as double bands due to the presence of faster migrating, singly N-glycosylated PrP forms. Marker: Molecular Weight Standard. PK: Proteinase K.

Prion infections could be established in permanent cell lines expressing PrP^C (Race *et al.*, 1987, Butler *et al.*, 1988). After inoculation of cells with brain homogenates from prion-infected animals, the cells continuously produce PrP^{Sc}. Pulse-chase studies in such cells provided evidence that PrP^C might be the precursor of PrP^{Sc} (Borchelt *et al.*, 1990). Furthermore, if PrP^C is removed from the cell surface of infected cells, synthesis of PrP^{Sc} is abrogated, indicating that PrP^C is needed as substrate in the conversion process. It is not clear, whether all PrP^C molecules can be converted into PrP^{Sc} or only distinct subsets of PrP^C molecules are suitable for PrP^{Sc} production. PrP^{Sc} and PrP^C have the same primary sequence and there does not seem to be a difference in covalent, posttranslational modifications between the two isoforms (Hope *et al.*, 1986).

3.3   Size of the infectious unit

Recent studies indicate that not all of the PrP^{Sc} molecules seem to be protease-resistant and protease-resistance is not a prerequisite for PrP^{Sc} to be infectious. Early calculations showed that ca. 100,000 PrP^{Sc} molecules are required to initiate scrapie infection in hamsters when inoculated by the intracerebral route (McKinley *et al.*, 1983). Several questions arise: Are all of these 100,000 molecules necessary to initiate the infection? Is there only a subset of PrP^{Sc} molecules which is infectious and the rest of the molecules is non-infectious but co-purifies with the infectious ones? In an extreme scenario, there could be only one infectious PrP^{Sc} molecule among the 99,999 non-infectious ones. Or is the minimal infectious unit an aggregate with a number of PrP^{Sc} molecules present in order to be infectious (Come *et al.*, 1993)? In the light of latest findings, the early calculations need to be handled with care since it is now known that more than 99% percent of the inoculum disappears from the brain within 24 hours after intracerebral inoculation. Is PrP^{Sc} being degraded in the brain? Is it cleared and degraded in the periphery? If all of the PrP^{Sc} molecules stayed in the brain, would much less of them be required to initiate an infection? Furthermore, PrP^{Sc} molecules immobilized on stainless steel wires have a high potential to initiate a prion infection in animal brains (Zobeley *et al.*, 1999). It will be interesting to learn whether the minimal infectious dose of PrP^{Sc} molecules is less under these conditions. Exact knowledge of the infectious unit of a prion is crucial to understand the molecular events initiating prion replication.

3.4   Amyloid in prion disease

Prion diseases have been described as infectious amyloidoses. Amyloid is a specific form of protein aggregate in the form of fibrils. These fibrils show high affinity for certain dyes such as congo red, similar to the binding characteristics of starch. There is no doubt that PrP27-30 forms aggregates which have the typical tinctorial properties of amyloid. PrP^{Sc} is a precursor for these amyloid fibrils: fibril formation occurs when PrP^{Sc} is treated with protease in the presence of detergent (McKinley *et al.*, 1991b). Two different kinds of fibrils with slightly different ultrastructures have been isolated from prion-infected tissues, prion rods (Prusiner *et al.*, 1983) and scrapie-associated fibrils (SAFs) described earlier (Merz *et al.*, 1981). Subtle differences in purification procedures might have favored formation of different ultrastructures of the aggregates identifiable by electron microscopy. Despite all the attention given to amyloid properties of prions in the past, amyloid fibril formation does not seem to be necessary for prion infection: after solubilization of rods, prions remain infectious (Gabizon *et al.*, 1987; Wille *et al.*, 1996).

4. BIOLOGICAL ACTIVITIES OF NORMAL AND INFECTIOUS PRION PROTEINS

Progress of most aspects of prion diseases has been greatly enhanced by the transmission of scrapie and other prion diseases to small laboratory animals, such as mice and hamsters (Chandler, 1961). In these animals, incubation times are significantly reduced in comparison with passaging prions in the original hosts. For example, scrapie in sheep shows an incubation time of one to two years, in mice it takes only half a year and in hamsters only seventy to eighty days. At the beginning, phenomena such as species barriers and strains lead the scientists to believe that the infectious agents were of viral nature. Transgenic and PrP knock out mice have given logical explanations for many of these phenomena. Furthermore, study of PrP^C function in prion replication and pathogenesis has been made possible.

4.1 PrP-deficient and transgenic mice

For the study of physiological function and importance of PrP^C in prion replication, knock-out mice, so called $Prnp^{0/0}$ mice, were constructed. They do not express PrP^C (Büeler *et al.*, 1992). Such mice have a normal life span, arguing that PrP^C has no essential function or that the function of PrP^C is compensated for in these PrP-deficient mice. When PrP-deficient mice are inoculated with prions they do not become scrapie-sick and they do not produce infectious prions (Büeler *et al.*, 1993; Prusiner *et al.*, 1993).

With the discovery of the prion protein and the prion protein gene, *Prnp*, transgenic (tg) mice could be produced which overexpress homologous or heterologous PrP^C (Scott *et al.*, 1989; Prusiner *et al.*, 1990) Overexpression of mouse PrP^C in mice, for example, leads to shortened incubation periods for mouse prions (Westaway *et al.*, 1991; Fischer *et al.*, 1996): thus, kinetics of prion replication are driven by the level of PrP expression. This finding is another strong evidence for the important role of prion protein in TSEs.

In tg mice expressing PrP^C of different species, a phenomenon called species barrier could be elegantly modelled and explained (Prusiner *et al.*, 1990). Transgenic mice were also used to study the involvement of different domains or single amino acids in PrP^{Sc} formation and pathogenesis by expressing mutated PrP^C molecules. Thus, mice expressing exclusively PrP^C molecules deleted for amino acids 32-80 can be infected with prions and develop prion disease (Fischer *et al.*, 1996). Tg mice expressing PrP^C with even more complex deletions could be infected with prions (Supattapone *et al.*, 1999). These mice produced so-called miniprions consisting of PrP^{Sc} molecules with only 106 amino acids compared to 209 in full length PrP^{Sc}.

4.2   Polymorphisms, pathogenic mutations and species barriers

4.2.1   Polymorphisms in the PrP gene

Four polymorphisms in the human PrP gene have been found. Polymorphism of amino acid 129 is studied best. In combination with a dominant mutation at position 178, it elicits two different forms of prion disease. If valine at position 129 is paired with aspartic acid instead of asparagine at position 178, the carrier develops familial CJD (fCJD) whereas individuals who carry the aspartic acid at position 178 in combination with methionine at position 129, develop FFI.

Amino acid 219 is polymorphic in Japanese individuals: interestingly, the polymorphic amino acid lysine at this position seems to protect against prion disease.

Individuals with serine instead of asparagine at position 171 or individuals with a deletion of one octarepeat seem to be healthy as well.

4.2.2   Dominant PrP mutations causing prion disease

Mutations have been found in humans suffering from familial prion disease. For several of these mutations, linkage analyses were performed which proved genetic linkage of the mutant PrP gene with familial prion disease (Gambetti *et al.*, 1999). The mutations do not seem to cluster at specific areas of the prion protein and they can be as diverse as conservative amino acid substitutions or insertions of several octarepeats. Despite a life-long expression of these mutant PrP^{C} molecules, carriers become sick only in the fourth or fifth decade of their lives. Thus, the mutant PrP does not seem to cause any harm in the PrP^{C} isoform, it seems to have a slightly increased tendency to refold into the PrP^{Sc} isoform, which someday, later in life, initiates prion infection. Interestingly, cell culture models seem to indicate that these mutant PrP^{C} molecules have inherent PrP^{Sc}-like biochemical and biophysical properties (Lehman and Harris, 1996). Inherited prion diseases can be transmitted by intracerebral inoculation of brain material from patients into animals which demonstrates that infectious prions were made *de novo* in these individuals.

4.2.3   Species barrier

When prions are transmitted from one species to another, the frequency of positive transmission is rare, only a few inoculated animals develop prion disease, and the incubation time in those can be significantly extended. Upon repeated passage in the same host, all of the inoculated animals develop prion disease in significantly shortened, similar incubation times. This phenomenon has been defined as species barrier (Pattison and Jones, 1967). Sequencing of many mammalian PrP cDNAs has revealed that PrP is highly conserved among mammals. In addition, it became evident that only a few amino acid differences between species might be responsible for the species barrier effect (Prusiner *et al.*, 1990). If PrP^{Sc} of a different species enters a new host, the endogenous PrP^C is not efficiently converted: homology between the amino acid sequence of the incoming PrP^{Sc} and the host PrP^C seems to be a prerequisite for efficient PrP^C-PrP^{Sc} interaction and PrP^{Sc} production. With the production of transgenic mice which express PrP^C molecules of a different species, in addition to endogenous mouse PrP, these phenomena could be explained in more detail. Thus, in mice which express hamster PrP^C, in addition to endogenous mouse PrP^C, the species barrier between mice and hamsters was abrogated: when such mice were infected with hamster prions, they became scrapie-sick and they produced hamster prions (Prusiner *et al.*, 1990). Normal mice were not susceptible for efficient hamster prion replication. On the other hand, when hamster PrP tg mice were infected with mouse prions, they produced mouse prions. These studies clearly demonstrated that the amino acid sequence of PrP determines species specificity and susceptibility. They also provided evidence that PrP^{Sc} acts as a template and interacts directly with PrP^C in the production of new PrP^{Sc} molecules (Prusiner *et al.*, 1990).

Recent studies suggest occurrence of subclinical prion disease when animals are infected with heterologous prions: very late after infection with hamster prions, the mice were not suffering from prion disease but had infectious prions in their brains (Hill *et al.*, 2000). In such cases, the incubation period seems to exceed the normal life span of the host animal. Similarily, during first passage of prions in a different host, PrP^{Sc} cannot be found in large quantities despite replication of infectious prions (Lasmézas *et al.*, 1997).

A species-barrier-like effect in intraspecies transmission seems to be created by polymorphisms in the PrP. Thus, heterozygosity seems to prolong incubation time in iatrogenic CJD, whereas individuals homozygous for methionine at this position contracted prion disease much earlier (Will *et al.*, 1999).

4.2.4   Protein X

When tg mice expressing human PrP, in addition to endogenous mouse PrP, are infected with human prions, they do not develop prion disease and do not produce human PrP^{Sc}(Telling *et al.*, 1995). In contrast, tg mice expressing human PrP on a $Prnp^{0/0}$ background develop prion disease and produce human PrP^{Sc}. From these findings, it was suggested that PrP^{C} must interact with a different cellular protein before it can be converted into PrP^{Sc} and such a protein was provisionally named protein X (Figure 5 A). Protein X could act as a molecular chaperone and assist in the refolding reactions by binding to PrP^{C}. Thus, tg mice that express both human and mouse PrP do not become sick by human prions because the endogenous mouse PrP^{C} has a higher affinity for the endogenous mouse protein X. It therefore blocks interaction between human PrP^{C} and mouse protein X. In tg $Prnp^{0/0}$ mice which express human PrP^{C}, there is no competition for protein X by endogenous mouse PrP^{C}. As a consequence, human PrP^{C} can interact with mouse protein X and it is then converted upon interaction with the inoculated human PrP^{Sc}. Several amino acid residues within the COOH-terminal half of the PrP sequence have been identified, which seem to contribute to a protein X binding site: polymorphic amino acid positions in sheep and humans seem to confer resistance to prion infection because they seem to have a high affinity to protein X and to block chaperone activity. This phenomenon could be simulated in scrapie-infected cell culture expression systems: mutant PrPs, that carry these protective polymorphic amino acid residues, act as dominant negative inhibitors in conversion of co-expressed wild type PrPs (Kaneko *et al.*, 1997b). Furthermore, the identified amino acid residues seem to be located close to each other in the folded PrP^{C} molecule. This is apparent in the three-dimensional structures of recombinant hamster, mouse, and human PrP determined by NMR spectroscopy (Donne *et al.*, 1997; Calzolai *et al.*, 2000; Figure 4 C). Protein X has not been identified so far. Despite the intriguing evidence for the existence of such a protein X, other explanations for the observed phenomena could still be envisioned: the identified amino acid residues could be crucial for PrP^{C} homo-oligomerization and a such dominant negative PrP molecule may either induce or abrogate oligomerization. This might interfere with conversion into PrP^{Sc}. Furthermore, mutations at amino acid positions different from the ones described above can also have a dominant negative activity (Priola *et al.*, 1994; Priola and Chesebro, 1995; Hölscher *et al.*, 1998). Taken together, the studies using transgenic mice expressing PrPs of two different species and phenomena observed with polymorphisms of the prion protein have further demonstrated the importance of PrP in prion infection. The elucidation of the protein X effect will reveal further important insights into replication of prions.

5. CONFORMATION OF PRION PROTEINS AND MODELS FOR PRION REPLICATION

5.1 Secondary structure elements of prion proteins

PrP^C and PrP^{Sc} have different biochemical and biophysical properties. In addition, PrP^{Sc} and PrP^C differ with respect to their folded structures as evidenced by Fourier Transformed Infrared spectroscopy (FTIR) and Circular Dicroism (CD) spectroscopy of the purified proteins (Pan *et al.*, 1992, 1993). PrP^C has high α-helical content (42%) and only 3% of the protein is folded into β-sheets. On the other hand, 43% of PrP^{Sc} is in a β-sheet structure and only 30% is folded into α-helices. These data provided strong evidence that PrP^C is considerably refolded during prion replication and the change in conformation is responsible for infectivity and pathogenicity of PrP^{Sc}. Thus, it is conceivable that chaperones might be involved in refolding of PrP^C into its infectious PrP^{Sc} isoform (Telling *et al.*, 1995).

5.2 Tertiary structure of recombinant prion proteins

Since replication of prions involves a change in PrP conformation by refolding, it is of great importance to know the detailed three-dimensional structures of PrP^C and PrP^{Sc}. As of today, crystal structures for neither of the two isoforms have been obtained. So far, NMR spectroscopy of recombinant PrP molecules expressed in E. coli has provided important information on a PrP^C-like conformation. Several fragments of PrP of different species have been analyzed by NMR and a canonical structure with minor differences in the lengths of α-helices was obtained (Riek *et al.*, 1996; James *et al.*, 1997; Donne *et al.*, 1997; Zahn *et al.*, 2000). According to these studies, PrP contains an unstructured NH2-terminus extending from amino acids 23 to 120 (Figure 4 A). The C-terminus contains a globular, folded domain with three α-helices and a small β-sheet structure. The two COOH-terminal α-helices are covalently linked by a disulfide bridge. The amino acids involved in protein X binding are positioned close to each other in the tertiary structure which supports the idea of a composite binding site for interactions with protein X (James *et al.*, 1997; Prusiner, 1998; Calzolai *et al.*, 2000; Figue 4 C). The short β-sheet structure could be a nucleation point from which refolding into β-sheet structures could be initiated in the conversion process. The octarepeat region was not found to be structured under the conditions used for spectroscopy. It is conceivable that, in the presence of Cu^{2+}-ions, this polypeptide region might adopt a more folded conformation.

Insolubility and hydrophobicity of PrP^{Sc} have not allowed NMR spectroscopy or crystallization. A computer-calculated model for PrP^{Sc} predicts β-sheet structures within region 90 to 145 of PrP and two α-helices seem to remain more or less unchanged at the COOH-terminus (Huang *et al.*, 1995; Figure 4 B). Several findings support this model. Thus, this region is not available for antibody binding or protease digestion without prior denaturation (Peretz *et al.*, 1997). The COOH-terminus, on the other hand, seems to be accessible to antibodies.

Figure 4. Three-dimensional structure of PrP^{C} and PrP^{Sc}. A. Structure of recombinant PrP. NMR structure of recombinant hamster PrP29-231 expressed in and purified from E. coli. The NH2-terminal half of the molecule is flexibly disordered and not folded in stable secondary structures under the conditions used. Colour coding indicates the degree of flexibility with red for highest and blue for lowest flexibility. The COOH-terminal globular domain consists of three α-helices and a short β-sheet structure. (Reprinted with permission from Donne *et al.*, 1997 [copyright 1997 National Academy of Sciences U.S.A.]) B. Proposed model for PrP^{Sc}. The computer-calculated structure illustrates that the NH2-terminal part of the PrP27-30 molecule is probably arranged in β-sheets whereas the COOH-terminal α-helices seem to remain unchanged during conversion. This structure is supported by several experimental studies. The four amino acid residues shown in ball and stick model are implicated in species barrier phenomena. (Reprinted with permission from Huang *et al.*, 1995 [copyright Elsevier Science] C. Structure of recombinant PrP. NMR structure of amino acids 121-231 of recombinant PrP highlights the proposed binding site for protein X, shown in blue. Residues in green might modulate transmission across species. (Reprinted, with permission from Prusiner 1998 [copyrights 1998 National Academy of Sciences, U.S.A.])

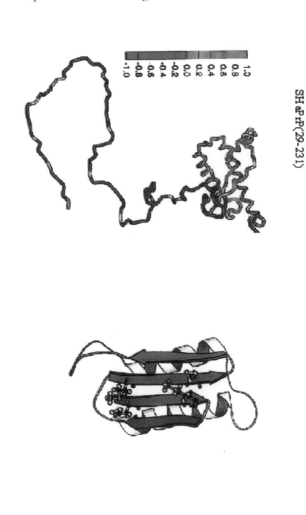

A. Structure of recombinant PrP
   SHaPrP(29-231)

B. Proposed model for PrPSc structure

C. Structure of recombinant PrP
   SHaPrP region 121-231

5.3   Mechanism of prion replication

Two theories for the conversion process are discussed: a template-assisted heterodimer hypothesis (Cohen and Prusiner, 1999) and a nucleation-dependent polymerization process (Come *et al.*, 1993; Eigen, 1996). In the template-assisted heterodimer model (Figure 5 A), it is suggested that monomers or dimers of PrP^C interact with a dimer of PrP^{Sc}. During the interaction, PrP^C is converted into PrP^{Sc} with the PrP^{Sc} in the heterodimer acting as a template. After conversion, the complex falls apart and the PrP^{Sc} dimers can be used again for the next conversion reaction. This leads to exponential prion replication. PrP^C seems to be a kinetically trapped folding intermediate (Huang *et al.*, 1995) and, thus, might require interaction with a chaperone in order to be in an activated state for conversion. It is then called PrP*. In this heterodimer model, prions replicate by repeated interaction of PrP^{Sc} dimers with PrP^C molecules where refolding of PrP^C into PrP^{Sc} is the rate-limiting step (Cohen and Prusiner, 1999).

In the nucleation-dependent polymerization theory (Figure 5 B), PrP^C oligomerization or oligomerization of a partially unfolded PrP^C, termed PrP^U, leads to the formation of a nucleus for polymerization: the PrP^C or PrP^U molecules in the nucleus refold into PrP^{Sc}. This then allows for attachment of further PrP^C or PrP^U molecules to the ends of the polymer. After binding to the polymer, PrP^C or PrP^U is refolded into PrP^{Sc} and the PrP^{Sc} polymer grows. In a prion infection by external prion sources, inoculated PrP^{Sc} aggregates serve as seeds for polymerization and thereby eliminate the need for formation of a nucleus made of PrP^C. In the nucleation-dependent polymerization process, formation of the nucleus is the rate-limiting step for sporadic PrP^{Sc} formation. Exponential growth occurs at some point during infection, when polymers reach a certain size and break (Eigen, 1996). When this happens, the newly released ends of the polymer fragments provide new polymerization sites. *In vitro* conversion systems seem to work by the seeding effect of PrP^{Sc} (Kocisko *et al.*, 1994)

Neither of the two models for prion replication can be unequivocally favoured at the moment.

A. Template-assisted Conversion

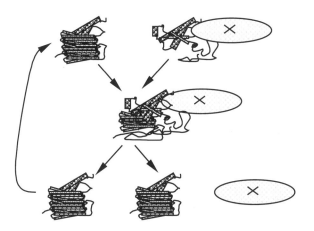

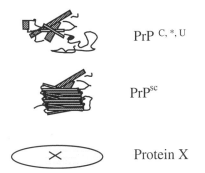

PrP $^{C, *, U}$

PrP^{sc}

Protein X

B. Nucleated Polymerization

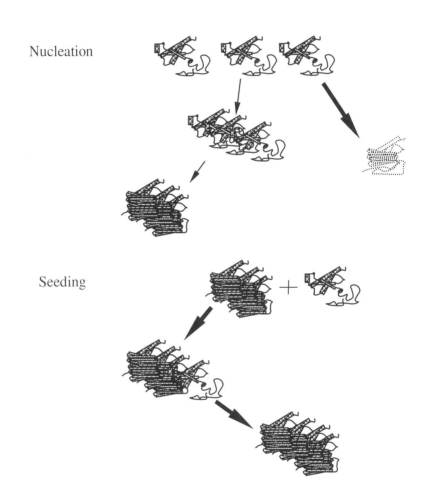

Nucleation

Seeding

Figure 5. Proposed mechanisms for prion conversion A. Template-assisted conversion. This model suggests that PrP^{Sc} and PrP^C or PrP*, an activated form of PrP^C, form a heterodimer. (PrP^{Sc} and PrP^C themselves may be present in the form of dimers). In order to be converted, PrP^C needs to interact with another cellular factor, designated protein X. After conversion, the heterodimer falls apart and PrP^{Sc} is again available as template for repeated conversion cycles (Cohen and Prusiner, 1999). B. Nucleated polymerization. Initial conversion requires oligomerization of PrP^C or PrP^U, a partially unfolded form of PrP^C, to form a nucleus for polymerization otherwise PrP^C is degraded. This formation of a nucleus is a rare event. Afterformation of the nucleus, PrP^C can convert into PrP^{Sc} which polymerizes. At the ends of the PrP^{Sc} polymer, PrP^C or PrP^{U}can be added and converted into PrP^{Sc}. In infectious prion disease, incoming PrP^{Sc} aggregates act as seeds for the growing polymers: this eliminates the need for initial PrP^C nucleation. Polymers break and thereby more catalytic sites for conversion are created allowing exponential replication of prions (Come *et al.*, 1993; Eigen 1996)

5.4 Prion formation *in vitro*?

5.4.1 *In vitro* conversion by mixing

Many attempts have been made to produce infectious prions *in vitro* but formation of infectious prions could not be demonstrated. Recombinant hamster PrP, expressed in E. coli, even when refolded into a β-sheet conformation is not infectious. This PrP is not glycosylated and does not contain a GPI-anchor which is necessary for PrP^C in order to be converted in mammalian cells. PrP^C isolated from non-infected cells was mixed with cell extracts from scrapie-infected cells but this did not result in newly formed PrP^{Sc} or infectivity (Raeber *et al.*, 1992). PrP^C incubated with infected hamster brain slices *in situ* was converted into a protease-resistant conformation (Bessen *et al.*, 1997). In a different attempt, metabolically labeled PrP^C could be converted into a protease-resistant PrP similar in size to PrP27-30 when mixed with almost equimolar amounts of purified PrP^{Sc}(Kocisko *et al.*, 1994). Whether the converted PrP^C was infectious could not be shown by animal infectivity bioassays because the preexisting PrP^{Sc}, which served as template for *in vitro* conversion, would have produced a high "background" infectivity such that newly formed infectious PrP^{Sc} would not have increased the overall infectivity significantly. In a different study using chimeric mouse hamster PrP^C, the so-called MH2M PrP, which allows trans-species transmission (Scott *et al.*, 1993), the newly formed protease-resistant PrP was not infectious (Hill *et al.*, 1999). So far, these studies have shown that PrP^{Sc} can indeed impart a new conformation into PrP^C evidenced by the formation of protease-resistant PrP even with strain characteristics (Bessen *et al.*, 1995). Finally, this *in vitro* conversion system also allows simulation of the species barrier: thus, hamster PrP^C can only be inefficiently converted by mouse PrP^{Sc} (Horiuchi *et al.*, 2000). These findings further underline the important role of amino acid sequence of PrP^{Sc} as template and PrP^C as substrate in the conversion process with regard to efficiency of intraspecies and inefficiency of interspecies transmission of prions. Once *in vitro* conversion or chemical synthesis of infectivity *in vitro* can be accomplished without the possibility of contamination by viruses or nucleic acids, the prion hypothesis could finally be accepted by everybody.

5.4.2   Animal model for *de novo* prion formation in familial prion disease

Transgenic mice expressing PrP^C with a mutation at the amino acid corresponding to human codon 102 that causes Gerstmann-Sträussler-Scheinker syndrome (Hsiao *et al.*, 1989) at high levels develop a neurological disease closely resembling prion disease: prion infectivity develops spontaneously by conversion of mutant PrP^C into PrP^{Sc} (Hsiao *et al.*, 1990). On the other hand, mouse lines expressing this mutant protein at low levels do not get sick spontaneously. When brain material from the sick high expresser animals is injected into the low expresser animals, these low expresser animals will develop prion disease: the disease has been transmitted from the high expressers to the low expressers (Hsiao *et al.*, 1994). Furthermore, when such low expresser mice are injected with a peptide containing the mutant PrP sequence folded into a β-sheet structure, the mice become sick as well (Kaneko *et al.*, 2000). This experiment indicates that β-sheet conformation in synthetic prion protein can induce prion formation. In these GSS mice, PrP^{Sc} is not protease-resistant further indicating that protease-resistance is not required for infectivity of PrP^{Sc}. Taken together, this transgenic mouse model lends further support to the prion hypothesis because it demonstrates that mutant PrP can induce prion formation and neurological disease spontaneously as it is the case in familial CJD, GSS, or FFI.

6.   PRION STRAINS

For a long time, prion research was challenged by the presence of different strains which seemed to be derived from mutation in an agent-specific genome (Scott *et al.*, 1999a). Prion strains can differ markedly with respect to length of incubation period in a host, the distribution of pathologic lesions in the brain of the host, in susceptibility to inactivation procedures etc. Today it seems to be clear that strains indeed represent distinct, stable variants of prions and the differences are not encoded in an agent-specific genome but in the conformation of the PrP^{Sc} molecule. The first evidence came from two different TME strains, "Hyper" and "Drowsy": both strains elicit prion disease but with different incubation times and different neuropathology in the brain of infected animals. Surprisingly, the protease-resistant PrP27-30 molecules of the two strains differed with respect to their size: the "Hyper" PrP27-30 was ca. 10 amino acids longer, indicating that a larger portion of "Hyper" PrP^{Sc} was inaccessible to the protease due to a different conformation (Bessen and Marsh, 1994).

Striking evidence for the inheritance of strain characteristics through protein conformation again was provided by a transgenic mouse system: tg mice expressing human PrP^C were either infected with prions from a patient who died of FFI or from a patient who died of sporadic or familial CJD (Telling *et al.*, 1996). All of these inocula transmitted prion disease to the infected mice and protease-resistant PrP^{Sc} was produced. As a first hint, the incubation times for the different prion inocula were different: they might have been caused by different titers of prions present in the inocula or differences between the amino acid sequences of the mutant PrP^{Sc} molecules in the inocula and the non-mutated human PrP^C molecules in the tg mice. Evidence for the existence of two different strains then arose when the sizes of the protease-resistant cores of PrP^{Sc} produced in the animals were compared: FFI produced a protease-resistant core with a molecular weight of 21kD after deglycosylation, sporadic CJD produced a protease-resistant core of 19kD (Telling *et al.*, 1996). These sizes matched exactly the different sizes observed for the protease-resistant cores of the respective PrP^{Sc} molecules produced in the patient brains (Monari *et al.*, 1994). Even after repeated passaging in the tg mice expressing human PrP, the strains kept their distinct characteristics: the strains bred true. This example clearly demonstrated that PrP^{Sc} imposes a distinct conformation onto a PrP^C molecule. This conformation is replicated with high fidelity and the differences in the conformation of strains are most probably responsible for different biological properties such as length of incubation period, lesion profile, susceptibility to inactivation procedures etc. There seem to be more than two human prion strains, especially with the unusual properties observed for vCJD prions. Reports on strain typing based different levels of glycosylation of PrP^{Sc} in combination with the different sizes of PrP27-30 for different human strains seems to be helpful for differential diagnosis of human prion strains (Gambetti *et al.*, 1999). As of today, there is no consensus on the number of distinct human strains that can be identified by this method.

Recent studies provided further evidence that strains of hamster scrapie are differently folded as evidenced by their unfolding characteristics in the presence of guanidinium (Peretz *et al.*, 2001). Furthermore, in infected brains, not only protease-resistant PrP^{Sc} is made but there seem to be differing amounts of protease-sensitive PrP^{Sc} molecules to be produced and degraded or cleared by other mechanisms. Thus, different strains seem to accumulate different amounts of protease-sensitive PrP^{Sc} in the hamster brains. It was suggested that the rate of PrP^{Sc} clearance determines the length of the incubation period of a certain strain (Safar *et al.*, 1998)

7.   PHYSIOLOGICAL AND PATHOLOGICAL FUNCTIONS OF NORMAL PRION PROTEINS

7.1   Copper binding activity of PrP^C

Since PrP^C is a highly conserved protein among mammals, it is tempting to believe that it serves an important physiological function. On the other hand, $Prnp^{0/0}$ mice seem to be healthy and do not have shortened life spans (Büeler *et al.*, 1992). Some electrophysiological studies in isolated hippocampus slices seemed to show some failure in long term potentiation (Collinge *et al.*, 1994) but these results could not be reproduced by others (Lledo *et al.*, 1996). In a different report, PrP-deficient mice seemed to show anomalies in the circadian rhythm (Tobler *et al.*, 1996).

Recent studies provided evidence for a Cu^{2+}-binding activity of PrP^C(Brown *et al.*, 1997; Stöckel *et al.*, 1998). The highly conserved octarepeat region in the NH2-terminal part of PrP^C binds Cu^{2+}-ions and membranes isolated from $Prnp^{0/0}$ brains had reduced Cu^{2+}-levels (Brown *et al.*, 1997). In addition, an enzyme activity dependent on the presence of Cu^{2+} seems to be reduced in the brain of PrP-deficient mice. Whether Cu^{2+}-binding is important for PrP^C refolding into PrP^{Sc} is not clear since a mutant PrP^C containing only one of the four octarepeats is converted into PrP^{Sc}(Fischer *et al.*, 1996). On the other hand, scrapie incubation times in these transgenic mice are significantly prolonged, suggesting that reduced Cu^{2+}-binding capability might slow down PrP^C conversion.

7.2   Pathological function by overexpression

Transgenic mice overexpressing high levels of PrP^C develop ataxia later in life indicating that high PrP^C levels can induce pathologic processes even in an uninfected animal (Westaway *et al.*, 1994). Tg mice expressing shorter versions of PrP^C, which do not support prion replication, were shown to have a severe disease and die soon after birth (Shmerling *et al.*, 1998). Interestingly, this deleterious effect of the shortened PrP^C could be abolished by co-expression of full length PrP^C in these animals. Taken together, these studies show that although PrP^C seems to be dispensable in knock out mice, the expression of PrP^C needs to be regulated to avoid neurological disease.

8.   MECHANISMS OF PRION PATHOGENESIS

The exact mechanism, by which prion-infected neurons die, has not yet been elucidated but several studies have expanded the understanding of the disease process. Prion-infected cells seem to die because apoptosis is triggered by prion infection (Schätzl *et al.*, 1997). It is not clear whether neighbouring cells are activating the apoptotic cascade or whether processes on the surface of the infected cells themselves are responsible. Oxydative stress elicited by reactive oxygen species produced by activated microglia is favored by some authors (Brown *et al.*, 1996). Furthermore, peptides containing a highly conserved, alanine-rich part of the PrP molecule which also has amyloidogenic properties is toxic for cells in culture (Forloni *et al.*, 1993). Toxicity of such peptides is dependent on the presence of PrP^C on the cells indicating that PrP^C may act as a receptor for the peptide or even as effector of toxicity (Brown *et al.*, 1998). In a recent study on expression of a GPI-anchored, amyloidogenic sequence of PrP in transgenic mice, toxicity of such a membrane-bound peptide lead to death of the animals soon after birth (Supattapone *et al.*, 2001a).

Since there is no replication of prions in PrP-deficient mice, it is clear that PrP^C is important for prion replication because it is the substrate. Elegant grafting experiments also showed the role of PrP^C in pathogenesis of prions: a graft containing PrP^C expressing cells was implanted into $PrP^{0/0}$ mice (Brandner *et al.*, 1996; Weissmann *et al.*, 1999). When these mice were infected with prions, PrP^{Sc} was made in the graft and vacuolar changes in the graft were indicative of typical prion pathology. Surprisingly, cells adjacent to the graft were not affected and showed no pathologic changes although PrP^{Sc} deposits had leaked into the regions close to the grafted cells. These results clearly demonstrated that PrP^C is necessary for prion pathogenesis and PrP^{Sc} *per se* is not toxic for cells. It is conceivable that, during the conversion reaction itself, pathologic processes are triggered.

Studies of the pathologic processes involved in GSS-like prion diseases showed an interesting correlation between a transmembrane form of mutant PrP and the potential of the mutant PrP to elicit neurologic disease: the PrP^C domain comprising the amyloidogenic peptide seems to have a high tendency to immerse into cellular membranes and trigger pathologic processes (Hegde *et al.*, 1998). Whether this transmembrane form or a similar membrane association of PrP^{Sc} is responsible for prion pathogenesis during prion infection, although intriguing (Hegde *et al.*, 1999), remains to be shown.

Taken together, it can be stated that key players in prion pathology have been identified but the molecular events and processes which cause cell death need to be understood in more detail. Identification of the mechanism by which neurons are killed during prion infection might one day yield targets for effective therapeutic treatment.

9.   PrP-LIKE PROTEIN DOPPEL

Not all lines of PrP$^{0/0}$ mice are completely healthy. Knock out mice which were produced by a different cloning strategy developed Purkinje cell loss and ataxia (Sakaguchi et al., 1996) in contrast to the line of PrP$^{0/0}$ mice constructed by Büeler and colleagues (Büeler et al., 1992). This phenomenon could be explained by the ectopic expression of an additional open reading frame (ORF) located downstream of the PrP ORF. This ORF is present in all mammalian species examined so far and it is usually not transcribed in brain. On the contrary, it is expressed in the atactic knock-out mice due to the deletion in the upstream PrP ORF. The ectopic expression of this protein caused cell death in the cerebellum resulting in ataxia (Moore et al., 1999). This protein, although shorter than PrP, is homologous to the COOH-terminal, globular domain of PrP but lacks the octarepeat region. It was named doppel because of the German word "Doppelgaenger" meaning double. Doppel is expressed at high levels mainly in testis. Similar to PrP, it is a membrane protein, glycosylated and tethered to the cell by a GPI-anchor (Silverman et al., 2000). NMR spectroscopy of a recombinant doppel protein showed very similar folding to PrP (Mo et al., 2001). Since doppel is not expressed during prion infection, it does not seem to be necessary for prion replication and pathogenesis. Grafting experiments, in which doppel-deficient, PrP^C expressing grafts were infected, showed that indeed doppel expression is not necessary for prion replication and pathogenesis (Behrens et al., 2001).

10.   REPLICATION OF PRIONS IN THE PERIPHERY AND NEUROINVASION

Familial prion diseases are caused by a dominant mutation in the PrP gene which leads to production of prions in the tissue of highest PrP expression, which is brain. Mutant PrP^C is refolded into the PrP^{Sc} isoform and initiates prion replication. Similarly, some of the iatrogenic transmissions have occured because prion-contaminated tissues or surgical instruments were in close contact with the brain of the recipient. The prions could then easily replicate in the new organism because they had already reached the organ with high expression levels of PrP^C. On the other hand, many transmitted prion diseases are initiated when prions enter the body via injection or ingestion. In such cases, prions have to find their way through the body into the CNS where prion infection finally leads to neuropathogenesis and neurodegenerative disease. This process has been termed neuroinvasion. It has been known for a long time that prions can replicate in the spleen of infected animals (Fraser and Dickinson, 1970; Kimberlin and Walker,

1979) and many other organs show prion infectivity to a limited extent (Hadlow *et al.*, 1982). The question arose then how prions enter the CNS after they have entered the body through the periphery. Many studies which aimed at identifying the mechanism for neuroinvasion used immunocompromised mice or mice which had congenital deficiencies of certain cell-types of the immune system, such as B- or T-lymphocytes. B cells, although not capable of replicating prions (Montrasio *et al.*, 2001), are essential for prion propagation in the periphery and invasion of central nervous system (Klein *et al.*, 1997) probably because B-lymphocytes secrete cytokines which are important for the differentiation of another cell type which supports prion replication in lymphoid tissues. Follicular dendritic cells (FDCs) express PrP^C at detectable levels and they seem to be responsible for replication of prions in lymphoid tissues of mice (Brown *et al.*, 1999a). Recent reports provide evidence that complement might play a role in early prion replication outside the CNS: uptake of PrP^{Sc} into FDCs via complement receptors can speed up peripheral infection with low amounts of prions (Klein *et al.*, 2001). Despite this progress in identifying cell types capable of prion replication in the lymphoid tissues, the route by which prions enter the central nervous system has not been established. The notion that peripheral nerves are in contact with lymphoid tissues suggests that prions might invade the CNS by infecting peripheral nerves which innervate the lymphoid organs. Finally, inhibiting spread of prions in the periphery before neuroinvasion seems to be a promising strategy to avoid fatal infection of the brain.

11.   CELL BIOLOGY OF PRIONS

PrP^C seems to be most efficiently converted in living cells. *In vitro* attempts using cell extracts or cell membranes have failed to produce infectious PrP^{Sc} (Raeber *et al.*, 1992; Hill *et al.*, 1999). Much progress in the understanding of cell biology of prions has been made on scrapie-infected neuroblastoma cells (Race *et al.*, 1987; Butler *et al.*, 1988). Recently, non-neuronal cell lines have been established which allow propagation even of heterologous prions (Vilette *et al.*, 2001). Several aspects of cell biology of prion conversion seem to explain the fact that cell-free conversion has been inefficient: conversion of PrP^C is a post-translational process (Borchelt *et al.*, 1990; Caughey and Raymond, 1991; Borchelt *et al.*, 1992). PrP^C has to leave the endoplasmic reticulum in order to be converted (Taraboulos *et al.*, 1992). N-linked glycosylation does not seem to be necessary for conversion as the presence of non-glycosylated PrP^{Sc} in infected cells demonstrates. Furthermore, tunicamycin treatment does not interfere with conversion and PrP mutants with defective N-linked glycosylation sites are converted (Taraboulos *et al.*, 1990; Korth *et al.*, 2000).

When PrP^C is removed from the surface of scrapie-infected neuroblastoma cells, PrP^{Sc} formation is abrogated, indicating that conversion takes place on the cell surface or after PrP^C has re-entered the cell (Caughey and Raymond, 1991). PrP conversion could take place anywhere on the way from the surface to the lysosomes because PrP^{Sc} has been found on the cell surface as well as in late endosomes and lysosomes (Vey et al., 1996; Arnold et al., 1995; McKinley et al., 1991a). Because of its protease-resistance, PrP^{Sc} cannot be easily degraded by cells: this leads to intracellular and surface deposition of PrP^{Sc} and PrP27-30 (McKinley et al, 1991a; Arnold et al., 1995; Vey et al., 1996). Most of the PrP^C molecules in prion-infected neuroblastoma cells are not converted. After PrP^C arrives at the cell surface, proteolytic cleavage close to the amino acid 112 (Chen et al., 1995) renders PrP non-convertible and initiates PrP^C degradation.

PrP^C and PrP^{Sc} are associated with specialized domains within cellular membranes (Taraboulos et al., 1995; Vey et al., 1996; Naslavsky et al., 1997). These domains are formed at the Golgi apparatus and are enriched for GPI-anchored proteins, cholesterol and glycosphingolipids (Brown and Rose, 1992). These domains were operationally named rafts or caveolae-like domains (CLDs) because they resemble caveolae with respect to their composition. Attachment of a GPI-anchor is essential for targeting of PrP^C to CLDs and it is prerequisite for efficient conversion into PrP^{Sc} (Taraboulos et al., 1995, Kaneko et al., 1997a). Integrity of CLD structure and function is important for conversion because substances which lower cholesterol content or modify lipids in CLDs interfere with prion conversion (Taraboulos et al., 1995; Kaneko et al., 1997a). Although the exact subcellular compartment(s) for prion conversion have not been identified, CLDs seem to be clearly involved. Identification and purification of the converting compartment(s) would not only advance understanding of the conversion process but it might also lead to identification of other cellular factors involved in conversion, such as the putative protein X. Furthermore, chemicals modifying CLDs, such as cholesterol-reducing drugs, might be promising therapeutics to stop prion infection.

Whether chaperones are involved in the conversion has not been clearly shown but scrapie-infected cells have an abnormal response to heat shock treatment (Tatzelt et al., 1995). Many studies have identified potential cellular binding proteins for interaction with PrP^C and PrP^{Sc} but physiological binding or even binding in vivo has never been demonstrated (Oesch et al., 1990; Edenhofer et al., 1996; Yehiely et al., 1997; Rieger et al., 1997).

A different class of molecules, namely sulfated glycosaminoglycans, have attracted similar interest because they have been found in amyloid plaques (Snow *et al.*, 1989) and they seem to bind prion proteins (Gabizon *et al.*, 1993; Caughey *et al.*, 1994). They deserve special attention, because they have been successfully used to stop PrP^{Sc} formation in cell culture (Caughey and Raymond, 1993) and they even proved to have protective effects against scrapie infection when administered before inoculation of animals (Kimberlin and Walker, 1986; Diringer and Ehlers, 1991; Ladogana *et al.*, 1992). It is suggested that prion proteins might interact with endogenous glycosaminoglycans and that inhibition of this interaction might lead to the observed interference with formation of PrP^{Sc}. Sulfated glycans also have stimulating effects in *in vitro* conversion reactions (Wong *et al.*, 2001). Further studies might yield more interesting facts about these molecules and their involvement in physiological or pathological functions of prion proteins.

12. INHIBITORS OF PRION CONVERSION

Since the successful infection of cell lines with prions, candidate inhibitors of PrP conversion could be studied much faster than in infectivity bioassays. Several drugs which interfered with PrP conversion in cell culture such as glycosaminoglycans, also proved to delay prion infection in infected animals. This again demonstrated the close link between PrP conversion and prion replication. Many compounds with different chemical properties and differing mecanisms of action have been shown to inhibit PrP conversion. Inhibitors of enzymes involved in cholesterol or glycosphingolipid biosynthesis reduce PrP^{Sc} formation (Taraboulos *et al.*, 1995; Kaneko *et al.*, 1997a). Dominant negative inhibition of PrP conversion by certain mutant prion proteins and determination of tertiary structure of PrP^{C} have paved the way to structure-based drug design: chemicals adopting a tertiary structure similar to the protein X binding site on PrP were shown to abrogate PrP^{Sc} formation in scrapie-infected cells (Perrier *et al.*, 2000). Attempts to unfold PrP^{Sc} have also shown initial success: short peptides which can break up β-sheets in proteins can make PrP^{Sc} sensitive to protease and delay scrapie incubation period (Soto *et al.*, 2000). A similar unfolding of PrP^{Sc} seems to be mediated by polyamines (Supattapone *et al.*, 2001b). Upon incubation of scrapie-infected cells with lysosomotropic reagents and inhibitors of cysteine proteases, PrP^{Sc} disappeared from cells (Doh-Ura *et al.*, 2000). With increasing numbers of such candidate therapeutic lead compounds, the symptomatic or prophylactic treatment of prion diseases seems to be a realistic goal.

13.   VARIANT CJD AND BSE

In 1996, a new form of Creutzfeldt-Jakob disease was described which seemed to affect mainly younger persons. Not only clinical symptoms but also neuropathologic findings indicated that this was a novel prion disease (Will *et al.*, 1996). The appearance of this new form of CJD ca. ten years after the diagnosis of BSE in the UK (Wells et al, 1987; Hope *et al.*, 1988) further lead to the conclusion that vCJD may be linked to BSE. In addition, this particular prion strain did not seem to be restricted by species barriers because several other mammals such as cats or exotic animals in zoos, which were fed with BSE-contaminated food, also developed prion disease (Kirkwood and Cunningham, 1994; Wyatt *et al.*, 1991). When macaques were injected with BSE brain material, they developed a BSE-like disease as well, showing that there is no effective species barrier between primates and cattle either (Lasmézas *et al.*, 1996). In the meantime, nearly 100 cases of vCJD have been diagnosed and the numbers are still increasing. Several independent experimental approaches have further indicated that vCJD is the human form of BSE and both resemble a distinct prion strain. Thus, when BSE and vCJD are transmitted to tg mice expressing bovine PrP, there does not seem to be a species barrier for vCJD because both transmit efficiently to these mice with short incubation times even in the first passage (Scott *et al.*, 1999b). Pathologic lesions in the brain of these mice are very similar between vCJD and BSE but distinct from mice infected with a sheep scrapie strain.

Finally, the size of the PrP27-30 in combination with a typical glycosylation pattern of PrP^{Sc} shared by BSE and vCJD clearly differentiated them from the already known human prion strains (Collinge *et al.*, 1996). In contrast to other human prion diseases and to BSE, vCJD prions can be detected in lymphoid organs such as tonsils, lymph nodes and appendix (Hill *et al.*, 1999). All of the vCJD victims were homozygous for methionine at position 129: this might indicate that methionine at this position renders human PrP more sensitive to conversion by the BSE strain but it does, by no means, show that individuals homozygous for valine or heterozygous Met/Val are protected against BSE or vCJD. The advent of specific *post mortem* screening tests for cattle have now been installed into routine veterinary inspection for cattle older than 30 months within the European Union. These *post mortem* tests are sensitive enough to detect BSE-infected cattle in an early phase of disease and even some animals which had not shown signs of disease could be prevented from entering the human food. All of these tests are based on the detection of PrP^{Sc} in material derived from the medulla oblongata, where BSE prions can be found earliest. Despite the obvious disadvantages of these tests, there is no specific *ante mortem* test for the diagnosis of prion diseases available today, mainly due to the low prion levels in easily accessible tissues or bodily fluids such as blood.

Latest studies on the differences of gene expression in some erythroid cells of scrapie-infected animals give rise to the hope that there may be surrogate markers which could indicate presence of prions soon after infection. The presently used BSE tests in combination with many other protective measures, such as removal of risk material in the production of food and medicines, already help to limit spread of BSE and fatal vCJD epidemic. Similar to the disappearance of the prion disease Kuru, which was accomplished by deserting cannibalistic rituals, minimalizing exposure to BSE will then probably lead to an end of vCJD transmission in the future.

14. FUTURE PERSPECTIVES

The discovery of prions has paved the way to diagnostic screening tests to combat spread of BSE and vCJD epidemics. PrP biology and disease have been studied in detail and significant progress has been made in elucidating the mechanism by which pathogenic properties of prions are replicated by prion proteins without the amplification of a pathogen-specific genome. Important questions about the mechanism of nerve cell destruction in prion disease and the PrP conversion itself remain to be answered. With the improvement of protein analytics, high throughput crystallization technologies and powerful animal model systems, answers to these questions might be found in the near future. With more sensitive detection systems on the horizon, safety of food and therapeutics will also reach higher standards which seems necessary with respect to the recent cases of BSE in many European countries.

REFERENCES

Alper, T., Cramp, W. A., Haig, D. A. and Clarke, M. C. 1967. Does the agent of scrapie replicate without nucleic acid ? *Nature (London)* **214:** 764-766

Arnold, J.E., Tipler, C., Laszlo, L., Hope, J., Landon, M., and Mayer, R.J., 1995, The abnormal isoform of the prion protein accumulates in late-endosome-like organelles in scrapie-infected mouse brain. *J.Pathol.* **176**: 403-411

Baron, H., Safar, J., Groth, D.G., DeArmond, S.J., and Prusiner, S.B., 1999, Biosafety issues in prion diseases. In *Prion biology and disease* (S.B. Prusiner, ed.), Cold Spring Harbor Laboratory Press, Cold Spring Harbor, New York, 743-777

Basler, B., Oesch, B., Scott, M., Westaway, D., Wälchli, M., Groth, D.F., McKinley, M.P., Prusiner, S.B., and Weissmann, C., 1986, Scrapie and cellular PrP isoforms are encoded by the same chromosomal gene. *Cell* **46**:417-428

Behrens, A., Brandner, S., Genoud, N., and Aguzzi, A., 2001, Normal neurogenesis and scrapie pathogenesis in neural grafts lacking the prion protein homologue doppel. *EMBO Rep.* **2**:347-352

Bellinger-Kawahara, C., Cleaver, J. E., Diener, T. O. and Prusiner, S. B. 1987. Purified scrapie prions resist inactivation by UV irradiation. *J. Virol.* **61**: 159-166

Bellinger-Kawahara, C., Kempner, E., Groth, D., Gabizon, R., and Prusiner, S. B. 1988. Scrapie prion liposomes and rods exhibit target sizes of 55,000 Da. *Virology.* **164**: 537-541

Bendheim, P.E., Brown, H.R., Rudelli, R.D., Scala, L.J., Goller, N.L., Wen, G.Y., Kascsak, R., Cashman, N., and Bolton, D.C., 1992, Nearly ubiquitous tissue distribution of the scrapie agent precursor protein. *Neurology* **42**:149-156

Bessen, R. A. and Marsh, R.F. 1994. Distinct PrP properties suggest the molecular basis of strain variation in transmissible mink encephalopathy. *J. Virol.* **68**: 7859-7868

Bessen, R. A., Kocisko, D. A., Raymond, G. J., Nandan, S., Lansbury, P. T. and Caughey, B., 1995, Non-genetic propagation of strain-specific properties of scrapie prion protein. *Nature (London)* **375**: 698-700

Bessen, R.A:, Raymond, G.J., and Caughey, B., 1997, *In situ*-formation of protease-resistant prion protein in transmissible spongiform encephalopathy-infected brain slices. *J.Biol.Chem.* **272**: 15227-15231

Billette de Villemeur, T., Deslys, J.P., Pradel, A, Soubrie, C., Alperovitch, A., Tardieu, M, Chaussain, J.L., Hauw, J.J., Dormont, D., Ruberg,M, and Agid, Y., 1996, Creutzfeldt-Jakob disease from contaminated growth hormone extracts in France. *Neurology* **47**: 690-695

Bolton, D. C., McKinley, M. P. and Prusiner, S. B., 1982., Identification of a protein that purifies with the scrapie prion. *Science* **218**: 1309-1311

Bolton, D.C., Meyer, R.K., and Prusiner, S.B., 1985, Scrapie PrP27-30 is a sialoglycoprotein. *J. Virol.* **53**: 596-606

Bolton, D.C., Bendheim, P.E., Marmorstein, A.D., and Potempska, A., 1987, Isolation and structural studies of the intact scrapie agent protein. *Arch. Biochem.Biophys.* **258**:579-590

Borchelt, D. R., Scott, M., Taraboulos, A., Stahl, N. and Prusiner, S. B. 1990. Scrapie and cellular prion proteins differ in their kinetics of synthesis and topology in cultured cells. *J. Cell Biol.* **110**: 743-752

Borchelt, D.R., Taraboulos, A., and Prusiner, S.B., 1992, Evidence for synthesis of scrapie prion proteins in the endocytic pathway. *J.Biol.Chem.* **267**:16188-16199

Brandner, S., Isenmann, S., Raeber, A., Fischer, M., Sailer, A., Kobayashi, Y., Marino, S., Weissmann, C., and Aguzzi, A., 1996. Normal host prion necessary for scrapie-induced neurotoxicity. *Nature (London)* **379**: 339-343.

Brown, D.A. and Rose, J.K., 1992, Sorting of GPI-anchored proteins to glycolipid-enriched membrane subdomains during transport to the apical cell surface. *Cell* **68**:533-544

Brown, D.R., Schmidt, B., and Kretzschmar, H.A. 1996. Role of microglia and host prion protein in neurotoxicity of a prion protein fragment. *Nature (London)* **380**: 345-347

Brown, D. R., Qin, K., Herms, J. W., Madlung, A., Manson, J., Strome, R., Fraser, P. E., Kruck, T., von Bohlen, A., Schulz-Schaeffer, W., Giese, A., Westaway, D., and Kretzschmar, H. 1997. The cellular prion protein binds copper *in vivo*. *Nature (London)* **390**: 684-687

Brown, D.R., Schmidt, B., and Kretzschmar, H.A. 1998. Prion protein fragments interact with PrP-deficient cells. *J. Neurosci. Res.* **52**: 260-267

Brown, K. L., Stewart, K., Ritchie, D. L., Mabbott, N. A., Williams, A., Fraser, H., Morrison, W. I. and Bruce, M. E. 1999a. Scrapie replication in lymphoid tissues depends on prion protein-expressing follicular dendritic cells. *Nature Medicine.* **5**: 1308-1312

Brown, P., Cervenakova, L., McShane, L.M., Barber, P., Rubenstein, R., and Drohan, W.N., 1999b, Further studies of blood infectivity in an experimental model of transmissible spongiform encephalopathy, with an explanation of why blood components do not transmit Creutzfeldt-Jakob disease in humans. *Transfusion* **39**: 1169-1178

Büeler, H., Fischer, M., Lang, Y., Bluethmann, H., Lipp, H.-P., DeArmond, S. J., Prusiner, S. B., Aguet, M. and Weissmann, C. 1992. Normal development and behaviour of mice lacking the neuronal cell-surface PrP protein. *Nature (London)* **356**: 577-582

Büeler, H., Aguzzi, A., Sailer, A., Greiner, R.-A., Autenried, P., Aguet, M. and Weissmann, C. 1993. Mice devoid of PrP are resistant to scrapie. *Cell* **73**: 1339-1347

Butler, D.A., Scott, M.R.D., Bockman, J.M., Borchelt, D., Taraboulos, A., hsiao, K., Kingsbury, D.T., and Prusiner, S.B., 1988, Scrapie-infected neuroblastoma cells produce protease-resistant prion proteins. *J.Virol.* **62**:1558-1564.

Calzolai, L., Lysek, D.A., Güntert, P., von Schroetter, C., Riek, R., Zahn, R., and Wüthrich, K., 2000, NMR structure of three single-residue variants of the human prion protein. *Proc.Natl.Acad.Sci.USA* **97**:8340-8345

Caughey, B., and Raymond, G.J., 1991, The scrapie-associated form of PrP is made from a cell surface precursor that is both protease- and phospholipase-sensitive. *J.Biol.Chem.* **266**:18217-18223.

Caughey, B., and Raymond, G., 1993, Sulfated polyanion inhibition of scrapie-associated PrP accumulation in cultured cells. *J.Virol.* **67**:643-650

Caughey, B., Brown, K., Raymond, G., Katzenstein, G.E., and Tresher, W., 1994, Binding of the protease-sensitive form of prion protein PrP to sulfated glycosaminoglycan and congo red. *J. Virol* **68**:2135-2141

Chandler, R. L. 1961. Encephalopathy in mice produced by inoculation with scrapie brain material. *Lancet* **I**: 1378-1379.

Chen, S.G., Teplow, D.B., Parchi, P., Teller, J.K., Gambetti, P., and Autilio-Gambetti, L., 1995, Truncated forms of the human prion protein in normal brain and in prion diseases. *J.Biol.Chem.* **270**:19173-19180

Chesebro, B., Race, R., Wehrly, K., Nishio, J., Bloom, M., Lechner, D., Bergstrom, S., Robbins, K., Mayer, L., Keith, J. M., Garon C. and Haase A. 1985. Identification of scrapie prion protein-specific mRNA in scrapie-infected and uninfected brain. *Nature (London)* **315**: 331-333

Cohen, F.E., and Prusiner, S.B., 1999, Pathologic conformation of prion proteins. *Annu.Rev.Biochem.* **67**:793-819

Collinge, J., Whittington, M. A., Sidle, K. C., Smith, C. J., Palmer, M. S., Clarke, A. R. and Jefferys, J. G. R. 1994. Prion protein is necessary for normal synaptic function. *Nature (London)* **370**: 295-297

Collinge, J., Sidle, K. C. L., Meads, J., Ironside, J. and Hill, A. F. 1996. Molecular analysis of prion strain variation and the aetiology of "new variant" CJD. *Nature (London)* **383**: 685-690

Come, J.H., Fraser, P.E., and Lansbury, P.T., 1993, A kinetic model for amyloid formation in prion dieases: importance of seeding. *Proc.Nal.Acad.Sci.USA* **90**: 5959-5963

DeArmond, S.J., and Ironside, J.W., 1999, Neuropathology of prion disease. In *Prion biology and disease* (S.B. Prusiner, ed.), Cold Spring Harbor Laboratory Press, Cold Spring Harbor, New York, 585-652

Diringer, H., and Ehlers, B., 1991, Chemoprophylaxis of scrapie in mice. *J.Gen.Virol.* **72**: 457-460

Donne, D. G., Viles, J. H., Groth, D., Mehlhorn, I., James, T. L., Cohen, F. E., Prusiner, S. B., Wright, P. E. and Dyson, H. J. 1997. Structure of the recombinant full-length hamster prion protein PrP (29-231): The N-terminus is highly flexible. *Proc. Natl. Acad. Sci. USA* **94**: 13452-13457

Doh-Ura, K., Iwaki, T., and Caughey, B., 2000, Lysosomotropic agents and cysteine protease inhibitors inhibit scrapie-associated prion protein accumulation. *J.Virol.* **74**: 4894-4897

Edenhofer, F., Rieger, R., Famulok, M., Wendler, W., Weiss, S., and Winnacker, E.-L., 1996, Prion Protein C interacts with molecular chaperones of the Hsp60 family. *J.Virol.* **70**:4724-4728

Eigen, M., 1996, Prionics or the kinetic basis of prion diseases. *Biophys.Chem.* **63**:118

Endo, T., Groth, D., Prusiner, S. B. and Kobata, A. 1989. Diversity of oligosaccharide structures linked to asparagines of the scrapie prion protein. *Biochemistry* **28**: 8380-8388

Fischer, M., Rülicke, T., Raeber, A., Sailer, A., Moser, M., Oesch, B., Brandner, S., Aguzzi, A. and Weissmann, C. 1996. Prion protein (PrP) with amino-proximal deletions restoring susceptibility of PrP knockout mice to scrapie. *EMBO J.* **15**: 1255-1264

Forloni, G., Angeretti, N., Chiesa, R., Monzani, E., Salmona, M., Bugiani, O. and Tagliavini, F.,1993. Neurotoxicity of a prion protein fragment. *Nature (London)* **362**: 543-546

Fraser, H. and Dickinson, A. G. 1970. Pathogenesis of scrapie in the mouse: the role of the spleen. *Nature (London)* **226**: 462-463

Gabizon, R., McKinley, M.P., and Prusiner, S.B. 1987. Purified prion proteins and scrapie infectivity copartition into liposomes. *Proc.Natl.Acad.Sci. USA* **84**:4017-4021

Gabizon, R., Meiner, Z., Halimi, M., and Ben-Sasson, S.A., 1993, Heparin-like molecules bind differentially to prion-proteins and change their intracellular metabolic fate. *J.Cell. Physiol.* **157**:319-325

Gajdusek, D. C. 1977. Unconventional viruses and the origin and disappearance of kuru. *Science* **197**: 943-960

Gajdusek, D. C., Gibbs, C. J., Jr., and Alpers, M. 1966. Experimental transmission of a kuru-like syndrome to chimpanzees. *Nature (London)* **209**: 794-796

Gambetti, P., Petersen, R.P., Parchi, P., Chen, S.G., Capellari, S., Goldfarb, L., Gabizon, R., Montagna, P., Lugarese, E., Piccardo, P., and Ghetti, B., 1999, Inherited prion diseases. In *Prion biology and disease* (S.B. Prusiner, ed.), Cold Spring Harbor Laboratory Press, Cold Spring Harbor, New York, 509-583.

Gibbs, C. J., Jr., Gajdusek, D. C., Asher, D. M., Alpers, M. P., Beck, E., Daniel, P. M. and Matthews, W. B. 1968. Creutzfeldt-Jakob disease (spongiform encephalopathy): Transmission to the chimpanzee. *Science* **161**: 388-389

Griffith, J. S. 1967. Self-replication and scrapie. *Nature (London)* **215**: 1043-1044

Hadlow, W.J., Kennedy, R.C., and Race, R.E., 1982, Natural infection of Suffolk sheep with scrapie virus. *J. Infect.Dis.***146**: 657-664

Hegde, R. S., Mastrianni, J. A., Scott, M. R., DeFea, K. A., Tremblay, P., Torchia, M., DeArmond, S. J., Prusiner, S. B. and Lingappa, V. R. 1998, A transmembrane form of the prion protein in neurodegenerative disease. *Science* **279**: 827-834

Hegde, R. S., Tremblay, P., Groth, D., DeArmond, S. J., Prusiner, S. B., and Lingappa, V. R., 1999, Transmissible and genetic prion disease share a common pathway of neurodegeneration. *Nature (London)* **402**: 822-826

Hill, A.F., Butterworth, R.J., Joiner, S., Jackson, G., Rossor, N.M., Thomas, D.J., Frosh, N., Tolley, N., Bell, J.E., Spencer, M., King, A., Al-Sarraj, S., Ironside, J.W., Lantos, P.L., and Collinge, J., 1999, Investigation of variant Creutzfeldt-Jakob disease and other human prion diseases with tonsil biopsy samples. *Lancet* **353**:183-189.

Hill, A.F., Joiner, S., Linehan, J., Desbruslais, M., Lantos, P.L., and Collinge, J., 2000, Species-barrier-independent prion replication in apparently resistant species. *Proc.Natl.Acad.Sci.USA* **97**:10248-53.

Hölscher, C., Deljus, H., and Bürkle, A., 1998, Overexpression of nonconvertible PrPC delta114-121 in scrapie-infected mouse neuroblastoma cells leads to trans-dominant inhibition of wild-type PrPSc accumulation. *J.Virol.* **72**:1153-1159

Hope, J., Morton, L.J.D., Farquhar, C.F., Multhaup, G., Beyreuther, K., and Kimberlin, R.H., 1986, The major polypeptide of scrapie-associated fibrils (SAF) has the same size, charge distribution and N-terminal protein sequence as predicted for the normal brain protein (PrP). *EMBO J.* **5**:2591-2597.

Hope, J., Reekie, L. J. D., Hunter, N., Multhaup, G., Beyreuther, K., White, H., Scott, A. C., Stack, M. J., Dawson, M. and Wells, G. A. H. 1988. Fibrils from brains of cows with new cattle disease contain scrapie-associated protein. *Nature (London)* **336**: 390-392

Horiuchi, M., Priola, S.A., Chabry, J., and Caughey, B., 2000, Interactions between heterologous forms of prion protein: binding, inhibition of conversion, and species barriers. *Proc.Natl.Acad,Sci.USA* **97**:5836-5841

Hsiao, K., Baker, H. F., Crow, T. J., Poulter, M., Owen, F., Terwilliger, J. D., Westaway, D., Ott, J. and Prusiner, S. B. 1989. Linkage of a prion protein missense variant to Gerstmann-Straussler syndrome. *Nature (London)* **338**: 342-345

Hsiao, K., Scott, M., Foster, D., Groth, D.F., DeArmond, S.J., and Prusiner, S.B., 1990, Spontaneous neurodegeneration in transgenic mice with mutant prion protein. *Science* **250**:1587-1590

Hsiao, K., Groth, d., Sc*ott, M.,* Yang, S.-L., Serban, H., Rapp, D., Foster, D., Torchia, M., DeArmond, S.J., and Prusiner, S.B., 1994, Serial transmission in rodents of neurodegeneration from transgenic mice expressing mutant prion protein. *Proc.Natl.Acad.Sci.USA* **91**:9126-9130.

Huang, Z., Prusiner, S.B., and Cohen, F.E., 1995, Scrapie prions: a three-dimensional model of an infectious fragment. *Fold.Des.* **1**:13-19

James, T. L., Liu, H., Ulyanov, N. B., Farr-Jones, S., Zhang, H., Donne, D. G., Kaneko, K., Groth, D., Mehlhorn, I., Prusiner, S. B. and Cohen, F. E. 1997. Solution structure of a 142-residue recombinant prion protein corresponding to the infectious fragment of the scrapie isoform. *Proc. Natl. Acad. Sci. USA* **94**: 10086-10091

Kaneko, K., Vey, M., Scott, M., Pilkuhn, S., Cohen, F. E. and Prusiner, S. B. 1997a. COOH-terminal sequence of the cellular prion protein directs subcellular trafficking and controls conversion into the scrapie isoform. *Proc. Natl. Acad. Sci. USA* **94**: 2333-2338

Kaneko, K., Zulianello, L., Scott, M., Cooper, C. M., Wallace, A. C., James, T. L., Cohen, F.E., and Prusiner, S. B. 1997b. Evidence for protein X binding to a discontinuous epitope on the cellular prion protein during scrapie prion propagation. *Proc. Natl. Acad. Sci. USA* **94**: 10069-10074

Kaneko, K, Ball, H., Wille, H, Zhang, H., Torchia, M., Tremblay, P., Safar, J., Prusiner, S.B., DeArmond, S.J., Baldwin, M.A., and Cohen, F.E., 2000, A synthetic peptide initiates Gerstmann-Sträussler-Scheinker disease in transgenic mice. *J. Mol.Biol.* **295**:997-1007

Kascsak, R. J., Rubenstein, R., Merz, P. A., Tonna-DeMasi, M., Fersko, R., Carp, R. I., Wisniewski, H. M. and Diringer, H. 1987. Mouse polyclonal and monoclonal antibody to scrapie-associated fibril proteins. *J. Virol.* **61**: 3688-3693

Kellings, K., Meyer, N., Mirenda, C., Prusiner, S. B. and Riesner, D. 1992. Further analysis of nucleic acids in purified scrapie prion preparations by improved return refocussing gel electrophoresis (RRGE). *J. Gen. Virol.* **73**: 1025-1029

Kimberlin, R.H., and Walker, C.A., 1979, Pathogenesis of mouse scrapie: Dynamics of agent replication in spleen, spinal cord and brain after infection by different routes. *J.Comp.Pathol.* **89**: 551-562

Kimberlin, R., and Walker, C.A., 1986, Suppression of scrapie infection in mice by heteropolyanion 23, dextran sulfate, and some other polyanions. *Antimicrobial Agents and Chemotherapy* **30**:409-413

Kirkwood, J.K., and Cunningham, A.A., 1994, Epidemiological observations on spongiform encephalopathies in captive wild animals in the British isles. *Vet.Rec.* **135**:296-303

Klein, M.A., Frigg, R., Flechsig, E., Raeber, A.J., Kalinke, U., Bluethmann, H., Bootz, F., Suter, M., Zinkernagel, R.M., and Aguzzi, A, 1997, A crucial role for B-cells in neuroinvasive scrapie. *Nature (London)* **390**:687-690.

Klein, M. A., Kaeser, P. S., Schwarz, P., Weyd, H., Xenarios, I., Zinkernagel, R. M., Carroll, M. C., Verbeek, J. S., Botto, M., Walport, M. J., Molina, H., Kalinke, U., Acha-Orbea, H. and Aguzzi, A. 2001. Complement facilitates early prion pathogenesis. *Nature medicine* **7**: 488-492

Kocisko, D.A., Come, J.H., Priola, S.A., Chesebro, B, Raymond, G.J., Lansbury, P.T., and Caughey, B., 1994, Cell-free formation of protease-resistant prion protein. *Nature (London)* **370:** 471-474

Kopp, N., Streichenberger, N., Deslys, J.P., Laplanche, J.L., and Chazot, G., 1996, Creutzfeldt-Jakob disease in a 52 year old woman with florid plaques. *Lancet* **348:** 1239-1240

Korth, K., Kaneko, K., and Prusiner, S.B., 2000, Expression of unglycosylated mutated prion protein facilitates PrPSc formation in neuroblastoma cells infected with different prion strains. *J.Gen.Virol.* **81:**2555-2563

Ladogana, A., Casaccia, P., Ingrosso, L., Cibati, M., Salvatore, M., Xi, Y., Masullo, C., and Pocchiari, M., 1992, Sulphate polyanions prolong the incubation period of scrapie-infected hamsters. *J.Gen.Virol.* **73:** 661-665

Laplanche, J.-L., Hunter, N., Shinagawa, M., and Williams, E., 1999, Scrapie, chronic wasting disease, and mink encephalopathy. In *Prion biology and disease* (S.B. Prusiner, ed.), Cold Spring Harbor Laboratory Press, Cold Spring Harbor, New York, 393-429

Lasmézas, C. I., Deslys, J.-P., Demaimay, R., Adjou, K. T., Lamoury, F., Dormont, D., Robain, O., Ironside, J. and Hauw, J.-J. 1996. BSE transmission to macaques. *Nature (London)* **381:** 743-744

Lasmézas, C. I., Deslys, J.-P., Robain, O., Jaegly, A., Beringue, V., Peyrin, J.-M., Fournier, J.-G., Hauw, J.-J., Rossier, J. and Dormont, D. 1997. Transmission of the BSE agent to mice in the absence of detectable abnormal prion protein. *Science* **275:** 402-405

Lehman, S., Harris, D.A., 1996, Mutant and infectious prion proteins display common biochemical properties in cultured cells. *J.Biol.Chem.* **271:**1633-1637

Lledo, P.-M., Tremblay, P., DeArmond, S. J., Prusiner, S. B. and Nicoll, R. A. 1996. Mice deficient for prion protein exhibit normal neuronal excitability and synaptic transmission in the hippocampus. *Proc. Natl. Acad. Sci. USA* **93:** 2403-2407

McKinley, M. P., Bolton, D. C. and Prusiner, S. B. 1983. A protease-resistant protein is a structural component of the scrapie prion. *Cell* **35:** 57-62

McKinley, M.P., Taraboulos, A., Kenaga, L., Serban, D., Stieber, A., DeArmond, S.J., Prusiner, S.B., and Gonatas, N., 1991a, Ultrastructural localization of scrapie prion proteins in cytoplasmic vesicles of infected cultured cells. *Lab.Invest.* **65:** 622-630

McKinley, M.P., Meyer, R.K., Kenaga, L., Rahbar, F., Cotter, R., Serban, A., and Prusiner, S.B., 1991b, Scrapie rod formation requires both detergent extraction and limited proteolysis. *J.Virol.* **65:**1340-1351.

Merz, P.A., Somerville, R.A., Wisniewski, H.M., and Iqbal, K., 1981, Abnormal fibrils from scrapie-infected brain. *Acta Neuropathol.* **54:** 63-74

Meyer, R. K., McKinley, M. P., Bowman, K. A., Braunfeld, M. B., Barry, R. A. and Prusiner, S. B. 1986. Separation and properties of cellular and scrapie prion proteins. *Proc. Natl. Acad. Sci. USA* **83:** 2310-2314

Mo, H., Moore, R. C., Cohen, F. E., Westaway, D., Prusiner, S. B., Wright, P. E., Dyson, H.J. 2001. Two different neurodegenerative diseases caused by proteins with similar structures. *Proc. Natl. Acad. Sci. USA* **98:** 2352-2357

Monari, L., Chen, S.G., Brown, P., Parchi, P., Petersen, R.B., Mikol, J., Gray, F., Cortelli, P., Montagna, P., Ghetti, B. *et al.*, 1994, Fatal familial insomnia and familial Creutzfeldt-Jakob disease: different prion proteins determined by a DNA polymorphism. *Proc.Natl.Acad.Sci.USA* **91:** 2839-2842

Montrasio, F., Cozzio, A., Flechsig, E., Rossi, D., Klein, M.A., Rulicke, T., Raeber, A.J., Vosshenrich, C.A., Proft, J., Aguzzi, A., and Weissmann, C., 2001, B lymphocyte-restricted expression of prion protein does not enable prion replication in prion protein knockout mice. *Proc. Natl.Acad. Sci. USA* **98:** 4034-4037.

Moore, R.C., Lee, I.Y., Silverman, G.L., Harrison, P.M., Strome, R., Heinrich, C., Karunaratne, A., Pasternak, S.H., Chishti, M.A., *et al.*, 1999, Ataxia in prion protein (PrP)-deficient mice is associated with upregulation of the novel PrP-like protein doppel. *J. Mol.Biol.* **292**: 797-817

Naslavsky, N., Stein, R., Yanai, A., Friedlander, G. and Taraboulos, A. 1997. Characterization of detergent-insoluble complexes containing the cellular prion protein and its scrapie isoform. *J. Biol. Chem.* **272**: 6324-6331

Oesch, B., Westaway, D., Wälchli, M., McKinley, M. P., Kent, S. B. H., Aebersold, R., Barry, R. A., Tempst, P., Teplow, D. B., Hood, L. E., Prusiner, S. B. and Weissmann C. 1985. A cellular gene encodes scrapie PrP 27-30 protein. *Cell* **40**: 735-746

Oesch, B., Teplow, D.B., Stahl, N., Serban, D., Hood, L.E., and Prusiner, S.B., 1990, Identification of cellular proteins binding to the scrapie prion protein. *Biochemistry* **29**:5848-5855

Pan, K.-M., Stahl, N., and Prusiner, S.B., 1992, Purification and properties of the cellular prion protein from Syrian hamster brain. *Protein Science* **1**:1343-1352

Pan, K.-M., Baldwin, M., Nguyen, J., Gasset, M., Serban, A., Groth, D., Mehlhorn, I., Huang, Z., Fletterick, R. J., Cohen, F. E. and Prusiner, S. B. 1993. Conversion of α-helices into β-sheets features in the formation of the scrapie prion proteins. *Proc. Natl. Acad. Sci. USA* **90**: 10962-10966

Pattison, I. H. and Jones, K. M. 1967. The possible nature of the transmissible agent of scrapie. *Vet. Rec.* **80**: 1-8

Peretz, D., Williamson, R. A., Matsunaga, Y., Serban, H., Pinilla, C., Bastidas, R., Rozenshteyn, R., James, T. L., Houghten, R. A., Cohen, F. E., Prusiner, S. B. and Burton , D. R. 1997. A conformational transition at the N terminus of the prion protein features in formation of the scrapie isoform. *J. Mol. Biol.* **273**: 614-622

Peretz, D., Scott, M., Groth, D., Williamson, A., Burton, D.R., Cohen, F.E., Prusiner, S.B., 2001, Strain-specified relative conformational stability of the scrapie prion protein. *Protein Science* **10**: 854-863

Perrier, V., Wallace, A.C., Kaneko, K., Safar, J., Prusiner, S.B., and Cohen, F.E., 2000, Mimicking dominant negative inhibition of prion replication through structure-based drug design. *Proc.Natl.Acad.Sci.USA* **97**: 6073-6078

Priola, S.A., Caughey, B., Race, R., and Chesebro, B., 1994, Heterologous PrP molecules interfere with accumulation of protease-resistant PrP in scrapie-infected murine neuroblastoma cells. *J.Virol.* **68**:4873-4878

Priola, S.A, and Chesebro, B., 1995, A single hamster PrP amino acid blocks conversion to protease-resistant PrP in scrapie-infected mouse neuroblastoma cells. J.Virol. **69**:7754-7758

Prusiner, S. B., Groth, D. F., Cochran, S. P., Masiarz, F. R., McKinley, M. P. and Martinez, H. M. 1980. Molecular properties, partial purification, and assay by incubation period measurements of the hamster scrapie agent. *Biochemistry* **19**: 4883-4891

Prusiner, S.B., McKinley, M.P., Groth, D.F., Bowman, K.A., Mock, N.I., Cochran, P., and Masiarz, F.R., 1981a, Scrapie agent contains a hydrophobic protein. *Proc.Natl.Acad.Sci.USA* **78**:6675-6679.

Prusiner, S.B., Groth, D.F., McKinley, M.P., Cochran, P., Bowman, K.A., and Kasper, K.C., 1981b, Thiocyanate and hydroxyl ions inactivate the scrapie agent. *Proc.Natl.Acad.Sci.USA* **78**:4606-4610

Prusiner, S. B. 1982. Novel proteinaceous infectious particles cause scrapie. *Science* **216**: 136-144

Prusiner, S. B., Bolton, D. C., Groth, D. F., Bowman, K. A., Cochran, S. P. and McKinley, M. P. 1982. Further purification and characterization of scrapie prions. *Biochemistry* **21**: 6942-6950

Prusiner, S. B., McKinley, M. P., Bowman, K. A., Bolton, D. C., Bendheim, P. E., Groth, D. F. and Glenner, G. G. 1983. Scrapie prions aggregate to form amyloid-like birefringent rods. *Cell* **35**: 349-358

Prusiner, S.B., Groth, D., Bolton, D.C., Kent, S.B., and Hood, L.E., 1984, Purification and structural studies of a major scrapie prion protein. *Cell* **38**:127-134

Prusiner , S.B., Scott, M., Foster, D., Pan, K.-M., Groth, D., Mirenda, C., Torchia, M., Yang, S.-l., Serban, D., Carlson, G.A:, Hoppe, P.C., Westaway, D., and DeArmond, S.J., 1990, Transgenetic studies implicate interactions between homologous PrP isoforms in scrapie prion replication. *Cell* **63**:673-686

Prusiner, S. B., Groth, D., Serban, A., Koehler, R., Foster, D., Torchia, M., Burton, D., Yang, S.-L. and DeArmond, S. J. 1993. Ablation of the prion protein (PrP) gene in mice prevents scrapie and facilitates production of anti-PrP antibodies. *Proc. Natl. Acad. Sci. USA* **90**: 10608-10612

Prusiner, S.B., Tremblay, P., Safar, J., Torchia, M., and DeArmond, S.J., 1999, Bioassays of prions. In *Prion biology and disease* (S.B. Prusiner, ed.), Cold Spring Harbor Laboratory Press, Cold Spring Harbor, New York pp.113-145

Prusiner, S.B., 1998, Prions. *Proc.Natl.Acad.Sci.USA* **95**:13363-13383.

Race, R.E., Fadness, L.H., and Chesebro, B., 1987, Characterization of scrapie infection in mouse neuroblastoma cell. *J.Gen.Virol.* **68**:1391-1399.

Raeber, A.J., Borchelt, D.R., Scott, M., and Prusiner, S.B., 1992, Attempts to convert the cellular prion protein into the scrapie isoform in cell-free systems. *J.Virol.* **66**: 6155-6163

Rieger, R., Edenhofer, F., Lasmézas, C.I., and Weiss, S., 1997, The human 37kDa laminin receptor precursor interacts with the prion protein in eukaryotic cells. *Nat.Med.* **3**:1383-1388

Riek, R., Hornemann, S., Wider, G., Billeter, M., Glockshuber, R. and Wüthrich, K. 1996. NMR structure of the mouse prion protein domain PrP (121-131). *Nature (London)* **382**: 180-182

Safar, J., Wille, H., Itri, V., Groth, D., Serban, H., Torchia, M., Cohen, F. E. and Prusiner, S. B. 1998. Eight prion strains have PrP^Sc molecules with different conformations. *Nat. Med.* **4**: 1157-1165

Sakaguchi, S., Katamine, S., Nishida, N., Moriuchi, R., Shigematsu, K., Sugimoto, T., Nakatani, A., Kataoka, Y., Houtani, T., Shirabe, S., Okada, H., Hasegawa, S., Miyamoto, T. and Noda, T. 1996. Loss of cerebellar Purkinje cells in aged mice homozygous for a disrupted PrP gene. *Nature (London)* **380**: 528-531

Schätzl, H.M., Laszlo, L., Holtzmann, D.M., Tatzelt, J., DeArmond, S.J., Weiner, R.I., Mobley, W.C., and Prusiner, S.B., 1997, A hypothalamic neuronal cell line persistently infected with scrapie prions exhibits apoptosis. *J.Virol* **71**:8821-8831.

Scott, M., Foster, D., Mirenda, C., Serban, D., Coufal, F., Wälchli, M, Torchia, M., Groth, D., Carlson, G., DeArmond, S.J., Westaway, D., and Prusiner, S.B., 1989, Transgenic mice expressing hamster prion protein produce species-specific scrapie infectivity and amyloid plaques. *Cell* **59**: 847-857

Scott, M., Groth, D., Foster, D., Torchia, M., Yang, S.-L., DeArmond, S. J. and Prusiner, S. B. 1993. Propagation of prions with artificial properties in transgenic mice expressing chimeric PrP genes. *Cell* **73**: 979-988

Scott, M., DeArmond, S.J., Prusiner, S.B., Ridley, R.M., and Baker, H.F., 1999a, Transgenetic investigations of the species barrier and prion strains. In *Prion biology and disease* (S.B. Prusiner, ed.), Cold Spring Harbor Laboratory Press, Cold Spring Harbor, New York, 307-342

Scott, M., Will, R., Ironside, J, Ngyen, H.-O. B., Tremblay, P., DeArmond, S.J., and Prusiner, S.B., 1999b, Compelling transgenetic evidence for transmission of bovine spongiform encephalopathy prions to humans. *Proc.Natl.Acad.Sci.USA* **96**:15137-15142

Shmerling, D., Hegyi, I., Fischer, M., Blättler, T., Brandner, S., Götz, J., Rülicke, T., Flechsig, E., Cozzio, A., von Mering, C., Hangartner, C., Aguzzi, A., and Weissman, C., 1998, Expression of amino-terminally truncated PrP in the mouse leading to ataxia and specific cerebellar lesions. *Cell* **93**: 203-214.

Sigurdsson, B. 1954. Rida, a chronic encephalitis of sheep with general remarks on infections which develop slowly and some of their special characteristics. *Br. Vet. J.* **110**: 341-354

Silverman, G.L., Qin, K., Moore, R.C., Yang, Y., Mastrangelo, P., Tremblay, P., Prusiner, S.B., Cohen, F.E., and Westaway, D., 2000, Doppel is an N-glycosylated, GPI-anchored protein: expression in mice and ectopic production in the brains of PrP0/0 mice predisposed to Purkinje cell loss. *J.Biol.Chem.* **275**: 26834-26841

Snow, A.D., Kisilevsky, R., Willmer, J., Prusiner, S.B., and DeArmond, S.J., 1989, Sulfated glycosaminoglycans in amyloid plaques of prion diseases. *Acta Neuropathol.* **77**:337-342

Soto, C., Kascsak, R.J., Saborio, G.P., Aucouturier, P., Wisniewski, T., Prelli, F., Kascsak, R., Mendez, E., Harris, D.A., Ironside, J., Tagliavini, F., Carp, R.I., and Frangione, B., 2000, Reversion of prion protein conformational changes by synthetic beta-sheet breaker peptides. *Lancet* **355**:192-197

Stahl, N., Borchelt, D. R., Hsiao, K. and Prusiner, S. B. 1987. Scrapie prion protein contains a phosphatidylinositol glycolipid. *Cell* **51**: 229-240

Stahl, N., Borchelt, D.R., and Prusiner, S.B., 1990, Differential release of cellular and scrapie prion proteins from cellular membranes by phosphatidylinositol-specific phospholipase C. *Biochemistry* **29**:5405-5412

Stöckel, J., Safar, J., Wallace, A. C., Cohen, F. E. and Prusiner, S. B. 1998. Prion protein selectively binds copper (II) ions. *Biochemistry* **37**: 7185-7193

Supattapone, S., Bosque, P., Muramoto, T., Wille, H., Aagaard, C., Peretz, D., Nguyen, H.-O., Heinrich, C., Torchia, M., Safar, J., Cohen, F.E., DeArmond, S.J., Prusiner, S.B., and Scott, M., 1999, Prion protein of 106 residues creates an artificial transmission barrier for prion replication in transgenic mice. *Cell* **96**: 869-878

Supattapone, S., Bouzamondo, E., Ball, H. L., Wille, H., Nguyen, H.-O. B., Cohen, F. E., DeArmond, S. J., Prusiner, S. B. and Scott, M. 2001a. A protease-resistant 61-residue prion peptide causes neurodegeneration in transgenic mice. *Molecular and cellular biology.* **21**: 2608-2616

Supattapone, S., Wille, H., Uyechi, L., Safar, J., Tremblay, P., Szoka, F.C., Cohen, F.E., Prusiner, S.B., and Scott, M.R., 2001b, Branched polyamines cure prion-infected neuroblastoma cells. *J.Virol.* **75**: 3453-3461

Taraboulos, A., Rogers, M., Borchelt, D.R., McKinley,M.P., Scott, M., Serban, D., Prusiner, S.B., 1990, Acquisition of protease-resistance by prion proteins in scrapie-infected cells does not require asparagine-linked glycosylation. *Proc.Natl.Acad.Sci.USA* **87**:8262-8266

Taraboulos, A., Raeber, A.J., Borchelt, D.R., Serban, D., and Prusiner, S.B., 1992, Synthesis and trafficking of prion proteins in cultured cells. *Mol.Biol.Cell* **3**:851-863

Taraboulos, A., Scott, M., Semenov, A., Avrami, D., Laszlo, L., and Prusiner, S.B., 1995, Cholesterol depletion and modification of COOH-terminal targeting sequence of the prion protein inhibits formation of the scrapie isoform. *J. Cell Biol.* **129**:121-132

Tatzelt, J., Zuo, J., Voellmy, R., Scott, M., Hartl, U., Prusiner, S. B. and Welch, W. J. 1995. Scrapie prions selectivity modify the stress response in neuroblastoma cells. *Proc. Natl. Acad. Sci. USA* **92**: 2944-2948

Taylor, D.M., 2000, Inactivation of transmissible degenerative encephalopathy agents: a review. *Vet J.* **159**:3-4

Telling, G. C., Scott, M., Mastrianni, J., Gabizon, R., Torchia, M., Cohen, F. E., DeArmond, S. J. and Prusiner, S. B. 1995. Prion propagation in mice expressing human and chimeric PrP transgenes implicates the interaction of cellular PrP with another protein. *Cell* **83**: 79-90

Telling, G. C., Parchi, P., DeArmond, S. J., Cortelli, P., Montagna, P., Gabizon, R., Mastrianni, J., Lugaresi, E., Gambetti, P. and Prusiner, S. B. 1996. Evidence for the conformation of the pathologic isoform of the prion protein enciphering and propagating prion diversity. *Science* **274**: 2079-2082

Tobler, I., Gaus, S. E., Deboer, T., Achermann, P., Fischer, M., Rülicke, T., Moser, M., Oesch, B., McBride, P. A. and Manson, J. C. 1996. Altered circadian activity rhythms and sleep in mice devoid of prion protein. *Nature (London)* **380**: 639-642

Verity, C.M., Nicoll, A., Will, R.G., Devereux, G., and Stellitano, L., 2000, Variant Creutzfeldt-Jakob disease in UK children: a national surveillance study. *Lancet* **356**:1224-1227

Vey, M., Pilkuhn, S., Wille, H., Nixon, R., DeArmond, S. J., Smart, E. J., Anderson, R. G., Taraboulos, A. and Prusiner, S. B. 1996. Subcellular colocalization of the cellular and scrapie prion proteins in caveolae-like membranous domains. *Proc. Natl. Acad. Sci. USA* **93**:14945-14949

Vilette, D., Andreoletti, O., Archer, F., Madelaine, M.F., Vilotte, J.L., Lehman, S., and Laude, H., 2001, Ex vivo propagation of infectious sheep scrapie agent in heterologous epithelial cells expressing ovine prion protein. *Proc.Natl.Acad.Sci.USA* **98**:4055-4059

Wells, G. A. H., Scott, A. C., Johnson, C. T., Gunning, R. F., Hancock, R. D., Jeffrey, M., Dawson, M. and Bradley, R. 1987. A novel progressive spongiform encephalopathy in cattle. *Vet. Rec.* **121**: 419-420

Weissmann, C., Raeber, A.J., Shmerling, D., Aguzzi, A., and Manson, J.C., 1999, Knockouts, transgenics, and transplants in prion research. In *Prion biology and disease* (S.B. Prusiner, ed.), Cold Spring Harbor Laboratory Press, Cold Spring Harbor, New York, 273-305.

Westaway, D., and Prusiner, S.B., 1986, Conservation of the cellular gene encoding the scrapie prion protein. *Nucl.Acid.Res.* **14**:2035-2044.

Westaway, D., Mirenda, C.A., Foster, D., Zebarjadian, Y., Scott, M., Torchia, M., Yang, S.-L., Serban, H., DeArmond, S.J., Ebeling, C., Prusiner, S.B., and Carlson, G.A., 1991. Paradoxical shortening of scrapie incubation times by expression of prion protein transgenes derived from long incubation period mice. *Neuron* **7**: 59-68

Westaway, D., DeArmond, S. J., Cayetano-Canlas, J., Groth, D., Foster, D., Yang, S-L., Torchia, M., Carlson, G. A. and Prusiner, S. B. 1994. Degeneration of skeletal muscle, peripheral nerves, and the central nervous system in transgenic mice overexpressing wild-type prion proteins. *Cell* **76**: 117-129

Will, R. G., Ironside, J. W., Zeidler, M., Cousens, S. N., Estibeiro, K., Alperovitch, A., Poser, S., Pocchiari, M., Hofman, A. and Smith, P. G. 1996. A new variant of Creutzfeldt-Jakob disease in the UK. *Lancet* **347**: 921-925

Will, R.G., Alpers, M.P., Dormont, D., Schonberger, L.B., and Tateishi, J., 1999, Infectious and sporadic prion diseases. In *Prion biology and disease* (S.B. Prusiner, ed.), Cold Spring Harbor Laboratory Press, Cold Spring Harbor, New York, 465-507

Wille, H., Zhang, G.-F., Baldwin, M.A., Cohen, F.E., and Prusiner, S.B., 1996, Separation of scrapie prion infectivity from PrP amyloid polymers. *J.Mol.Biol.* **259**:608-621.

Wong, C., Xiong, L.W., Horiuchi, M., Raymond, L., Wehrly, K., Chesebro, B., and Caughey, B., 2001, Sulfated glycans and elevated temperature stimulate PrP^{Sc}-dependent cell-free formation of protease-resistant prion protein. *EMBO J.* **20**: 377-386

Wyatt, J.M., Pearson, G.R., and Smerdon, T.N., 1991, Naturally occurring scrapie-like spongiform encephalopathy in five domestic cats. *Vet.Rec.* **129**: 233-236

Yehiely, F., Bamborough, P., DaCosta, M., Perry, B.J., Thinakaran, G., Cohen, F.E., Carlson, G.A., and Prusiner, S.B., 1997, Identification of candidate proteins binding to prion protein. *Neurobiology of Disease* **3**:339-355

Zahn, R., Liu, A., Kührs, T., Riek, R., von Schroetter, C., Garcia, F.L., Billeter, M., Calzolai, L., Wider, G., and Wüthrich, K., 2000, NMR solution structure of the human prion protein. *Proc.Natl.Acad.Sci.USA* **97**:145-150

Zobeley, E., Flechsig, Cozzio, A., Enari, M., and Weissmann, C., 1999, Infectivity of scrapie prions bound to a stainless steel surface. *Molecular Medicine* **5**:240-243

Chapter 5.2

Human Immunodeficiency Virus:
From Virus Structure to Pathogenesis

HANS R. GELDERBLOM[*] and KLAUS BOLLER[#]
*Robert Koch-Institut, Nordufer 20, D-13353 Berlin, Germany
#Paul-Ehrlich-Institut, Paul-Ehrlich-Str. 51 — 59, D-63225 Langen, Germany

1.   AIDS, FACTS ON THE EPIDEMIC

Only two decades ago, mankind became aware of a hitherto unknown, deadly, transmissible health risk - observed first in homosexually active young man and in recipients of blood and blood products (CDC, 1981). Epidemiological and virological studies soon established a retrovirus as the common cause (Barré-Sinoussi et al., 1983; Feorino et al., 1984; Levy et al., 1984; Popovic et al., 1984) leading after years of incubation to an acquired immunodeficiency syndrome (AIDS), and ultimately to death.

When serological tests suitable for screening large populations became available (Sarngadharan et al., 1984), an exponential spread of the infection within defined high risk groups - highly promiscous homosexual men, i.v.-drug users and hemophiliacs - was uncovered, initially in the USA and Western Europe. In contrast to the restricted pattern in the Northern hemisphere, a heterosexual spread of the infection became prominent on other continents (reviewed in Piot et al., 2001). The threat of a pandemy promoted intense research on the nature of the etiological agent and on means to control the infection. In 1986, the designation human immunodeficiency virus (HIV) was adopted for the differently named isolates, e.g. LAV (lymphadenopathy-associated virus, Barré-Sinoussi et al., 1983; Feorino et al., 1984); ARV (AIDS-associated retrovirus, Levy et al.,

Structure-Function Relationships of Human Pathogenic Viruses, Edited by
Holzenburg and Bogner, Kluwer Academic/Plenum Publishers, New York, 2002   295

1984); HTLV (Coffin et al., 1986; Popovic et al., 1984). Additional HIV isolates from Westafrica were shown to form another type of HIV and designated therefore HIV-2 (Clavel et al., 1986). Today, the global spread of HIV has exceeded all predictions: 36 million of people are living with the virus – worldwide one out of 167 persons is infected, and about 20 million have already died from AIDS (Piot et al., 2001).

To fight HIV more successfully, both socio-economic changes and an improved understanding of basic viral properties are essential. This review focuses mainly on structural properties of HIV and, based on detailed structural information, functional and pathogenetic mechanisms are deduced. Knowing that HIV is today the best studied virus, we are aware that this article - reflecting also personal experiences and views of the authors - inevitably will miss important aspects, we apologize for that. For readers interested in further information, we have included a number of authoritative reviews in the references.

2.   THE ORIGINS OF HIV, VIRUS CLASSIFICATION AND DIVERSITY

2.1   The origins of HIV

Phylogenetic studies supplied clear evidence that HIV-1 and HIV-2 came to humans from the chimpanzee and the sooty mangabey monkey, respectively. The simian species are naturally infected with simian immunodeficiency viruses (SIV), which are not pathogenic in their natural simian hosts. When the chimpanzee virus (SIVcpz) and the mangabey virus (SIVsm) crossed the species border, the zoonosis adapted to men because of favourable conditions resulting in a dramatic rise in pathogenicity and transmission rate. The original trans-species transmission happened in equatorial Africa, probably several times through hunting and husbandry (for review see Weiss et al., 1999). The later worldwide spread of HIV, however, is man-made, facilitated by dense populations and behavioural changes of humans.

HIV-1 and HIV-2 do not consist of two or a few "wild-type" species. Because of the high mutation frequency - inherent to all RNA viruses – HIV has evolved rapidly into a virus population consisting of many related genotypes. These "quasi-species" (for review see Coffin, 1995; Domingo, 1995; Korber et al., 1998) are able to adapt rapidly by continuous mutations to new environments or new types of cells. In that respect HIV represents a rapidly moving target, able to escape specific immune responses and inappropriate therapeutic attacks.

HIV-1 isolates form three genetically closely related groups: the main (M) group comprising the great majority of virus isolates with subtypes or clades A to K, the outlier group (O) and a very small N (non-M, non-O) group. All three are found in Central Africa in regions corresponding to the distribution of the chimpanzee subspecies Pan *tryglodytes tryglodytes* which is harbouring SIVcpz as a clinically latent virus. HIV groups O and N are still confined to Central Africa, whereas group M has spread worldwide developing into distinct subtypes present on different continents. E.g. subtype B clade prevails in the Americas and Europe, while subtype C and E viruses are devastating sub-Saharan Africa and Thailand, respectively. The initially clear geographic distribution of HIV1 subtypes is increasingly lost due to worldwide trafficking. Furthermore, when different subtypes replicate in one and the same individual, "mosaic" viruses may arise following genetic recombination.

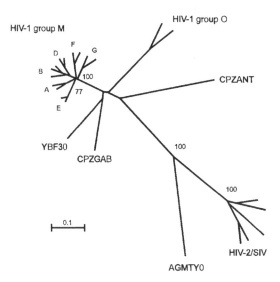

Figure 1: Phylogenetic tree of HIV and SIV, based on env gene alignment, showing a cloud of group M subtypes of HIV-1 and the relation to the HIV/SIV complex. Scale bar indicates genetic relatedness (according to Los Alamos HIV Database, Korber et al., 1998, 2000).

In the analysis of the phylogenetic development, groups M, O and N occur together with respective SIVcpz isolates indicating a shared evolution of these groups and multiple transmissions from monkey to man (Figure 1). By comparing great numbers of sequence data of envelope and gag genes of

HIV-1 group M subtypes and assuming a molecular clock (reflecting a constant rate of genetic changes) the origin of the expanding M group was estimated to the year 1931 (Korber et al., 2000).

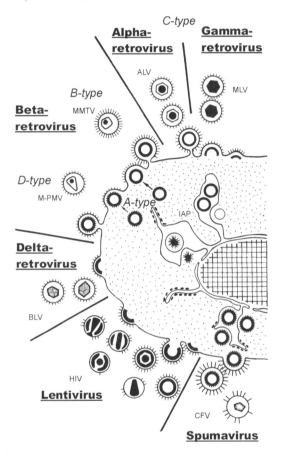

Figure 2. Retrovirus taxonomy reflects morphology. All retroviruses are assembled at cellular membranes and show characteristic structural features during assembly, maturation and ageing. Thus they can be classified into different groups using morphological criteria, e.g. the position and form of the core in the mature virion, or the dimensions of the glycoprotein knobs. The former structure-based classification into A-, B-, C- and D-type has been replaced recently by a genetically based taxonomy (van Regenmortel et al., 2000). The genus Epsilon-Retrovirus, comprising the retroviruses of fishes, has been omitted in this scheme, they ressemble morphologically mammalian C-type viruses. Type species of retrovirus families: MMTV – mouse mammary tumor virus; MuLV – mouse leukemia virus; ALV – avian leukosis virus M-PMV – Mason-Pfizer monkey virus; BLV – bovine leukemia virus; HIV – human immunodeficiency virus; CFV – chimpanzee foamy virus (IAP – intracisternal A particles).

2.2 Taxonomic classification of HIV

Members of the retrovirus family occur in many vertebrates as exogenous viruses, basically with a fairly restricted, species-specific distribution, though cross-species transmissions may occur. Vertebrates also harbour endogenous retroviruses often representing genetically incomplete DNA-proviruses that have arisen from exogenous retroviruses that have integrated into germ line cells. While endogenous viruses are transmitted vertically from parents to offspring like Mendelian genes, the exogenous agents are spread as infectious particles by a number of routes: horizontally by infectious blood, saliva or sperm, or vertically via infection of the offspring, by perinatal infection or by infectious milk. Retroviruses cause a variety of diseases: some members induce malignancies (carcinomas, sarcomas or lymphomas), others cause immunodeficiencies or central nervous system disorders (for review see Coffin et al., 1986; Coffin et al., 1997; Levy, 1986).

Retroviruses share a number of basic characteristics. The virion is an enveloped particle, 110 to 140 nm in diameter, surrounded by a fringe of virus-coded glycoproteins. It contains a core encasing the viral genome which consists of two identical copies of single-stranded RNA 7 to 11 kb in size. During replication, the viral reverse transcriptase (RT) copies the genomic RNA into double-stranded DNA. This provirus inserts into the host genome with the help of the viral integrase (IN) and, after activation, utilizes the cell's protein machinery to generate virus progeny (for review see Coffin et al., 1997).

Retroviruses are classified in seven genetically distinct groups (van Regenmortel, 2000). These groups show typical morphological peculiarities caused by differences in virus structural proteins, i.e. basically by genomic differences (Figure 3). During their extracellular life, retroviruses pass through three functional states: assembly, maturation, and ageing (Figures 5, 6). The possibility to clearly differentiate developmental steps is unique to retroviruses. These states as well as two early steps in virus-cell interaction, i.e. virus adsorption and entry, can be assessed with high resolution by transmission electron microscopy (TEM). Therefore, starting with the pioneering work of Wilhelm Bernhard in the late 1950s, EM has been used to assign "new" retroviruses according to morphology into the proper genus (Bernhard, 1960); for details see (Gelderblom, 1991; Gelderblom, 1997; Nermut et al., 1996). Based on both, comparative sequencing and morphology, HIV was classified in 1985 as a member of the lentivirus group of retroviruses (Gelderblom et al., 1985a, 1988; Gonda et al., 1985; Rabson et al., 1985; Wain-Hobson et al., 1985).

Because of the close structure to function relations, structural studies including immuno-EM, have contributed much to the current understanding of viral function and pathogenesis and the development of antiretroviral strategies (for review see Coffin et al., 1997; Frankel et al., 1998; Gelderblom, 1991; Gelderblom, 1997; Hunter, 1994; Jones et al., 1998; Nermut et al., 1996; Nermut et al., 2001). The recent application of cryo-EM and 3D-reconstruction has added further insights into the structural organisation of HIV (for review on technical aspects see Baker et al., 1999). Cryo-EM, in particular when applied in combination with other structural techniques, offers promising new high resolution information. Nuclear magnetic resonance and x-ray structural studies have established the oligomer-status and 3D-organization of many HIV components down to atomic resolution (Turner et al., 1999). These physical data help to complement and interpret observations from EM in a successful, "divide and conquer" approach (Wilk et al., 1999a; Wilk et al., 2001).

3.   STRUCTURE AND COMPOSITION OF THE MATURE VIRION

3.1   Components of HIV

HIV-coded proteins are designated according to function and localization within the virion by 2-letter acronyms (Leis et al., 1988); for details see Figures 3, 4). During maturation, Pr55 Gag is cleaved sequentially by the viral protease (PR) into the proteins of the mature virion: Membrane associated or matrix protein (MA, p17), capsid, (CA, p24), nucleocapsid (NC, p10), and the C-terminal core-envelope link (LI, p6) (Gorelick et al., 1988; Henderson et al., 1992; Höglund et al., 1992). Similarly, the two viral envelope proteins, surface (SU, gp120) and transmembrane (TM, gp41) protein, are processed from the Env gp160 precursor. In addition, all replication competent retroviruses code for three essential enzymes, reverse transcriptase (RT), integrase (IN) and protease (PR) that are also contained in the virion. Complex retroviruses, like HIV, code also for accessory proteins that regulate virus life cycle and expression (Table 1; for review see (Coffin et al., 1997; Cullen, 1998; Frankel and Joung, 1998).

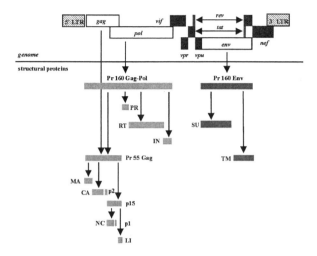

Figure 3. Genomic map of HIV provirus and the coded proteins. The genome comprises 9400 base pairs and is flanked by the LTR's, two identical non translated sequences with regulatory functions. The genome is read in 3 different reading frames to produce the mRNA for the structural proteins, enzymes and accessory proteins. The former are translated as two large precursor proteins, Gag-Pol and Env, that are subsequently, during maturation, cleaved at multiple sites to produce the definitive structural proteins.

3.2   Constituents and organization of the mature core

While MA in the mature virion remains tightly bound to the inner leaflet of the phospholipid-bilayer, CA and NC are found with the cone-shaped core typical of members of the lentivirus group (Figures 4, 5, 6; Gelderblom et al., 1985; Gonda et al., 1988; for review see Gelderblom, 1991). The core capsid is made up of CA (Accola et al., 2000; Chrystie et al., 1989; Gelderblom et al., 1987a). Cleavage of Pr155 Gag-Pol liberates in addition multiple copies of RT, IN, and PR. A virion contains about 1200 molecules of CA and 80 of RT (Layne et al., 1992). Also several accessory viral proteins (see Table 1), i.e. Nef, Vif, Vpr (and Vpx) are present in the virion. Vpr is found in all lentiviruses, while Vpx is restricted to HIV-2 and SIV. Vpr and Vpx had been localized by immuno EM (IEM) between the core shell and the viral envelope inside the virion (Liska et al., 1994), however, the indirect IEM technique used turned out not to allow a precise localization. Recent studies on isolated HIV-1 cores have revealed Vpr being enclosed in the core capsid (Accola et al., 2000; Welker et al., 2000). Nevertheless, high resolution IEM analysis is still required to clarify the composition of both, the electron-opaque proteins of the CEL and the "lateral bodies" along the axis of the cone-shaped cores (Figures 4 - 6). The fine organization of the viral genome inside the core appears complex (Accola et al., 2000; Höglund et al., 1992; Takasaki et al., 1997).

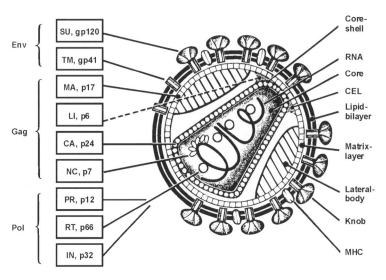

Figure 4. Diagram of HIV-1 relating viral proteins to the morphological units of the virion. The localization of LI in the CEL is hypothetic, based on morphological evidence as well as on studies of deletion mutants and isolated cores.

The regular HIV core resembles a cone, about 100 nm in length, about 60 nm at the wide end and 20 nm at the narrow end in diameter, showing often an angular cap (Figures 4, 5, 6; Gelderblom et al., 1987a). Recently, a curved lattice of hexamers with pentamers included on both ends was deduced from cryo-EM observations of in vitro assembled CA-NC cores as the probable underlying symmetry (Ganser et al., 1999). Occasionally, also tubular cores are observed in HIV (Figure 6d; Gelderblom, 1988). Synthetic tubular and cone-shaped cores are scrutinized presently to determine symmetry principles (Ganser et al., 1999; Gross et al., 2000).

3.3   Host constituents of the viral core

In addition to virus coded proteins, a number of host cell constituents e.g. actin, ubiquitin, and cyclophilin A are also incorporated. The presence of cytosceletal proteins points to their functional role in the transport of viral proteins to the plasma membrane (Ott et al., 2000; Wilk et al., 1999b). Cyclophilin A, a cellular chaperone, is found only in HIV-1, not in HIV-2. It binds to CA of HIV-1 and was shown to be required for the infectivity of HIV-1 strains of the M-group viruses (Franke et al., 1994; Grättinger et al.,

1999; Luban et al., 1993; Ott et al., 1996; Thali et al., 1994). Ubiquitin is present in the virion in two forms: as free molecules and specifically conjugated to the p6 portion of the Gag precursor. In vitro studies were unable to detect a function of this posttranslational modification in viral replication (Ott et al., 2000).

Table 1. Accessory proteins of HIV

Gene regulatory and other accessory proteins of HIV-1 adopted from Cullen (1998) and Frankel and Young (1998)

Protein, name and size	Localization	Function
Tat (transactivator of transcription) 86 amino acids, 2 exons, 14 kDalton	nucleus, nucleolus early protein	essential for replication: binds to TAR in the 5'-LTR region of the HIV provirus and several cellular trans-cription factors; transactivates viral mRNA expression and modulates cytokin- and receptor-expression
Rev (regulator of virus protein expression), 116 amino acids, 2 exons, 19 kDalton	nucleus, early protein shuttling between nucleolus and cytoplasm	essential for replication. Binds to RRE of viral RNA and cellular factors. RRE-binding promotes export of un- or singly- splized viral mRNA to the cytoplasm; modulates transcription + expression of structural proteins Gag, Pol, Env and of Vif, Vpu, Vpr
Vif (viral infectivity factor) 192 amino acids 23 kDalton, late protein	cytoplasm, also membrane associated, a few molecules encased in the mature virion	essentiell for high infectivity and replication in primary cells, not essential in permanent, permissive cells; role in early steps in HIV replication, i.e. "stabilization" of viral progeny-DNA and virus maturation
Vpr (viral protein R) 96-106 amino acids 10-15 kDalton, late protein	constituent of the viral core and pre-integration complex (including Vpr, RT, IN, RNA and MA) membrane associated forming ion channels	essential for nuclear uptake of the pre-integration complex mediating replication in nondividing cells, e.g. resting monocytes /macrophages; not essential for infection of T cell lines. arrests cell proliferation at G2, leading to apoptosis of T cells
Vpx (virusprotein X) 112 amino acids 12-16 kDalton, HIV-2 specific	late protein constituent of the virion	functionally similar to Vpr (absent in HIV-1, sequence homologies)
Vpu (virus protein U) specific for HIV-1 81 amino acids (phosphorylated) 9,2 kDalton, 16 kD	late protein localized to the ER and as a trans-membrane protein at the inner leaflet of the membranes of infected cells	degradation of CD4 in gp 160-CD4 complexes in ER, liberating Env gp160 for further processing and assembly; down-regulates cell surface expression of MHC class I; increases HIV release, also of other retroviruses; in vitro not essential; integral membrane protein forming oligomeric ion-channels
Nef (negativ factor) 206 amino acids 27 kDalton, early protein (N-terminally myristoylated)	in the virion (up to 70 molecules) localized also at cellular membranes in both cytoplasm and nucleus	increases infectivity/production of HIV in early stages by supporting viral DNA synthesis and binding to tyrosine protein kinases. down-regulates expression of MHC class I and CD4; in vitro not essential. In vivo-pathogenesis: long-term nonprogressors often show defects in Nef

3.4   Components and organization of the viral envelope

Viral lipids are derived from the plasma membrane during budding (Aloia et al., 1993). The two virus coded envelope proteins SU and TM are proteolytically processed from the Env precursor Pr160 in the Golgi region (Hallenberger et al., 1992). Both are glycosylated with SU containing 50 percent of carbohydrate. The incorporation of the glycoprotein knobs into the budding virion (Figures 5-8) depends on the N-terminal part of MA (Dorfman et al., 1994; Freed et al., 1996; Yu et al., 1992). There is ample evidence for a close proximity of the MA and TM (Bugelski et al., 1995; Pepinsky et al., 1984) for review see (Hunter, 1994). The membrane-spanning region of TM can be replaced by that of an unrelated TM protein without changing viral functions, and also the C-terminus of TM appears to be dispensable (Mammano et al., 1994; Reil et al., 1998; Wilk et al., 1996). There are 72 glycoprotein knobs inserted into the viral envelope forming a regular T = 7 laevo pattern (Özel et al., 1988).

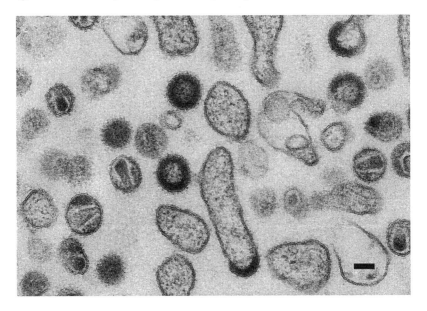

Fig. 5. Ultrathin section of a HIV-1 producing cell showing viral assembly, maturation and ageing. Between cellular processes and villi, HIV particles reveal different stages in development. Besides many mature virions - characterized by the cone-shaped cores - also budding viruses and cell-released immature virions are shown. Bar represents 100 nm.

The knobs are oligomers of loosely connected TM-SU heterodimers. A trimer organization was early on deduced from electron microscopy (Gelderblom et al., 1987a, 1988; Grief et al., 1989), though heavily debated and only much later generally accepted (Chan et al., 1997). By comparing SU sequences of different HIV-1 isolates, five variable (V1 - V5) and five conserved regions (C1 - C5) have been identified (Modrow et al., 1987). SU consists basically of two major domains, the outer and inner domain that are connected by a mini-domain, consisting of four antiparallel beta-sheets. The outer domain is heavily glycosylated, while the inner domain is required for interaction with TM and trimerization (Parren et al., 1999).

3.5 Host constituents of the viral envelope

Like other enveloped viruses, HIV acquires the lipids of its envelope from the virus-producing cell (Aloia et al., 1993). Also part of the glycoproteins exposed on the viral envelope are host-encoded. Besides the virus-coded SU and TM, a number of cellular proteins are tightly inserted during viral budding. MHC class I, class II, ICAM-1, LFA-1 and other proteins, i.e. genuine cellular proteins evolved for specific cell-cell-interaction, are present on the envelope (for review see Tremblay et al., 1998). By IEM and quantitative immunological techniques, a selective, nonrandom inclusion was shown (Figure 8c; Gelderblom et al., 1987b; Hoxie et al., 1987; Meerloo et al., 1993). The host proteins exceed in number the viral SU and TM (Arthur et al., 1992; Henderson et al., 1987) and might have a functional role by facilitating infection besides the well established gp120-CD4 interaction. While the viral SU is shed spontaneously (Gelderblom et al., 1985b; Layne et al., 1992), the host proteins are firmly inserted (Figure 8 c). By binding to receptor proteins on prospective host cells, these "viral" constituents definitely enhance viral infectivity (discussed by Tremblay et al., 1998), indeed, functionally they might partly replace the gp120-CD4 interaction.

Host components can also induce a number of unexpected phenomena. Some are masking HIV-1 as a "wolf in a sheep's clothing" from antibody-, complement- or soluble CD4-mediated neutralization (Marschang et al., 1995; Montefiori et al., 1994; Moog et al., 1997; Saifuddin et al., 1995). On the other hand, infectivity of a virus grown in MHC class II expressing cells can be neutralized by anti-ICAM-1, -LFA-1 and class II antibody. This observation raises the possibility of inducing protective immunity by allo-immunization (Arthur et al., 1995). The receptor-ligand interaction might induce further biological responses of the host cell and the immune system, e.g. with regard to intracellular signalling and possible cellular responses ranging from activation to anergy and/or apoptosis (for review see Tremblay et al., 1998). A caveat for biochemical studies should be mentioned here: even highly purified suspensions of HIV are contaminated by cell-derived microvesicles filled up with host-encoded proteins (Bess et al., 1997; Dettenhofer et al., 1999; Gluschankof et al., 1997).

4.   MORPHOGENESIS AND MATURATION OF HIV

Three maturation states of HIV can be studied in detail by conventional TEM (for review see (Gelderblom, 1991; Gelderblom, 1997; Gonda, 1988; Hunter, 1994; Kräusslich et al., 1996; Nakai et al., 1996; Nermut et al., 1996). (1) The virus is formed at the plasma membrane by an ordered self-assembly of virus structural Gag and Gag-Pol precurser polyproteins (Pr55 Gag, Pr160 Gag-Pol) together with genomic RNA. (2) After closure of the spherical ribonucleoprotein shell (RNP) the immature virion is released from the cell and undergoes functional and morphological maturation as a consequence of proteolytic cleavage of the precursor proteins by the viral PR. The cleavage occurs during assembly or after release of the immature particle. (3) The mature, infectious virion ages, rapidly losing infectivity caused by shedding of the SU knobs (Gelderblom et al., 1985b; Layne et al., 1992).

4.1   The budding process

Like the majority of retroviruses, HIV assembles at the plasma membrane (Figures 5, 6a, 8a, c). As a first indication an electron-dense patch appears directly underneath the lipid bilayer, while on the bud, on the outside of the bilayer, individual virus envelope knobs can be seen. The patch is growing continuously by side-to-side assembly of the Gag- and Gag-Pol-polyproteins, ultimately forming a spherical, electron-dense ribonucleoprotein (RNP) shell. This immature core is surrounded by a tightly fitting lipid bilayer. High resolution EM of immature virions shows lateral interaction of Gag-rods to the RNP (Fuller et al., 1997; Gross et al., 2000; Hockley et al., 1988; Nermut et al., 1994; Wilk et al., 2001). HIV is often observed to bud in a polar manner (Figure 9e) at sites were also Env is expressed suggesting that Gag proteins are attracted to domains where the viral glycoproteins are inserted (Bugelski et al., 1995). Late stages of virus assembly also depend on energy and host cell factors (Tritel et al., 2001).

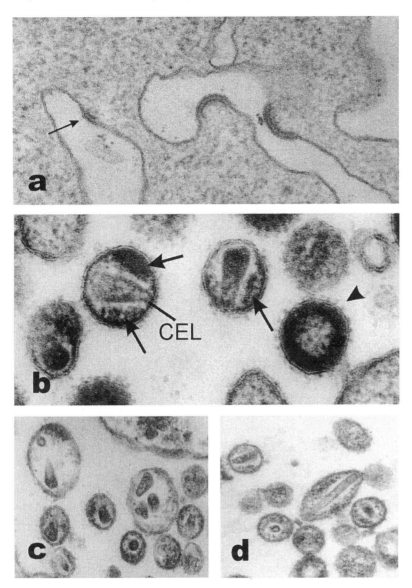

Figure 6. Ultrathin sections demonstrating assembly and maturation of HIV-1. (a) Two particles during early budding, and a third one (arrow) in a very early phase of formation (x 80.000). (b) Two mature virions revealing core-envelope-links (CEL), lateral bodies (arrows) and loss of envelope knobs. The immature virion is still densely covered with SU knobs (arrow head) (x 160.000). (c) Aberrant virions with two or three cone-shaped cores - sometimes with distinctly angular core-caps - or (d) with tubular cores (x 80.000).

Only after complete sealing of the RNP shell, the concentrically organized immature virion separates from the plasma membrane and pinches off into the extracellular space (Gelderblom, 1991; Göttlinger et al., 1991). The immature virion is slightly bigger than the mature particle (120-140 nm versus 110-130 nm; Gentile et al., 1994; Nermut et al., 1993), studded with glycoprotein knobs (Figures 5 - 8) and appears to be more rounded than the latter which often shows an angular outline (Renz, 1993). The angular shape of the virion early on has led to the conclusion that SIV and HIV represent isometric, probably icosahedral structures (Goto et al., 1990; Marx et al., 1988; Nermut et al., 1993; Nermut et al., 1994; Nermut et al., 1996; Özel et al., 1988), a view supported also by modelling (Forster et al., 2000). Recent combined data on both Gag-constructs and immature virions from cryo-electron microscopy, image processing and 3D reconstruction, however, have weakened the arguments for a strict icosahedral symmetry of the immature shell and pointed to just locally ordered domains (Fuller et al., 1997; Wilk et al., 1999a; Wilk et al., 2001). Another interesting model for the symmetry of the cone-shaped core was described recently: Synthetic CA-NC structures assembled in the presence of various RNA templates resembled closely the authentic HIV core. The asymmetric, though highly regular appearance of the core was proposed to follow a curved p6 lattice (Ganser et al., 1999). No doubt, the discussion on the symmetries governing the architecture of immature and mature cores will go on (Nermut et al., 2001; Wilk et al., 1999a).

Virus formation and release depend on a number of requirements. Using recombinant techniques, intact Pr55 Gag was shown to contain all the information necessary for oligomerization, assembly and release (Gheysen et al., 1989; Wills et al., 1991). The authentic RNP core of HIV is formed from two proteins, Pr55 Gag and Pr160 Gag-Pol, co-assembled side-to-side at a rate of approximately 20 to 1. Formation of infectious virus, however, involves besides a sequential and controlled cleavage also the rearrangement of the polyprotein products. Thus, virus formation essentially represents a very constrained process (Freed, 1998; Hunter, 1994; Jones et al., 1998; Nermut et al., 1994; Tritel et al., 2000; Wiegers et al., 1998; Zhang et al., 1996). Here, a few factors shall be mentioned.

4.1.1   Requirements for a functional MA

In the virion, MA is associated with the lipid bilayer of the envelope (Bugelski et al., 1995; Gelderblom et al., 1987a). Co-translational myristylation of the MA domain of Pr55 Gag and a highly basic region of the N-terminal MA domain, binding to acidic membrane phospholipids, are required for this interaction (Bryant et al., 1990; Göttlinger et al., 1989;

Zhou et al., 1994). Other regions of MA determine the targeting of Pr55 Gag to the plasma membrane, the regular site of viral assembly (Fäcke et al., 1993; Yuan et al., 1993). MA functions are also required for both, the assembly of the Gag- and the Gag-Pol-polyproteins and the insertion of the viral Env protein into the prospective envelope (Dorfman et al., 1994; Ono et al., 2000; Yu et al., 1992).

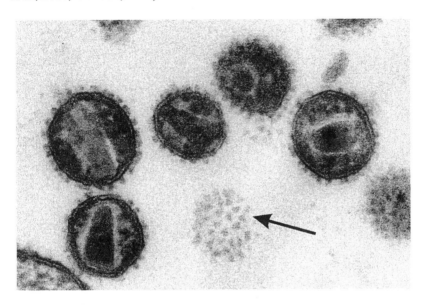

Figure 7. Ultrathin section showing 4 mature virions and a circular array of regularly arranged „spots", each of a distinct tri-angular outline. The triangles represent SU knobs, cut perpendicular to the envelope's surface (x 180.000).

During replication, immediately after viral uptake, MA is required by forming, together with the DNA provirus, Vpr and IN, the viral pre-integration complex. At its N-termal basic region MA contains putative nuclear localization signals which, together with similar signals of Vpr, mediate the transport of the complex through nuclear pores into the nucleus of quiescent cells, thus permitting the infection of non-dividing cells (Bukrinsky et al., 1993; Heinzinger et al., 1994). However, large parts of MA are dispensable for its proper function or can be replaced by a heterologous myristic anchor (Reil et al., 1998).

4.1.2 Properties of a functional CA

CA comprises 211 amino acids, domains mainly at the C-terminus are essential for proper assembly of the RNP shell (Zhang et al., 1996), while N-terminal domains are required for the formation of the mature, cone-shaped core (Fitzon et al., 2000). The middle region containing the "major homology motif" is well conserved among different retroviruses and essential for the co-incorporation of Pr160 Gag -Pol into the growing core shell (Mammano et al., 1994; Smith et al., 1993; for review see Freed, 1998; Hunter, 1994; Jones et al., 1998; Morikawa et al., 2000).

The immature virion, characterized by its spherical core (Figures 5, 6b), is not infectious and represents a morphologically stable intermediate in the extracellular viral life cycle (Gross et al., 2000). Virus-like particles produced by the expression of Pr55 Gag as well as immature HIV particles, generated in greater quantities in vitro by blocking the viral RT (Schätzl et al., 1991), consistently reveal regular side-to-side assembly of Gag polyprotein rods (Fuller et al., 1997; Gross et al., 2000; Nermut et al., 1996; Nermut, 2001; Wilk et al., 2001). The assembly of the Gag presursors was shown to involve multiple protein-protein interactions.

4.1.3   On the roles of p2 and p1

At the C-terminus of CA a spacer peptide of 14 amino acids (p2) is situated; another spacer, (p1) is located between NC and p6 (Figure 3; Henderson et al., 1992): both are virtually absent from the mature virion, but detectable in lysates of infected cells. The C-terminal portion of CA including part of p2 are crucial for the ordered assembly of the spherical core shell. Proteolytic removal of p2 from the C-terminus of CA occurs late during the controlled Pr55 and Pr160 cleavage kinetics and is essential for virus maturation, i.e. the transition from the spherical to the cone-shaped core (Göttlinger et al., 1989; Kräusslich et al., 1995; Pettit et al., 1994). Governing the shift from spherical to conical assembly are both the pH and the absence of p2 as shown recently by in vitro assembly studies (Gross et al., 2000). Similarly, the presence of p1 prevents maturation (Wiegers et al., 1998).

4.1.4   On the role of NC

The 55 amino acids of NC form strictly conserved domains that interact in many aspects with the viral RNA. E.g. for efficient binding of the viral genome to the packaging signal during assembly of the Gag precursor the interaction of two zinc-finger-like motifs with the viral RNA is required (Bess et al., 1992; Clavel et al., 1990; Gorelick et al., 1990). Functional Zn^{2+} binding proteins are required also for a number of chaperone activities as well as for early steps in the infection and in Gag precursor processing and particle formation (Gorelick et al., 1999; Rein et al., 1998).

4.1.5 On the function of p6

The 51 C-terminal amino acids of Pr55 gag form p6. This protein is required for the incorporation of Vpr (Müller et al., 2000; Paxton et al., 1993) and late step in viral budding (Göttlinger et al., 1991). Deletions in p6 prohibit both the closure of the spherical RNP shell and particle release. From thin section EM, p6 was supposed to also form in mature particles the so-called core-envelope-link (CEL) connecting the narrow end of the core to the MA shell (Figures 4, 5, 6b; Höglund et al., 1992). This assumption is corroborated by recent studies on isolated HIV-1 cores. The cores obtained after mild detergent treatment and sucrose cushion centrifugation were morphologically intact and essentially devoid of p6 (and of the morphologically defined CEL), while the accessory protein Vpr was co-purifying with the cores (Accola et al., 2000; Welker et al., 2000). This "central" localization of Vpr points again to its role in the viral pre-integration complex (Bukrinsky et al., 1993; Heinzinger et al., 1994).

4.2 Virus ageing and comparative aspects

Because of non-covalent, i.e. weak SU-TM interactions in the trimeric envelope knobs, viral SU is shed spontaneously from the virion (Gelderblom et al., 1985b; Poignard et al., 1996; Renz, 1993). For HIV-1 strain IIIB a shedding half-life of 30 hrs was determined. This rate easily explains the rapid decay of viral infectivity of IIIB (Layne et al., 1992). The SU knobs on HIV-2, however, and in particular on primary HIV isolates reportedly are more tightly associated with the viral envelope, though a direct proof is still missing. Except for possible differences in anchorage stability of the knobs, no consistent morphological differences are apparent between HIV-1, -2, and SIV (Chrystie et al., 1988; Gelderblom, 1991; Gonda, 1988; Goto et al., 1998; Nakai et al., 1996; Nermut et al., 1996; Palmer et al., 1988).

Here it should be emphasized, that hitherto structural studies on HIV, also the most recent ones that combine image reconstruction techniques and refined freeze-preparation techniques (Müller, 1992) or cryo-electron microscopy (Baker et al., 1999; Wilk et al., 1999a), are based on either cell culture grown HIV or mutants of HIV, or on virus structures obtained in artificial expression systems. Though the present structural models show an overall resemblance, details of the assembly process and the rules governing viral symmetry are controversely discussed (Fuller et al., 1997; Nermut et al., 1994; Nermut et al., 1998; Nermut et al., 2001; Wilk et al., 1999a). Presently it is not clear whether and how these differences influence also functional properties of HIV. However, it might be rewarding to structurally characterize HIV also ex vivo, i.e. to study morphological and antigenic properties of HIV derived from viremic plasma as a more close to reality object.

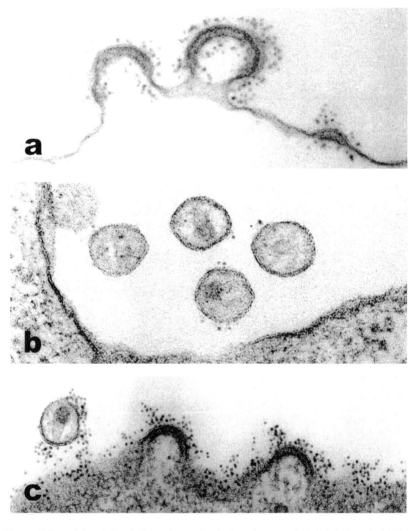

Figure 8. Reactivity of the viral envelope using indirect immuno-ferritin labeling. (a, b) The viral buds are densely studded with viral SU antigen, while the mature particles have lost most of the label (x 160.000). (c) MHC class I labelling of budding and muture HIV-1 particles (x 100.000).

5. ASPECTS OF VIRAL PATHOGENESIS

5.1 Virus cell interactions

HIV infects mainly cells of the immune and the central nervous system utilizing sequentially the CD4 receptor and one of a series of co-receptors. CD4 is expressed on T-helper lymphocytes, monocytes/macrophages, dendritic cells and microglia in the central nervous system (for summary see Fauci, 1988; Levy, 1994). The CD4-tropism primarily determines the range of infectable cells and the clinical course of the infection. Several early stages of the infection can be studied and differentiated by EM: (1) adsorption, mediated by electrostatic forces, followed by (2) specific binding of SU to cellular receptors and (3) virus uptake, either at the cell surface by a pH independent direct fusion of the viral and the host membrane or endosomal fusion following receptor-mediated uptake using clathrin-coated pits and vesicles. Both ways can lead to infection (Figure 9a,b; Grewe et al., 1990; Stein et al., 1987). Virus uptake is controlled by a number of sequential steps: Briefly, the primary SU-CD4-interaction causes a conformational change in SU, exposing the co-receptor binding site, while the following SU-coreceptor reaction induces further structural changes that ultimately allow for fusion.

The SU monomer, due to its beta-sheet bridging domain and high degree of glycosylation, shows a high degree of conformational flexibility. Accordingly, depending on the specific domain arrangement, also the SU trimer displays only certain sites, e.g. for virus neutralization or co-receptor binding (for review see Parren et al., 1999). Entry is a multivalent process requiring formation of several gp120-CD4 bonds. The CD4-receptor binding site of SU is formed by a groove at the interface of the 3 domains. The high affinity interaction with the amino-terminal domain of CD4 results in conformational changes in both molecules. While in SU the previously hidden co-receptor binding site becomes accessible, the flexible CD4 bends away thus bringing viral and cellular membranes in close proximity so that cellular co-receptors can interact with SU. Co-receptor binding mediates the destabilization of the SU-TM structure and the formation of a pre-hairpin intermediate of TM that inserts into the membrane of the target cell, while SU is released as monomeric gp120. The amino-terminal glycine rich portion of the membrane-inserted TM undergoes further rearrangements: a transformation into the fusogenic "hairpin structure" allows the hydrophobic "fusion peptide" to penetrate the cellular membrane ultimately leads to fusion of both membranes and release of the viral core into the cytoplasm (for review see Chan et al., 1998; Dragic, 2001; Sattentau et al., 1991; Wyatt and Sodroski, 1998).

As coreceptor HIV utilizes one of the seven-transmembrane-domain G protein-coupled receptors (for review see Dragic, 2001; Horuk, 1999; Littman, 1998). More than 10 different co-receptors have been identified for HIV-1, HIV-2 and SIV with only two of them, CCR5 and CXCR4, being the major co-receptors in vivo. Co-receptor binding to the viral SU protein is multivalent (Kuhmann et al., 2000). At the cellular level both, the amino acid composition and charging of SU and the differential expression of co-receptors are thought responsible for viral tropism. Because of this key function in HIV infectivity, CD4 and co-receptors offer a promising drug target (for review see Richman, 2001).

5.2   Clinical Aspects of HIV

According to biological and in vitro-criteria, two groups of HIV-1 are to be differentiated in the clinical course. Early after transmission and during the first years of the disease, the large majority of clinical HIV-1 isolates (tested as primary, early passages of HIV in PBMC) replicates at low levels in primary monocytes/macrophages and primary T lymphocytes, but not in immortalized T cells. Because of the preferential use of CCR5 as coreceptor, the primary macrophage-tropic HIV is designated R5 virus, while T-cell adapted strains - using CXCR4 - are named R4 viruses (Berger et al., 1998).

Infection by HIV is most often transmitted sexually. Here the first crucial step in the pathogenesis of AIDS is the infection of monocytes/macrophages and Langerhans cells. These front-line cells in the network of the specific and unspecific immune defense are situated in mucous membranes and the skin. As antigen presenting cells (APC) they take up and react with foreign antigens and microbes. After migration into lymphnodes and antigen processing, they present respective determinants to CD4+ T helper cells. Being the primary target of R5 viruses, APC can live up to several months with a productive HIV infection. In the lymph node, during antigen presentation, however, they may transmit the infection across cell-cell contacts directly to CD4 T helper cells.

Late in the course of the disease, the R4 type of HIV predominates: it grows in vitro in PBMC to higher titers, but can also productively infect T cell lines and induce syncytia (Figures 9 c, d). When adapted to high titer growth in permanent T cell lines, R4 viruses lose their ability to replicate in primary macrophage/monocyte cultures concomitant with the change in co-receptor usage (Connor et al., 1997; for review see Horuk, 1999; Littman, 1998). Infection by R5 viruses can induce also early neuropathogenesis based on the co-expression of CD4 and CCR5 plus some CXCR4 (Albright et al., 1999).

After membrane fusion, the viral core is liberated into the cytoplasm of the infected cell were the viral genome is transcribed into the DNA provirus. In non-dividing cells the provirus, together with MA, IN and Vpr as the pre-integration complex, enters the interphase nucleus where it becomes chromosomally integrated. Integration is at random, without any specific chromosomal preference. In dividing cells, the preintegration complex may reach the nucleus during mitotic division when nuclear membranes have dissolved. After integration, HIV enters a state of latency. There are a number of factors that can wake the provirus from latency, the main being activation of the host cell by e.g. immune phenomena, cytokine action and/or chemical or physical stress (Fauci, 1996).

5.3 Virus function and pathogenesis

Clinically HIV causes a profound immuno deficiency that results in susceptibility for opportunistic infections and malignancies (for details consult clinical textbooks, e.g. Merigan et al., 1999, and Wormser, 1996). Pathognomonic for the HIV disease is the progressive loss of CD4+ T-helper cells in the peripheral blood. These cells are crucial in initiating and organizing the specific immune defence against foreign antigens and become infected in the lymph node during their interaction with infected APC. The depletion of CD4+ cells in the periphery apparently is due to the destruction of both mature T helper cells and the death of progenitor cells in bone marrow and thymus. Over years 10^{10} HIV particles are produced per day, and 10^9 infected cells destroyed. The two parameters virus load and CD4 counts form a dynamic steady state: The virus is readily inactivated and the T-cell loss can be compensated over years without overt clinical signs by increased cell proliferation until the system ultimately is exhausted (Ho et al., 1995). However, also changes in the distribution of CD4 cells between blood and the lymph nodes contribute to it (for review see McCune, 2001).

Clinically HIV, like other lentiviruses, shows a bi-phasic course of expression. The primary disease, characterized by mild clinical signs and a marked viremia, is followed by an immune-mediated viral clearance and clinical latency. After 6 to 10 years clinical symptoms reemerge, accompanied by a rising viremia and progressive immune failure. The primary, acute syndrome can be observed in the majority of patients between weeks three to six after infection. It lasts about one week and is caused by a burst of HIV with titers of up 10^3 and 10^7 genome equivalents per ml of plasma (for review see Coffin et al., 1997; McCune, 2001). A transitional mononucleosis-like syndrome with fever, pharyngitis, lymph-adenopathy, T-cell activation and skin rash prevail. The primary disease is accompanied by a clearcut drop in CD4+ cells. This loss, also later in the clinical course, is directly related to viral load as measured in the plasm. The early symptoms, however, disappear often completely during weeks 5 or 6 p.i., when a vigorous cell-mediated immunity (CMI) against HIV is mounted. Antibody-mediated immunity (AMI), i.e. neutralizing antibody is detected usually a few weeks later. CMI and AMI in addition to endogenous cytokines help to control viremia by destroying the majority of virus-infected cells. Another pool of HIV, however, is integrated in resting CD4+ cells and remains undetectable by the immune system. Ultimately, when resting T cells become activated, the provirus will often be co-expressed. The resulting virus progeny leads to new rounds of HIV infections and an initially hidden and lateron overt progressive immunopathogenesis

Because of close relations between HIV and the immune system, the clinical course of the HIV infection will be affected by all factors that interact with the patient´s immune system (for review see Fauci, 1996; Merigan et al., 1999). Primarily, acute and chronic infections will activate the immune system (Wahl et al., 1998): Common opportunistic agents, like herpes simplex virus, human herpesvirus type 6, cytomegalovirus, mycobacteria, and Pneumocystic carinii will cause chronic activation, cytokine excretion and the activation of HIV from latently infected cells. Cell and virus activation and new rounds of HIV infections start primarily in lymph nodes because the close cell-cell contacts facilitate an effective virus transfer. In the placenta, on the other hand, the cell contacts allow the transfer of HIV from the mother to the trophoblast of the embryo by transcytosis and free virus (Lagaye et al., 2001). In the closed milieu of the lymphnode, HIV-infected macrophages often show a dramatic "upregulation" of virus production (Orenstein et al., 1997). The resulting viremia will lead to additional rounds of infection. Similarly any vaccination of HIV patients will enhance viremia: the virus burst inevitably will lead to T-cell losses, pointing to the deleterious effects of "immune stress".

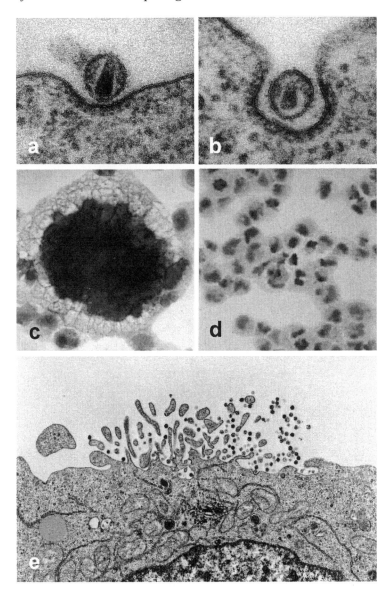

Figure 9. (a, b) Ultrathin sections of HIV-1 with two sequential stages of receptor-mediated uptake (x 100.000), showing the transformation from clathrin-coated pit to a -coated groove. The virions interact by a fringe of SU knobs with cellular counterparts. (c, d) Light microscopy showing an HIV induced syncytium and respective control cells (x 1.000). (e) EM of an HIV-1 infected cell. The polar virus burst induces membrane destabilization (x 10.000).

5.4   Molecular mechanisms of HIV pathogenesis

At the molecular level, a host of direct and indirect mechanisms is involved in T-cell depletion. Besides viral factors, the immune system itself, its cytokine network and the interaction with accompanying opportunistic infections contribute to the T-cell depletion (McCune, 2001). Due to limited space we will restrict this outline mainly to the impact of HIV itself. Readers interested in the wide fields of immune and cytokine regulation and dysbalances in the clinical course are referred to respective reviews (Fauci, 1988; Fauci, 1996; Littman, 1998; McCune, 2001; Merigan, 1999; Wormser, 1996).

A productive infection kills the T helper host cell within less than two days. This direct cell death may be caused by a number of mechanisms which have been observed mainly in cell culture experiments: Early after infection an abundance of full length proviral DNA is found in the cytoplasm of HIV infected cells. The accumulation of viral genomes apparently is cytotoxic. HIV is often produced in a polar manner. When virus formation occurs as a burst with a high concentration of budding events in time, the lipid composition of the plasma membrane will change resulting in changes in membrane fluidity (Figure 9e). A major part of the cytopathic potential of HIV resides in the viral Env protein. Besides the labilization of the plasma membrane also the fusion activity of the viral SU-TM protein will ultimately destroy the HIV infected helper cell (LaBonte et al., 2000).

High amounts of HIV will directly lyse CD4+ cells (Grewe et al., 1990), while low concentrations of isolated SU protein by intracellular signalling may induce T-cell apoptosis. Also Tat and other regulatory proteins like Vpr and Nef are reported to be involved in HIV mediated T cell apoptosis (summarized by Levy, 1994; McCune, 2001; see Table 2). When expressed on the surface of HIV infected cells, SU-TM will bind to receptors of uninfected cells thus inducing fusion of the cell membranes and formation of syncytia (Figure 9 c, d; Gelderblom at al., 1988; Lifson et al., 1986).

The viral SU may be shed spontaneously from the viral surface (Gelderblom et al., 1985b; Layne et al., 1992). When free SU binds to uninfected CD4 positive cells, these cells will acquire the antigenic make-up of HIV and may be recognized and destroyed by strong antiviral immune mechanisms (Wyatt and Sodroski, 1998).

Table 2. Pathomechanisms of HIV

Pathogenic properties of HIV
(adapted from Fauci, 1996; Levy, 1994; McCune, 2001)

Factors promoting virus survival

- High mutability of the genome: escape from immune defence and therapy
- Infection also of resting cells: advantage before other viruses
- Integration into host cell chromosomes: HIV inaccessible by immune attackes
- Heavy glycosilation of viral ENV protein and conformational
  activity of viral SU domains leading to: relative protection against virus neutralization
- Insertion of host cell constituents in the viral envelope
  functional role in viral entry
  protection against complement mediated lysis
  shedding of viral SU: binds to and abrogates neutralizing antibodies

Mechanisms for direct killing of infected CD4+ T cells

- Accumulation of unintegrated proviral DNA in cytoplasm
- Envelope-mediated lysis or apoptosis of infected cells
- Vpr-induced G2 arrest and apoptosis
- Membrane destabilization and/or syncytia formation

Indirect mechanisms leading to dysfunction or death of uninfected T cells

- SU shedding: lysis of SU-decorated, uninfected cells by NK/specific cytolytic cells
- Weakening of the immune system by viral TM
- Auto-immune reactions of humoral or cellular nature
- Incorporation of uninfected cells into syncytia by HIV-producing cells
- Apoptosis upon cell activation or cross-linking of CD4 bound gp 120
- Enhanced HIV transmission from APC to T-helper cells in lymphnodes
- De-regulation of immune system by enhanced TNF- alpha and IL-1beta secretion

Both, the heavy glycosylation and the conformational changes during virus cell interaction of the viral SU help HIV to evade neutralization by specific antibody, complement or soluble CD4 (discussed in detail by Parren et al., 1999; Wyatt and Sodroski, 1998). Finally, animal retroviruses are known to be lysed in vitro by the complement system present in human serum. When HIV was tested for its susceptibility to complement-mediated lysis, it was resistant. When studied in more detail, the protection from lysis turned out to be species-specific and mediated by membrane-bound complement inhibitors, i.e. host proteins CD46, CD55 and CD59 acquired during budding from human cells and present on the envelope of HIV (Marschang et al., 1995; Montefiori et al., 1994).

Moreover, the viral envelope proteins are assumed to interfere in several, detrimental ways with the network of normal immune regulation (for review see Dalgleish et al., 1999; McCune, 2001). Possibly, also the TM of HIV itself is immuno-suppressive. This has been deduced from sequence homologies of the TM of HIV with the immuno-supressive TM protein of murine C-type retroviruses (Denner, 2000).

Presently it is not clear whether all these factors – and to what a degree – contribute to in vivo pathogenicity. This deficit in information appears understandable, because this field is not accessible to experimental work. Nevertheless, a restricted number of mechanisms was shown in HIV patients involved in the decline of CD4-positive cells (for review see Fauci, 1996; McCune, 2001). By immuno EM of lymphnodes of HIV-infected patients, SU was detected on apoptotic CD4+ T cells. Because no free HIV was detected, it appears very likely that premature cell death was caused by free gp120 bound to non-infected cells and immune elimination of the SU-decorated cells (Sunila et al., 1997). Likewise, in lymphnodes, the transmission of HIV from APC to T-helper cells and the lysis of HIV-producing helper cells and uptake by macrophages have been shown (Orenstein, 2000; Tenner-Racz et al., 1998). Finally, there is also evidence on HIV induced giant cell formation from macrophages in tonsils and in the central nervous system (Koenig et al., 1986; Orenstein et al., 1999).

5.5   Outlook

No doubt, today the perspective for a worldwide eradication of HIV, comparable to the successful campaign against smallpox, appears dim. HIV utilizes biological niches for its survival. By its high mutation rate it can evade immune defence and develop resistance against chemo-therapeutic approaches. The conformational activity of the viral Env proteins, the high degree of glycosylation and the integration of host proteins in the viral envelope offer the virus an efficient camouflage. Indeed, HIV behaves like a wolf in sheep's clothing. Nevertheless, research on the virus has revealed already significant parts of its complexity and complex relation to the immune system resulting also in a number of therapeutic approaches. As outlined in the begin of this chapter, to combat this moving target successfully and to control the pandemic spread of HIV, the efforts have to come from many different angles. In view of the global catastrophe there is definitely the need for additional painstakingly detailed basic research.

REFERENCES

Accola, M. A., Öhagen, A., and Göttlinger, H. G., 2000, Isolation of human immunodeficiency virus type 1 cores: retention of Vpr in the absence of p6(gag). *J. Virol.* **74**: 6198-6202

Albright, A. V., Shieh, J. T., Itoh, T., Lee, B., Pleasure, D., O'Connor, M. J., Doms, R. W., and Gonzalez-Scarano, F., 1999, Microglia express CCR5, CXCR4, and CCR3, but of these, CCR5 is the principal coreceptor for human immunodeficiency virus type 1 dementia isolates. *J. Virol.* **73**:205-213

Aloia, R. C., Tian, H., and Jensen, F. C., 1993, Lipid composition and fluidity of the human immunodeficiency virus envelope and host cell plasma membranes. *Proc.Natl.Acad.Sci.U.S.A* **90**:5181-5185

Arthur, L. O., Bess, J. W., Sowder, R. C., Benveniste, R. E., Mann, D. L., Chermann, J. C., and Henderson, L. E., 1992, Cellular proteins bound to immunodeficiency viruses: implications for pathogenesis and vaccines. *Science* **258**:1935-1938

Arthur, L. O., Bess, J. W., Jr., Urban, R. G., Strominger, J. L., Morton, W. R., Mann, D. L., Henderson, L. E., and Benveniste, R. E., 1995, Macaques immunized with HLA-DR are protected from challenge with simian immunodeficiency virus. *J. Virol.* **69**:3117-3124

Baker, T. S., Olson, N. H., and Fuller, S. D., 1999, Adding the third dimension to virus life cycles: three-dimensional reconstruction of icosahedral viruses from cryo-electron micrographs. *Microbiol.Mol.Biol.Rev.* **63**:862-922, table

Barré-Sinoussi, F., Chermann, J. C., Rey, F., Nugeyre, M. T., Chamaret, S., Gruest, J., Dauguet, C., Axler-Blin, C., Vezinet-Brun, F., Rouzioux, C., Rozenbaum, W., and Montagnier, L., 1983, Isolation of a T-lymphotropic retrovirus from a patient at risk for acquired immune deficiency syndrome (AIDS). *Science* **220**:868-871

Berger, E. A., Doms, R. W., Fenyo, E. M., Korber, B. T., Littman, D. R., Moore, J. P., Sattentau, Q. J., Schuitemaker, H., Sodroski, J., and Weiss, R. A., 1998, A new classification for HIV-1. *Nature* **391**:240

Bernhard, W., 1960, The detection and study of tumor viruses with the electron microscope. *Cancer Res.* **20**:712-727

Bess, J. W., Gorelick, R. J., Bosche, W. J., Henderson, L. E., and Arthur, L. O., 1997, Microvesicles are a source of contaminating cellular proteins found in purified HIV-1 preparations. *Virology* **230**:134-144

Bess, J. W., Powell, P. J., Issaq, H. J., Schumack, L. J., Grimes, M. K., Henderson, L. E., and Arthur, L. O., 1992, Tightly bound zinc in human immunodeficiency virus type 1, human T-cell leukemia virus type I, and other retroviruses. *J. Virol.* **66**:840-847

Bryant, M. and Ratner, L., 1990, Myristoylation-dependent replication and assembly of human immunodeficiency virus 1. *Proc.Natl.Acad.Sci.U.S.A* **87**:523-527

Bugelski, P. J., Maleeff, B. E., Klinkner, A. M., Ventre, J., and Hart, T. K., 1995, Ultrastructural evidence of an interaction between Env and Gag proteins during assembly of HIV type 1. *AIDS res.hum.Retroviruses* **11**:55-64

Bukrinsky, M. I., Haggerty, S., Dempsey, M. P., Sharova, N., Adzhubel, A., Spitz, L., Lewis, P., Goldfarb, D., Emerman, M., and Stevenson, M., 1993, A nuclear localization signal within HIV-1 matrix protein that governs infection of non-dividing cells. *Nature* **365**:666-669

CDC, 1981, Kaposi's sarcoma and Pneumocystis pneumonia among homosexual men - New York City and California. *MMWR Morb.Mortal.Wkly.Rep.* **30**:305-308

Chan, D. C., Fass, D., Berger, J. M., and Kim, P. S., 1997, Core structure of gp41 from the HIV envelope glycoprotein. *Cell* **89**:263-273

Chan, D. C. and Kim, P. S., 1998, HIV entry and its inhibition. *Cell* **93**:681-684

Chrystie, I. L. and Almeida, J. D., 1988, Further studies of HIV morphology by negative staining . *AIDS* **2**:459-464

Chrystie, I. L. and Almeida, J. D., 1989, Recovery of antigenically reactive HIV-2 cores. *J.Med.Virol.* **27**:188-195

Clavel, F., Guetard, D., Brun-Vezinet, F., Chamaret, S., Rey, M. A., Santos-Ferreira, M. O., Laurent, A. G., Dauguet, C., Katlama, C., Rouzioux, C., Klatzmann, D., Champalimaud, J. L. and Montagnier, L., 1986, Isolation of a new human retrovirus from West African patients with AIDS. *Science* **233**:343-346

Clavel, F. and Orenstein, J. M., 1990, A mutant of human immunodeficiency virus with reduced RNA packaging and abnormal particle morphology. *J.Virol.* **64**:5230-5234

Coffin, J., Haase, A., Levy, J. A., Montagnier, L., Oroszlan, S., Teich, N., Temin, H., Toyoshima, K., Varmus, H., Vogt, P., and Weiss, R., 1986, Human immunodeficiency viruses. *Science* **232**:697

Coffin, J. M., 1995, HIV population dynamics in vivo: implications for genetic variation, pathogenesis, and therapy. *Science* **267**:483-489

Coffin, J. M., Hughes, S.H.J., and Varmus, H.E., 1997, Retroviruses. Cold Spring Harbor, Cold Spring Harbor Laboratory Press.

Connor, R. I., Sheridan, K. E., Ceradini, D., Choe, S., and Landau, N. R., 1997, Change in coreceptor use coreceptor use correlates with disease progression in HIV-1--infected individuals. *J.Exp.Med.* **185**:621-628

Cullen, B. R., 1998, HIV-1 auxiliary proteins: making connections in a dying cell. *Cell* **93**:685-692

Dalgleish, A. G., Marriott, J. B., Souberbielle, B., Montesano, C., Sheikh, J., Sidebottom, D., Westby, M., and Austen, B., 1999, The role of HIV gp120 in the disruption of the immune system. *Immunol.Lett.* **66**:81-87

Denner, J., 2000, How does HIV induce AIDS? The virus protein hypothesis. *J.Hum.Virol.* **3**:81-82

Dettenhofer, M. and Yu, X. F., 1999, Highly purified human immunodeficiency virus type 1 reveals a virtual absence of Vif in virions. *J.Virol.* **73**:1460-1467

Domingo, E., Holland, J.J., Biebricher, C. and Eigen, M. 1995. In: Molecular Basis of Virus Evolution. Eds. A.J. Gibbs, C.H. Calisher, F. Garcia-Arenal. Cambridge Univ Press, UK, pp 181-191.

Dorfman, T., Mammano, F., Haseltine, W. A., and Göttlinger, H. G., 1994, Role of the matrix protein in the virion association of the human immunodeficiency virus type 1 envelope glycoprotein. *J.Virol.* **68**:1689-1696

Dragic, T., 2001, An overview of the determinants of CCR5 and CXCR4 co-receptor function. *J.Gen.Virol.* **82**:1807-1814

Fäcke, M., Janetzko, A., Shoeman, R. L., and Kräusslich, H. G., 1993, A large deletion in the matrix domain of the human immunodeficiency virus gag gene redirects virus particle assembly from the plasma membrane to the endoplasmic reticulum. *J.Virol.* **67**:4972-4980

Fauci, A. S., 1988, The human immunodeficiency virus: infectivity and mechanisms of pathogenesis. *Science* **239**:617-622

Fauci, A. S., 1996, Host factors and the pathogenesis of HIV-induced disease. *Nature* **384**:529-534

Feorino, P. M., Kalyanaraman, V. S., Haverkos, H. W., Cabradilla, C. D., Warfield, D. T., Jaffe, H. W., Harrison, A. K., Gottlieb, M. S., Goldfinger, D., Chermann, J. C., Spira, T. T., McDougal, J. S., Curran, J. W., Montagnier, L., Murphy, F. A. and Francis, D. P., 1984, Lymphadenopathy associated virus infection of a blood donor-recipient pair with acquired immunodeficiency syndrome. *Science* **225**: 69-72

Fitzon, T., Leschonsky, B., Bieler, K., Paulus, C., Schröder, J., Wolf, H., and Wagner, R., 2000, Proline residues in the HIV-1 NH2-terminal capsid domain: structure determinants for proper core assembly and subsequent steps of early replication. *Virology* **268**:294-307

Forster, M. J., Mulloy, B., and Nermut, M. V., 2000, Molecular modelling study of HIV p17gag (MA) protein shell utilising data from electron microscopy and X-ray crystallography. *J.Mol.Biol.* **298**:841-857

Franke, E. K., Yuan, H. E., and Luban, J., 1994, Specific incorporation of cyclophilin A into HIV-1 virions. *Nature* **372**:359-362

Frankel, A. D. and Young, J. A., 1998, HIV-1: fifteen proteins and an RNA. *Annu. Rev. Biochem.* **67**:1-25

Freed, E. O., 1998, HIV-1 gag proteins: diverse functions in the virus life cycle. *Virology* **251**:1-15

Freed, E. O. and Martin, M. A., 1996, Domains of the human immunodeficiency virus type 1 matrix and gp41 cytoplasmic tail required for envelope incorporation into virions. *J. Virol.* **70**:341-351

Fuller, S. D., Wilk, T., Gowen, B. E., Kräusslich, H. G., and Vogt, V. M., 1997, Cryo-electron microscopy reveals ordered domains in the immature HIV-1 particle. *Curr.Biol.* **7**:729-738

Ganser, B. K., Li, S., Klishko, V. Y., Finch, J. T., and Sundquist, W. I., 1999, Assembly and analysis of conical models for the HIV-1 core. *Science* **283**:80-83

Gelderblom, H., Reupke, H., Winkel, T., Kunze, R., and Pauli, G., 1987b, MHC-antigens: constituents of the envelopes of human and simian immunodeficiency viruses. *Z.Naturforsch.[C.]* **42**:1328-1334

Gelderblom, H. R., 1991, Assembly and morphology of HIV: potential effect of structure on viral function [editorial]. *AIDS* **5**:617-637

Gelderblom, H. R., 1997, Fine structure of HIV and SIV. III-31-III-44, Los Alamos National Laboratory.

Gelderblom, H. R., Hausmann, E. H., Özel, M., Pauli, G., and Koch, M. A., 1987a, Fine structure of human immunodeficiency virus (HIV) and immunolocalization of structural proteins. *Virology* **156**:171-176

Gelderblom, H. R., Özel, M. and Pauli., G., 1985a, T-Zell-spezifische Retroviren des Menschen: Vergleichende morphologische Klassifizierung und mögliche funktionelle Aspekte. *Bundesgesundheitsblatt* **28**:161-171

Gelderblom, H. R., Reupke, H., and Pauli, G., 1985b, Loss of envelope antigens of HTLV-III/LAV, a factor in AIDS pathogenesis? *Lancet* **2**:1016-1017

Gelderblom, H. R., Özel, M., Hausmann, M., Winkel, T., Pauli, G., and Koch., M.A., 1988, Fine structure of human immunodeficiency virus (HIV), immunolocalization of structural proteins and virus-cell relation. *Micron Microscopica* **19**:41-60

Gentile, M., Adrian, T., Scheidler, A., Ewald, M., Dianzani, F., Pauli, G., and Gelderblom, H. R., 1994, Determination of the size of HIV using adenovirus type 2 as an internal length marker. *J. Virol.Methods* **48**:43-52

Gheysen, D., Jacobs, E., de Foresta, F., Thiriart, C., Francotte, M., Thines, D., and De Wilde, M., 1989, Assembly and release of HIV-1 precursor Pr55gag virus-like particles from recombinant baculovirus-infected insect cells. *Cell* **59**:103-112

Gluschankof, P., Mondor, I., Gelderblom, H. R., and Sattentau, Q. J., 1997, Cell membrane vesicles are a major contaminant of gradient-enriched human immunodeficiency virus type-1 preparations. *Virology* **230**:125-133

Gonda, M. A., Wong-Staal, F., Gallo, R. C., Clements, J. E., Narayan, O., and Gilden, R. V., 1985, Sequence homology and morphologic similarity of HTLV-III and visna virus, a pathogenic lentivirus. *Science* **227**:173-177

Gonda, M.A., 1988, Molecular genetics and structure of the human immunodeficiency virus. J. Electron Microsc. Techn. 8: 17-40.

Gorelick, R. J., Gagliardi, T. D., Bosche, W. J., Wiltrout, T. A., Coren, L. V., Chabot, D. J., Lifson, J. D., Henderson, L. E., and Arthur, L. O., 1999, Strict conservation of the retroviral nucleocapsid protein zinc finger is strongly influenced by its role in viral infection processes: characterization of HIV-1 particles containing mutant nucleocapsid zinc- coordinating sequences. *Virology* **256**:92-104

Gorelick, R. J., Henderson, L. E., Hanser, J. P., and Rein, A., 1988, Point mutants of Moloney murine leukemia virus that fail to package viral RNA: evidence for specific RNA recognition by a "zinc finger- like" protein sequence. *Proc.Natl.Acad.Sci.U.S.A* **85**:8420-8424

Gorelick, R. J., Nigida, S. M., Bess, J. W., Arthur, L. O., Henderson, L. E., and Rein, A., 1990, Noninfectious human immunodeficiency virus type 1 mutants deficient in genomic RNA. *J.Virol.* **64**:3207-3211

Goto, T., Ikuta, K., Zhang, J. J., Morita, C., Sano, K., Komatsu, M., Fujita, H., Kato, S., and Nakai, M., 1990, The budding of defective human immunodeficiency virus type 1 (HIV-1) particles from cell clones persistently infected with HIV-1. *Arch.Virol.* **111**:87-101

Goto, T., Nakai, M., and Ikuta, K., 1998, The life-cycle of human immunodeficiency virus type 1. *Micron* **29**:123-138

Göttlinger, H. G., Dorfman, T., Sodroski, J. G., and Haseltine, W. A., 1991, Effect of mutations affecting the p6 gag protein on human immunodeficiency virus particle release. *Proc.Natl.Acad.Sci.U.S.A* **88**:3195-3199

Göttlinger, H. G., Sodroski, J. G., and Haseltine, W. A., 1989, Role of capsid precursor processing and myristoylation in morphogenesis and infectivity of human immunodeficiency virus type 1. *Proc.Natl.Acad.Sci.U.S.A* **86**:5781-5785

Grättinger, M., Hohenberg, H., Thomas, D., Wilk, T., Müller, B., and Kräusslich, H. G., 1999, In vitro assembly properties of wild-type and cyclophilin-binding defective human immunodeficiency virus capsid proteins in the presence and absence of cyclophilin A. *Virology* **257**:247-260

Grewe, C., Beck, A., and Gelderblom, H. R., 1990, HIV: early virus-cell interactions. *J.Acquir.Immune.Defic.Syndr.* **3**:965-974

Grief, C., Hockley, D. J., Fromholc, C. E., and Kitchin, P. A., 1989, The morphology of simian immunodeficiency virus as shown by negative staining electron microscopy. *J.Gen.Virol.* **70**:2215-2219

Gross, I., Hohenberg, H., Wilk, T., Wiegers, K., Grättinger, M., Müller, B., Fuller, S., and Kräusslich, H. G., 2000, A conformational switch controlling HIV-1 morphogenesis. *EMBO J.* **19**:103-113

Hallenberger, S., Bosch, V., Angliker, H., Shaw, E., Klenk, H. D., and Garten, W., 1992, Inhibition of furin-mediated cleavage activation of HIV-1 glycoprotein gp160 . *Nature* **360**:358-361

Heinzinger, N. K., Bukinsky, M. I., Haggerty, S. A., Ragland, A. M., Kewalramani, V., Lee, M. A., Gendelman, H. E., Ratner, L., Stevenson, M., and Emerman, M., 1994, The Vpr protein of human immunodeficiency virus type 1 influences nuclear localization of viral nucleic acids in nondividing host cells. *Proc.Natl.Acad.Sci.U.S.A* **91**:7311-7315

Henderson, L. E., Bowers, M. A., Sowder, R. C., Serabyn, S. A., Johnson, D. G., Bess, J. W., Arthur, L. O., Bryant, D. K., and Fenselau, C., 1992, Gag proteins of the highly replicative

MN strain of human immunodeficiency virus type 1: posttranslational modifications, proteolytic processings, and complete amino acid sequences. *J. Virol.* **66**:1856-1865

Henderson, L. E., Sowder, R., Copeland, T. D., Oroszlan, S., Arthur, L. O., Robey, W. G., and Fischinger, P. J., 1987, Direct identification of class II histocompatibility DR proteins in preparations of human T-cell lymphotropic virus type III. *J. Virol.* **61**:629-632

Ho, D. D., Neumann, A. U., Perelson, A. S., Chen, W., Leonard, J. M., and Markowitz, M., 1995, Rapid turnover of plasma virions and CD4 lymphocytes in HIV-1 infection. *Nature* **373**:123-126

Hockley, D. J., Wood, R. D., Jacobs, J. P., and Garrett, A. J., 1988, Electron microscopy of human immunodeficiency virus. *J.Gen.Virol.* **69**: 2455-2469

Höglund, S., Öfverstedt, L. G., Nilsson, A., Lundquist, P., Gelderblom, H., Özel, M., and Skoglund, U., 1992, Spatial visualization of the maturing HIV-1 core and its linkage to the envelope. *AIDS Res.Hum.Retroviruses* **8**:1-7

Horuk, R., 1999, Chemokine receptors and HIV-1: the fusion of two major research fields. *Immunol.Today* **20**:89-94

Hoxie, J. A., Fitzharris, T. P., Youngbar, P. R., Matthews, D. M., Rackowski, J. L., and Radka, S. F., 1987, Nonrandom association of cellular antigens with HTLV-III virions. *Hum.Immunol.* **18**:39-52

Hunter, E., 1994, Macromolecular interactions in the assembly of HIV and other retroviruses. *Sem.Virol.* **5**:71-83

Jones, I. M. and Morikawa, Y., 1998, The molecular basis of HIV capsid assembly. *Rev. Med. Virol.* **8**:87-95

Koenig, S., Gendelman, H. E., Orenstein, J. M., Dal Canto, M. C., Pezeshkpour, G. H., Yungbluth, M., Janotta, F., Aksamit, A., Martin, M. A., and Fauci, A. S., 1986, Detection of AIDS virus in macrophages in brain tissue from AIDS patients with encephalopathy. *Science* **233**:1089-1093

Korber, B., Muldoon, M., Theiler, J., Gao, F., Gupta, R., Lapedes, A., Hahn, B. H., Wolinsky, S., and Bhattacharya, T., 2000, Timing the ancestor of the HIV-1 pandemic strains. *Science* **288**:1789-1796

Korber, B., Theiler, J., and Wolinsky, S., 1998, Limitations of a molecular clock applied to considerations of the origin of HIV-1. *Science* **280**:1868-1871

Kräusslich, H. G., Fäcke, M., Heuser, A. M., Konvalinka, J., and Zentgraf, H., 1995, The spacer peptide between human immunodeficiency virus capsid and nucleocapsid proteins is essential for ordered assembly and viral infectivity. *J. Virol.* **69**:3407-3419

Kräusslich, H. G. and Welker, R., 1996, Intracellular transport of retroviral capsid components. *Curr.Top.Microbiol.Immunol.* **214**:25-63

Kuhmann, S. E., Platt, E. J., Kozak, S. L., and Kabat, D., 2000, Cooperation of multiple CCR5 coreceptors is required for infections by human immunodeficiency virus type 1. *J. Virol.* **74**:7005-7015

LaBonte, J. A., Patel, T., Hofmann, W., and Sodroski, J., 2000, Importance of membrane fusion mediated by human immunodeficiency virus envelope glycoproteins for lysis of primary CD4-positive T cells. *J. Virol.* **74**:10690-10698

Lagaye, S., Derrien, M., Menu, E., Coito, C., Tresoldi, E., Mauclere, P., Scarlatti, G., Chaouat, G., Barré-Sinoussi, F., and Bomsel, M., 2001, Cell-to-cell contact results in a selective translocation of maternal human immunodeficiency virus type 1 quasispecies across a trophoblastic barrier by both transcytosis and infection. *J. Virol.* **75**: 4780-4791

Layne, S. P., Merges, M. J., Dembo, M., Spouge, J. L., Conley, S. R., Moore, J. P., Raina, J. L., Renz, H., Gelderblom, H. R., and Nara, P. L., 1992, Factors underlying spontaneous inactivation and susceptibility to neutralization of human immunodeficiency virus. *Virology* **189**:695-714

Leis, J., Baltimore, D., Bishop, J. M., Coffin, J., Fleissner, E., Goff, S. P., Oroszlan, S., Robinson, H., Skalka, A. M., Temin, H. M., and ., 1988, Standardized and simplified nomenclature for proteins common to all retroviruses. *J.Virol.* **62**:1808-1809

Levy, J. A., 1986, The multifaceted retrovirus. *Cancer Res.* **46**:5457-5468

Levy, J. A., 1994, HIV and the Pathogenesis of AIDS. Washington, ASM Press .

Levy, J. A., Hoffman, A. D., Kramer, S. M., Landis, J. A., Shimabukuro, J. M., and Oshiro, L. S., 1984, Isolation of lymphocytopathic retroviruses from San Francisco patients with AIDS. *Science* **225**:840-842

Lifson, J. D., Reyes, G. R., McGrath, M. S., Stein, B. S., and Engleman, E. G., 1986, AIDS retrovirus induced cytopathology: giant cell formation and involvement of CD4 antigen. *Science* **232**:1123-1127

Liska, V., Spehner, D., Mehtali, M., Schmitt, D., Kirn, A., and Aubertin, A. M., 1994, Localization of viral protein X in simian immunodeficiency virus macaque strain and analysis of its packaging requirements. *J.Gen.Virol.* **75**:2955-2962

Littman, D. R., 1998, Chemokine receptors: keys to AIDS pathogenesis? *Cell* **93**:677-680

Luban, J., Bossolt, K. L., Franke, E. K., Kalpana, G. V., and Goff, S. P., 1993, Human immunodeficiency virus type 1 Gag protein binds to cyclophilins A and B. *Cell* **73**:1067-1078

Mammano, F., Öhagen, A., Höglund, S., and Göttlinger, H. G., 1994, Role of the major homology region of human immunodeficiency virus type 1 in virion morphogenesis. *J.Virol.* **68**:4927-4936

Marschang, P., Sodroski, J., Würzner, R., and Dierich, M. P., 1995, Decay-accelerating factor (CD55) protects human immunodeficiency virus type 1 from inactivation by human complement. *Eur.J.Immunol.* **25**:285-290

Marx, P. A., Munn, R. J., and Joy, K. I., 1988, Computer emulation of thin section electron microscopy predicts an envelope-associated icosadeltahedral capsid for human immunodeficiency virus. *Lab Invest* **58**:112-118

McCune, J. M., 2001, The dynamics of CD4+ T-cell depletion in HIV disease. *Nature* **410**:974-979

Meerloo, T., Sheikh, M. A., Bloem, A. C., de Ronde, A., Schutten, M., van Els, C. A., Roholl, P. J., Joling, P., Goudsmit, J., and Schuurman, H. J., 1993, Host cell membrane proteins on human immunodeficiency virus type 1 after in vitro infection of H9 cells and blood mononuclear cells. An immuno-electron microscopic study. *J.Gen.Virol.* **74**:129-135

Merigan, T. C. Bartlett J. G. and Bolognesi D. P., 1999, Textbook of AIDS Medicine. Baltimore., Williams and Wilkins.

Modrow, S., Hahn, B. H., Shaw, G. M., Gallo, R. C., Wong-Staal, F., and Wolf, H., 1987, Computer-assisted analysis of envelope protein sequences of seven human immunodeficiency virus isolates: prediction of antigenic epitopes in conserved and variable regions. *J.Virol.* **61**:570-578

Montefiori, D. C., Cornell, R. J., Zhou, J. Y., Zhou, J. T., Hirsch, V. M., and Johnson, P. R., 1994, Complement control proteins, CD46, CD55, and CD59, as common surface constituents of human and simian immunodeficiency viruses and possible targets for vaccine protection. *Virology* **205**:82-92

Moog, C., Spenlehauer, C., Fleury, H., Heshmati, F., Saragosti, S., Letourneur, F., Kirn, A., and Aubertin, A. M., 1997, Neutralization of primary human immunodeficiency virus type 1 isolates: a study of parameters implicated in neutralization in vitro. *AIDS res.hum.Retroviruses* **13**:19-27

Morikawa, Y., Shibuya, M., Goto, T., and Sano, K., 2000, In vitro processing of human immunodeficiency virus type 1 Gag virus- like particles. *Virology* **272**:366-374

Müller, B., Tessmer, U., Schubert, U., and Kräusslich, H. G., 2000, Human immunodeficiency virus type 1 Vpr protein is incorporated into the virion in significantly smaller amounts than gag and is phosphorylated in infected cells. *J. Virol.* **74** :9727-9731

Müller, M., 1992, The integrating power of cryofixation-based electron microscopy in biology. *Acta Microscópia* **1**:37-44

Nakai, M. and Goto, T., 1996, Ultrastructure and morphogenesis of human immunodeficiency virus . *J.Electron Microsc.(Tokyo)* **45**:247-257

Nermut, M. V., Grief, C., Hashmi, S., and Hockley, D. J., 1993, Further evidence of icosahedral symmetry in human and simian immunodeficiency virus. *AIDS res.hum.Retroviruses* **9**:929-938

Nermut, M. V. and Hockley, D. J., 1996, Comparative morphology and structural classification of retroviruses. *Curr.Top.Microbiol.Immunol.* **214**:1-24

Nermut, M. V., Hockley, D. J., Bron, P., Thomas, D., Zhang, W. H., and Jones, I. M., 1998, Further evidence for hexagonal organization of HIV gag protein in prebudding assemblies and immature virus-like particles. *J.Struct.Biol.* **123**:143-149

Nermut, M. V., Hockley, D. J., Jowett, J. B., Jones, I. M., Garreau, M., and Thomas, D., 1994, Fullerene-like organization of HIV gag-protein shell in virus-like particles produced by recombinant baculovirus. *Virology* **198**:288-296

Nermut, M. V., Bron, P., Thomas, D., Zhang, W.H., Jones, J.M., Rumlova, M., and T. Ruml, T., 2001, Structural aspects of retrovirus assembly. *Virus Research* **77**: in press

Ono, A., Orenstein, J. M., and Freed, E. O., 2000, Role of the Gag matrix domain in targeting human immunodeficiency virus type 1 assembly. *J. Virol.* **74**:2855-2866

Orenstein, J. M., 2000, In vivo cytolysis and fusion of human immunodeficiency virus type 1- infected lymphocytes in lymphoid tissue. *J.Infect.Dis.* **182**:338-342

Orenstein, J. M., Fox, C., and Wahl, S. M., 1997, Macrophages as a source of HIV during opportunistic infections. *Science* **276**:1857-1861

Orenstein, J. M. and Wahl, S. M., 1999, The macrophage origin of the HIV-expressing multinucleated giant cells in hyperplastic tonsils and adenoids. *Ultrastruct.Pathol.* **23**:79-91

Ott, D. E., Coren, L. V., Chertova, E. N., Gagliardi, T. D., and Schubert, U., 2000, Ubiquitination of HIV-1 and MuLV Gag. *Virology* **278**:111-121

Ott, D. E., Coren, L. V., Kane, B. P., Busch, L. K., Johnson, D. G., Sowder, R. C., Chertova, E. N., Arthur, L. O., and Henderson, L. E., 1996, Cytoskeletal proteins inside human immunodeficiency virus type 1 virions. *J. Virol.* **70**:7734-7743

Özel, M., Pauli, G., and Gelderblom, H. R., 1988, The organization of the envelope projections on the surface of HIV. *Arch.Virol.* **100**:255-266

Palmer, E. and Goldsmith, C. S., 1988, Ultrastructure of human retroviruses. *J.Electron Microsc.Tech.* **8**:3-15

Parren, P. W., Moore, J. P., Burton, D. R., and Sattentau, Q. J., 1999, The neutralizing antibody response to HIV-1: viral evasion and escape from humoral immunity. *AIDS* **13** **Suppl A:** S137-S162

Paxton, W., Connor, R. I., and Landau, N. R., 1993, Incorporation of Vpr into human immunodeficiency virus type 1 virions: requirement for the p6 region of gag and mutational analysis. *J. Virol.* **67**:7229-7237

Pepinsky, R. B. and Vogt, V. M., 1984, Fine-structure analyses of lipid-protein and protein-protein interactions of gag protein p19 of the avian sarcoma and leukemia viruses by cyanogen bromide mapping. *J. Virol.* **52**:145-153

Pettit, S. C., Moody, M. D., Wehbie, R. S., Kaplan, A. H., Nantermet, P. V., Klein, C. A., and Swanstrom, R., 1994, The p2 domain of human immunodeficiency virus type 1 Gag regulates sequential proteolytic processing and is required to produce fully infectious virions. *J.Virol.* **68**:8017-8027

Piot, P., Bartos, M., Ghys, P. D., Walker, N., and Schwartlander, B., 2001, The global impact of HIV/AIDS. *Nature* **410**:968-973

Poignard, P., Fouts, T., Naniche, D., Moore, J. P., and Sattentau, Q. J., 1996, Neutralizing antibodies to human immunodeficiency virus type-1 gp120 induce envelope glycoprotein subunit dissociation. *J.Exp.Med.* **183**:473-484

Popovic, M., Sarngadharan, M. G., Read, E., and Gallo, R. C., 1984, Detection, isolation, and continuous production of cytopathic retroviruses (HTLV-III) from patients with AIDS and pre-AIDS. *Science* **224**:497-500

Rabson, A. B. and Martin, M. A., 1985, Molecular organization of the AIDS retrovirus. *Cell* **40**:477-480

Reil, H., Bukovsky, A. A., Gelderblom, H. R., and Göttlinger, H. G., 1998, Efficient HIV-1 replication can occur in the absence of the viral matrix protein. *EMBO J.* **17**:2699-2708

Rein, A., Henderson, L. E., and Levin, J. G., 1998, Nucleic-acid-chaperone activity of retroviral nucleocapsid proteins: significance for viral replication. *Trends Biochem.Sci.* **23**:297-301

Renz, H., Weinberg, G., Özel, M., Gelderblom, H.R., Pauli, G., and Müller, M., 1993, High resolution scanning electron microscopy of HIV after conventional- and cryo-preparation. Comparison with transmission electron microscopical results. *Beitr.Elektronenmikroskop.Direktabb.Oberfl.* **26**:209-219

Richman, D. D., 2001, HIV chemotherapy . *Nature* **410**:995-1001

Saifuddin, M., Parker, C. J., Peeples, M. E., Gorny, M. K., Zolla-Pazner, S., Ghassemi, M., Rooney, I. A., Atkinson, J. P., and Spear, G. T., 1995, Role of virion-associated glycosylphosphatidylinositol-linked proteins CD55 and CD59 in complement resistance of cell line-derived and primary isolates of HIV-1. *J.Exp.Med.* **182**:501-509

Sarngadharan, M. G., Popovic, M., Bruch, L., Schüpbach, J., and Gallo, R. C., 1984, Antibodies reactive with human T-lymphotropic retroviruses (HTLV-III) in the serum of patients with AIDS. *Science* **224**: 506-508

Sattentau, Q. J. and Moore, J. P., 1991, Conformational changes induced in the human immunodeficiency virus envelope glycoprotein by soluble CD4 binding. *J.Exp.Med.* **174**:407-415

Schätzl, H., Gelderblom, H. R., Nitschko, H., and von der, Helm K., 1991, Analysis of non-infectious HIV particles produced in presence of HIV proteinase inhibitor. *Arch.Virol.* **120**:71-81

Smith, A. J., Srinivasakumar, N., Hammarskjold, M. L., and Rekosh, D., 1993, Requirements for incorporation of Pr160gag-pol from human immunodeficiency virus type 1 into virus-like particles. *J.Virol.* **67**:2266-2275

Stein, B. S., Gowda, S. D., Lifson, J. D., Penhallow, R. C., Bensch, K. G., and Engleman, E. G., 1987, pH-independent HIV entry into CD4-positive T cells via virus envelope fusion to the plasma membrane. *Cell* **49**:659-668

Sunila, I., Vaccarezza, M., Pantaleo, G., Fauci, A. S., and Orenstein, J. M., 1997, gp120 is present on the plasma membrane of apoptotic CD4 cells prepared from lymph nodes of HIV-1-infected individuals: an immunoelectron microscopic study. *AIDS* **11**:27-32

Takasaki, T., Kurane, I., Aihara, H., Ohkawa, N., and Yamaguchi, J., 1997, Electron microscopic study of human immunodeficiency virus type 1 (HIV- 1) core structure: two RNA strands in the core of mature and budding particles. *Arch.Virol.* **142**:375-382

Tenner-Racz, K., Stellbrink, H. J., van Lunzen, J., Schneider, C., Jacobs, J. P., Raschdorff, B., Grosschupff, G., Steinman, R. M., and Racz, P., 1998, The unenlarged lymph nodes of HIV-1-infected, asymptomatic patients with high CD4 T cell counts are sites for virus replication and CD4 T cell proliferation. The impact of highly active antiretroviral therapy. *J.Exp.Med.* **187**:949-959

Thali, M., Bukovsky, A., Kondo, E., Rosenwirth, B., Walsh, C. T., Sodroski, J., and Göttlinger, H. G., 1994, Functional association of cyclophilin A with HIV-1 virions. *Nature* **372**:363-365

Tremblay, M. J., Fortin, J. F., and Cantin, R., 1998, The acquisition of host-encoded proteins by nascent HIV-1. *Immunol.Today* **19**:346-351

Tritel, M. and Resh, M. D., 2000, Kinetic analysis of human immunodeficiency virus type 1 assembly reveals the presence of sequential intermediates. *J. Virol.* **74**:5845-5855

Tritel, M. and Resh, M. D., 2001, The late stage of human immunodeficiency virus type 1 assembly is an energy-dependent process. *J. Virol.* **75**:5473-5481

Turner, B. G. and Summers, M. F., 1999, Structural biology of HIV. *J.Mol.Biol.* **285**:1-32

van Regenmortel, M.H. V., Fauquet, C.M., Bishop, D.H.L., Carstens, E.B., Estes, M.K., Lemon, S.M., Maniloff, J., Mayo, M.A., McGeoch, D.J., PringleC.R., and Wickner.,R.B., 2000, Virus taxonomy, classification and nomenclature of viruses, Seventh report of the International Committee on Taxonomy of Viruses. pp 369-387San Diego, London, Academic Press.

Wahl, S. M., Greenwell-Wild, T., Peng, G., Hale-Donze, H., Doherty, T. M., Mizel, D., and Orenstein, J. M., 1998, Mycobacterium avium complex augments macrophage HIV-1 production and increases CCR5 expression. *Proc.Natl.Acad.Sci.U.S.A* **95**:12574-12579

Wain-Hobson, S., Alizon, M., and Montagnier, L., 1985, Relationship of AIDS to other retroviruses. *Nature* **313**:743

Weiss, R. A. and Wrangham, R. W., 1999, From Pan to pandemic. *Nature* **397**:385-386

Welker, R., Hohenberg, H., Tessmer, U., Huckhagel, C., and Kräusslich, H. G., 2000, Biochemical and structural analysis of isolated mature cores of human immunodeficiency virus type 1. *J. Virol.* **74**:1168-1177

Wiegers, K., Rutter, G., Kottler, H., Tessmer, U., Hohenberg, H., and Kräusslich, H. G., 1998, Sequential steps in human immunodeficiency virus particle maturation revealed by alterations of individual Gag polyprotein cleavage sites. *J. Virol.* **72**:2846-2854

Wilk, T. and Fuller, S. D., 1999a, Towards the structure of the human immunodeficiency virus: divide and conquer. *Curr.Opin.Struct.Biol.* **9**:231-243

Wilk, T., Gowen, B., and Fuller, S. D., 1999b, Actin associates with the nucleocapsid domain of the human immunodeficiency virus Gag polyprotein. *J. Virol.* **73**:1931-1940

Wilk, T., Gross, I., Gowen, B. E., Rutten, T., de Haas, F., Welker, R., Kräusslich, H. G., Boulanger, P., and Fuller, S. D., 2001, Organization of immature human immunodeficiency virus type 1. *J. Virol.* **75**:759-771

Wilk, T., Pfeiffer, T., Bukovsky, A., Moldenhauer, G., and Bosch, V., 1996, Glycoprotein incorporation and HIV-1 infectivity despite exchange of the gp160 membrane-spanning domain. *Virology* **218**:269-274

Wills, J. W. and Craven, R. C., 1991, Form, function, and use of retroviral gag proteins. *AIDS* **5**:639-654

Wormser, G. P., Editor, 1996, A Clinical Guide to AIDS and HIV. Lippincott-Raven, Philadelphia

Wyatt, R. and Sodroski, J., 1998, The HIV-1 envelope glycoproteins: fusogens, antigens, and immunogens. *Science* **280**:1884-1888

Yu, X., Yuan, X., Matsuda, Z., Lee, T. H., and Essex, M., 1992, The matrix protein of human immunodeficiency virus type 1 is required for incorporation of viral envelope protein into mature virions. *J. Virol.* **66**:4966-4971

Yuan, X., Yu, X., Lee, T. H., and Essex, M., 1993, Mutations in the N-terminal region of human immunodeficiency virus type 1 matrix protein block intracellular transport of the Gag precursor. *J. Virol.* **67**:6387-6394

Zhang, W. H., Hockley, D. J., Nermut, M. V., Morikawa, Y., and Jones, I. M., 1996, Gag-Gag interactions in the C-terminal domain of human immunodeficiency virus type 1 p24 capsid antigen are essential for Gag particle assembly. *J. Gen. Virol.* **77**:743-751

Zhou, W., Parent, L. J., Wills, J. W., and Resh, M. D., 1994, Identification of a membrane-binding domain within the amino-terminal region of human immunodeficiency virus type 1 Gag protein which interacts with acidic phospholipids. *J. Virol.* **68**:2556-2569

Chapter 5.3

The Role of Hemagglutinin and Neuraminidase in Influenza Virus Pathogenicity

RALF WAGNER[1], ANKE FELDMANN[1], THORSTEN WOLFF[2], STEPHAN PLESCHKA[3], WOLFGANG GARTEN[1], AND HANS-DIETER KLENK[1]
Institut für Virologie, Klinikum der Philipps-Universität Marburg, Robert-Koch-Str. 17, 35037 Marburg, Germany. [2]Robert-Koch-Insititut, Division of Viral Infections, Nordufer 20, 13353 Berlin, Germany. [3]Institut für Virologie, Justus-Liebig-Universität Giessen, Frankfurter Str. 107, 35392 Giessen, Germany

1.   INTRODUCTION

The pathogenicity of influenza viruses is determined by many of its other biological properties, such as efficiency of replication, tissue tropism, host range, spread of infection, as well as response to and modulation of host defense. All of these properties are controlled by the complex interplay of viral and host factors at virtually each stage in the life cycle of the virus. Activation of the hemagglutinin by host cell proteases has been shown in many studies to have a dramatic effect on pathogenesis. We will therefore first briefly review the molecular details of proteolytic activation. We will then concentrate on recent studies in which we have analysed the interplay of hemagglutinin (HA) and neuraminidase (NA) in receptor binding and virus release and some of the factors that determine spread of infection in the organism.

Structure-Function Relationships of Human Pathogenic Viruses, Edited by
Holzenburg and Bogner, Kluwer Academic/Plenum Publishers, New York, 2002

2.   PROTEOLYTIC ACTIVATION OF HA IS A MAJOR FACTOR IN PATHOGENICITY

The hemagglutinin glycoprotein (HA) of influenza A virus plays an essential role in virus entry. After binding to the cellular receptor, the virus is internalized through receptor-mediated endocytosis. HA then induces fusion of the viral envelope with the endosomal membrane, thus allowing delivery of the nucleocapsid into the cytoplasm. To show fusion activity, HA has to undergo a biphasic activation process. The first step involves cleavage by host proteases into the fragments HA_1 and HA_2 (Klenk and Garten, 1994). Proteolytic cleavage renders HA in a metastable form susceptible to a conformational change that is triggered by low pH in the endosome. This second step in the activation process results in the exposure of a hydrophobic domain on HA_2 that can now interact with the lipid bilayer of the target membrane (Chen et al., 1998; Wiley and Skehel, 1987).

HA cleavage is a determinant of pathogenicity, because it is necessary for spread of infection through the organism. The mammalian and the non-pathogenic avian influenza virus strains have HAs that are usually cleaved only in a restricted number of cell types. Therefore these viruses cause local infection. In contrast, the pathogenic avian strains have different HAs that are activated in a broad range of different host cells and therefore cause systemic infection. The major structural property of HA that determines differential cleavability is the link between HA_1 and HA_2 in the uncleaved precursor. With HAs of restricted cleavability, the linker usually consists of a single arginine, whereas highly cleavable HAs mainly have multiple basic residues in this position showing the consensus sequence R-X-K/R-R (Horimoto et al., 1994; Klenk and Garten, 1994; Klenk and Rott, 1988).

Various proteases have been described recognizing monobasic sites. One of them is tryptase Clara, which is probably the most important protease activating human influenza viruses in their natural environment, the respiratory system (Kido et al., 1992). Bacterial proteases which have been shown to promote the development of influenza pneumonia after coinfection cleave also at monobasic sites (Tashiro et al., 1987). In contrast, HAs with multibasic sites are cleaved by furin and some furin related enzymes which all belong to the proprotein convertase family of eukaryotic subtilisin-like, calcium-dependent serine endoproteases. Furin is a constituent of the constitutive secretory pathway in practically all cells and accumulates in the trans-Golgi network, which is also the cellular compartment where this HA type is cleaved.

Highly pathogenic influenza virus strains with multibasic HA cleavage sites were previously thought to be strictly confined to birds. However, such a virus has now also been identified as the causative agent of an outbreak of human influenza with a high case-mortality rate that occurred 1997/1998 in Hong Kong (Subbarao et al., 1998). Thus the potential of these viruses as a threat to human health has become a realistic matter of concern.

3. BALANCE OF RECEPTOR BINDING AND RELEASE REGULATES VIRUS GROWTH

HA mediated attachment of influenza viruses to sialic acid containing receptors on the host cell surface is the initial step of infection. Influenza HA contains at the tip a narrow crevice lined with highly conserved amino acids. By its ability to specifically bind sialic acids this crevice has been identified as the receptor binding site (Eisen et al., 1997; Rogers et al., 1983; Weis et al., 1988). The precise structure of this HA domain is known to be of crucial importance for the process of virus binding to its receptor. Accordingly, single amino acid substitutions in the binding pocket can result in altered receptor binding specificity and altered host range of the respective viruses (Aytay and Schulze, 1991; Connor et al., 1994; Vines et al., 1998). Furthermore, employing vector expressed FPV HA we could show that oligosaccharides flanking the binding site modulate receptor affinity (Ohuchi et al., 1997). To evaluate the impact of each individual N-glycan at the FPV-HA tip on the growth of intact viruses, we generated recombinant influenza viruses containing the oligosaccharide-deleted HA mutants (Wagner et al., 2000) (Fig. 1). Our studies demonstrate that the glycans flanking the receptor binding pocket are potent regulators of virus growth in cell culture. The oligosaccharide attached to Asn 149 (absent in mutant G2) plays a dominant role in controlling virus spread while that attached to Asn 123 (absent in mutant G1) is less effective. Growth of viruses lacking both N-glycans was found to be reduced in cell culture due to a restricted release of progeny viruses from infected cells (Fig. 2). These findings on the growth of recombinant viruses are an important extension of our previous work investigating the receptor interaction of transiently expressed HA. There is now experimentally based evidence for a distinct regulatory function of individual N-glycans located at the HA tip on the viral life cycle. By sequentially removing N-glycans from the vicinity of the HA receptor binding site we have also delineated a novel approach to specifically generate influenza viruses with a gradual extension of attenuation in cell culture.

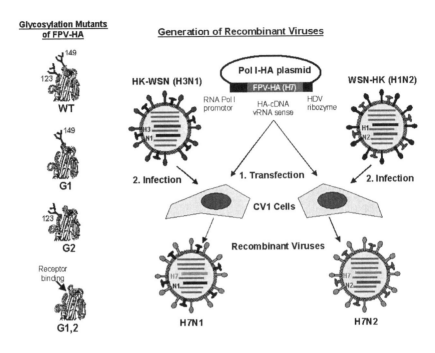

Figure 1: Inset on the left: The head region of FPV-HA is shown on the top. N-linked oligosaccharides adjacent to the receptor binding pocket are indicated. Mutants G1 and G2 lack the glycosylation sites at Asn 123 and Asn 149, respectively. Both sites are absent in mutant G1,2. The arrow marks the entrance to the receptor binding pocket. Body of figure: recombinant viruses were generated by a RNA-polymerase I-based reverse genetics system. Two reassortants of strain A/WSN/33 were used as helper viruses to obtain two series of HA-mutant viruses only differing in NA.

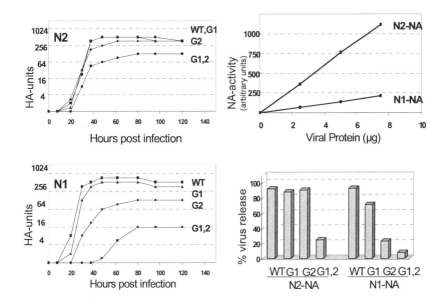

Figure 2. (A) Growth curves of recombinants in MDCK cells. Cell monolayers were infected at a MOI of 0.001 with recombinant viruses, and supernatants were monitored for HA titers at the time points indicated. The results obtained with viruses of the N2-series and of the N1-series are shown. ■ WT; σ G1; υ G2; λ G1,2. (B) Comparison of specific NA-activities of WT/N1 and WT/N2 viruses. Different amounts of purified virus were incubated with 4-methylumbelliferyl N-acetylneuraminic acid for 20 min at 37°C. The reaction was stopped and NA activity was calculated by measuring the fluorescence of the liberated methylumbelliferone. The data are means of three experiments. They indicate that WT/N2 has a higher NA activity than WT/N1. (C) Release of recombinant viruses from MDCK cells. MDCK cells were infected at a MOI of 5 with recombinant viruses and incubated at 37°C overnight. One hour before virus harvest VCNA was added to the culture media to one half of the samples. Titers of progeny viruses released into the media were determined by plaque assay. Levels of virus release in the absence of VCNA are presented as percent values relative to the virus titers released after VCNA treatment (from Wagner et al., 2000).

By removing terminal sialic acid residues from oligosaccharide side chains of glycoconjugates the viral neuraminidase (NA) acts as a receptor destroying enzyme in influenza viruses (Colman, 1998; Lamb and Krug, 1996). When NA activity was blocked by either antibodies (Compans et al., 1969), inhibitors (Gubareva et al., 1996; Palese and Compans, 1976), or temperature-sensitive mutations (Palese et al., 1974), formation of large viral aggregates on the surface of infected cells was observed as was with virus lacking NA either partly (Mitnaul et al., 1996) or completely (Liu et al., 1995). Accordingly, viral NA is regarded an important factor for the release of progeny virus from host cells promoting the efficient progression of an infection. In light of this it was of special interest to examine how different NA subtypes affect the attenuated phenotype of the recombinant viruses lacking N-glycans at the HA tip. Several N1-NAs have a deletion in the stalk region that is most extensive with FPV-NA (Hausmann et al., 1997). NA enzymatic activity has been reported to vary according to the length of the stalk region of the molecule with NA species containing a deletion in the stalk having a lower activity (Castrucci and Kawaoka, 1993; Els et al., 1985; Luo et al., 1993; Matrosovich et al., 1999). By choosing appropriate helper viruses we generated recombinants in which the HA mutants were combined with either the WSN virus NA (N1 subtype) containing a stalk deletion or the Hong Kong virus NA (N2 subtype) that has no deletion. When assayed for neuraminidase activity recombinant viruses carrying N2-NA exceeded those with N1-NA at least sixfold (Fig. 2). Thus, our set of recombinants was ideally suited to analyze in depth the impact of different NA activities on the growth of mutant influenza viruses specifically designed to show distinct receptor binding activities. Using this system we were able to demonstrate that the growth behavior of HA mutant viruses is governed by the nature of the accompanying viral NA. Among the viruses with the high activity N2-NA growth restriction was observed only when the G1,2 mutant was present showing the highest receptor affinity, while recombinants containing G1 and G2 grew essentially like virus carrying wildtype HA. Yet, the situation was different with viruses containing the low activity N1-NA. Here, the growth of G1,2 mutant viruses was significantly impeded in cell culture due to a restricted release from host cells. This effect was less pronounced with G2 mutant viruses, but still evident. Obviously, unlike N2-NA the lower activity N1-NA is not capable to overcome the high affinity interaction of G1,2 and G2 HA with its receptor (Fig. 2).

Hence, our data clearly point out that, for the establishment of productive infection, influenza viruses are strictly dependent on a highly balanced action of HA and NA. An increase in receptor binding affinity apparently needs to be accompanied by a concomitant increase in the receptor destroying activity of the viral NA. Otherwise, the enhanced receptor binding is a serious disadvantage in the late stage of infection by preventing the release of progeny viruses from host cells. The need for such a match of HA and NA activities had so far only been deduced from studies analyzing natural virus isolates or laboratory generated reassortants (Kaverin et al., 1998; Kaverin and Klenk, 1999; Matrosovich et al., 1999; Rudneva et al., 1996). Taken together, our work represents the first concise study of the functional interrelationship of distinct HA and NA species and provides experimental evidence for the strict requirement of a fine tuning of HA receptor binding and NA receptor destroying activity in order to allow for an efficient influenza virus propagation (Fig. 3).

Balance of Receptor Binding and Release

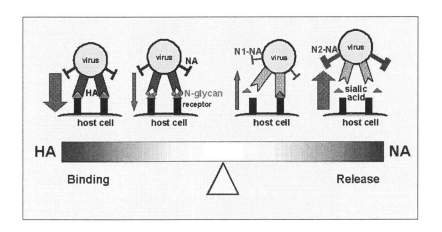

Figure 3: Regulation of virus binding and release by HA glycosylation and neuraminidase activity. Receptor affinity is controlled by the oligosaccharides adjacent to the receptor binding site on HA. The efficiency of release depends on the activity of NA.

There is evidence that N-glycans flanking the receptor binding site not only modulate receptor affinity, but control also receptor specificity. Thus, subtype H1 influenza strains with an oligosaccharide in such a position have been shown to bind preferentially to α2,3-linked neuraminic acid, whereas mutants lacking this oligosaccharide had a preference for the α2,6-linkage (Gambaryan et al., 1998; Günther et al., 1993). Furthermore, it has been shown recently that glycans carrying neuraminic acid in α2,3 or α2,6 linkages gain access to the receptor binding pocket from opposite sides (Eisen et al., 1997). Sterical hindrance by a glycan adjacent to the receptor binding site may therefore be a determinant of receptor specificity. Finally, the number and structure of N-glycans neighbouring the receptor binding pocket have been suggested to determine host range and pathogenicity of influenza viruses (Deom et al., 1986; Gambaryan et al., 1998; Perdue et al., 1995). In view of these findings, it will now be interesting to employ our panel of recombinant viruses to elucidate the contributions of individual HA tip glycans to tissue tropism and host range.

4.   FACTORS DETERMINING ENDOTHELIOTROPISM OF FPV

Although cleavage of HA is an important determinant of spread of infection in the organism, it is not the only factor involved. This is illustrated by a recent study in which spread of FPV in the chick embryo was investigated (Feldmann et al., 2000). We have found in this study that FPV shows strict endotheliotropism when infecting 11 days old chick embryos (Fig. 4). Since hemorrhages and edema are major symptoms in FPV infected chickens, it was not unexpected to see that the vasculature is an important target of infection. However, we were surprised when we could not detect viral replication in other cell types. Besides endothelia, myocytes and lymphatic tissues were found to be sites of virus replication, when hatched chickens were infected with other pathogenic H5 and H7 strains (Kobayashi et al., 1996; van Campen et al., 1989a; van Campen et al., 1989b). These differences in cell tropism may depend on the developmental stage of the host and on the virus strains used.

The results obtained with reassortants of FPV and virus N clearly indicate that endotheliotropism requires the presence of FPV HA. This observation is in line with the concept that, because of its susceptibility to ubiquitous proteolytic activation, FPV HA allows virus entry from the allantoic cavity into the highly vascularized mesenchymal layer of the chorioallantoic membrane and, thus, mediates hematogenic spread of infection. On the other hand, the restricted cleavability of virus N HA confines infection to the inner layer of the membrane and the allantoic cavity (Rott et al., 1980).

Reassortants containing virus N HA did therefore not have access to endothelia when infected through the chorioallantoic route.

Whereas cleavage activation of HA proved to be essential for targeting the virus to endothelia, it was not responsible for confining the infection to these cells. Furin and PC5/PC6, the activating proteases of FPV HA, were identified in all chicken tissues analyzed including endothelial cells. This observation indicates that the lack of spread of infection from endothelia to surrounding tissues cannot be attributed to the absence of activating proteases.

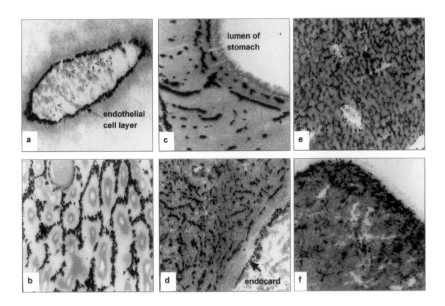

Figure 4: Localization of FPV-infected cells in embryonic tissues by *in situ* hybridization. Brightfield photomicrographs showing autoradiograms with black grains representing bound HA-specific riboprobe in organs of the chick embryo. After *in situ* hybridization, slides were covered by photoemulsion, exposed for 2 days, developed, and counter-stained with hematoxilin-eosin. a: blood vessel (arrowheads indicating infected blood cells), b: lung, c: stomach, d: heart, e: liver, f: spleen. Magnification x75 (from Feldmann et al., 2000).

In contrast, tissue-specific expression of virus receptors appears to be an important factor in restricting infection to endothelia. Using lectin binding assays a α-2,3-linked as well as a α-2,6-linked neuraminic acid could be detected. Both proved to be able to serve as FPV receptors on epithelial cells and cells of the reticulo-endothelial system only. However, receptor determinants on other cells, such as myocytes, fibroblasts, and hepatocytes. Thus, it appears that cells lacking a measurable amount of neuraminic acid receptors cannot be infected by FPV and are therefore a barrier for the spread of infection. This concept is nicely supported by observations made in lung tissue. When this organ is infected via the hematogenic route as is the case in the embryo at day 11, the virus is retained in the endothelial cells of the capillary vessels. Since neuraminic acid is present in α2,6 linkage on these cells, it is clear that this type of neuraminic acid can serve as FPV receptor, although it appears that binding of avian strains is generally determined by the α2,3 linkage. The alveolar epithelia, although expressing virus receptor in large amounts, are not infected because virus access is prevented by the connective tissue lacking neuraminic acid. On the other hand, when embryos are infected through the airways as can be done by inoculating virus 2 days before hatching into the now almost dry allantoic cavity, virus replication is readily detected in lung epithelia (data not shown).

It has to be pointed out that expression of neuraminic acid in the chick embryo depends on tissue differentiation (Codogno and Aubery, 1983). This may explain the absence of detectable amounts of neuraminic acid on fibroblasts *in situ*, whereas cultured fibroblasts readily express receptors as indicated by their ability to allow efficient virus replication. It has also to be assumed that the subendothelial connective tissue is not a very tight barrier in the choriallantoic membrane where it allows penetration of the virus from the allantoic epithelium into the mesodermal endothelia. In fact, low amounts of virus budding from mesodermal fibroblasts have been observed indicating that these cells may play a role in mediating spread of infection (Rott et al., 1980). Whether the spread through the mesodermal layer is driven by the particularly high virus replication rates in the allantoic epithelium and the presence of neuraminic acid on mesodermal fibroblasts or by some other mechanism remains to be seen.

Our data show also that the polarity of virus budding is another factor contributing to the confinement of infection to endothelial cells. Studies on Sendai virus in a mouse model have shown before that the sidedness of virus maturation has a distinct effect on spread of infection in the organism and on pathogenicity. Wild type Sendai virus released exclusively from the apical surface of lung epithelia is strictly pneumotropic, whereas the mutant F1-R which matures at the apical as well as the basolateral side causes pantropic infection (Tashiro et al., 1992; Tashiro et al., 1990). It has long been known that FPV matures preferentially at the luminal side of endothelia (Rott et al., 1980), and the observations made here on virus budding and HA transport support this concept. The luminal budding polarity of FPV supports therefore the hematogenic spread of the virus and prevents at the same time infection of subendothelial cells.

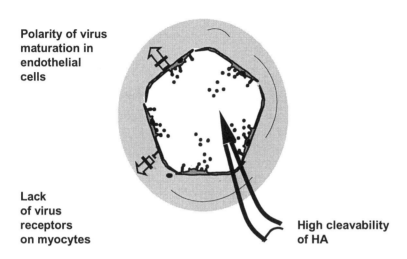

Figure 5: The factors determining endotheliotropism of FPV. HA cleavage is responsible for virus entry into the vascular system. Polar budding and lack of receptors in adjacent tissues are responsible for restricting infection to endothelia.

Taken together, our data indicate that endotheliotropism of FPV in the chick embryo is the result of an interplay of several factors determined by the virus and the host (Fig. 5). These include proteolytic activation of HA by ubiquitous proteases which is responsible for entry of the virus into the vascular system and at least 2 mechanisms contributing to the confinement of the virus to endothelia: the polarity of virus budding at the luminal side of endothelial cells and cell specific differences in the expression of neuraminic acid receptors. Endotheliotropism without doubt plays an important role in the generalization of FPV infection and in the generation of typical symptoms of the disease, such as hemorrhages and edema. Systemic infection and severe vascular injury are also the central pathogenetic mechanisms of hemorrhagic fevers in primates caused by filoviruses and other agents, and there is evidence that at least some of these viruses replicate also in endothelia (Ryabchikova et al., 1999; Schnittler et al., 1993; Zaki and Goldsmith, 1999). It will therefore be interesting to see if similar mechanisms as described here for FPV infection play also a role in the pathogenesis of hemorrhagic fevers in other species.

5.   CONCLUSION

The high variability of the influenza virus genome is reflected by a wide spectrum in host tropism, tissue specificity, and pathogenicity, ranging from local infection of the respiratory tract or the gut, as is the case with most mammalian strains and the apathogenic avian viruses, to systemic infection caused by fowl plague virus or other highly pathogenic avian strains. Pathogenicity is determined by the interaction of viral proteins with each other or with host factors. This concept is supported by many studies on the viral envelope proteins. Proteolytic activation of the hemagglutinin as a fusion protein has been known for a long time to be an important determinant of pathogenicity. Receptor specificity is another hemagglutinin trait responsible for host range and tissue tropism. There is a close interdependence between hemagglutinin and neuraminidase in controlling binding and release of virus. The multifactorial character of tissue tropism and pathogenicity is illustrated by the determinants of endotheliotropism: high cleavability of the hemagglutinin (mediating virus entry into the vascular system) on one hand, and restricted receptor expression and polar budding (to prevent spread of infection into tissues surrounding endothelia) on the other.

ACKNOWLEDGMENTS

Research in the author's laboratories was supported by grants from the Deutsche Forschungsgemeinschaft (SFB 286 and Kl 238/6-1) and from the Fonds der Chemischen Industrie.

REFERENCES

Aytay, S., and Schulze, I.T., 1991, Single amino acid substitutions in the hemagglutinin can alter the host range and receptor binding properties of H1 strains of influenza A virus. *J Virol.* **65:** 3022-8.

Castrucci, M. R., and Kawaoka, Y., 1993, Biologic importance of neuraminidase stalk length in influenza A virus. *J Virol.* **67:** 759-64.

Chen, J., K. H. Lee, D. A. Steinhauer, S. J. Stevens, J. J. Skehel, and D. C. Wiley, 1998. Structure of the hemagglutinin precursor cleavage site, a determinant of influenza pathogenicity and the origin of the labile conformation. Cell **95:** 409-417.

Codogno, P., and Aubery, M., 1983, Changes in cell-surface sialic acid content during chick embryo development. *Mech. Ageing Dev.* **23:** 307-314

Colman, P., 1998, Structure and function of the neuraminidase, p. 65-73. *In* K. G. Nicholson, R. G. Webster, and A. J. Hay (ed.), Textbook of influenza. Blackwell Science, London, England

Compans, R. W., Dimmock, N.J., and. Meier-Ewert, H., 1969, Effect of antibody to neuraminidase on the maturation and hemagglutinating activity of an influenza A2 virus. *J Virol.* **4:** 528-34.

Connor, R. J., Kawaoka, Y., Webster, R.G., and Paulson, J.C., 1994, Receptor specificity in human, avian, and equine H2 and H3 influenza virus isolates. *Virology* **205:** 17-23.

Deom, C. M., Caton, A.J., and Schulze, L.T., 1986, Host cell-mediated selection of a mutant influenza A virus that has lost a complex oligosaccharide from the tip of the hemagglutinin. *Proc Natl Acad Sci U S A.* **83:** 3771-5.

Eisen, M. B., Sabesan, S., Skehel, J.J., and Wiley, D.C., 1997, Binding of the influenza A virus to cell-surface receptors: structures of five hemagglutinin-sialyloligosaccharide complexes determined by X- ray crystallography. *Virology* **232:** 19-31.

Eisen, M. B., Sabesan, S., Skehel, J.J., and Wiley, D.C., 1997, Binding of the influenza A virus to cell-surface receptors: structures of five hemagglutinin-sialyloligosaccharide complexes determined by X- ray crystallography. *Virology* **232:** 19-31.

Els, M. C., Air, G.M., Murti, K.G., Webster, R.G., and Laver, W.G., 1985, An 18-amino acid deletion in an influenza neuraminidase. *Virology* **142:** 241-7.

Feldmann, A., Schäfer, M.K.-H., Garten, W., and Klenk, H.-D., 2000, Targeted infection of endothelial cells by the avian influenza virus A/FPV/ROSTOCK/34 (H7N1) in chicken embryos. *J. Virol.* **74,** 8018-27.

Gambaryan, A. S., Marinina, V.P., Tuzikov, A.B., Bovin, N.V., Rudneva, L.A., Sinitsyn, B.V., Shilov, A.A., and Matrosovich, M.N., 1998, Effects of host-dependent glycosylation of hemagglutinin on receptor- binding properties on H1N1 human influenza A virus grown in MDCK cells and in embryonated eggs. *Virology* **247:** 170-7.

Gubareva, L. V., Bethell,R., Hart, G.J., Murti, K.G., Penn, C.R., and Webster, R.G., 1996, Characterization of mutants of influenza A virus selected with the neuraminidase inhibitor 4-guanidino-Neu5Ac2en. *J Virol.* **70:** 1818-27.

Günther, I., Glatthaar, B., Doller, G., and Garten, W., 1993, A H1 hemagglutinin of a human influenza A virus with a carbohydrate- modulated receptor binding site and an unusual cleavage site. *Virus Res.* **27:** 147-60.

Hausmann, J., Kretzschmar, E., Garten, W., and Klenk, H.-D., 1997, Biosynthesis, intracellular transport and enzymatic activity of an avian influenza A virus neuraminidase: role of unpaired cysteines and individual oligosaccharides. *J Gen Virol.* **78:** 3233-45.

Horimoto, T., Nakayama, K., Smeekens, S.P., and Kawaoka, Y., 1994, Proprotein-processing endoprotease PC5 and furin both activate hemagglutinin of virulent avian influenza viruses. *J. Virol.* **68:** 6074-6078.

Kaverin, N. V., Gambaryan, A.S., Bovin, N.V., Rudneva, L.A., Shilov, A.A., Khodova, O.M., Varich, N.L., Sinitsin, B.V., Makarova, N.V., and Kropotkina, E.A., 1998, Postreassortment changes in influenza A virus hemagglutinin restoring HA-NA functional match. *Virology* **244:** 315-21.

Kaverin, N. V., and Klenk, H.D., 1999, Strain-specific differences in the effect of influenza A virus neuraminidase on vector-expressed hemagglutinin. *Arch Virol.* **144:** 781-6.

Kido, H., Y. Yokogoshi, Y., Sakai, K., Tashiro, M., Kishino Y., Fukutomi, A., Katunuma, A.N., 1992, Isolation and characterization of a novel trypsin-like protease found in rat bronchiolar epithelial Clara cells: A possible activator of the viral fusion glycoprotein. *J. Biol. Chem.* **267:** 13573-13579.

Klenk, H.-D., and Garten, W., 1994, Host cell proteases controlling virus pathogenicity. *Trends Micobiol.* **2:** 39-43.

Klenk, H.-D., and Rott, R., 1988, The molecular of influenza virus pathogenicity. *Adv. Virus Res.* **34:** 247-281.

Kobayashi, Y., Horimoto, T., Kawaoka, Y., Alexander, D., and Itakura, C., 1996, Pathological studies of chickens experimentally infected with two highly pathogenic avian influenza viruses. *Avian pathology* **25**: 285-304.

Lamb, R. A., and Krug, R.M., 1996, Orthomyxoviridae: The viruses and their replication. In *Fields Virology*, (B. N. Fields, D. M. Knipe, and P. M. Howley eds.),. Lippincott-Raven Publishers, Philadelphia, ISA, pp1353-1395.

Liu, C., Eichelberger, M.C., Compans, R.W., and Air, G.M., 1995, Influenza type A virus neuraminidase does not play a role in viral entry, replication, assembly, or budding. *J Virol.* **69**: 1099-106.

Luo, G., Chung, J., and Palese, P., 1993, Alterations of the stalk of the influenza virus neuraminidase: deletions and insertions. *Virus Res.* **29**: 321.

Matrosovich, M., Zhou, N., Kawaoka, Y., and Webster, R., 1999, The surface glycoproteins of H5 influenza viruses isolated from humans, chickens, and wild aquatic birds have distinguishable properties. *J Virol.* **73**: 1146-55.

Mitnaul, L. J., Castrucci, M.R., Murti, K.G., and Kawaoka, Y., 1996, The cytoplasmic tail of influenza A virus neuraminidase (NA) affects NA incorporation into virions, virion morphology, and virulence in mice but is not essential for virus replication. *J Virol.* **70**: 873-9.

Ohuchi, M., Ohuchi, R., Feldmann, A., and Klenk, H.D., 1997, Regulation of receptor binding affinity of influenza virus hemagglutinin by its carbohydrate moiety. *J Virol.* **71**: 8377-84.

Palese, P., and Compans, R.W., 1976. Inhibition of influenza virus replication in tissue culture by 2-deoxy- 2,3-dehydro-N-trifluoroacetylneuraminic acid (FANA): mechanism of action. *J Gen Virol.* **33**: 159-63.

Palese, P. Tobita, K., Ueda, M., and Compans, R.W., 1974, Characterization of temperature sensitive influenza virus mutants defective in neuraminidase. *Virolog.* **61**: 397-410.

Perdue, M. L., Latimer, J.W., and Crawford, J.M., 1995, A novel carbohydrate addition site on the hemagglutinin protein of a highly pathogenic H7 subtype avian influenza virus. *Virolog.* **213**: 276-81.

Rogers, G. N., Paulson, J.C., Daniels, R.S., Skehel, J.J., Wilson, L.A., and Wiley, D.C., 1983, Single amino acid substitutions in influenza haemagglutinin change receptor binding specificity. *Nature* **304**: 76-8.

Rott, R., Reinacher, M., Orlich, M., and Klenk, H.D., 1980, Cleavability of hemagglutinin determines spread of avian influenza viruses in the chorioallantoic membrane of chicken embryo. *Arch. Virol.* **65**: 123-133.

Rudneva, I. A., Sklyanskaya, E.L., Barulina, O.S., Yamnikova, S.S., Kovaleva, V.P., Tsvetkova, L.V., and Kaverin, N.V., 1996, Phenotypic expression of HA-NA combinations in human-avian influenza A virus reassortants. *Arch Virol.* **141**: 1091-9.

Ryabchikova, E. J., Kolesnikova, L.V., and Netesov, S.V., 1999, Animal pathology of filoviral infections. *Curr. Top. Microbiol. Immunol.* **235**: 145-173.

Schnittler, H. J., Mahner, F., Drenckhahn, D., Klenk, H.-D., and Feldmann, H., 1993, Replication of Marburg virus in human endothelial cells. A possible mechanism for the development of viral hemorrhagic disease. *J. Clin. Invest.* **91**: 1301-1309.

Subbarao, K., Klimor, A., Katz, J., Regnery, H., Lim, W., Hall, H., Perdue, M., Swayne, D., Bender, C., Huang, J., Hemphill, M., Rowe, Z., Shaw, M., Xu, X., Fukada, K., and Cox, N., 1998, Characterization of an avian influenza A (H5N1) virus isolated from a child with a fatal respiratory illness, *Science* **279**: 393-395.

Tashiro, M., Seto, J.T., Choosakul, S., Yamakawa, M., Klenk, H.D., and Rott, R., 1992, Budding site of Sendai virus in polarized epithelial cells is one of the determinants for tropism and pathogenicity in mice. *Virology* **187**: 413-422.

Tashiro, M., Ciborowski, P., Klenk, H.D., Pulverer, G., and Rott, R., 1987, Role of Staphylococcus protease in the development of influenza oneumonia, *Nature* **325:** 536-537.

Tashiro, M., Yamakawa, M., Tobita, K., Seto, J.T., Klenk, H.D., and Rott, R., 1990. Altered budding site of a pantropic mutant of Sendai virus, F1-R, in polarized epithelial cells. *J. Virol.* **64:** 4672-4677.

van Campen, H., Easterday, B.C., and Hinshaw, V.S., 1989a, Destruction of lymphocytes by a virulent avian influenza A virus. *J. Gen. Virol.* **70:** 467-472.

van Campen, H., Easterday, B.C., and Hinshaw, V.S., 1989b, Virulent avian influenza A viruses: their effect on avian lymphocytes and macrophages in vivo and in vitro. *J. Gen. Virol.* **70:** 2887-2895.

Vines, A., Wells, K., Matrosovich, M., Castrucci, M.R., Ito, T., and Kawaoka, Y., 1998, The role of influenza A virus hemagglutinin residues 226 and 228 in receptor specificity and host range restriction. *J Virol.* **72:** 7626-31.

Wagner, R., Wolff, T., Herwig, A., Pleschka, S., and Klenk, H.-D., 2000, Interdependence of hemagglutinin glycosylation and neuraminidase as regulators of influenza virus growth – A study by reverse genetics. *J. Virol.* **74:** 6316-23.

Weis, W., Brown, J.H., Cusack, S., Paulson, J.C., Skehel, J.J., and Wiley, D.C., 1988, Structure of the influenza virus haemagglutinin complexed with its receptor, sialic acid. *Nature* **333:** 426-31.

Wiley, D.C., and Skehel, J.J,. 1987,The structure and function of the hemagglutinin membrane glycoprotein of influenza virus. *Annu. Rev. Biochem.* **56:** 365-394.

Zaki, S. R., and Goldsmith, C.S.,. 1999, Pathologic features of filovirus infections in humans. *Curr. Top. Microbiol. Immunol.* **235:** 97-116.

6. Viral oncogenesis

Chapter 6.1

Rhadinovirus Pathogenesis

ARMIN ENSSER[1], FRANK NEIPEL[1], and HELMUT FICKENSCHER[2]

[1]*Institut für Klinische und Molekulare Virologie, Friedrich-Alexander-Universität Erlangen-Nürnberg, Schlossgarten 4, D-91054 Erlangen;* [2]*Abteilung Virologie, Hygiene-Institut, Ruprecht-Karls-Universität, Im Neuenheimer Feld 324, D-69120 Heidelberg, Germany*

1. INTRODUCTION

Rhadinoviruses or γ_2-herpesviruses have been known for many years as animal pathogens causing diseases such as malignant catarrhal fever in cattle or T-cell lymphoma in certain neotropical primates. The existence of a human member of this virus group had been assumed for a long time. However, only few years ago, the human herpesvirus type 8 (HHV-8) was discovered as the first human rhadinovirus which is strongly associated with Kaposi's sarcoma (KS), multicentric Castleman's disease, and primary effusion lymphoma (PEL). DNA of this virus is regularly found in all forms of KS, and particularly in the spindle cell which is a specific cell type for KS. In contrast, the virus DNA is not detected in most other malignancies. Thus, HHV-8 was also termed KS-associated herpesvirus (KSHV). Antibodies against this virus are rare in normal blood donors and more frequently found in risk groups for KS. KSHV seroconversion precedes the tumor development. A series of KSHV genes have been shown to exhibit transforming potential in cell culture systems. In addition, the virus encodes several cytokines and angiogenic factors, as well as a cytokine receptor. This is of particular interest because pathogenesis models for KS emphasize the importance of inflammatory cytokines produced in the lesion. In contrast to other virus-induced malignancies, the pathogenetic role of KSHV in KS seems to be complicated by various cellular and viral factors. Closely related herpesviruses have been discovered in various Old World primates, however

Structure-Function Relationships of Human Pathogenic Viruses, Edited by
Holzenburg and Bogner, Kluwer Academic/Plenum Publishers, New York, 2002

these rhadinoviruses are as far as known only loosely associated with pathogenicity.

A New World primate species, the squirrel monkey (*Saimiri sciureus*), is the natural host of the T-lymphotropic herpesvirus saimiri (HVS), the long-known rhadinovirus prototype. In its natural host, this persisting virus does not cause disease, whereas it induces fatal acute T-cell lymphoma in other monkey species after experimental infection. Similar to other rhadinoviruses, the HVS genome encodes a series of functional viral proteins with pronounced homology to cellular counterparts. Most of these viral genes have been shown to be dispensable for the transforming and pathogenic capability of this virus. However, they are considered relevant for the apathogenic persistence of HVS in its natural host. A terminal region of the non-repetitive coding part of the virus genome is essential for pathogenicity and T-cell transformation. Certain HVS strains such as C488 are capable of transforming human T lymphocytes to stable proliferation in culture. The transformed human T cells maintain multiple copies of the viral genome as non-integrated episomes and do not produce virus particles. The transformed cells retain the antigen specificity and many other essential functions of their parental T-cell clones. Based on the preserved functional phenotype of the transformed T cells, HVS provides useful tools for T-cell immunology, for gene transfer, and possibly also for experimental adoptive immunotherapy.

A series of other rhadinoviruses have been described in various domestic and feral animals, including ungulates and rodents. Some of these viruses exhibit relevant pathogenicity, such as the malignant catarrhal fever of cattle caused by rhadinoviruses of sheep or wildebeests. Thus, a family of genetically closely related viruses shows strongly divergent pathogenic phenotypes, including various forms of lymphoma.

2.   KAPOSI'S SARCOMA-ASSOCIATED HERPESVIRUS

2.1   Epidemiology and disease associations

2.1.1   Evidence for sexual transmission

Kaposi's sarcoma (KS) has been a rare disease in Europe and the United States. In its "classical" form, it afflicts elderly men of Mediterranean or Eastern European origin, where it may progress slowly over several years. The unusual occurrence of a more aggressive variant of KS in young homosexual men was one of the first signs of the beginning AIDS epidemic (Friedman-Kien *et al.*, 1982). Out of the 90,990 persons with AIDS reported until March 1989, 13,616 or 15% suffered from KS (Beral *et al.*, 1990). KS was thus at least 20,000 times more common in persons with AIDS than expected in the general population. The peculiar epidemiology of KS in different HIV-transmission groups clearly indicated that an environmental or transmissible agent other than HIV must be involved in KS pathogenesis (Beral *et al.*, 1990; Siegal *et al.*, 1990). Most notably, whereas about 20% of homosexual and bisexual AIDS patients developed KS, only 1% of age- and sex-matched men with hemophilia suffered from this uncommon tumor. This triggered a broad search for behavioral factors (Beral *et al.*, 1992) and infectious agents such as known viruses which would be associated with an increased risk for KS. However, none of these factors or viruses were found to be consistently linked to KS (Lifson *et al.*, 1990; Kempf *et al.*, 1994, 1995, 1997). Application to KS of a PCR-based method for the identification of differences between two complex genomes (representational difference analysis; Lisitsyn *et al.*, 1993) resulted in the identification of two short DNA fragments from a new herpesvirus (Chang *et al.*, 1994). This virus, now termed human herpesvirus 8 (HHV-8) or KS-associated herpesvirus (KSHV), is the first known human member of the genus rhadinovirus or γ_2-herpesviruses.

2.1.2 Association of Kaposi's sarcoma with KSHV

Consistent detection of viral DNA in tumor tissue is not sufficient to prove that a virus is involved in oncogenesis. In theory, different roles are possible for KSHV in KS pathogenesis (Neipel *et al.*, 1997b), ranging from a mere "passenger model", where KSHV would be an innocent bystander, to the role of a directly transforming virus. Most human tumor viruses are relatively widespread when compared to the frequency of the malignancy they cause. For example, this is true for Epstein-Barr Virus (EBV), the human herpesvirus most closely related to KSHV, and for human papilloma viruses. The epidemiology of AIDS-associated KS in the metropolitan areas of the Northern hemisphere suggested that a relatively infrequent, sexually transmitted agent is involved in KS pathogenesis.

This poses three important epidemiological questions: First, is the KSHV prevalence significantly increased in KS risk groups, as compared to the general population? Second, does the infection precede KS-development? Third, are the modes of transmission compatible with the claimed transmissible agent? DNA PCR is not sufficient to answer these questions, as the site of KSHV latency in non-KS patients is unknown. Several serological assays are available. Immunofluorescence techniques based on KSHV-positive cell lines derived from primary effusion lymphomas (PEL; Cesarman *et al.*, 1995) are widely used. Alternatively, purified virus or recombinant antigen is utilized in ELISA techniques. The validity of these assays, especially when it comes to the crucial question of seroprevalence in persons not at risk for KS, is still a hotly debated issue (Rabkin *et al.*, 1998). This is essentially due to the high inter-assay variability observed with low-titered sera from non-KS patients. However, as justified as it may be, this debate should not obscure three important findings: first, the prevalence of high anti-KS antibody titers correlates very well with KS or KS risk. This is true both when various AIDS-transmission groups are compared within one geographic area, and when areas of low KS prevalence (*e.g.*, Northern Europe or the United States) are compared with regions where KS is endemic, *e.g.*, Mediterranean Europe or parts of Africa (Gao *et al.*, 1996a, b; Kedes *et al.*, 1997; Miller *et al.*, 1996; Raab *et al.*, 1998; Simpson *et al.*, 1996). Second, KSHV seroconversion or at least a marked increase in anti-KSHV antibody titer precedes KS-development in homosexual AIDS patients (Martin *et al.*, 1998; Oksenhendler *et al.*, 1998). Third, high anti-KSHV antibody titers were also associated with the number of homosexual partners and a history of other sexually transmitted diseases (Martin *et al.*, 1998). In summary, serological data largely agree that the titer of anti-

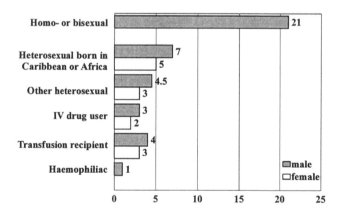

Figure 1. Percentage of AIDS patients with Kaposi's sarcoma in different HIV transmission groups. A total of 88,739 persons with AIDS were included in the study. Fifteen percent of 13,616 of these had KS. Thus the risk of KS is at least 20,000 times greater in AIDS patients than in the general population. Among persons with AIDS the percentage ranges from 23% (= 21,612) in homo- or bisexual men to 1% (= 9) in men with hemophilia (Beral *et al.*, 1990).

KSHV antibodies reflects KS or the risk of KS development. So far, only one exception has been noted: KSHV infection, as measured by an immunofluorescence assay, appears to be very common in The Gambia. However, in contrast to some areas of Central Equatorial Africa, KS is rarely seen in Gambian AIDS patients. In contrast, one study reported an epidemiological link between HIV-1 and KS. Although HIV-2 infection is much more common than HIV-1 infection in The Gambia, the few cases of KS that were reported occurred in AIDS patients infected with HIV-1 (Ariyoshi *et al.*, 1998). Although some uncertainty still remains with the current serological assays and reactivation of KSHV cannot be excluded as the reason for the markedly increased anti-KSHV titers in KS patients, the high levels of antibodies directed against both lytic and latent antigens of KSHV clearly hint at a pathogenetic role of the active KSHV infection in KS.

2.2   Genome structure, persistence and replication

KSHV is a member of the γ_2-herpesvirus or rhadinovirus subfamily (Chang *et al.*, 1994; Roizman *et al.*, 1992; Russo *et al.*, 1996). The rhadinoviruses have a so-called M genome with intermediate density in CsCl gradients (M-DNA). The γ_2-herpesviruses were termed "rhadino" viruses utilising the ancient greek word for fragile (Roizman *et al.*, 1992), because the M-DNA splits into two DNA molecules of highly different density, the L-DNA containing all the virus genes (low density, low G+C content) and the terminal repetitive H-DNA (high density, high G+C content) without coding capacity.

The almost complete nucleotide sequence of KSHV was determined from both a PEL cell line (Russo *et al.*, 1996) and from a KS biopsy specimen (Neipel *et al.*, 1997c). The central low-GC (53.3% G+C, L-DNA) coding fragment is flanked by numerous tandem repeats of high GC content (84.5% G+C, H-DNA). The L-DNA contains at least 89 open reading frames, most of which have homologs in other herpesviruses, especially in the closely related γ_2-herpesvirus prototype herpesvirus saimiri (HVS). Consequently, KSHV is the first human member of this virus group that clearly exhibits oncogenic potential.

The KSHV genome is present in all PEL cells and, at least at later stages of the disease, in virtually all spindle cells in KS lesions (Stürzl *et al.*, 1997, 1999b). In the vast majority of these cells the genome is present in a latent state with only few genes being expressed (Stürzl *et al.*, 1999a, b). On the transcript level, the best characterized cluster of latently expressed genes maps to one genomic segment close to the right end of the L-DNA

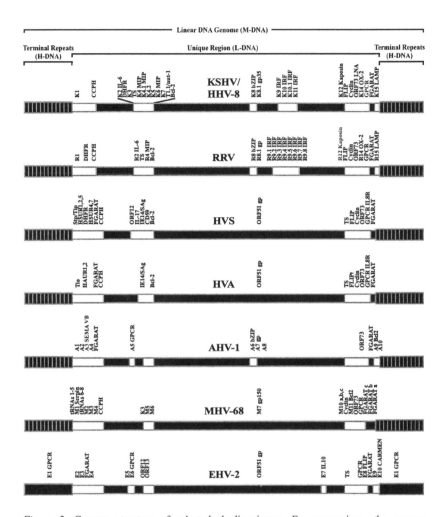

Figure 2. Genome structure of selected rhadinoviruses. For comparison, the genome structures of the rhadinoviruses KSHV/HHV-8 (Russo *et al.*, 1996), rhesus rhadinovirus (RRV; Alexander *et al.*, 2000; Searles *et al.*, 1999), herpesvirus saimiri (HVS; Albrecht *et al.*, 1992a), herpesvirus ateles (HVA, Albrecht, 2000), alcelaphine herpesvirus type 1 (AHV-1, Ensser *et al.*, 1997), murine herpesvirus 68 (MHV-68; Virgin *et al.*, 1997), and equine herpesvirus type 2 (EHV-2, Telford *et al.*, 1995) are depicted. Rather conserved genomic regions of virus genes with typical herpesvirus functions are shown in black. As white boxes, the variable areas are marked harboring non-conserved genes or genes with homology to cellular counterparts. The abbreviations are explained in the text.

comprising open reading frames (orf) *orf71* to *orf73* (Sarid *et al.*, 1998, 1999). Whereas *orf71* and *orf72* code for proteins with homology to cellular FLICE/caspase-8 inhibitors (FLIP, Thome *et al.*, 1997) or cyclin D (Li *et al.*, 1997; Swanton *et al.*, 1997), respectively, *orf73* encodes another protein conserved amongst the rhadinoviruses only. Although transcription of all three genes seems to be driven by one single latent promoter (Dittmer *et al.*, 1998), only the protein encoded by *orf73*, termed latent nuclear antigen-1 (LANA-1, LNA-1) has invariably been found to be expressed in all cells latently infected by KSHV (Kellam *et al.*, 1997; Rainbow *et al.*, 1997). One of the functions of this gene can be seen in analogy to the Epstein-Barr virus nuclear antigen-1 (EBNA-1). By tethering viral DNA to the chromosomes, LANA-1 mediates the persistence of extrachromosomal KSHV DNA (Ballestas *et al.*, 1999). Surprisingly, the cis- acting DNA elements involved in this function seem to be located within the GC-rich H-DNA (Ballestas *et al.*, 2001; Garber *et al.*, 2001). In addition to its function in episomal DNA maintenance, LANA-1 seems also to modulate cellular and viral gene expression (Garber *et al.*, 2001; Lim *et al.*, 2001; Renne *et al.*, 2001), although there are a few conflicting results (Hyun *et al.*, 2001; Renne *et al.*, 2001). Furthermore, LANA-1 of KSHV may also be involved in cell growth transformation as it has been shown to interact with the cellular p53 and retinoblastoma proteins (Friborg *et al.*, 1999; Radkov *et al.*, 2000).

Although it is possible to propagate KSHV on endothelial cells in the laboratory to a limited extent (Ciufo *et al.*, 2001; Sakurada *et al.*, 2001), an efficient lytic cell culture system is not yet available. It is thus not surprising that our knowledge on the molecular biology of KSHV lytic replication is still very limited. It is clear, however, that the protein encoded by *orf50*, which due to its homology with the EBV R-transactivator has been termed RTA or ART, is a key regulator for KSHV replication sufficient for initiation of the KSHV lytic cycle (Lukac *et al.*, 1998, 1999). More recently, the first molecules involved in KSHV target-cell binding have been identified. At least two viral proteins, the glycoprotein B and the highly immunogenic glycoprotein K8.1, mediate attachment of KSHV to target cells by binding to heparansulfate (Akula *et al.*, 2001; Birkmann *et al.*, 2001).

2.3   Models of Pathogenesis

2.3.1   Candidates for Transforming Virus Genes

The membership of KSHV to a class of oncogenic viruses and the close epidemiological association of KSHV and KS, immediately triggered a search for putative oncogenes of KSHV. However, due to the lack of an KSHV animal model and the inability to transform cultured cells by KSHV only (Kliche *et al.*, 1998) researchers are still relying on heterologous over expression systems and rodent fibroblast transformation assays. A series of putatively transforming genes, namely K1, K9 encoding a viral interferon (IFN)-response factor homolog (vIRF), K12, and the viral interleukin-8 receptor homolog (vIL-8R), have been identified in these systems to date.

Known transformation-relevant genes of other γ-herpesviruses are transmembrane proteins that map towards the ends of the non-repetitive L-DNA portion of the genome. Examples are the oncoproteins StpA, StpC, and Tip of HVS, Tio of herpesvirus ateles (HVA) (see section 3.4), and LMP-1 (latent membrane protein-1) of the more distantly related EBV. Remarkably, none of these transforming proteins is conserved in KSHV. However, a non-conserved transmembrane glycoprotein termed K1 was identified in an homologous position. In lack of an efficient permissive cell culture or animal model system for KSHV, HVS was utilized as an expression vector system for K1. In these experiments, the transforming gene *stpC* of HVS strain C488 by was replaced by K1 of KSHV. This revealed that K1 was able to substitute for the transforming function of StpC both in cell culture with respect of T-cell transformation, and in an animal model system. K1 was able to induce T-cell lymphoma in a common marmoset (*Callithrix jacchus*) when expressed from the HVS genome lacking *stpC* (Lee *et al.*, 1998b). However, the second transformation relevant gene of HVS C488 (*tip*), was still present in the recombinant HVS vector genome. Its contribution to the transformation remains to be clarified. K1 has been found to associate with non-receptor protein tyrosine kinases (Lee *et al.*, 1998a).

The KSHV open reading frame K9 encodes a viral IFN regulatory factor (vIRF-1), a protein with low, but significant homology to cellular IFN regulatory factors (IRF). The gene encoding vIRF-1 is located within a 12 kb non-conserved region of the viral genome. It is noteworthy that there are at least two other genes in this non-conserved area which are significantly homologous both to each other and to K9. IFNs are multifunctional cytokines with marked antiviral activity. IFN regulatory factors are a family of transcription factors linked to the IFN-induced signal transduction pathway. They bind to conserved *cis*-acting DNA elements, termed IFN-stimulated response elements (ISRE) or IFN-γ activated sequence (GAS), within the promoters of IFN regulated genes. IRFs can act as inducers or repressors of IFN-regulated genes by binding to ISRE or GAS elements. Examples for cellular IRFs are IRF-1 and IRF-2 (Fujita *et al.*, 1988; Itoh *et al.*, 1989). Cellular IRF-1 is an IFN-β inducible transcription factor with an N-terminal DNA-binding domain and a C-terminal transactivation domain (Lin *et al.*, 1994). IRF-2 also has a N-terminal DNA-binding domain which binds to the same ISRE as IRF-1. However, IRF-2 is a negative regulator of IRF-1 induced transcription. It has been shown that IRF-1 and IRF-2 can act as antagonists in cell proliferation regulation (Nguyen *et al.*, 1995). vIRF-1 was shown to inhibit the IFN-β signaling pathway (Gao *et al.*, 1997; Li *et al.*, 1998b; Zimring *et al.*, 1998). In addition, vIRF was able to transform rodent fibroblasts (Gao *et al.*, 1997; Li *et al.*, 1998b) which formed tumors in nude mice (Gao *et al.*, 1997). The molecular mechanisms of transcription regulation by vIRF-1 are partially understood. In contrast to many cellular IRFs, a direct DNA binding of vIRF-1 could not be shown. Instead, vIRF-1 interacts directly with several proteins. This includes the key tumor suppressor p53. By binding to a central region of p53, vIRF-1 inhibits phosphorylation of and thus transcriptional activation by p53 (Nakamura *et al.*, 2001). In addition, vIRF-1 interacts with the coadapter CREB-binding protein (CBP) which results in a 15 fold increase of *myc* transcription (Burysek *et al.*, 1999; Jayachandra *et al.*, 1999). The same interaction also prevents cellular IRF-3 from binding to CBP which in turn reduces the antiviral response to IFNs (Lin *et al.*, 2001). Thus, vIRF-1 may not only inhibit the hosts response against viral infections, but also shows transforming potential.

Similarly to most other rhadinoviruses, KSHV encodes a protein with homology to the family of G-protein coupled receptors (GPCR) with seven transmembrane regions. The KSHV gene is most homologous to the human IL-8R and hence termed vIL-8R. This viral receptor is capable of activating the protein kinase C pathway, regardless of whether a suitable chemokine is present or not. This constitutive activity was sufficient to enhance the proliferation of transfected primary rat fibroblasts (Arvanitakis *et al.*, 1997). Moreover, the transfection of vIL-8R induced focus formation in fibroblasts with an efficiency comparable to the m1 muscarinic receptor, but independently of an extracellular ligand. vIL-8R transformed cells efficiently formed tumors in nude mice (Bais *et al.*, 1998). Transgenic mice were generated that expressed vIL-8R under control of the CD2 promoter on lymphoid cells. Interestingly, this resulted in angioproliferative lesions in multiple organs. Histologically, these lesions strikingly resembled KS (Yang *et al.*, 2000). Thus, the constitutively signaling viral IL-8R homolog is another potential KSHV-encoded oncogene.

In contrast to K1, K9/vIRF, K12, and vIL-8R, the KSHV-encoded cyclin-D homolog (k-cyclin) has not yet been shown to transform cultured cells or to induce tumors in nude mice. However, it deserves to be listed amongst the potential KSHV oncogenes. Briefly, k-cyclin has been shown to associate with the catalytic subunit cdk6 to form an active holoenzyme (Li et al., 1997; Godden-Kent et al., 1997). This functional complex phosphorylates the retinoblastoma protein and stimulates cell-cycle progression from G1 to S in quiescent fibroblasts. However, the activity of the k-cyclin/cdk6 holoenzyme was resistant to inhibition by cdk-inhibitory proteins p21^{cip1} and p27^{Kip1} (Swanton et al., 1997). Moreover, the k-cyclin/cdk6 complex has a broader substrate range including p27^{kip1} and specific substrates of cdk2. Phosphorylation of p27^{kip1} results in the proteolytic degradation of this cdk-inhibitor and progression of cells arrested in G1 to S (Ellis et al., 1999; Mann et al., 1999). Deregulation of the G1/S restriction point is a central mechanism present in many tumor cells types and induced by various DNA tumor viruses (reviewed in Pines, 1995). It is therefore intriguing to speculate that k-cyclin might be involved in KSHV pathogenesis. However, it must be noted that at present there is no direct evidence for a transforming ability of k-cyclin.

Table 1. Non-conserved and cell-homologous genes of KSHV

ORF	Product	Hypothetical mechanism	Expression
K1	glycoprotein	B-cell signalling, transformation	Lytic
K2	IL-6	B-cell proliferation, paracrine stimulation	Lytic
K3		MHC-I downregulation	
K4	MIP	Inflammation, angiogenesis	Lytic
K4.1	MIP	Inflammation, angiogenesis	
K4.2			
K5		MHC-I downregulation	
K6	MIP	Inflammation, angiogenesis	Lytic
K7			
16	Bcl-2	Apoptosis resistance, enhanced productivity	Lytic
K8	bZIP		
K8.1	glycoprotein	Receptor ligand	
K9	IRF	Block of interferon action, transformation	Lytic
K10	IRF	Block of interferon action, transformation	
K10.5	IRF	Block of interferon action, transformation	Latent
K11	IRF		
K12	Kaposin	Transformation	Latent
71/K13	FLIP	Caspase inhibition, apoptosis resistance	Latent
72	k-cyclin	Proliferation, cell-cycle stimulation	Latent
73	LNA-1	Inhibition of p53	Latent
K14	OX-2		
74	GPCR/IL-8R	Angiogenesis, transformation	Lytic
K15	LAMP	Hypothetical homolog of EBV LMP-2	Lytic

In summary, five candidate KSHV oncogenes have been identified so far. This includes K1, K9/vIRF, vIL-8R, and K12. K9/vIRF and vIL-8R exhibited oncogenic potential in rodent fibroblasts. In contrast, fibroblast transformation by K12 was by far less efficient (Muralidhar *et al.*, 1998). Rodent fibroblasts expressing K9/vIRF or vIL-8R were able to induce tumors in nude mice, and K1 was shown to be a functional oncogene for T-cell transformation. However, there is a special irony in KSHV research: although many potential oncogenes have been identified in KSHV, probably more than in any other tumor virus, it has not been possible to transform cultured cells with KSHV, and only a small number of KSHV-carrying tumor cell lines could be isolated from patients. This is in sharp contrast to EBV, which efficiently transforms human B lymphocytes in culture. The poor replication of KSHV in cell culture and the lack of an efficient permissive system makes it difficult to assess the significance of the candidate oncogenes of KSHV.

2.3.2 Angiogenesis, B-cell stimulation, and inflammation

A closer look at the pathology of KS raises doubts about the relevance of classical oncogenes for the pathogenesis of this peculiar tumor. At least in its early stages, KS is not a typical sarcoma. Firstly, KS develops in a polyclonal fashion. Multiple lesions appear simultaneously, often in an symmetrical distribution. This is not compatible with metastatic tumor growth. In contrast to other sarcomas, classical KS may wax and wane over decades, and single lesions may disappear completely. This was also observed for the more aggressive AIDS-associated form of KS, where complete remissions were seen during antiretroviral combination therapy (Lebbe *et al.*, 1998). Although molecular data on the clonality of several KS lesions in a single patient are conflicting, it appears that at least in its early stages KS is not a monoclonal tumor (Delabesse *et al.*, 1997; Gill *et al.*, 1997). However, late nodular lesions may eventually progress to true sarcoma as indicated by clonality (Rabkin *et al.*, 1997; Gill *et al.*, 1998). Secondly, the histology of early KS lesions is not typical for other sarcomas. The lesions contain many different cell types. The so-called KS spindle cells are considered the tumor-defining, malignant cell type of KS. KS spindle cells are thought to be derived from lymphatic endothelial cells (Weninger *et al.*, 1999). However, they are not the dominant cell type until late stage disease. The KS lesions begin as subtle neovascularisation (Ruszczak *et al.*, 1987a) and the pathognomonic spindle cells are not yet detectable. In addition, KS-lesions contain infiltrating inflammatory cells and multicentrically arising vascular endothelia forming capillaries and other vessels (Ruszczak *et al.*, 1987b). Thirdly, the proliferation of cultured spindle cells is highly dependent on inflammatory cytokines and, with a few exceptions, KS spindle cells are not tumorigenic in nude mice (Nakamura *et al.*, 1988; Salahuddin *et al.*, 1988). Based on clinical observations and on data derived from these culture systems, models of KS-pathogenesis were already developed long before KSHV was discovered. Several groups agreed upon the notion that KS develops as an interplay of inflammatory cytokines and angiogenic factors (reviewed in Ensoli and Stürzl, 1998; Samaniego *et al.*, 1998), although the cytokines focused on varied in different reports. Amongst them are basic fibroblast growth factor, vascular endothelial growth factor (VEGF), IL-6, tumor necrosis factor (TNF) α and β, and oncostatin M. It is remarkable that several of the cytokines thought to be involved in KS pathogenesis in the pre-KSHV era have now been shown to be induced or encoded by this virus.

The open reading frame K2 of KSHV encodes viral IL-6 (vIL-6), a protein with moderate yet significant homology to the multifunctional human cytokine IL-6 (hIL-6) (Neipel *et al.*, 1997a). The vIL-6 cytokine is functional and supports the proliferation of both murine (Molden *et al.*, 1997; Moore *et al.*, 1996; Nicholas *et al.*, 1997) and human (Burger *et al.*, 1998) IL-6 dependent cells. Both human and viral IL-6 are produced in KS lesions and the proliferation of cultured spindle cells can be stimulated with oncostatin M and hIL-6 (Miles *et al.*, 1992; Masood *et al.*, 1994). Nevertheless, the relevance of IL-6 for KS pathogenesis has been questioned as some groups were not able to detect expression of the IL-6 receptor α-chain (IL-6Rα) by KS spindle cells (Murakami-Mori *et al.*, 1995). As a consequence, data on the proliferation stimulation of cultured KS cells by hIL-6 are controversial. The IL-6R consists of two chains. The β-chain, also termed gp130, is almost ubiquitously expressed. It is responsible for ligand-dependent signal transduction and is shared by several receptors of four-helical cytokines such as IL-6, IL-11, ciliary neurotrophic factor, cardiotrophin-1, oncostatin M, and leukemia inhibitory factor. The β-chain alone is not sufficient for hIL-6 binding. hIL-6 needs to bind IL-6Rα first. This complex associates with a homodimer of two gp130 molecules before signal transduction occurs. Similarly to many other transmembrane proteins, IL-6Rα exists also in a soluble form (sIL-6Rα). Soluble cytokine receptors maintain their affinity to the ligand, compete with the membrane-bound form for binding of the cytokine and thus usually act as antagonists. Soluble IL-6Rα, however, has been shown to act as an agonist. Whereas cells which do not express IL-6Rα are not able to respond to this cytokine, the presence of soluble IL-6Rα renders these cells responsive to hIL-6 (reviewed in Peters *et al.*, 1998). Stimulation or not of the proliferation of cultured KS spindle cells can therefore be explained by the presence or absence of sIL-6Rα in the culture medium, respectively. sIL-6Rα is present in the blood of healthy individuals in low concentrations and increases with certain infectious and malignant disorders. Stimulation of cells expressing both chains of the IL-6R by vIL-6 can be reduced by antibodies against IL-6Rα, the IL-6specific chain of the IL-6R (Burger *et al.*, 1998). This indicates that vIL-6 is able to bind IL-6Rα, albeit with reduced affinity. However, there is compelling evidence that IL-6Rα is not required for binding of vIL-6 to the signal transducer gp130 (Chow *et al.*, 2001; Li *et al.*, 2001; Molden *et al.*, 1997). Thus, vIL-6 may act in a way similar to the highly active "hyper-IL6", a unimolecular fusion of hIL-6 and sIL-6Rα (Fischer *et al.*, 1997).

Three genes with homology to the family of CC chemokines are located in the same non-conserved region of the KSHV genome as vIL-6. The proteins encoded by the open reading frames K4 and K6 are homologous to the macrophage inflammatory protein 1α (MIP-1α). They have been termed vMIP- II and vMIP-I, respectively (Russo *et al.*, 1996; Nicholas *et al.*, 1997). The amino acid sequence homology of these viral proteins to their cellular counterpart is only about 30%, but they are 48% homologous to each other. This could indicate that rather a gene duplication event occurred than the genes were independently acquired from the host cell genome. Another KSHV reading frame, K4.1, is also homologous to the CC chemokine family, with similarity to both MIP-1β and macrophage chemoattractant protein. Functional data are available for vMIP-I (K6) and vMIP-II (K4). vMIP-II binds efficiently to the CC-chemokine receptor CCR3 (Boshoff *et al.*, 1997). In contrast to cellular MIP-1α, it does also bind to CXC-chemokine receptors such as CXCR4, albeit with lower affinity. The CC-chemokine receptor CCR3 is known to be relevant for the chemoattraction of both eosinophil granulocytes and Th2 lymphocytes. Correspondingly, a potent chemoattractive effect of vMIP-II on eosinophils was described (Boshoff *et al.*, 1997). In contrast to cellular MIP-1α and to the CC-chemokine RANTES, both vMIP-I and vMIP-II were highly angiogenic in the chorioallantoic assay (Boshoff *et al.*, 1997). The chemokine homologs encoded by KSHV could thus contribute to two features of KS histology: the presence of inflammatory infiltrates containing Th2 lymphocytes, and the prominent angiohyperplasy. In addition to vMIP-I and -II, the viral IL-8R homolog is also capable of inducing angiogenesis (Bais *et al.*, 1998).

As discussed above, vIL-8R is a constitutively signaling CXC-chemokine receptor capable to alter rodent fibroblasts to an oncogenic phenotype. It engages the JNK/SAPK and p38-MAPK protein kinase pathways characteristic of inflammatory cytokines (Bais *et al.*, 1998). Most interestingly, this was associated with an increased secretion of vascular endothelial growth factor (VEGF). vIL8R-transfected fibroblasts induced angiogenic responses both in cell culture and in nude mice. This effect could be blocked by antibodies against VEGF. Receptors for VEGF are not only expressed on vascular endothelium, but also on KS spindle cells (Jussila *et al.*, 1998). VEGF has been regarded as one of the key players in paraendocrine models of KS-pathogenesis.

2.3.3   Latent gene expression

KSHV belongs to a family of viruses that frequently exhibit oncogenic potential in latently infected cells. At least four potentially transforming genes have been identified in the KSHV genome. In addition, several cytokines, chemokines, and cytokine receptors have been identified that fit the paracrine pathogenesis model of KS as an initially semi-malignant disease initiated by inflammatory and angiogenic signals. HVS has been successfully used as a paradigm for KSHV oncogenesis (reviewed by Jung *et al.*, 1999), and data obtained through HVS or other lymphotropic γ-herpesviruses might be useful for understanding the pathogenesis of KSHV-associated lymphomas. However, there is still no convincing animal model for KS. Thus, data on the expression pattern of viral genes in KS are required for an initial evaluation of their pathogenetic relevance.

The first basic observation is that KSHV is clearly present in the majority of KS spindle cells of both early and late KS lesions (Boshoff *et al.*, 1995; Kennedy *et al.*, 1998). Thus, the virus is present at the scene of the crime (Ganem, 1997). The virus is apparently in a latent state in the vast majority of these spindle cells, as judged from the inability to detect mRNA coding for structural proteins (Blasig *et al.*, 1997; Staskus *et al.*, 1997). Only few cells are interspersed amongst these latently infected cells that express RNAs mapping to structural genes, implying that these cells are lytically infected and produce viral particles. This is compatible with a classical transformation model where viral genes expressed in latently infected cells are responsible for the genetic alterations resulting in tumor growth, in a similar way as well known for other oncogenic viruses, such as EBV and papillomaviruses.

The second basic question asks whether the putative oncogenes are indeed expressed in KS spindle cells. However, the most attractive KSHV oncogenes such as K1, vIRF, and vIL-8R are obviously not expressed in latency, but belong to a class of genes only expressed in the productive replication cycle. Regulation of KSHV gene expression has been studied essentially in cultured PEL cells (Lagunoff and Ganem, 1997; Dittmer *et al.*, 1998; Sarid *et al.*, 1998, 1999; Sun *et al.*, 1999). Caution is certainly appropriate when applying data derived from cultured PEL B-cell lines to the *in vivo* situation of KS spindle cells. So far, *in situ* hybridization studies revealed that viral gene expression in KS essentially reflects the situation found in PEL cells (Blasig *et al.*, 1997; Davis *et al.*, 1997; Dittmer *et al.*, 1998; Staskus *et al.*, 1997; Stürzl *et al.*, 1997; Sun *et al.*, 1999). K1 transcription was detected only after phorbolester (TPA) stimulation of BCBL-1 cells, making it unlikely as a candidate for a latency-associated protein and for transformation of KS spindle cells (Lagunoff and Ganem, 1997), although one might argue that low-level transcription was not detectable and a stable protein could still be present in amounts of functional significance. The study of KSHV transcription regulation and the definition of latent transcription units is complicated by the spontaneous reactivation of KSHV in most PEL cell lines. Depending on the culture conditions, KSHV spontaneously enters lytic replication in 0.5-3% of BCBL-1 cells (Renne *et al.*, 1996). Spontaneous KSHV replication seems to be less common in the EBV/KSHV dually positive PEL cell line BC-1 (Miller *et al.*, 1997).

In a comprehensive transcription analysis, three kinetic classes of KSHV transcripts were defined in BC-1 cells. Class I describes truly latent viral transcripts that are detectable in the absence of TPA and not inducible by TPA. To class II, those viral RNAs were assigned which are detectable in the absence of TPA, but are induced by TPA. Class III transcripts were only detectable after TPA induction which is typical for gene expression during lytic replication (Sarid *et al.*, 1998). These data were essentially confirmed recently in a study using BC-3 cells and DNA arrays (Jenner *et al.*, 2001). The viral IL-8R homolog encoded by *orf74* was clearly identified as class III, which excludes it from the list of candidate latency genes. Transcripts for K1, vIRF, and K12 were assigned to class II. In contrast to BCBL-1 cells, late gene expression was not seen in BC-1 cells in the absence of chemical induction (Miller *et al.*, 1997). This is often interpreted as a sign for tight latency control in BC-1 cells. If so, this would imply that genes of class II (Sarid *et al.*, 1998) are expressed in latency. However, close inspection of class II genes identified in BC-1 cells shows that this is most likely not the case. The pattern of genes that are detectable without and inducible by TPA (class II) goes far beyond the usual latency program. It includes several conserved herpesvirus genes. Most notably, all genes with homology to identified immediate-early genes of other herpesviruses are expressed (*orf50, K8, orf57, K3, K5*), in addition several conserved members of the early class (major DNA binding protein, helicase, uracil DNA glucosidase, *orf59*, but not DNA polymerase). This pattern is more compatible with an abortive lytic infection blocked in the early phase, than with true latency. This view is supported by the inability to detect several class II transcripts in KS biopsy specimens by *in situ* hybridization (Sun *et al.*, 1999). Thus, neither K1, nor vIRF, nor vIL-8R are likely to be expressed in latently infected KS spindle cells. But which virus genes are then really latently expressed? None of the typical latent genes of EBV or HVS is conserved in KSHV. A KSHV transcript of 0.7kb (T0.7) had originally been described as being abundantly expressed in non-stimulated PEL cells. T0.7 mRNA, which contains the putative open reading frame K12, is also detectable in the majority of KS spindle cells. However, the oncogenic potential of K12 in rodent fibroblast transformation assays was very weak (Muralidhar et al., 1998) and remains controversial. It is also not clear which protein is translated in vivo from the transcripts originating from the K12 area. It is therefore not yet possible to assess the oncogenic potential of K12 or T0.7. Only one specific area of the KSHV genome was consistently found to be transcriptionally active in latency. As described above, it is located close to the right end of the genome (Sarid et al., 1998; Sun et al., 1999) and comprises orf71/vFLIP, orf72/k-cyclin, and orf73/LANA-1. LANA-1 (LNA-1) is important for maintenance of the episomal KSHV genome in latently infected cells (Ballestas et al., 1999; Ballestas et al., 2001). Moreover LANA-1 has been described to inhibit p53 and to provide protection from cell death (Friborg et al., 1999; Radkov et al., 2001). The viral FLIP (FLICE/caspase-8 inhibiting protein) which is homologous to a cellular apoptosis inhibitor, appears to be expressed in latency and may play an important role to block antiviral cytotoxic T cell-induced apoptosis (Thome et al., 1997). A similar transcription pattern was observed in KS biopsies by in situ hybridization: only transcripts for k-cyclin and/or T0.7 were detectable in the vast majority of KS spindle cells (Blasig et al., 1997; Davis et al., 1997; Dittmer et al., 1998; Staskus et al., 1997; Stürzl et al., 1997; Sun et al., 1999). In contrast to the latently infected cells, the full transcription program was active in a few interspersed cells representing only up to 3% of the total cell population in KS (Blasig et al., 1997; Davis et al., 1997; Sun et al., 1999).

2.3.4   Multifactorial tumorigenesis of Kaposi's sarcoma

Several lines of evidence strongly suggest that KSHV plays an essential role in KS pathogenesis: the virus is invariably present in KS, it is present in the tumor cells themselves, and it is rather infrequently found outside the population at risk of KS. Moreover, KSHV belongs to a group of herpesviruses known to be oncogenic. It is thus intriguing to assume that a mechanism similar to those identified in EBV and HVS might be relevant for the pathogenesis of KS by KSHV. Starting from the analogy to EBV and exploiting the close relationship to HVS, several putative oncogenes were identified. However, the probable lack of expression of directly transforming genes in KS spindle cells indicates that the mechanisms of lymphocyte transformation by other γ-herpesviruses may not apply to KS pathogenesis.

The oncogenic potential of the membrane protein K15 is not yet known. Transcripts mapping to the K15 area were not detectable in BC-1 cells without TPA induction (Sarid *et al.*, 1998). The peculiar pathology of the neoplasia KS is another hint at a more complex, indirect mechanism of pathogenesis. Pre-KSHV models of KS pathogenesis have emphasized the role of inflammatory cytokines. Interestingly, KSHV encodes or induces several cytokines with intriguing similarity to the cellular factors shown to be required for *in vitro* models of KS. An example is VEGF which is secreted by cells expressing the constitutively active vIL-8R. This leads to an alternate model of KS pathogenesis, where increased secretion of both viral and cellular cytokines, the latter in part induced by KSHV, promote inflammatory infiltrates (vMIP, vIL-6), angiogenesis (vMIP, vIL-8R), and enhance spindle cell proliferation (vIL-6, vIL-8R via VEGF). The reliable presence of KSHV in the spindle cells points to additional, more direct viral factors beyond those induced through the 1-3% productively infected cells. The viral k-cyclin is possibly one of them. Deregulation of the G1/S checkpoint is a very common feature of malignancy and of human tumor viruses. k-cyclin has been shown to overcome arrest in G1 and to force progression of the cell cycle. This property is most likely required for the virus to overcome cell cycle arrest. The latter is one of the means by which cells fight viral infection, as replication of the viral genome requires proteins present in dividing, but not in quiescent cells. One physiological function of the restriction point in G1/S is to block cell division in the case of DNA damage. This is usually achieved through the activation of p53 which in turn results in an increased synthesis of cdk inhibitors such as p21. However, the k-cyclin/cdk6 complex is resistant to inhibition by several cdk inhibitors (Swanton *et al.*, 1997). Thus, k-cyclin may not only be responsible for increased proliferation of KSHV-infected endothelial and spindle cells, but sustained expression of k-cyclin may also result in the unrestricted acquisition of somatic mutations, eventually resulting in the malignant phenotype and monoclonality associated with advanced stage disease. Thus, KS may result from a complex interplay of both viral and cellular cytokines and angiogenic factors, induced by a sustained inflammatory reaction initiated by up to 3% productively infected cells. The viral cyclin and perhaps other latency-associated proteins such as LANA-1 might further enhance the proliferation of KS cells and favor the development of truly malignant cells by indirect means, *e.g.*, the reduced control of accidental DNA damage. As KS is an unusual malignancy, resembling hyperplastic, angioproliferative inflammatory changes rather than true sarcoma, such a multifactorial model might be more compatible with KS pathogenesis than the classical transformation model by viral oncogenes, as described for EBV, HVS, and possibly even for KSHV-associated B-cell malignancies (PEL).

2.4   Related Old-World primate rhadinoviruses

The discovery of KSHV as the first human rhadinovirus in 1994 stimulated the search for rhadinoviruses in other Old World primates. Serological studies using KSHV-derived antigens indicated that a herpesvirus closely related to KSHV may exist in rhesus monkeys. Co-cultivation of lymphocytes from seropositive rhesus monkeys with fibroblasts resulted in the isolation of a novel rhesus monkey rhadinovirus (RRV) (Desrosiers et al., 1997). The complete genomic sequences from two independent RRV isolates are already available (Alexander et al., 2000; Searles et al., 1999). This revealed that RRV is indeed more closely related to KSHV than the 'prototypic' rhadinovirus herpesvirus saimiri (HVS). The RRV genome organisation is essentially colinear with KSHV. It encodes (at least) 79 genes, 67 of which are homologous to genes found in both KSHV and HVS. Of the remaining 12 genes, 8 are similar to genes identified in KSHV. This includes the viral IL-6 and several genes with homology to the family of IFN regulatory factors. In contrast to KSHV, RRV can be readily propagated in cell culture, and at least amongst monkeys kept in captivity, RRV seems to be very widespread. It is thus not surprising that no clear disease association has been identified for RRV in otherwise healthy macaques as yet. However, when rhesus macaques were immunosuppressed due to an infection with the simian immunodeficiency virus, inoculation with RRV resulted in the development of a multifocal lymphoproliferative disease resembling multicentric Castelman's disease observed in KSHV infected patients (Wong et al., 1999).

In a second approach targeted at the identification of Old World primate rhadinoviruses, T. Rose and coworkers studied tissue specimens from simian retroperitoneal fibromatosis (RF). RF has been identified as an infrequent disease syndrome occurring in various immunosuppressed macaques (Giddens et al., 1985). RF lesions consist of an aggressively proliferating fibrous tissue with a high degree of vascularisation and thus remotely resemble KS. Transmission studies indicated that an infectious agent may be involved in RF pathogenesis (Giddens et al., 1985). By using a degenerated PCR technique, fragments of a herpesvirus DNA polymerase gene were identified in RF tissues from two macaque species, Macaca nemestrina and Macaca mulatta (Rose et al., 1997). Sequence comparisons indicated, that at least the DNA polymerase genes of these two novel rhadinoviruses, tentatively termed RFHVMm and RFHVMn, are more closely related to KSHV than the DNA polymerase genes of RRV. However, attempts to isolate these viruses on cultured cells or to obtain additional sequence information have not been successful so far.

By using the degenerate PCR technique, rhadinoviral DNA polymerase fragments could be identified in various Old World primates. This includes African green monkeys (Greensill *et al.*, 2000), chimpanzees (Lacoste *et al.*, 2000a, 2001; Greensill *et al.*, 2000), gorillas (Lacoste *et al.*, 2000a), and mandrills (Lacoste *et al.*, 2000b). The interesting point here is: phylogenetic analysis of the DNA polymerase fragments revealed that Old World primate rhadinoviruses may be segregated into two groups. One group of sequences is more closely related to KSHV, the other group is more closely related to RRV (Schultz *et al.*, 2000; Strand *et al.*, 2000). The finding that two different rhadinoviruses appear to co-exist in almost any Old World primate examined so far may hint at an as yet unidentified second, more RRV-like, human rhadinovirus.

3. HERPESVIRUS SAIMIRI AND ATELES

3.1 Natural occurrence and pathogenicity

3.1.1 Apathogenic persistence of or leukemogenesis by herpesvirus saimiri

The rhadinovirus herpesvirus saimiri (HVS; saimiriine herpesvirus type 2) is regularly found in squirrel monkeys (*Saimiri sciureus*) which naturally occur in South American rain forests. As natural hosts of HVS, squirrel monkeys are infected via saliva usually within the first two years of life. HVS does not cause disease or tumors and establishes lifelong persistence in this species (Melendez *et al.*, 1968).

The infection with HVS can lead to distinct results in different monkey species (Fleckenstein and Desrosiers, 1982; Fickenscher and Fleckenstein, 2001). Whereas this virus is apathogenic in its natural host, it reproducibly induces lymphoma and leukemia in different simian species. Within less than two months after experimental infection, HVS causes acute peripheral T-cell lymphoma in other New World primate species such as tamarins (*Saguinus* species), common marmosets (*Callithrix jacchus*), or owl monkeys (*Aotus trivirgatus*) (Melendez *et al.*, 1969; Wright *et al.*, 1976; reviewed in Fleckenstein and Desrosiers, 1982). These monkey species are presumably not infected by HVS under wildlife conditions. The experimental infection is usually performed intramuscularly or intravenously with a typical dose of 10^6 tissue culture-infectious doses per ml. Purified virion DNA was also infectious and oncogenic after intramuscular injection into susceptible primates (Fleckenstein et al., 1978b). Depending on the pathogenic properties and on the sequence divergence in the terminal non-repetitive genomic region (see below), HVS strains were classified into the three subgroups A, B, and C (Desrosiers and Falk 1982; Medveczky et al., 1984, 1989b). The major representative strains are A11 (Falk et al., 1972a) for subgroup A, B-S295C (Melendez et al., 1968) and B-SMHI (Daniel et al., 1976a) for subgroup B, and C488 (Biesinger et al., 1990) and C484 (Medveczky et al., 1984) for subgroup C. Generally, subgroup B viruses have the weakest and subgroup C strains the strongest oncogenic properties. Tamarins (Saguinus spec.) are most susceptible to all subgroups, whereas subgroup B viruses did not cause disease in adult common marmosets (Callithrix jacchus) (reviewed in detail in Fleckenstein and Desrosiers, 1982; Bröker and Fickenscher, 1999; Fickenscher and Fleckenstein, 1998).

The pathogenicity of viral deletion mutants based on HVS C488 has been investigated in common marmosets (Callithrix jacchus) and in cottontop tamarins (Saguinus oedipus) in which the acute T-cell lymphoma developed within a few weeks only (Duboise et al., 1998a, b; Knappe et al., 1998a, b). Virus strain C488 induced a similar fulminant disease even in Old World monkeys (Macaca mulatta, rhesus monkeys; Macaca fascicularis, cynomolgus) within a few weeks after experimental infection with a high virus dose (Alexander et al., 1997; Knappe et al., 2000a). The pathological findings in these macaques were similar to those in New World primates (Hunt et al., 1972; Knappe et al., 1998b, 2000a). The wide-spread CD3+ infiltrates could also be designated as pleomorphic peripheral T-cell lymphoma or pleomorphic T-lymphoproliferative disorder showing similarities to human EBV-induced post-transplantation B-lympho-proliferative disorders (Knappe et al., 2000a). Since the discovery of the oncogenic properties of HVS thirty years ago, research has concentrated on the acute lymphoma induced after experimental infection in various monkey species other than squirrel monkeys. Although it was initially expected that this disease would serve as a tumor model for human EBV disease, it soon became obvious that T cells are the major cell type involved, in contrast to B cells in EBV pathogenesis (Wright et al., 1976). Thus, the HVS-induced disease remained rather a general model for tumor development than a specific model for a human disease, since a comparable form of acute peripheral pleomorphic T-cell lymphomas is not known in humans.

Similarly to the situation in persistently infected squirrel monkeys, HVS can be isolated from the peripheral blood of leukemic animals, presumably from infected T cells, by cocultivation with permissive epithelial owl monkey kidney (OMK) cells (Daniel et al., 1976b; Falk et al., 1972a; Wright et al., 1976). HVS replicates productively in OMK cells and induces lysis after three to 20 days. It reaches titers of 10^9 particles and of 10^6 to 10^7 plaque forming units per ml and is relatively stable at +4°C (Fickenscher and Fleckenstein, 1998).

3.1.2 Semi-permissive persistence in transformed monkey T-cell lines

A limited series of transformed T-cell lines was derived from leukemias or tumors of subgroup A or B infected tamarins during the early phase of HVS research (reviewed in Fleckenstein and Desrosiers, 1982). Such cell lines have been cultivated continuously for up to many years. While most of these cultures initially produced virus particles, the virion production was lost after prolonged culture time. The cell lines 1670 (Marczynska *et al.*, 1973; Fleckenstein and Desrosiers 1982) and 70N2 (Falk *et al.*, 1972b) are derived from virus strain A11, and L77/5 from B-S295C (Fleckenstein *et al.*, 1977). Some of these cell lines carry rearrangements or large deletions in the episomal HVS genomes (Kaschka-Dierich *et al.*, 1982). Moreover, the episomal DNA is heavily methylated in such cell lines (Desrosiers *et al.*, 1979). *In vitro*, marmoset and tamarin T cells can be transformed by HVS to stable proliferation (Chou *et al.*, 1995; Desrosiers *et al.*, 1986; Kiyotaki *et al.*, 1986; Schirm *et al.*, 1984; Szomolanyi *et al.*, 1987). These transformed T-cell lines are semi-permissive: They become transformed while still releasing virus particles. Specifically HVS subgroup C strains such as C488 are able to transform also human T cells to stable proliferation *in vitro* (Biesinger *et al.*, 1992). In contrast to semi-permissive T-cell lines from New World monkeys, the C488-transformed human T cells do not produce virus particles (Biesinger *et al.*, 1992; Fickenscher *et al.*, 1996a, 1997). In macaques which are, as Old World monkeys, more closely related to humans, a reproducible weak semi-permissivity of C488-transformed T cells has been described (Alexander *et al.*, 1997; Knappe *et al.*, 2000a). Whereas HVS pathogenicity has not been reported in rodents, a non-permissive infection and tumor induction was described in New Zealand white rabbits, however with variable efficiency (Ablashi *et al.*, 1985; Medveczky *et al.*, 1989b). As described, the features of infection can vary considerably between closely related species (*e.g.*, productivity of C488-transformed T cells from macaques versus humans). Not even the macaque model seems optimal for the non-permissive persistence conditions in human T cells (Knappe *et al.*, 2000a). Thus, it is difficult to predict the behaviour of HVS in another species.

3.1.3   The closely related herpesvirus ateles

A closely related rhadinovirus, herpesvirus ateles (HVA), can be isolated at a high rate from spider monkeys (*Ateles* species) (reviewed in Fleckenstein and Desrosiers, 1982). Isolate #810 from *Ateles geoffroyii* (Melendez *et al.*, 1972) is officially classified as ateline herpesvirus type 2, whereas isolate #73 and related strains (#87, 93, 94) from *Ateles paniscus* are designated as ateline herpesvirus type 3 (Falk *et al.*, 1974). HVA replicates in OMK cells (Daniel *et al.*, 1976b), but remains strictly cell-associated with syncytia formation. As a result, supernatants of such cultures have low and unstable HVA titers. Similarly to HVS, HVA is not pathogenic in its natural host, but causes acute T-cell lymphoma in various New World primate species including cottontop tamarins (*Saguinus oedipus*) and owl monkeys (*Aotus trivirgatus*) (Hunt *et al.*, 1972). The pathological changes are similar to those observed after HVS infection. In addition, HVA transforms T cells of certain New World monkey species such as cottontop tamarins in culture yielding cytotoxic T-cell lines (Falk *et al.*, 1978; Johnson and Jondal, 1981a, b; Kiyotaki *et al.*, 1988; reviewed in Fleckenstein and Desrosiers, 1982). Human T cells have not been susceptible to transformation with various HVA strains. In comparison to HVS far less is known about HVA biology.

3.2   Genome structure, persistence, and replication

3.2.1   Genome organization

As a member of the γ-herpesvirus family, HVS is the prototype of the rhadinovirus or γ_2-herpesvirus subfamily (Albrecht *et al.*, 1992a; Roizman *et al.*, 1992). HVA was confirmed by sequencing to have a similar genome to HVS (Albrecht, 2000; Fleckenstein *et al.*, 1978a). In the case of HVS A11, the terminal repetitive H-DNA (high density, high G+C content) contains multiple tandem repeats of 1,444 bp with 70.8 % G+C and a density of 1.729 g/ml, whereas the long unique L-DNA (low density, low G+C content) has 112,930 bp with 34.5 % G+C and 1.695 g/ml (Albrecht *et al.*, 1992a; Fleckenstein and Desrosiers, 1982). The size of the total M-DNA (1.705 g/ml) genome of approximately 140,000 bp is variable due to the varying copy numbers of H-DNA segments attached to both ends of the linear virion genome.

3.2.2 Conserved and divergent genome regions in herpesvirus saimiri

The HVS L-DNA genome carries up to 77 open reading frames (Albrecht, 2000; Albrecht *et al.*, 1992a; Biesinger *et al.*, 1990; Ensser *et al.*, 1997). There are gene blocks of typical herpesvirus genes which are highly conserved also in viruses of other herpesvirus families (Albrecht and Fleckenstein, 1990; Gompels *et al.*, 1988; Nicholas *et al.*, 1992a). In addition, certain genes of HVS do usually not occur in other herpesvirus families. Among these genes are transforming oncogenes and a series of viral homologs of cellular genes described below. While most genes are well conserved between different HVS strains (Knappe *et al.*, 1997), there is extensive sequence variation at the so-called left end of the L-DNA. Together with the different pathogenicity in different hosts, this divergence formed the basis for the classification of HVS strains into the subgroups A, B, and C (Medveczky *et al.*, 1984, 1989b; Szomolanyi *et al.*, 1987). Deletion mutants of this region lose their transforming potential (Desrosiers *et al.*, 1985a, 1986; Duboise *et al.*, 1996, 1998a; Knappe *et al.*, 1997; Medveczky *et al.*, 1993). Whereas virus strains of all subgroups were capable of inducing T-cell lymphoma in susceptible non-human primates, only subgroup C strains of HVS were able to transform human T cells *in vitro* (Biesinger *et al.*, 1992). The transformation-relevant genes encoded in the respective left-terminal L-DNA region and the functions of the derived gene products are described below. In addition, the region of the R transactivator gene *orf50* occurs as two strongly divergent alleles in the strains A11 and C488 and in related virus isolates (Thurau *et al.*, 2000).

3.2.3 Viral functions for lytic replication and episomal persistence

In comparison to the herpesvirus prototype herpes simplex virus (HSV), the DNA replication mechanism is simplified because there is only a single unique genomic region and the terminal repetitions are simple tandem-repeated segments without coding capacity. Based on the hypothesis of the rolling circle mechanism for replication, virus mutants were generated by homologous recombination into the right-terminal L- to H-DNA transition (Grassmann and Fleckenstein, 1989). In this case, a single crossing over event was sufficient to integrate the recombination plasmid into the viral genome. Because H-DNA segments were not included in this plasmid, it is obvious that the terminal repeats are automatically attached to the right L-DNA terminus during rolling circle replication. The origin of lytic replication was mapped to the untranslated region upstream of the thymidylate synthase gene (Lang and Fleckenstein, 1990; Schofield, 1994).

In transformed human T cells, HVS persists as stable non-integrated episomes at high copy number (Biesinger *et al.*, 1992). Unexpectedly, an isolate termed C139 was found to persist episomally at low copy number in transformed human T cells (Fickenscher *et al.*, 1997). The genomic correlates of a plasmid-like origin of replication and of the viral factors involved are yet unknown. By analogy to observations made for KSHV, the *orf73* product was suggested to be involved because it shares homology to the latent nuclear antigen of KSHV. However, this is unlikely, because transcripts of this gene were not detectable by Northern blotting and by subtractive cDNA cloning from C488-transformed human T cells (Fickenscher *et al.*, 1996a; Knappe *et al.*, 1997). In contrast, ORF73 is expressed in the nuclei of a persistently HVS-infected human lung carcinoma cell line (Hall *et al.*, 2000a, b). A DNA fragment of strain C484 was described to mediate plasmid maintenance (Kung and Medveczky, 1996). However, this fragment maps to the variable region of the left-terminal L-DNA which had been previously shown to be dispensable for episomal persistence (Medveczky *et al.*, 1989b).

3.2.4   Viral regulatory genes

In contrast to HSV, the infection of tissue culture cells by HVS is asynchronous (Randall *et al.*, 1985). Therefore, the assignment of HVS genes to the immediate-early (IE) phase of infection was contradictory, mostly based on using cycloheximide to inhibit protein synthesis. The IE gene *ie57* codes for a nuclear phosphoprotein of 52 kDa (Hoyle *et al.*, 1990; Nicholas *et al.*, 1988, 1990; Randall *et al.*, 1984). IE57 shares structural and functional homology with ICP27/IE63 of HSV. Correspondingly, IE57 acts as a post-transcriptional regulator stimulating the expression of unspliced and repressing the expression of spliced transcripts (Goodwin *et al.*, 2000; Whitehouse *et al.*, 1998b). IE57 redistributes nuclear splicing factors and is involved in nuclear RNA export, binding to importin α1 and α5 (Cooper *et al.*, 1999; Goodwin and Whitehouse, 2001; Goodwin *et al.*, 1999).

Besides *ie57* only two other genes were assigned to the IE phase. Contradictory results were published about transcripts for the thymidylate synthase gene *orf70* which were detectable or not in presence of cycloheximide (Bodemer *et al.*, 1984, 1986; Nicholas *et al.*, 1990). Presumably, the unusual transcription behavior resulted from the fact that the lytic replication origin is localized in the promoter region of this gene (Lang and Fleckenstein, 1990; Schofield, 1994). Another IE-transcript was mapped to the gene *ie14* (Hoyle *et al.*, 1990; Nicholas *et al.*, 1990). The derived gene product has sequence homology to murine superantigens (Sag) (Nicholas *et al.*, 1990; Thomson and Nicholas, 1991). However, *ie14/sag* was shown to be dispensable for virus replication (Duboise *et al.*, 1998b; Knappe *et al.*, 1997, 1998a, b) which largely excludes a contribution of this gene to the regulation of virus replication. Thus, *ie57* appears to be the sole regulatory viral IE-gene of HVS.

A strong viral transactivator was mapped to the delayed-early *orf50* gene showing a limited sequence homology to the R transactivator of EBV (Nicholas *et al.*, 1991). Due to differential splicing and promoter usage, the gene from HVS A11 codes for a larger protein ORF50A and for a smaller C-terminal variant ORF50B (Whitehouse *et al.*, 1997a). Only ORF50A transactivated reporters carrying the promoters of responsive virus genes, such as *orf6* and *orf57* (Whitehouse *et al.*, 1997b, 1998a). The C-terminal amino acids of the ORF50 proteins represent the transactivation domain and were shown to bind to the TATA binding protein in the basal transcription complex (Hall *et al.*, 1999). Surprisingly, the genomic *orf50* region was found to be strongly divergent between the virus strains A11 from subgroup A and C488 from subgroup C (Thurau *et al.*, 2000). Moreover, a strong antisense transcription was detected for C488 *orf50* during later phases of lytic replication. In contrast to strain A11, the ORF50B protein of C488 had full transactivation properties (Thurau *et al.*, 2000). Because the IE57-mediated post-transcriptional inhibition is obviously not relevant in this context, this suggests that *orf50* exerts the dominant function for replication regulation, at least in HVS strain C488.

3.3 Virus genes with homology to cellular counterparts

3.3.1 Molecular piracy of cellular genes

Rhadinoviruses carry sets of genes which seem to be pirated from cellular DNA. Remarkably, the viral gene copies are typically not disrupted by introns which would suggest a role for reverse transcription, possibly involving endogenous retroviruses. This exciting hypothesis was discussed extensively, however, experimental results have not yet helped to understand this. The fact that several cellular gene copies are common to different rhadinoviruses or even other γ-herpesviruses including EBV, suggests that the uptake of cellular genes is a rare event during evolution. Similarly to KSHV and other rhadinoviruses, HVS has acquired a series of genes from its host cells. The transformation-associated gene products StpC and Tip share limited homology to cellular proteins. Because these genes are required for T-cell transformation and pathogenicity, they are described in a separate section below. Whereas StpC and Tip are constitutively expressed, most of the cell-homologous genes are not transcribed in C488-transformed human T cells which are non-permissive for virus replication (Fickenscher *et al.*, 1996a; Knappe *et al.*, 1997). Most of these cell-homologous virus genes have been found expressed during lytic virus replication. Thus, they could play a role in the natural host, the squirrel monkey (*Saimiri sciureus*), either to enhance lytic replication and spreading, or to ensure the apathogenic persistence of HVS while evading the host's immune defense mechanisms. Some of the cell-homologous gene products exert stimulatory effects on T cells. However, T-cell transformation or HVS-induced leukemia have not been reported for the natural host.

3.3.2   Cell-homologous genes of the nucleotide metabolism and cell-cycle control

A first group of viral proteins are homologous to enzymes of the nucleotide metabolism or to cell cycle-promoting factors. *orf2* codes for a dihydrofolate reductase (Ensser *et al.*, 1999; Trimble *et al.*, 1988) and *orf70* for a functional thymidylate synthase (Bodemer *et al.*, 1984, 1986; Honess *et al.*, 1986). *orf2* and the viral U-RNA genes are dispensable for virus replication and T-cell transformation (Ensser *et al.*, 1999). Both *orf3* and *orf75* encode large tegument proteins which share local homology to formylglycineamide ribotide amido transferase (FGARAT) (Ensser *et al.*, 1997). These enzymatic functions may play a role in augmenting the free nucleotide pool and thus facilitate DNA synthesis and virus replication. *orf72* codes for a functional viral cyclin D version (Jung *et al.*, 1994; Nicholas *et al.*, 1992a, b). In contrast to cellular cyclin D, the viral cyclins of HVS and KSHV are not inactivated by the cyclin-dependent kinase inhibitors p16, p21, and p27 (Swanton *et al.*, 1997). This deregulation pushes the cell cycle towards the S phase and thereby may support virus replication in permissive cells (Swanton *et al.*, 1997). Moreover, enhanced proliferation could secondarily promote transformation or tumor induction. The expression of the viral cyclin in semi-permissive HVS-transformed marmoset T cells might contribute to their activated and transformed phenotype (Jung *et al.*, 1994). However, these functions would be rather auxiliary, since the viral cyclin has been shown not to be required for replication, T-cell transformation, and lymphoma induction (Ensser *et al.*, 2001).

Table 2. Non-conserved and cell-homologous genes of HVS

ORF	Product	Hypothetical mechanism	Role for the growth transformation of human T cells
01b	StpC	Transformation	essential
01a	Tip	Transformation, Lck interaction	essential
	5 URNAs	Splice regulation	dispensable
04	CCPH	Complement inhibition, immune evasion	not tested
13	IL-17	Cytokine	dispensable
14	IE14/vSag	Mitogen, proliferation	dispensable
15	CD59	Complement inhibition, immune evasion	not tested
16	Bcl-2	Apoptosis block, immune evasion	not tested
71	FLIP	Apoptosis block, immune evasion	dispensable
72	Cyclin	Proliferation, cell-cycle progression	dispensable
74	IL-8R/GPCR	Cytokine receptor	dispensable

3.3.3   Viral homologs to complement-regulatory proteins

Two other cell-homologous HVS genes target the complement system which is relevant for pathogen elimination. *orf4* codes for a complement control protein homolog (CCPH) which inhibits C3 convertase in the initiation of complement activation (Albrecht and Fleckenstein, 1992; Fodor *et al.*, 1995). *orf15* is a viral variant of CD59 and blocks the terminal complement cascade (Albrecht *et al.*, 1992b; Bramley *et al.*, 1997; Rother *et al.*, 1994). Both proteins may be utilized for the escape from immune surveillance *in vivo*.

3.3.4   Viral inhibitors of apoptosis

A third class of lytically expressed viral genes are inhibitors of induced cell death or apoptosis. The *orf71* product is a viral FLICE-inhibitory protein (FLIP) sharing the death-effector domains with FLICE (caspase-8) and cellular FLIP (Thome *et al.*, 1997). The rhadinoviral FLIP inhibited death receptor-dependent apoptosis mediated by Fas ligand, TNF-α, TRAIL, or TRAMP. The viral FLIP was shown to interact with FADD and FLICE via their death-effector domains. Thus, the formation of the death signal-induced signal complex was blocked, caspase-8 (FLICE) was not activated, and cell death did not occur. Moreover, HVS infection partially protected permissive OMK cells from Fas-dependent apoptosis at a late stage of infection before the cells are lysed (Thome *et al.*, 1997). However, the anti-apoptosis activity of FLIP is dispensable for virus replication, T-cell transformation, and lymphoma induction (Glykofrydes *et al.*, 2000). The product of *orf16* contains the Bcl-2 homology domains BH1 and BH2 and shows an anti-apoptotic activity similar to cellular Bcl-XL. Both Bcl-XL and viral Bcl-2 inhibit cell death induced by either cell-autonomous (independent of death receptors) or death receptor-mediated mechanisms, depending on the cell type studied. The viral Bcl-2 homolog was shown to act at the level of mitochondrial stabilization and to inhibit apoptosis induced by Sendai virus or by treatment with Fas ligand, dexamethasone, menadione, or irradiation (Bellows *et al.*, 2000; Derfuss *et al.*, 1998; Nava *et al.*, 1997). Thus, two HVS-encoded proteins appear to render infected cells resistant both to death-receptor mediated and cell-autonomous apoptosis resulting in enhanced virion production from permissive cells (Meinl *et al.*, 1998).

3.3.5   Viral cytokine genes

A fourth group of cell-homologous viral genes is related to the cytokine network. This includes viral homologs of the cytokine IL-17 and of the IL-8 chemokine receptor. The HVS gene *orf13* led to the discovery of its cellular homolog which was initially termed *ctla8* (Rouvier *et al.*, 1993). This gene codes for IL-17, a cytokine produced specifically by CD4+ T cells. IL-17 induces the secretion of IL-6, IL-8, G-CSF, and PG-E2 by stroma cells such as fibroblasts, endothelial, or epithelial cells, and promotes the proliferation and maturation of CD34+ hematopoetic progenitor cells into neutrophils. Among various other functions, IL-17 was shown to support T-cell proliferation (Fossiez *et al.*, 1996, 1998; Kennedy *et al.*, 1996; Spriggs, 1997; Yao *et al.*, 1995a, b, 1996b). The viral IL-17 is functionally not distinguishable from its cellular counterpart. Viral null mutants for *orf13*/IL-17 had full replicative, transforming, and pathogenetic capabilities (Knappe *et al.*, 1998a). Similarly to KSHV, the product of *orf74* is a viral IL-8 receptor (IL-8R), a seven transmembrane G protein-coupled receptor of the low-affinity B type of IL-8R (Ahuja and Murphy, 1993; Murphy, 1994; Nicholas *et al.*, 1992b). A paracrine stimulation model was postulated for *in vivo* virus replication, whereby viral IL-17, produced by virus-infected cells, would induce IL-8 on neighboring stroma cells. The induced IL-8 might then bind to the viral IL-8R on the virus-infected cell and lead to further activation. However, the existence of such a paracrine stimulation mechanism has not yet been substantiated by experimental data.

3.3.6   The viral homolog to murine superantigens

Finally, *orf14* was one of the few viral IE genes assumed to play a role in the regulation of virus replication (Nicholas *et al.*, 1990). A protein homology to the superantigen (Sag) of mouse mammary tumor virus (MMTV) and to murine *mls* superantigens led to the designation IE14/vSag for the *orf14* product. Superantigens are characterized by their efficient antigen-independent stimulation of T cells if they express the superantigen-specific Vβ families of T-cell receptor molecules. Superantigens crosslink the Vβ chains of the T-cell receptor with major histocompatibility complex (MHC) class II molecules on accessory cells leading to T-cell stimulation. Recombinant viral IE14/vSag bound to MHC-II molecules and stimulated T-cell proliferation. However, as yet there is no evidence for a selective advantage for certain V*b*-families, neither after stimulation of human T cells with IE14/vSag *in vitro*, nor after infection and transformation with HVS (Duboise *et al.*, 1998b; Knappe *et al.*, 1997; Yao *et al.*, 1996a). Therefore, the IE14 should not be termed a superantigen, but rather a superantigen homolog or simply a mitogen. The deletion of *ie14/vsag* from HVS C488 did neither impair its ability to induce tumors in cottontop tamarins (*Saguinus oedipus*), nor to transform human and simian T cells *in vitro* (Knappe *et al.*, 1997, 1998b). In another study, however, similar deletion mutants did not induce T-cell lymphoma in common marmosets (*Callithrix jacchus*) and were unable to transform their T cells *in vitro*. Surprisingly, this virus did not even persist in infected animals (Duboise *et al.*, 1998b). This contradiction could either be explained by species-specific differences between New World primate species, or by comparably lower virus titers used for the experimental infection in the latter study.

Possibly *ie14/vsag* plays a role in the apathogenic persistence of HVS in the squirrel monkey, the natural host of the virus (Knappe *et al.*, 1997, 1998a, b). A similar situation as known for the MMTV superantigen in mice is conceivable. In the latter case, the superantigen facilitates infection and spread of the virus in B cells through attracting and stimulating T helper cells, followed by apoptosis and specific deletion of the superantigen-stimulated lymphocytes from the T-cell repertoire (Held *et al.*, 1993, 1994).

3.3.7 General role for the cell-homologous genes of herpesvirus saimiri

In summary, besides the essential genes *stpC* and *tip*, there is little evidence for a functional role of the cell-homologous genes of HVS in the transformation of human T cells. Growing knowledge from deletion mutagenesis demonstrates that at least most of these genes are dispensable for virus replication, T-cell transformation in culture, and pathogenesis. In susceptible New World monkeys, where the tumorigenic HVS establishes a (semi-)permissive infection, most cell-homologous genes are transcribed

(Fickenscher *et al.*, 1996a), and proteins such as the viral cyclin could possibly contribute to the transforming and pathogenic phenotype. However, in the natural host, the squirrel monkey, the cell-homologous genes are most likely involved in the efficient infection, replication, spreading, and in the establishment of apathogenic persistence.

3.4   Transforming functions

3.4.1   The transformation-associated genomic region of herpesvirus saimiri

The viral functions responsible for the induction of T-cell leukemia *in vivo* and the T-cell transformation *in vitro* were mapped to the variable region at the left end of the HVS L-DNA (Desrosiers *et al.*, 1985a, 1986; Koomey *et al.*, 1984; Murthy *et al.*, 1989). In subgroup A strains, only one gene is found at this position, termed *stpA* (saimiri transformation-associated protein of subgroup A strains) (Murthy *et al.*, 1989). The position-homologous *stpB* of subgroup B strains seems to have a low transforming activity (Choi *et al.*, 2000; Hör *et al.*, 2001). At the homologous location, the virus strains of subgroup C carry two open reading frames (Biesinger *et al.*, 1990) which were termed *stpC* (saimiri transformation-associated protein of subgroup C strains) and *tip* (tyrosine kinase interacting protein) (Biesinger *et al.*, 1990; Biesinger *et al.*, 1995; Jung and Desrosiers, 1991). StpA shares limited sequence homology with StpC, but is structurally unrelated to Tip. The deletion of any of these genes abolishes the transformation capacity of HVS in vivo and *in vitro*, while none of them is required for viral replication (Duboise *et al.*, 1996, 1998a; Knappe *et al.*, 1997; Medveczky *et al.*, 1993). The closely related HVA encodes the protein Tio at the homologous genomic position (Albrecht *et al.*, 1999). All these transformation-related proteins of HVS and HVA carry a hydrophobic carboxy-terminal membrane-anchor sequence.

3.4.2   The saimiri transformation-associated protein Stp

The *stpC* gene product is a perinuclear membrane-associated phosphoprotein with a predicted molecular mass of 10 kDa and an apparent molecular weight of 20 kDa. The specific N-terminal domain of StpC consists of 17 mostly charged aminoacids. The C-terminal hydrophobic region probably serves as an anchor to perinuclear membranes. In between, there are approximately 17 collagen tripeptide repeats, (GPX)n, which may mediate trimerization of the protein (Biesinger *et al.*, 1990; Fickenscher *et al.*, 1997; Jung and Desrosiers, 1991, 1992, 1994). StpA of strain A11 and StpC of strain C488 are oncoproteins: Transfected rodent fibroblasts formed foci *in vitro* and induced tumors in nude mice (Jung *et al.*, 1991). *stpA*-transgenic mice developed polyclonal peripheral T-cell lymphoma, while an *stpC* transgene induced various epithelial tumors (Kretschmer *et al.*, 1996; Murphy *et al.*, 1994). StpA was reported to bind to and to be phosphorylated by the non-receptor tyrosine kinase Src (Lee *et al.*, 1997). The less transforming StpB also associates with Src (Choi *et al.*, 2000; Hör *et al.*, 2001). StpC was shown to interact with Ras favoring its active GTP-bound state and stimulating MAP kinase activity (Jung and Desrosiers, 1995). StpA and StpC interact with TNF-α receptor associated factors (TRAF) leading to NF-κB activation (Lee *et al.*, 1999). Although the precise biochemical mechanisms still have to be resolved, the transforming potential of StpA and StpC is well established.

3.4.3 The tyrosine-kinase interacting protein Tip

The reading frame *tip*, which only exists in subgroup C strains, is localized downstream of *stpC* at the left genomic end of the L-DNA. Both *stpC* and *tip* are transcribed as a single bicistronic mRNA from a common activation-dependent promoter complex. In C488-transformed human T cells, StpC and Tip are the only viral proteins which have been shown to be constitutively expressed (Biesinger *et al.*, 1995; Fickenscher *et al.*, 1996a, 1997; Knappe *et al.*, 1997). The phosphoprotein Tip associates with the T-cell specific non-receptor tyrosine kinase p56^{lck} in C488-transformed T cells as demonstrated in kinase assays after immunoprecipitation with Lck antibodies (Biesinger *et al.*, 1995). Tip of HVS strain C488 has a predicted molecular mass of 29 kDa with an apparent size of 40kDa. The Tip proteins of different virus strains carry one or two N-terminal glutamate-rich regions, one or two serine-rich regions, and a C-terminal hydrophobic domain which anchors the molecule at the inside of the plasma membrane (Biesinger *et al.*, 1990, 1995; Fickenscher *et al.*, 1997; Lund *et al.*, 1995, 1996). Only five aminoacids are assumed to project into the extracellular space (Lund *et al.*, 1996). Three tyrosine residues which are conserved in all strains serve as substrates for the kinase Lck. The association between Tip and Lck is mediated by two sequence motifs in the C-terminal third of Tip: The C-terminal Src kinase homology domain (CSKH) of nine amino acids which is similar to the regulatory regions of various Src kinases, and a proline-rich Src homology 3 (SH3) domain binding (SH3B) sequence. Both motifs are required for the interaction with the kinase (Biesinger *et al.*, 1995; Hartley *et al.*, 2000; Jung *et al.*, 1995a).

The binding of herpesviral Tip to the tyrosine kinase Lck resulted in a strong activation of the enzyme in various approaches (Fickenscher *et al.*, 1997; Hartley *et al.*, 1999; Lund *et al.*, 1997b; Wiese *et al.*, 1996). Recombinant Lck was activated by recombinant Tip also *in vitro*, showing that their direct interaction does not depend on other molecules (Wiese *et al.*, 1996). The Lck-Tip interaction is highly specific: The other Src-family members which are active in transformed T cells, Fyn and Lyn, did neither phosphorylate Tip, nor were activated by Tip (Fickenscher *et al.*, 1997; Wiese *et al.*, 1996). Lck was activated by Tip even if the regulatory tyrosines of Lck had been mutated, suggesting a novel mechanism of Lck activation (Hartley *et al.*, 1999). In addition, both the SH3B and the CSKH motif were shown to be required for binding and activating Lck (Hartley *et al.*, 2000). When compared with their non-infected parental clones, HVS C488-transformed human T-cell clones had increased basal levels of tyrosine phosphorylation (Wiese *et al.*, 1996). Lck has a central role in T-cell signalling, and constitutively active Lck mutants have a transforming potential. Conditional *tip*-transgenic mice even developed T-cell lymphoma (Wehner *et al.*, 2001). Thus, the Tip-induced activation of Lck-signalling would contribute to the activated and transformed phenotype of HVS-transformed T cells. Accordingly, Tip and StpC would act in synergy, which was further confirmed by the induction of IL-2 production in *stpC*- and *tip*-cotransfected Molt-4 leukemia cells (Merlo and Tsygankov, 2001). This would allow a simple model for the HVS-transformation of human T cells, supported by the observation that either *stpC* or *tip* or both together are essential for T-cell transformation and pathogenicity (Duboise *et al.*, 1996, 1998a; Knappe *et al.*, 1997; Medveczky *et al.*, 1993).

Virtually contradictory results were obtained by stable transfection of *tip* and drug selection. Jurkat T cells with high levels of Tip had low basal levels of tyrosine phosphorylation, and their response to T-cell receptor activation was impaired (Jung *et al.*, 1995b). This effect was even more pronounced when Tip was mutated to enhance its binding to Lck (Guo *et al.*, 1997). In addition, Tip partially reversed the transformed phenotype of fibroblasts which had lost contact inhibition after transfection with a constitutively active Lck mutant (Guo *et al.*, 1997; Jung *et al.*, 1995b). Consequently, an alternative model assigned to Tip an inhibitory role: the function of Tip in T-cell transformation by HVS seemed similar to that of LMP-2A in EBV infection. LMP-2A inhibits B-cell receptor-mediated signalling in EBV-transformed B cells and favors the latent infection by blocking EBV replication (Brielmeier *et al.*, 1996; Fruehling and Longnecker, 1997; Miller *et al.*, 1994). This hypothesis suggested Tip as an inhibitory antagonist of StpC. The contradiction to the cooperative hypothesis could possibly be explained by negative feed back regulation after long-lasting Lck-activation by Tip in stably transfected cells.

Besides its interaction with Lck, Tip has been shown to associate with STAT factors (signal transducer and activator of transcription). An activation of STAT1 and STAT3 was initially observed after transfection of Tip-C484 into Jurkat T cells. The phosphorylated STAT factors were found in complex with Lck and Tip (Lund *et al.*, 1997a). The 48 aa Lck-binding domain of Tip, including the CSKH and SH3B domains, was sufficient for activation of Lck and STAT3 in Jurkat cells (Lund *et al.*, 1999). In addition, tyrosine phosphorylation in the N-terminal portion of Tip-C484 by Lck was required for STAT factor binding and activation of STAT factor-dependent transcription (Hartley and Cooper, 2000). The respective motif YXPQ of Tip is a consensus binding site for STAT factors. These observations further support the cooperative model for Tip function which is based on Lck activation by Tip.

A third cellular Tip-interacting factor, termed Tap (Tip associated protein), was identified by a yeast two hybrid screen using a C-terminal deletion mutant of Tip (Yoon *et al.*, 1997). Tap, a protein of 65kDa, is an RNA export factor (Grüter *et al.*, 1998), of which no T-cell specific function is known as yet. Coexpression of Tip and Tap resulted in strong aggregation of Jurkat T-leukemia cells in parallel with overexpression of adhesion molecules and NF-κB activation. Transfection of *tap* into Tip-expressing Jurkat cells reversed the signalling inhibition caused by *tip* alone (Yoon *et al.*, 1997). The role of Tap in T-cell transformation has not yet been investigated.

3.4.4 The herpesvirus ateles gene Tio from the transformation-associated genome region

In HVA strain 73, a split gene was detected in the left-terminal L- to H-DNA transition region. The derived viral protein, termed Tio ("two in one"), shares local sequence homology with StpC and Tip of HVS. Tio is expressed in HVA-transformed simian T cells. After cotransfection, Tio bound to and was phosphorylated by the Src kinases Lck or Src (Albrecht *et al.*, 1999).

3.5 Growth-transformation of human T cells

3.5.1 Scientific background of T-cell transformation

Rapidly proliferating T-lymphoblastic tumor cell lines are frequently used as a model for primary human T cells, which is limited because tumor-derived cell lines such as Jurkat (Schneider *et al.*, 1977) have a strongly altered phenotype in comparison to primary T cells with respect to signal transduction, gene regulation, and proliferation control. In contrast, primary T-cell cultures are limited in their natural lifespan. It is laborious and frequently impossible to grow primary T cells to large numbers, and it requires considerable effort to amplify them in periodic response to a specific antigen in presence of accessory cells with the appropriate MHC restriction elements. The immortalization of functional human T cells should be the ideal way to solve such problems. Virus-mediated transformation is a routine method for human B cells which are efficiently immortalized *in vitro* by EBV. This method has been instrumental for the analysis of the human B-cell repertoire and function. EBV-transformed lymphoblastoid B cells retain their antigen-presenting capability and are widely used to study T-cellular antigen specificity. An established method to transform human T cells in culture utilizes HTLV-1 (Miyoshi *et al.*, 1981; Popovic *et al.*, 1983; Yamamoto *et al.*, 1982). For numerous applications, this approach proved useful, although the transformation is largely confined to CD4+ T cells. Retrovirus-transformed T lymphocytes produce HTLV-1 virions regularly. However, the HTLV-1 transformed T cells tend to lose their T-cell receptor complex, their cytotoxic activity, and their dependence on IL-2 after prolonged cultivation (Inatsuki *et al.*, 1989; Yssel *et al.*, 1989).

Another method of T-cell transformation in cell culture became available through the observation that cell-free HVS strain C488 was capable of stimulating human T lymphocytes to stable antigen-independent proliferation in culture (Biesinger *et al.*, 1992). These growth-transformed human T cells did not shed virus particles and retained many essential T-cell functions including the MHC-restricted antigen-specific reactivity of their parental T-cell clones (Bröker *et al.*, 1993; De Carli *et al.*, 1993; Mittrücker *et al.*, 1993; Weber *et al.*, 1993). These observations have opened up a novel research direction which links T-cell biology, signal transduction pathways, and transforming functions to HVS, an oncogenic herpesvirus which has thus been converted to an immunological tool.

3.5.2   Infection of human cells by a monkey herpesvirus

The infection of human cells by HVS is a prerequisite of the growth transformation of human T lymphocytes. The HVS strain B-SMHI revealed a weak productive activity in primary human fetal cells (Daniel *et al.*, 1976a). Selectable HVS A11 recombinants were utilized to study the spectrum of cells which can be infected by the virus (Simmer *et al.*, 1991). A broad range of epithelial, mesenchymal, and hematopoetic cell types became persistently infected and carried non-integrated episomal DNA of the recombinant viruses. The pancreatic carcinoma line PANC-1 and human foreskin fibroblasts even produced infectious virus under selection conditions (Simmer *et al.*, 1991; Stevenson *et al.*, 2000c). These findings suggested that the unknown HVS receptor is widely distributed among various tissues. The receptor seems to be well conserved among the species because rabbit T cells can also be infected and transformed by HVS strains (Ablashi *et al.*, 1985; Medveczky *et al.*, 1989b). Cell lines which had been infected with recombinant viruses under selection pressure retained the non-integrated viral episomes after withdrawal of the selecting drug for long time periods. The lack of counter selection against persisting viral episomes suggested that the virus persists with suppressed viral gene expression (Simmer *et al.*, 1991). This model is also supported by the observation that the persisting non-integrated viral episomes were heavily methylated in simian tumor cells (Desrosiers *et al.*, 1979) and may carry extensive genomic rearrangements or deletions (Kaschka-Dierich *et al.*, 1982; Schirm *et al.*, 1984). It is likely that these results for strain A11 are also valid for subgroup C strains and for transformed human T cells. Although monkey T lymphocytes produce HVS particles in many cases, it was not possible to isolate virus from transformed human T-cell cultures which carry non-integrated viral episomes in high copy number (Biesinger *et al.*, 1992). Even after treatment with phorbol esters, nucleoside analogs and other drugs known to reactivate other viruses such as EBV, or after specific or non-specific stimulation of the T cells, virion production could not be demonstrated (Fickenscher *et al.*, 1996a). Nevertheless, it will be difficult to provide formal proof that the virus can never be reactivated from transformed human T lymphocytes.

3.5.3   Preserved functional phenotype of transformed human T Cells

The growth-transformation procedure of human T lymphocytes by HVS C488 (Biesinger *et al.*, 1992) was described in methods reviews (Fickenscher and Fleckenstein, 1998; Meinl and Fickenscher, 2000). The infection of peripheral blood mononuclear cells, cord blood cells, thymocytes, or established T-cell clones by HVS C488 results in continuously proliferating T-cell lines which do not release virus particles and show the morphology of T blasts with irregular shape. Optimal proliferation of HVS-transformed T cells depends on a high cell density and on exogenous IL-2. However, in contrast to primary T cells, regular restimulations with antigen or mitogen in presence of feeder or antigen-presenting cells are not required (Biesinger et al., 1992). CD4+ or CD8+ T cells carrying αβ or γδ T-cell receptors have been transformed by HVS. Mixed populations may occur when polyclonal populations are infected. After transformation of established T-cell clones the phenotype of the parental T cells is conserved (Bröker et al., 1993; De Carli et al., 1993; Fickenscher et al., 1997; Klein et al., 1996; Pacheco-Castro et al., 1996; Weber et al., 1993; Yasukawa et al., 1995). The karyotypes of HVS-transformed T-cell lines were normal (Troidl et al., 1994). While the Vβ-family expression profile in polyclonally derived T-cell lines is initially not altered, a few clones usually overgrow in long term culture (Fickenscher et al., 1996b; Knappe et al., 1997).

HVS-transformed T lymphocytes have the surface phenotype of mature, activated T cells (Biesinger et al., 1992; Fickenscher and Fleckenstein, 1998; Meinl et al., 1995b; Meinl and Fickenscher, 2000). The typical surface markers are the T-cell receptor, CD3, CD2, CD4 or CD8, CD5, CD7, CD11a/CD18 (LFA-1), CD45, CD54 (ICAM-1), and CD58 (LFA-3). As typical activation markers, CD25 (IL-2Rα), CD26, CD30, CD40 ligand, CD69, CD86 (B7.2), and MHC-II are detectable on HVS-transformed T cells. The surface antigen CD56 (an NK cell marker) is typically expressed, while the the NK markers CD16 and CD57 are lacking. CD34+ triple negative thymocytes matured to either αβ or γδ T cells after HVS transformation depending on the cytokines added (Pacheco-Castro et al., 1996).

The MHC-restricted antigen-specificity of the T cells is conserved after transformation. Characterized T helper-cell clones reacting specifically to a series of defined antigens retained their MHC-restricted antigen specificity as measured by proliferation and cytokine production (Bröker et al., 1993; De Carli et al., 1993; Meinl et al., 1995a; Mittrücker et al., 1993; Weber et al., 1993). Another study reported that EBV-specific cytotoxic T-cell lines maintained their antigen-specific cytotoxicity after viral transformation (Berend et al., 1993). Transformed CD4+ T cells secreted IL-2, IL-3, IFN-γ, TNF-α, TNF-β, and GM-CSF after stimulation (Bröker et al., 1993; De Carli et al., 1993; Meinl et al., 1995a; Mittrücker et al., 1992; Weber et al., 1993). IL-4 and IL-5 were produced only at low rates by transformed Th2 cells (De Carli et al., 1993). HVS-transformed T cells are capable of delivering non-specific B-cell help via membrane-bound TNF-α or via CD40 ligand (Del Prete *et al.*, 1994; Hess *et al.*, 1995; Saha *et al.*, 1996). Transformed CD8+ lines and to a lesser extent CD4+ cells showed a non-specific NK-like cytolytic activity on K562 cells (Biesinger *et al.*, 1992). The lectin-dependent cytolytic activity of Th1 clones against P815 target cells was enhanced after transformation, while Th2 clones showed this activity only in the transformed state (De Carli *et al.*, 1993). The cytotoxic activity of a transformed γδ T-cell clone on K562 was inducible by stimulation with IL-12 (Klein *et al.*, 1996). HVS-transformed T cells have a similar sensitivity to apoptosis signals as their parental cells indicating that the growth-transformation is not based on a resistance to apoptosis (Kraft *et al.*, 1998). HVS-transformed T cells can be driven into activation-dependent death by a strong stimulation with phorbol ester or CD3-specific monoclonal antibodies. This cell death is apparently independent of CD95/CD95 ligand interactions (Bröker *et al.*, 1997).

The early signal transduction properties such as tyrosine phosphorylation and calcium mobilization patterns of the transformed cells after IL-2, anti-CD3 and/or anti-CD4 stimulation were similar to those of the uninfected parental cells (Bröker *et al.*, 1993; Mittrücker *et al.*, 1993; Tsygankov *et al.*, 1992). The functionality of CD3, CD4, and the IL-2 receptor was shown by signalling, by proliferation, and by IFN-γ production (Bröker *et al.*, 1993; Weber *et al.*, 1993). The IL-2 dependent proliferation of transformed lymphocytes and the activity and abundance of the CD4-bound fraction of the tyrosine kinase p56^{lck} were strongly inhibited by soluble CD4 antibodies (Bröker *et al.*, 1994).

3.5.4 Changed T-cellular features after T-cell transformation

Whereas HVS-transformed T lymphocytes retain multiple normal T-cell functions, only a few specific cellular and biochemical features are typically changed in comparison to their parental cells. First, the protein tyrosine kinase Lyn, which is usually expressed in B cells but not in T cells, is enzymatically active in HVS-transformed T cells (Fickenscher *et al.*, 1997; Wiese *et al.*, 1996), as it is in T cells immortalized by HTLV-1 (Yamanashi *et al.*, 1989).

Second, HVS shifts the range of cytokines secreted by stimulated T cells towards a Th1 profile: secretion of IL-2 and IFN-γ is increased, IL-4 and IL-5 production are diminished in comparison with parental cells (De Carli *et al.*, 1993). Particularly, the IFN-γ secretion can be stimulated to very high levels (Bröker *et al.*, 1993; De Carli *et al.*, 1993; Weber *et al.*, 1993). In addition, many transformed clones secrete large amounts of the Th1-type

CC-chemokines MIP-1α, MIP-1β, and RANTES after stimulation (Mackewicz *et al.*, 1997; Saha *et al.*, 1998; Schrum *et al.*, 1996).

Third, the most relevant HVS-induced functional alteration in transformed T cells is a hyper-responsiveness to CD2 ligation (Mittrücker *et al.*, 1992). CD2 is a T-cell surface molecule transmitting mitogenic signals in a complex fashion. Binding to its ligand CD58, which can be mimicked by antibodies directed against the CD2.1 epitope, is not sufficient for the activation of primary T cells which require the simultaneous ligation of a different epitope, CD2.2 (Meuer *et al.*, 1984). In contrast, ligation of the CD2.1 epitope alone suffices for inducing proliferation and IL-2 production of HVS-transformed T cells (Mittrücker *et al.*, 1992). Since the transformed T cells express the CD2 ligand CD58 at high density on their surface, cell contact leads to autostimulation. Thus, the hyper-responsiveness to CD2 ligation contributes to the spontaneous proliferation and to the transformed phenotype of HVS T cells. However, transformed T cells from a patient lacking CD18/LFA-1 expression could not be stimulated via the CD2 pathway, suggesting that the CD2 hyperreactivity, though typically observed, might not be essential for HVS transformation (Allende *et al.*, 2000).

Differentially expressed cellular genes were cloned from HVS-transformed T cells by representational difference analysis (Knappe *et al.*, 1997, 2000b). By this approach, a novel cellular gene, *ak155*, was identified which is a sequence homolog of cellular IL-10. Specifically HVS-transformed T cells overexpress cellular *ak155* and secrete the protein into the supernatant. In other T-cell lines and in native peripheral blood cells, but not in B cells, *ak155* is transcribed at low levels. The AK155 protein was shown to form homodimers similarly to IL-10. As a lymphokine, AK155 (= IL-26) may contribute to the transformed phenotype of human T cells after infection by HVS (Knappe *et al.*, 2000b).

3.5.5   State of herpesvirus saimiri in transformed human T cells

Numerous genomic regions of the virus have been used to search for viral transcripts derived from the persisting multicopy episomes in the non-permissive transformed human T cells. With rare exceptions, viral transcription could not be demonstrated. In the left-terminal H-/L-DNA transition region which is essential for transformation, the viral gene *stpC/tip* is strongly and inducibly transcribed as a bicistronic message (Fickenscher *et al.*, 1996a; Medveczky *et al.*, 1993). Whereas large amounts of StpC are produced (Fickenscher *et al.*, 1996a), Tip is expressed at low levels and is detectable only after immunoprecipitation with Lck antibodies in Lck kinase assays (Biesinger *et al.*, 1995; Lund *et al.*, 1997b; Wiese *et al.*, 1996). StpC and Tip are the sole viral proteins which have been demonstrated in HVS-transformed human T cells.

The non-coding viral U-RNA genes (HSUR, HVS U-RNA) were abundantly expressed in the presence of episomal viral DNA, in a similar way to the EBER RNAs of EBV (Ensser *et al.*, 1999). The deletion of the respective genes did not influence virus replication or T-cell transformation. In addition, after chemical stimulation of transformed human T cells with phorbol ester, the gene *ie14/vsag* was abundantly transcribed for a few hours only (Knappe *et al.*, 1997). By this observation, *ie14/vsag* can be considered the third transformation-associated gene of HVS C488. The mitogenic properties of the secreted protein would fit well into a model of growth transformation caused by a combination of different virus functions. However, the functional importance of this is unclear because this gene is dispensible for the transformation of human T cells (Knappe *et al.*, 1997). A few other virus genes, mainly with regulatory function, were found transcribed, however at low abundance and only after additional T-cell stimulation: the IE-gene *ie57*, the early gene *orf50*, and the viral thymidylate synthase gene (Knappe *et al.*, 1997; Thurau *et al.*, 2000). This and the fact, that virus replication could not be induced by various means in C488-transformed human T cells, argues strongly for a block of virus replication downstream of the level of the expression of the regulatory genes *orf50* and *ie57*.

In particular the HVS strain C488 has been used as a tool for the targeted transformation of human T cells (Biesinger *et al.*, 1992). In addition, various subgroup C virus strains are able to transform human T cells, but to a varying extent. Virus strain C484 was reported to transform human T cells to short term, IL-2 independent growth (Medveczky *et al.*, 1993). Other closely related subgroup C strains of HVS can cause unexpected differences in the functional phenotype of growth-transformed human T cells (Fickenscher *et al.*, 1997).

3.5.6 Transformed T-cell lines for studying primary human immunodeficiencies

The method of T-cell transformation by HVS C488 has also been successfully applied for the analysis of primary human immunodeficiencies, in which specific T-cell functions are disturbed or missing. Surprisingly, T-cells with defects in T-cell receptor-dependent signalling could also be transformed by HVS. The defect of the CD3γ chain did not prevent transformation (Rodriguez-Gallego *et al.*, 1996). The transformed CD3γ-deficient cells were utilized for biochemical studies of the structure of the T-cell receptor complex (Zapata *et al.*, 1999). The transformed T cells from a patient with an atypical X-linked severe combined immunodeficiency (SCID) showed a spontaneous partial reversion of the genetic defect affecting the IL-2Rγ chain (Stephan *et al.*, 1996). Transformed CD95-deficient T cells of a human SCID patient were useful for studying CD95-independent activation-dependent cell death (Bröker *et al.*, 1997). Transformed T-cell lines were established from a series of further patients with genetic T-cell defects involving the IL-12R (Altare *et al.*, 1998), MHC class II (Alvarez-Zapata *et al.*, 1998), and with Wiskott-Aldrich syndrome (Gallego *et al.*, 1997). HVS-transformed T cells from a patient with leucocyte adhesion deficiency lacking CD18/LFA-1 expression were unexpectedly not reactive to CD2 stimulation (Allende *et al.*, 2000). HVS-transformation seems promising for studying T cells from SCID patients. In many cases, this has been the only way to cultivate and amplify the patients' cells for further research.

3.5.7   HIV infection of transformed human T cells

HVS-transformed human CD4+ T cells provide a productive lytic system for T-lymphotropic viruses such as human herpesvirus type 6 (F. Neipel, unpublished observation) and human immunodeficiency virus (HIV) (Nick *et al.*, 1993; Vella *et al.*, 1997). The prototype viruses HIV-1$_{IIIB}$ and HIV-2$_{ROD}$ replicated rapidly causing cell death within two weeks. Also a poorly replicating HIV-2 strain and primary clinical isolates replicated to high titers. HVS-transformed human CD4+ T cells can be used for poorly growing HIV strains with narrowly restricted host cell range (Nick *et al.*, 1993). Moreover, HVS-transformed T cells can be persistently and productively infected with HIV. In comparison to conventional T-cell lines, the down-regulation of surface CD4 molecules is delayed (Vella *et al.*, 1997). Similarly to cultures of primary peripheral blood cells, HVS-transformed T cells allow the propagation of macrophage-tropic HIV isolates without selecting for subtypes with changed phenotype or cell tropism (Vella *et al.*, 1999a, b). HVS-transformed CD8+ T cells were reported to secrete soluble HIV-inhibiting factors different to the known inhibitory cytokines (Copeland *et al.*, 1995; Copeland *et al.*, 1996; Lacey *et al.*, 1998; Leith *et al.*, 1999). Transformed CD4+ and CD8+ T cells produced varying amounts of the cytokines IL-8, IL-10, TNF-α, TNF-β, RANTES, MIP-1α, and MIP-1β (Mackewicz *et al.*, 1997; Saha *et al.*, 1998). Surprisingly, transformed CD4+ T-cell clones from AIDS patients produced no RANTES and little or no MIP-1α or MIP-1β and were more readily infectable with HIV in comparison to T cells from non-progressors which produced high amounts of chemokines and were less infectable (Saha *et al.*, 1998). HVS-transformed human CD4+ T cells expressing CCR5 and CXCR4 were fully functional as antigen-presenting target cells for HIV-specific, MHC class I-restricted cytotoxic T-cell activity (Bauer *et al.*, 1998).

3.5.8 Herpesvirus saimiri-transformed T cells in animal models

Animal models are valuable for understanding the status of growth-transformation which may be either a rather mild phenomenon restricted to cell-cultures or alternatively may correspond to neoplastic transformation *in vivo*. When C488-transformed human T cells were tested for tumorigenesis in *nude* or SCID mice in conventional implantation experiments, tumor formation could not be observed, whereas xenogeneic graft-versus-host disease could be induced in a similar way as with primary human T cells (Huppes *et al.*, 1994).

The behavior of HVS C488 in various monkey systems is of interest, because on the one hand, HVS is a tumor virus of New World monkeys, and on the other hand, Old World monkeys such as macaques are the most used animal model for the close-to-human situation. In New World primate T cells, HVS establishes a semi-permissive infection *in vitro* and *in vivo*: the cells do produce virus for long time periods, and they become transformed as well (summarized in Fickenscher *et al.*, 1996a). Similarly to human T cells, the T lymphocytes from macaque monkeys can be growth-transformed by HVS (Akari *et al.*, 1996, 1999; Feldmann *et al.*, 1997; Knappe *et al.*, 2000a; Meinl *et al.*, 1997). Some researchers observed IL-2 dependence, others IL-2 independence. In many respects, the transformed macaque T cells resembled their human counterparts. The phenotype of activated T cells was preserved, and antigen-specific T-cell lines against myelin basic protein or streptolysin O retained their reactivity after transformation. The MHC-II expressing transformed cells were able to present the antigen to each other even in the absence of autologous presenter cells (Meinl *et al.*, 1997). One major difference is the pronounced frequency of double-positive CD4+/CD8+ T cells, which are uncommon in humans. T-cell immunology and T-cell transformation from macaques is greatly hampered by reactivation of the foamy virus with which most rhesus monkeys in primate centers are infected (Feldmann *et al.*, 1997; Knappe *et al.*, 2000a). Initially, HVS-transformed macaque T cells seemed to be non-permissive for HVS similarly to human T cells (Feldmann *et al.*, 1997; Meinl *et al.*, 1997). However, in contrast to their human counterparts, these cells were shown to shed low amounts of virus particles in many cases (Alexander *et al.*, 1997; Knappe *et al.*, 2000a).

In order to study the behavior *in vivo* in a close-to-human situation, HVS-transformed autologous T cells of macaque monkeys were infused intravenously into the respective donor. The animals remained healthy, without occurrence of lymphoma or leukemia for an observation period of more than one year. Over several months virus genomes were detectable in peripheral blood cells and in cultured T cells (Knappe *et al.*, 2000a). Monkeys, which had previously received autologous T-cell transfusions, were protected from lymphoma after challenge infection with the wild-type virus C488. In naive control animals, a high-dose intravenous infection rapidly induced pleomorphic peripheral T-cell lymphoma (Alexander *et al.*, 1997; Knappe *et al.*, 2000a). Thus, HVS-transformed T cells were well tolerated after autologous reinfusion. This may form the basis for a new concept of experimental T-cell mediated adoptive immunotherapy.

3.5.9   Non-transforming herpesvirus saimiri as an oncogene trap

Non-transforming and non-oncogenic HVS deletion variants were used as eukaryotic expression vectors in order to investigate heterologous oncogenes. HVS deletion mutants without the left-terminal transformation-associated L-DNA region neither caused malignant disease in animals, nor transformed simian lymphocytes in culture (Desrosiers *et al.*, 1984, 1986). Homologous recombination was used for the insertion of transgenes into the virus genome. The transforming functions of HTLV-1 were defined by using an HVS vector containing the HTLV-1 X region or mutants thereof, instead of the homologous transformation-associated genes from HVS A11. These assays revealed Tax and not Rex as the transforming principle of HTLV-1 for human T cells (Grassmann and Fleckenstein, 1989; Grassmann *et al.*, 1989, 1992).

Overexpression of the proto-oncogene c-*fos* is known to induce transformation in various systems. c-*fos* recombinant HVS-vectors expressed large amounts of the oncoprotein upon persistent infection of human neonatal fibroblasts. However, these primary mesenchymal cells did not show any sign of transformation (Alt *et al.*, 1991; Alt and Grassmann, 1993). The transforming function of *stpC* for T-cell transformation *in vitro* was successfully substituted by cellular *ras* (Guo *et al.*, 1998), by the *K1* gene of KSHV (Lee *et al.*, 1998b), or by *R1* of rhesus monkey rhadinovirus (RRV; Damania *et al.*, 1999) in HVS. In KSHV the gene *K1* is located at the respective transformation-associated left-terminal L-DNA position and its gene product interacts with various cellular signalling molecules via its immunoreceptor activation motif and prevents the transport of the B-cell receptor complex to the cell surface (Lee *et al.*, 1998b, 2000). The ability of

K1 to complement an *stpC* deletion mutant of HVS to a transforming and tumorigenic phenotype identified it as an oncogene (Lee *et al.*, 1998a). In the case of RRV, the oncogene *R1* is situated at the homologous position. *R1* was also sufficient to substitute for *stpC* for T-cell transformation (Damania *et al.*, 1999, 2000). This system of complementing a transformation-deficient HVS strain to a transforming phenotype by heterologous oncogenes, is applicable as an oncogene trap for novel cellular or viral transforming genes.

3.5.10   Herpesvirus saimiri vectors for heterologous gene expression

HVS has been further used as a vector for growth hormone, for secreted alkaline phosphatase, and an autofluorescent protein (Desrosiers *et al.*, 1985b; Duboise *et al.*, 1996; Stevenson *et al.*, 1999). An early preclinical gene therapy trial was performed with a non-transforming replication-competent HVS vector expressing the bovine growth hormone from an intron-containing gene. Persistently infected simian T cells produced high amounts of the bovine hormone. Experimentally infected New World primates produced circulating bovine growth hormone and developed a humoral immune response (Desrosiers *et al.*, 1985b). These observations suggested that persisting HVS vectors could be used to replace missing or defective genes in hereditary genetic disorders.

The original method of generating HVS expression vectors utilized homologous recombination via a single stretch of viral terminal L-DNA in the recombination construct (Desrosiers *et al.*, 1985b; Grassmann and Fleckenstein, 1989). A more elaborated procedure for the isolation of HVS mutants was developed by the insertion of an autofluorescent reporter gene flanked by single-cutter restriction endonuclease recognition sites into the viral genome instead of the transforming *stpC* gene. Thus, the reporter gene cassette could easily be replaced by other transgenes after simple restriction enzyme digestion and ligation (Duboise *et al.*, 1996). This vector has mainly been used for the expression of heterologous oncogenes in order to assay transforming activity in monkey T cells. An alternative approach has been developed by cloning HVS C488 into cosmid vectors. The co-transfection of overlapping cosmids into permissive epithelial owl monkey kidney cells led to the reconstitution of recombinant replication-competent virus (Ensser *et al.*, 1999). This approch is valuable for generating expression vectors for foreign genes, because a contamination with wild-type virus is excluded.

Non-selectable recombinant viruses expressing an autofluorescent protein were able to transduce human hematopoietic progenitor cells in culture, but at low efficiency and with a tendency towards differentiation (Stevenson *et al.*, 1999, 2000b). Moreover, active HVS replication was observed in certain human cell types (Daniel *et al.*, 1976a; Simmer *et al.*, 1991; Stevenson *et al.*, 1999, 2000c). HVS efficiently infected totipotent mouse embryonic stem cells under drug selection pressure. The infected cells stably maintained the viral episomal genome and could be differentiated into mature haematopoietic cells, while the heterologous gene was rather stably expressed (Stevenson *et al.*, 2000a). This system may be of particular interest for studying transgene effects during cell differentiation *in vitro* and *in vivo*.

3.5.11   Episomal vectors for adoptive immunotherapy

Gene transfer into primary human T cells by transfection or retroviral transduction remains difficult and unreliabe with respect to long-term transgene expression. The maintained functional phenotype of HVS-transformed T cells suggested the use of HVS as a vector for human T cells at least for cell culture experiments. A reactivation of recombinant or wild-type virus from transformed human T cells has not been observed, but cannot be formally excluded. The techniques of homologous recombination and cosmid complementation are applied for constructing replication-defective, but transformation-competent deletion variants which preclude reactivation. Furthermore, additional genes are introduced into HVS vectors to study the gene products in human T cells.

HVS might be useful as a gene vector for targeted amplification of functional human T cells even for therapeutic applications if a series of biosafety issues are clarified. By reinfusion of autologous transformed T cells into the donor macaques, an intrinsic oncogenic phenotype could be excluded, because the animals did not develop disease while the infused T cells persisted for extended periods (Knappe *et al.*, 2000a). To improve the biosafety of such vectors, the prodrug activating gene thymidine kinase (TK) of HSV was inserted into the genome of HVS. TK-expressing transformed T cells were efficiently eliminated in the presence of low concentrations of ganciclovir over an observation period of one year. At any time during the course of a therapeutic application, TK-expressing transformed human T cells might be eliminated after administration of ganciclovir. In principle, this function could be useful for the T-cell dependent immunotherapy of resistant blood cancer while avoiding the risk of uncontrolled graft-versus-host disease (Hiller *et al.*, 2000). Replication-deficient vector variants are another step towards applicability. The use of HVS vectors for redirecting the antigen specificity of primary human T cells may provide an important tool for experimental cancer therapy.

4. RHADINOVIRUSES IN OTHER ANIMALS

4.1 Malignant Catarrhal Fever by alcelaphine and ovine rhadinoviruses

4.1.1 Symptoms and pathology

Malignant Catarrhal Fever (MCF) is a disease of domestic and wild ungulates with a high case fatality rate and a worldwide distribution. The pathology is characterized by a combination of lymphoproliferation and degenerative symptoms in the affected animals. The disease can be categorized into peracute, intestinal, head and eye, and mild forms (Götze, 1930). The head and eye and intestinal form frequently combine to an unfavourable prognosis. Symptoms of MCF in cattle and other susceptible ruminants include marked depression, persistently high body temperature (39.5-42°C), mucopurulent nasal and ocular discharges, dyspnea, bilateral kerato-conjunctivitis with corneal opacity, enlargement of the superficial lymph nodes and marked erythema and/or superficial mucosal erosions and necrosis. At necropsy, generalized swelling of lymph nodes, and mild enlargement of spleen and liver are present in all forms, whereas the extent of erosive or necrotic lesions in the upper respiratory and digestive tracts, brain, eyes, liver, kidneys and the urinary bladder is variable. Histopathologically fibrinoid vasculitis accompanied by lymphocytic infiltration in the parenchyma of the affected tissues is evident (reviewed by Reid and Buxton, 1989). A certain variation of the predominant morphological changes has been noted between different affected species, from lymphosarcoma-like changes in deer (Blake *et al.*, 1990) to vasculitis and necrosis predominating in other species (Reid and Buxton, 1989). Though the occurrence of mild forms is still subject to discussions, with the advent of modern PCR- and ELISA-based diagnostics it is now increasingly recognized that nonfatal cases occur. Surviving animals have a high prevalence of obliterative arteriopathy and remain PCR-positive for viral DNA in peripheral blood (O'Toole *et al.*, 1995, 1997). Infiltrating lymphocytes and IL-2 dependent lymphoblastoid cell lines derived from peripheral blood of MCF diseased cattle and deer express CD2, CD5 and CD25 and CD4+ and/or CD8+ (Schock *et al.*, 1998). Thus, both the biological properties and phenotype of transformed cells resemble the pathology caused by HVS in New World primate species (Fleckenstein and Desrosiers, 1982).

4.1.2   Epidemiology

MCF occurs in two basic epizootical forms. The first form is the African wildebeest-associated MCF (WA-MCF), with outbreaks in timely relationship to the wildebeest calving season. Contact of susceptible cattle to wildebeest or other antelopes during this period of increased shedding of the causative agent by the wildebeest calves is linked with disease (Reid and Buxton, 1989). The second form is the sheep-associated MCF (SA-MCF) occurring when sheep and cattle are housed together (Götze and Liess, 1930), which accounts for sporadic cases or small enzootics of MCF with worldwide distribution (Reid and Buxton, 1989). Either form may be responsible for outbreaks in zoos. Though experimental transmission of SA-MCF by cross-inoculation of blood from diseased cattle had been demonstrated (Götze and Liess, 1929), no infectious agent had been associated with the disease, until an etiological agent, termed alcelaphine herpesvirus 1 (AHV-1), was isolated from blue wildebeest, *Connochaetes taurinus taurinus* (Plowright *et al.*, 1960). Various species of wild and captive gnu antelopes (*Connochaetes taurinus* spp.) are latently infected with AHV-1, and closely related herpesviruses were isolated from topi and hartebeest (Seal *et al.*, 1989a, b) and roan antelopes (*Hippotragus equinus*) (Reid and Bridgen, 1991). Attempts to isolate the agent of SA-MCF were unsuccessful, but partial genomic sequences of a virus with close relationship to AHV-1 have been characterized from lymphoblastoid cells of diseased cattle, deer and rabbits (Bridgen and Reid, 1991). This agent has been designated ovine herpesvirus type 2 (OHV-2). It is readily demonstrated by serology (Li *et al.*, 1994) and PCR (Wiyono *et al.*, 1994) in samples from SA-MCF.

Ungulate species susceptible to MCF include members of the subfamily *Bovinae* such as cattle, buffaloes, bison, Indian gaur (*Bos gaurus*) (Zimmer *et al.*, 1981), greater kudu (*Tragelaphus strepsiceros*) (Castro *et al.*, 1982), the family Cervidae (many deer species; Peer David's, Sika, white-tailed, or Rusa deer are highly susceptible) (Reid and Buxton, 1989), but also pigs (*Suidae*) (Loken *et al.*, 1998). Experimental small animal models for MCF are based on the inoculation of infectious blood or tissue into rabbit, rat, hamster, and guinea pig, whereas mice are refractory (Buxton and Reid, 1980; Jacoby *et al.*, 1988).

Infection with OHV-2 is widespread in sheep (*Ovis aries*), and highly sensitive detection can be achieved by PCR, or a slightly less sensitive competitive inhibition ELISA (Li *et al.*, 1994, 1995; Müller-Doblies *et al.*, 1998). OHV-2 is detectable by PCR from the B-cell fraction of adult sheep PBMC (Baxter *et al.*, 1997). Horizontal transmission efficiently occurs when uninfected and OHV-2 seropositive adult sheep are combined. In contrast, transfusion of blood or purified PBMC from infected to uninfected sheep is inefficient, arguing for a mostly nonproductive latent infection of blood cells (Li *et al.*, 2000a). Infection of lambs can be prevented by their separation from infected animals at the age of two months; nonseparated lambs usually are all PCR-positive at an age of six months (Li *et al.*, 1998a; 1999a). OHV-2 DNA was detectable from nasal secretions of most lambs in with peak levels and incidence (88%) at an age of 7.5 months (Li *et al.*, 1998a). Another study found shedding of OHV-2 DNA from perinatal lambs (n=9) within the first seven weeks (Baxter *et al.*, 1997). In addition to sheep, mouflon (*Ovis musimon*) and pygmy goats (*Capra hircus*) also have a high seroprevalence of an OHV-2 like virus; interestingly, fallow deer (*Dama dama*) can also be infected asymptomatically (Li *et al.*, 1999b; Frölich *et al.*, 1998).

Table 3. Rhadinovirus species

Designation, Abbreviation	Host	Associated pathogenicity
Kaposi's sarcoma associated herpesvirus, KSHV	Human	Kaposis's sarcoma, multicentric Castleman's disease, primary effusion lymphoma
Retroperitoneal fibromatosis-associated herpesvirus, RFHV	Rhesus monkey	Retroperitoneal fibromatosis ?
Rhesus monkey rhadinovirus, RRV	Rhesus monkey	B-cell hyperplasia ?
Herpesvirus saimiri, HVS	Squirrel monkey	T-cell lymphoma in other neotropical monkeys
Herpesvirus ateles, HVA	Spider monkey	T-cell lymphoma in other neotropical monkeys
Alcelaphine herpesvirus 1, AHV-1	Wildebeest	Malignant cararrhal fever in cattle
Alcelaphine herpesvirus 2, AHV-2	Hartebeest, topi	
Ovine herpesvirus 2, OHV-2	Sheep	Malignant cararrhal fever in cattle and deer
Caprine herpesvirus 2	Goat	Chronic disease in deer ?
Bovine herpesvirus 4, BHV-4	Cattle	None reported
Equine herpesvirus 2, EHV-2	Horse	Mononucleosis in horses?
Equine herpesvirus 5, EHV-5	Horse	Mononucleosis in horses?
Porcine lymphotropic herpesviruses, PLHV-1/2	Pig	None reported
Herpesvirus sylvilagus	Cottontail rabbit	Mononucleosis ?
Murine γ-herpesvirus, 68 MHV-68	Bank vole	Mononucleosis in mice

DNA sequences corresponding to new γ-herpesviruses from healthy feral and domestic pigs, termed pig lymphotropic herpesvirus (PLHV-1 and 2, respectively) (Ehlers *et al.*, 1999a, b; Ulrich *et al.*, 1999), and from cattle, termed bovine lymphotropic herpesvirus (BLHV) (Rovnak *et al.*, 1998), were recently amplified by consensus PCR with degenerate primers. The limited phylogenetic analysis that was possible with these partial sequences suggested that these viruses are most closely related to AHV-1 and OHV-2, but they do not seem to be associated with MCF. DNA-sequences with close homology to AHV-1 and OHV-2 were also amplified from a MCF case in white-tailed deer (Li *et al.*, 2000b). However, the natural host of this virus is unknown, and the virus seems to be different from goat, sheep, and antelope isolates. Recently, a new rhadinovirus has been found in goats, and phylogenetic analysis of the available amplified short DNA sequences places this virus in close relationship to AHV-1 and OHV-2 (Li *et al.*, 2001), and may account for sporadic cases of MCF which are not associated with contact to wildebeest or sheep. Thus, MCF viruses seem to be derived from apparently healthy host animals of the subfamilies *Caprinae* or *Alcelaphinae*.

4.1.3   Genome structure and pathogenic factors

The genome organisation of AHV-1 strain C500 (Ensser *et al.*, 1997) is essentially collinear with the well characterized rhadinovirus prototype, HVS (Albrecht *et al.*, 1992a), but marked differences were found in the regions interspersed between the blocks of conserved herpesvirus genes. The coding 130,608 bp L-DNA region of the genome has a low (46.17%) GC-content and marked suppression of CpG dinucleotide frequency. The L-DNA is flanked by approximately 20 to 25 GC-rich (71.83%) H-DNA repeats of 1113 to 1118 bp. The analysis of the L-DNA sequence revealed 70 open reading frames; 61 thereof showed homology to other herpesviruses and are arranged in 4 blocks collinear to other rhadinovirus genomes. These gene blocks are flanked by non-conserved regions, containing genes without similarities to the known cytokine-related or to the transformation-associated genes of KSHV and HVS (reviewed in Jung *et al.*, 1999). The positionally equivalent reading frames *A1* and *A2* show no similarity but are in homologous position to the *stp/tip* and *tio* genes of HVS and HVA, and may exert similar activating functions (Ensser *et al.*, 1997). It is encoded by a spliced transcript and has a motif similar to nuclear localization signals and the basic domain of the stress-induced transcription factor ATF3 (Chen *et al.*, 1996). *A3* is a gene with homology to the semaphorin family and a related gene of poxviruses (Ensser and Fleckenstein, 1995). It is homologous to cellular Sema7A/CDw108 (Lange *et al.*, 1998) which is expressed on

B- and T-lymphocytes (Mine *et al.*, 2000). Semaphorins are a growing gene family of chemoattractant and/or repulsive factors with important roles in neuronal and lymphocyte development. The T-cellular semaphorin Sema4D/CD100 is upregulated after CD40 stimulation and turns off negative signals by CD72; it has important non-redundant functions in the immune system (Kumanogoh *et al.*, 2000; Shi *et al.*, 2000). The high-affinity receptor for Sema7A is Plexin 1-C, and Plexin-1B for Sema4D. The transforming genes of AHV-1 have not been defined. The region coding for ORF50/RTA was reported to be associated with pathogenicity in rabbits (Handley *et al.*, 1995), however the described clone seems to be strongly rearranged and differs from the AHV-1 genomic sequence (Ensser *et al.*, 1997). The putative transcriptional regulators encoded by *orf50* and *orf57* are only weakly conserved to other rhadinoviruses, presumably reflecting the changes necessary in the process of adaption of regulatory genes to different host environments. This may also apply to the ORF73 protein, sharing structural similarity with HVS and KSHV LNA-1/ORF73. Similarly to the MHV-68 M11 (Wang *et al.*, 1999), the AHV-1 Bcl2 homolog (A9) is located at a different genomic position, but is similar in size and contains a conserved BH1 domain and a hydrophobic carboxy-terminus.

The lymphoblastoid T-cell lines established from cattle with MCF harbour multiple copies (10 to > 500) of the OHV-2 genome, and we have continuously propagated such cells lines for more than 18 months *in vitro*. The lymphoid proliferations observed in experimentally infected rabbits are also of T-cell origin (Schock and Reid, 1996). Similarly to KSHV and RRV, propagation and transmission of the infectious agent associated with SA-MCF is only possible by the cultivation of latently infected lymphoid cells, whereas attempts to virus isolation in lytic culture succeeded only for AHV-1 and closely related WA-MCF strains.

Despite common biological and epidemiological properties, AHV-1 differs from HVS and from all other γ-herpesviruses in its viral cell homologs, suggesting that different viral effector molecules achieve a similar phenotype in T-cell transformation. MCF and the HVS-induced T-cell lymphoma in marmosets show remarkable similarities. Both causative agents belong to the genus rhadinovirus, both are apathogenic in their natural hosts, and both cause lymphoproliferative disease in other species. Both viruses can induce a similar pathology after infection of rabbits. The involved lymphocytes have an analogous phenotype, and lymphocytic cell lines can be established from naturally and experimentally infected animals.

4.2   Rhadinoviruses of other domestic animals

4.2.1   Bovine herpesvirus type 4

Bovine herpesvirus 4 (BHV-4), or bovine cytomegalovirus, is not associated with a defined clinical disease in cattle (Goltz and Ludwig, 1991), and has been recognized as a member of the rhadinoviruses (Bublot *et al.*, 1992). It has a low GC content of the coding L-DNA, and the arrangement of conserved herpesviral genes is similar to HVS, but lacks most homologs to cellular genes found in HVS or KSHV (Lomonte *et al.*, 1996; Zimmermann *et al.*, 2001). BHV-4 can replicate on human cells *in vitro* with a lytic cycle including cytolytic activity (embryonic lung cell lines MRC-5 and Wistar-38, and giant-cell glioblastoma cells), thus human BHV-4 infection seems possible in theory (Egyed, 1998). Transduction of human RD4 rhabdomyosarcoma cells by the recombinant virus BHV-4neo was also accompanied by the production of infectious virus (Donofrio *et al.*, 2000). BHV-4 encodes functional vBcl2 (Bellows *et al.*, 2000) and vFLIP (E2) homologs (Wang *et al.*, 1997). However, virus yield in BHV-4 lytic replication is increased by host cell apoptosis (Sciortino *et al.*, 2000). At the left terminal region, an open reading frame initially missed due to an sequencing error codes for a β-1,6-N-acetylglucosaminyl transferase homolog, an enzyme with a crucial role in glycan synthesis (Vanderplasschen *et al.*, 2000). The sequence of BHV-4 has been published recently (Zimmermann *et al.*, 2001).

A virus termed herpesvirus aotus 2 with genomic properties of a γ-herpesvirus similar to HVS had been isolated from blood of asymptomatic owl monkeys on OMK cells (Fuchs *et al.*, 1985); however, this virus was later found to have almost identical antigenic and genomic properties to BHV-4, and most likely represented a contamination from bovine serum (Bublot *et al.*, 1991). The term BHV-3 was ambiguously used for different herpesvirus species, including BHV-4 and AHV-1 strain WC11; therefore this designation should not be used any more (Roizman *et al.*, 1992).

4.2.2   Equine herpesviruses type 2 and 5

Equine herpesvirus (EHV)-2 and EHV-5 were provisionally classified as equine cytomegaloviruses due to their biological properties (Plummer *et al.*, 1969), but genomic analysis later showed they are γ-herpesviruses (Telford *et al.*, 1993). Both viruses are ubiquitously found in horses with seropositivity rates exceeding 90% (Borchers *et al.*, 1999; Edington *et al.*,

1994; Murray *et al.*, 1996). Infection of foals with EHV-2 is highly prevalent even at young age, it can be asymptomatic or be associated with mild respiratory symptoms, and the virus can be isolated with increased frequency from tracheal aspirates of foals with respiratory symptoms (Murray *et al.*, 1996). Foals are seronegative at birth and prior to sucking, and most seem to seroconvert within the first two months. EHV-2 respiratory infection has been identified as a predisposing factor for *Rhodococcus (R.) equi* pneumonia in foals, and vaccination reduces mortality by *R. equi* (Nordengrahn *et al.*, 1996; Varga *et al.*, 1997). The virus can be frequently isolated from equine peripheral blood cells by cultivation on monolayer cultures of equine fetal kidney (EFK) cells or monocytes, and seems to be persist in and reactive from latently infected B-lymphocytes (Drummer *et al.*, 1996). It was further possible to transmit EHV-2 to mice, resulting in a nonletal productive infection with weight loss and conjunctivitis, and the virus was detectable in respiratory tissues, trigeminal ganglia, and olfactory bulbs reflecting the tissue tropism in ponies (Rizvi *et al.*, 1997). DNA-polymerase sequences amplified from peripheral blood cells of zebra and wild ass revealed sequence identities exceeding 80% to EHV-2 and -5 (Ehlers *et al.*, 1999a).

The physical mapping (Agius *et al.*, 1992; Browning and Studdert, 1989) and the genomic sequence of EHV-2 (Telford *et al.*, 1995) revealed that the coding sequence of EHV-2 is not flanked by multiple noncoding GC-rich tandem repeats like in HVS or KSHV. EHV-2 has evolved two single large GC-rich repeats instead, each containing an open reading frame encoding a vGPCR homolog. This EHV-2 E1 encodes a functional GPCR that binds Eotaxin (Camarda *et al.*, 1999). Of particular interest are homologs to genes that regulate apoptosis, such as the Bcl2 (ORF16), vFLIP (E8; Bertin *et al.*, 1997; Hu *et al.*, 1997; Thome *et al.*, 1997), and vCarmen/CLAP (E10). vCarmen/CLAP is a unique CARD (caspase-recruitment domain) containing protein that has been shown to induce both apoptosis and NFκB activation in mammalian cells. NFkB activation seems to be mediated by E10 CARD domain interaction with the carboxyterminal region of IKKκ (Thome *et al.*, 1999; Yan *et al.*, 1999).

The virus lacks homologs to the other putative transforming or transformation associated genes of the transforming rhadinoviruses. Though it had been suggested that EHV-2 may belong to a new γ-herpesvirus subfamily (Telford *et al.*, 1993, 1995), the organization of homologous genes places EHV-2 in closer relationship to HVS, KSHV, and the other rhadinoviruses; most EHV-2 genes have counterparts in HVS, and only the IL-10 homolog encoded by E7 relates to EBV.

Herpesvirus sylvilagus is a lymphotropic herpesvirus of cottontail rabbits (*Sylvilagus floridanus*) with a genome organization similar to the rhadinovirus prototype, HVS. Infected animals showed a transient lymphoproliferation, but the virus is neither oncogenic nor transforming *in vitro* (Medveczky *et al.*, 1989a).

4.3   Murine γ-herpesvirus 68

Murine γ-herpesvirus 68 (MHV-68) is the best studied of several murine γ-herpesvirus isolates and was initially obtained from asymptomatic bank voles (*Clethrionomys glareolus*) in Slovakia by passage in neonatal mouse brain (Blaskovic *et al.*, 1980). Other closely related isolates include MHV-60, -72, -76, and -78 (Reichel *et al.*, 1991). Since MHV-68 was subsequently reisolated from the trigeminal ganglia of naturally and experimentally infected mice, it was first classified as a murine a-herpesvirus (Rajcani *et al.*, 1985). Early reports found severe pneumonia and generalized infection with considerable letality after infection of 5-21 d old outbred mice (Rajcani *et al.*, 1985), and virus could be recovered from lungs and trigeminal ganglia of surviving animals up to six month after infection. Later reports noted that infection of immune competent mice results in less severe respiratory symptoms, accompanied by weight loss, and thymic and splenic atrophy; lymphoproliferation in lungs and spleen was noted in asymptomatic mice with prolonged persistence of virus (Sunil-Chandra *et al.*, 1992). MHV-68 is a valuable small animal model for mononucleosis-like diseases and for studying immune response to γ-herpesvirus infection (Brooks *et al.*, 1999; Cardin *et al.*, 1996; Doherty *et al.*, 1997; Sangster *et al.*, 2000), and even vascular disease (Alber *et al.*, 2000). Mice deficient in CD4+ or CD8+ T cells or both T cell subsets, B cells, IFN-γ, or inducible nitric oxide synthase (iNOS) can not appropriately control MHV-68 infection, resulting in a frequently lethal phenotype (Dutia *et al.*, 1997; Kulkarni *et al.*, 1997; Usherwood *et al.*, 1996b; Weck *et al.*, 1996). A study in IL-6 deficient mice shows that IL-6 is not required for T-cell activation in MHV-68 infection (Sarawar *et al.*, 1998).

The DNA sequence of MHV-68 confirmed the classification as a rhadinovirus (Virgin *et al.*, 1997) with the typical genome structure of conserved herpesvirus genes and interspersed virus-specific genes. Similarly to the other rhadinoviruses, MHV-68 encodes several modulators of the host immune response and sequestered cellular genes. M1 encodes a protein similar to poxvirus serpins. M1 and 4 of 8 tRNA-like sequences encoded in the same region are not required for replication, establishment of, or reactivation from latency *in vivo* (Simas *et al.*, 1998). The deletion of M1 may even increase reactivation from latency (Clambey *et al.*, 2000). M2 is latency-associated as are M11(vBcl2) and *orf73* (Husain *et al.*, 1999; Virgin *et al.*, 1999). The M3 gene encodes an abundantly secreted 44-kD chemokine binding protein (hvCKBP) protein that blocks the interaction of CC, CXC, C, and CX(3)C chemokines and inhibits chemokine-induced elevation of intracellular calcium levels. hvCKBP does neither bind to human B cell-specific, nor to murine neutrophil-specific CXC chemokines (Parry *et al.*, 2000). Moreover, this virus produces a functional complement regulator (Kapadia *et al.*, 1999). The viral cyclin encoded by B-lymphotropic MHV-68 (M-Cyclin) is expressed as a lytic leaky-late or early-late transcript (van Dyk *et al.*, 1999, 2000). M-Cyclin acts as an oncogene when overexpressed in transgenic mice which then develop T-cell lymphoma (van Dyk *et al.*, 1999). On the other hand, the M-Cyclin has a restricted cdk-preference as it binds only to cdk2 in a similar way to Cyclin-A, and this binding is partially inhibited by $p27^{Kip1}$ (Card *et al.*, 2000). Although it was shown that the MHV-68 M-Cyclin is dispensable for viral replication and pathogenicity in immuno deficient mice and that it is necessary for efficient reactivation from latency (Hoge *et al.*, 2000; van Dyk *et al.*, 2000), this model does not allow to address the question of oncogenesis and transformation. Pathogenesis of this recombinant virus was assessed in immune-deficient SCID mice lacking lymphocytes, thus lymphoma induction can not be studied in this model (Hoge *et al.*, 2000; van Dyk *et al.*, 2000). Most likely, though not precisely defined, death in this model is the result of the overwhelming lytic viral replication and the associated cell destruction. Although MHV-68 infection was reported to be associated with lymphoproliferative disease and lymphoma in aging mice (Sunil-Chandra *et al.*, 1994), tumor cell lines established from such lesions can only rarely be cultivated and typically do not harbor MHV-68 (Usherwood *et al.*, 1996a), and lymphocyte growth transforming properties are absent *in vitro* (Dutia *et al.*, 1999). Thus, there is an obvious contrast between an oncogenic phenotype in a transgenic system and uncertain tumorgenicity ofMHV-68 infection in mice. In contrast, after induction of B- or T-cell lymphoma in immune competent cottontop tamarins by EBV or HVS, respectively, growth transformed tumor derived cell lines harboring thetransforming virus can be regularly expanded.

5. CONCLUSIONS

Several genetic and infectious factors are known to contribute to the pathogenesis of KS, an unusually complex tumor. These include genetic predisposition, male gender, immunodeficiency, HIV-1 and KSHV infection. Whereas all of the above factors are relevant for KS pathogenesis, none of them is sufficient, and most are not required. The importance of KSHV in KS development is underscored by the notion that it is the only known factor that is required for KS pathogenesis. Two major models of KS pathogenesis have been proposed: First, the direct transformation induced by viral oncogenes, similarly to EBV- or HVS-induced lymphomas, and second, caused by KSHV-encoded or -induced cytokines and angiogenic factors, a chronic inflammatory reaction. In the first model, malignant growth transformation would be the initial event. In contrast, according to the alternative model, this would occur rather late in the pathogenesis process, again facilitated through specific KSHV gene products. It will be difficult to assess these two hypotheses experimentally, since at present there is no animal or tissue culture model for KS. However, it will be of clinical relevance to discrimate between these two models: KS may respond to inhibitors of KSHV replication such as phosphonoformic acid or ganciclovir if the second model is more valid, but the therapeutic benefit of these antivirals is doubtful if KS represents a tumor mainly induced by transforming latent gene products of KSHV. HVS leukemogenesis has been discussed as a model for tumor induction by KSHV. Although the specific biology of the respective virus-associated tumors (KS versus T-cell lymphoma) is not directly comparable, it is rather convincing that these viruses have many genomic features in common. Since the putative disease-relevant genes are variable between HVS and KSHV, HVS was proposed as a vector for KSHV genes in order to study their pathogenic potential. A main argument in favor of this application is that HVS replicates in permissive epithelial cells, whereas it is difficult to manipulate KSHV in absence of a classical permissive system.

Most rhadinoviruses have sequestered a specific set of homologs to cellular genes which is variable in different rhadinoviruses. The integration sites are concentrated in the terminal L-DNA regions which could be facilitated by the process of viral replication that seems to originate in these areas of the genome. Whereas functional tests for such pirated genes are not possible in the case of KSHV in absence of a replicative system, cell-homologous genes of HVS were shown by deletion mutagenesis to be dispensable for virus replication, T-cell transformation, and pathogenicity. This has led to a new hypothesis, which puts the function of the cell-homologous genes closer to the mechanisms of persistence than to transformation or pathogenesis. However, such work will be hampered by the limited availability of seronegative squirrel monkeys.

In HVS, the basis for further research into molecular oncology was the definition of an oncogenic, transformation-associated region in the HVS genome. Later, the complete genome sequence of the HVS prototype A11 - the first sequenced rhadinovirus- facilitated functional studies on individual virus genes. Subsequently, the genome sequences of a series of other rhadinoviruses were completed, including the closely related HVA. The impression is rather convincing that HVA resembles an ancient variant of HVS which has collected a smaller set of cell-homologous genes. A new chapter of HVS research was initiated by the observation that the HVS strain C488 is capable of transforming human T lymphocytes to stable proliferation in culture. Thus, the transforming virus functions became interesting in the context of signal transduction in human T cells, such as the specific interaction between Tip of HVS C488 and the T-cellular tyrosine kinase Lck. HVS-transformed human T cells are a promising tool for laboratory studies in T-cell immunology, including inherited and acquired immunodeficiencies. Surprisingly, the transformed human T cells retained many essential features including the MHC-restricted antigen-specificity of their non-transformed progenitor clones. Because transformed autologous T cells were well tolerated in macaque monkeys, and moreover, HVS was useful as an efficient vector for delivering foreign genes into primary human T lymphocytes, a novel concept for possible therapeutic applications of HVS as a vector for T cells becomes conceivable.

ACKNOWLEDGMENTS

Original work included in this review article has been supported by the Deutsche Forschungsgemeinschaft (Sonderforschungsbereich 466), the Bundesministerium für Bildung und Forschung, the Bayerische Forschungs-Stiftung, the German-Israeli Foundation, and the Wilhelm Sander-Stiftung. We are grateful to all the colleagues from our department and from cooperating groups for their scientific contributions, and especially to Bernhard Fleckenstein for continuous support.

REFERENCES

Ablashi, D.V., Schirm, S., Fleckenstein, B., Faggioni, A., Dahlberg, J., Rabin, H., Loeb, W., Armstrong, G., Peng, J.W., and Aulahk, G., 1985, Herpesvirus saimiri-induced lymphoblastoid rabbit cell line: growth characteristics, virus persistence, and oncogenic properties. *J. Virol.* **55**:623-633

Agius, C.T., Nagesha, H.S., and Studdert, M.J., 1992, Equine herpesvirus 5: comparisons with EHV2 (equine cytomegalovirus), cloning, and mapping of a new equine herpesvirus with a novel genome structure. *Virology* **191**:176-186

Ahuja, S. K. and Murphy, P.M., 1993, Molecular piracy of mammalian interleukin-8 receptor type B by herpesvirus saimiri. *J. Biol. Chem.* **268**:20691-20694

Akari, H., Mori, K., Terao, K., Otani, I., Fukasawa, M., Mukai, R., and Yoshikawa, Y., 1996, In vitro immortalization of Old World monkey T lymphocytes with herpesvirus saimiri: its susceptibility to infection with simian immunodeficiency viruses. *Virology* **218**:382-388

Akula, S.M., Pramod, N.P., Wang, F.Z., and Chandran, B., 2001, Human herpesvirus 8 envelope-associated glycoprotein B interacts with heparan sulfate-like moieties. *Virology* **284**:235-249

Alber, D.G., Powell, K.L., Vallance, P., Goodwin, D.A., and Grahame-Clarke, C., 2000, Herpesvirus infection accelerates atherosclerosis in the apolipoprotein E-deficient mouse. *Circulation* **102**:779-785

Albrecht, J.C., 2000, Primary structure of the herpesvirus ateles genome. *J. Virol.* **74**:1033-1037

Albrecht, J.C. and Fleckenstein, B., 1990, Structural organization of the conserved gene block of herpesvirus saimiri coding for DNA polymerase, glycoprotein B, and major DNA binding protein. *Virology* **174**:533-542

Albrecht, J.C. and Fleckenstein, B., 1992, New member of the multigene family of complement control proteins in herpesvirus saimiri. *J. Virol.* **66**:3937-3940

Albrecht, J.C., Friedrich, U., Kardinal, C., Koehn, J., Fleckenstein, B., Feller, S.M., and Biesinger, B., 1999, Herpesvirus ateles gene product Tio interacts with nonreceptor protein tyrosine kinases. *J. Virol.* **73**:4631-4639

Albrecht, J.C., Nicholas, J., Biller, D., Cameron, K.R., Biesinger, B., Newman, C., Wittmann, S., Craxton, M.A., Coleman, H., Fleckenstein, B., and Honess, R.W., 1992a, Primary structure of the herpesvirus saimiri genome. *J. Virol.* **66**:5047-5058

Albrecht, J.C., Nicholas, J., Cameron, K.R., Newman, C., Fleckenstein, B., and Honess, R.W., 1992b, Herpesvirus saimiri has a gene specifying a homologue of the cellular membrane glycoprotein CD59. *Virology* **190**:527-530

Alexander, L., Du, Z., Rosenzweig, M., Jung, J.U., and Desrosiers, R.C., 1997, A role for natural simian immunodeficiency virus and human immunodeficiency virus type 1 nef alleles in lymphocyte activation. *J. Virol.* **71**:6094-6099

Alexander, L., Denekamp, L., Knapp, A., Auerbach, M.R., Damania, B., and Desrosiers, R.C., 2000, The primary sequence of rhesus monkey rhadinovirus isolate 26-95: sequence similarities to Kaposi's sarcoma-associated herpesvirus and rhesus monkey rhadinovirus isolate 17577. *J. Virol.* **74**:3388-3398

Allende, L.M., Hernandez, M., Corell, A., Garcia-Perez, M.A., Varela, P., Moreno, A., Caragol, I., Garcia-Martin, F., Guillen-Perales, J., Olive, T., and Espanol, T., 2000, A novel CD18 genomic deletion in a patient with severe leukocyte adhesion deficiency: a possible CD2/lymphocyte function-associated antigen-1 functional association in humans. *Immunology* **99**:440-450

Alt, M. and Grassmann, R., 1993, Resistance of human fibroblasts to c-fos mediated transformation. *Oncogene* **8**:1421-1427

Alt, M., Fleckenstein, B., and Grassmann, R., 1991, A pair of selectable herpesvirus vectors for simultaneous gene expression in human lymphoid cells. *Gene* **102**:265-269

Altare, F., Durandy, A., Lammas, D., Emile, J.F., Lamhamedi, S., Le Deist, F., Drysdale, P., Jouanguy, E., Doffinger, R., Bernaudin, F., Jeppsson, O., Gollob, J. A., Meinl, E., Segal, A. W., Fischer, A., Kumararatne, D., and Casanova, J.L., 1998, Impairment of myco-bacterial immunity in human interleukin-12 receptor deficiency. *Science* **280**:1432-1435

Alvarez-Zapata, D., de Miguel, O.S., Fontan, G., Ferreira, A., Garcia-Rodriguez, M.C., Madero, L., van den Elsen, P., and Regueiro, J.R., 1998, Phenotypical and functional characterization of herpesvirus saimiri-immortalized human major histocompatibility complex class II-deficient T lymphocytes. *Tissue Antigens* **51**:250-257

Ariyoshi, K., Schim van der Loeff, M., Cook, P., Whitby, D., Corrah, T., Jaffar, S., Cham, F., Sabally, S., O'Donovan, D., Weiss, R.A., Schulz, T.F., and Whittle, H., 1998, Kaposi's sarcoma in the Gambia, West Africa, is less frequent in human immunodeficiency virus type 2 than in human immunodeficiency virus type 1 infection despite a high prevalence of human herpesvirus 8. *J. Hum. Virol.* **1**:193-199

Arvanitakis, L., Geras Raaka, E., Varma, A., Gershengorn, M.C., and Cesarman, E., 1997, Human herpesvirus KSHV encodes a constitutively active G-protein-coupled receptor linked to cell proliferation. *Nature* **385**:347-350

Bais, C., Santomasso, B., Coso, O., Arvanitakis, L., Raaka, E.G., Gutkind, J.S., Asch, A.S., Cesarman, E., Gerhengorn, M.C., and Mesri, E.A., 1998, G-protein-coupled receptor of Kaposi's sarcoma-associated herpesvirus is a viral oncogene and angiogenesis activator. *Nature* **391**:86-89

Ballestas, M.E., Chatis, P.A., and Kaye, K.M., 1999, Efficient persistence of extra-chromosomal KSHV DNA mediated by latency-associated nuclear antigen. *Science* **284**:641-644

Ballestas, M.E. and Kaye, K.M., 2001, Kaposi's sarcoma-associated herpesvirus latency-associated nuclear antigen 1 mediates episome persistence through cis-acting terminal repeat (TR) sequence and specifically binds TR DNA. *J. Virol.* **75**:3250-3258

Bauer, M., Lucchiari-Hartz, M., Fickenscher, H., Eichmann, K., McKeating, J., and Meyerhans, A., 1998, Herpesvirus saimiri-transformed human CD4+ T-cell lines: an efficient target cell system for the analysis of human immunodeficiency virus-specific cytotoxic CD8+ T-lymphocyte activity. *J. Virol.* **72**:1627-1631

Baxter, S.I., Wiyono, A., Pow, I., and Reid, H.W., 1997, Identification of ovine herpesvirus-2 infection in sheep. *Arch. Virol.* **142**:823-831

Bellows, D.S., Chau, B.N., Lee, P., Lazebnik, Y., Burns, W.H., and Hardwick, J.M., 2000, Antiapoptotic herpesvirus Bcl-2 homologs escape caspase-mediated conversion to proapoptotic proteins. *J. Virol.* **74**:5024-5031

Beral, V., Bull, D., Darby, S., Weller, I., Carne, C., Beecham, M., and Jaffe, H., 1992, Risk of Kaposi's sarcoma and sexual practices associated with faecal contact inhomosexual or bisexual men with AIDS. *Lancet* **339**:632-635

Beral, V., Peterman, T.A., Berkelman, R.L., and Jaffe, H.W., 1990, Kaposi's sarcoma among persons with AIDS: a sexually transmitted infection? *Lancet* **335**:123-128

Berend, K.R., Jung, J.U., Boyle, T.J., DiMaio, J.M., Mungal, S.A., Desrosiers, R.C., and Lyerly, H.K., 1993, Phenotypic and functional consequences of herpesvirus saimiri infection of human CD8+ cytotoxic T lymphocytes. *J. Virol.* **67**:6317-6321

Bertin, J., Armstrong, R.C., Ottilie, S., Martin, D.A., Wang, Y., Banks, S., Wang, G.H., Senkevich, T.G., Alnemri, E.S., Moss, B., Lenardo, M.J., Tomaselli, K.J., and Cohen, J.I., 1997, Death effector domain-containing herpesvirus and poxvirus proteins inhibit both Fas-and TNFR1-induced apoptosis. *Proc. Natl. Acad. Sci. USA* **94**:1172-1176

Biesinger, B., Müller-Fleckenstein, I., Simmer, B., Lang, G., Wittmann, S., Platzer, E., Desrosiers, R.C., and Fleckenstein, B., 1992, Stable growth transformation of human T lymphocytes by herpesvirus saimiri. *Proc. Natl. Acad. Sci. USA* **89**:3116-3119

Biesinger, B., Trimble, J.J., Desrosiers, R.C., and Fleckenstein, B., 1990, The divergence between two oncogenic Herpesvirus saimiri strains in a genomic region related to the transforming phenotype. *Virology* **176**:505-514

Biesinger, B., Tsygankov, A.Y., Fickenscher, H., Emmrich, F., Fleckenstein, B., Bolen, J.B., and Bröker, B.M., 1995, The product of the herpesvirus saimiri open reading frame 1 (tip) interacts with T cell-specific kinase p56lck in transformed cells. *J. Biol. Chem.* **270**:4729-4734

Birkmann, A., Mahr, K., Ensser, A., Yaguboglu, S., Titgemeyer, F., Fleckenstein, B., and Neipel, F., 2001, Cell surface heparan sulfate is a receptor for human herpesvirus-8 and interacts with envelope glycoprotein K8.1. *J. Virol.* **75**:11583-11593

Blake, J.E., Nielsen, N.O., and Heuschele, W.P., 1990, Lymphoproliferation in captive wild ruminants affected with malignant catarrhal fever: 25 cases (1977-1985). *J. Am. Vet. Med. Assoc.* **196**:1141-1143

Blasig, C., Zietz, C., Haar, B., Neipel, F., Esser, S., Brockmeyer, N.H., Tschachler, E., Colombini, S., Ensoli, B., and Stürzl, M., 1997, Monocytes in Kaposi's sarcoma lesions are productively infected by human herpesvirus 8. *J. Virol.* **71**:7963-7968

Blaskovic, D., Stancekova, M., Svobodova, J., and Mistrikova, J., 1980, Isolation of five strains of herpesviruses from two species of free living small rodents. *Acta Virol.* **24**:468

Bodemer, W., Knust, E., Angermüller, S., and Fleckenstein, B., 1984, Immediate-early transcription of herpesvirus saimiri. *J. Virol.* **51**:452-457

Bodemer, W., Niller, H. H., Nitsche, N., Scholz, B., and Fleckenstein, B., 1986, Organization of the thymidylate synthase gene of herpesvirus saimiri. *J. Virol.* **60**:114-123

Borchers, K., Frölich, K., and Ludwig, H., 1999, Detection of equine herpesvirus types 2 and 5 (EHV-2 and EHV-5) in Przewalski's wild horses. *Arch. Virol.* **144**:771-780

Boshoff, C., Endo, Y., Collins, P.D., Takeuchi, Y., Reeves, J.D., Schweickart, V.L., Siani, M.A., Sasaki, T., Williams, T.J., Gray, P.W., Moore, P.S., Chang, Y., and Weiss, R.A., 1997, Angiogenic and HIV-inhibitory functions of KSHV-encoded chemokines. *Science* **278**:290-294

Boshoff, C., Schulz, T.F., Kennedy, M.M., Graham, A.K., Fisher, C., Thomas, A., McGee, J.O., Weiss, R.A., and O'Leary, J.J., 1995, Kaposi's sarcoma-associated herpesvirus infects endothelial and spindle cells. *Nat. Med.* **1**:1274-1278

Bramley, J.C., Davies, A., and Lachmann, P.J., 1997, Herpesvirus saimiri CD59 - baculovirus expression and characterisation of complement inhibitory activity. *Biochem. Soc. Trans.* **25**:354S.

Bridgen, A. and Reid, H.W., 1991, Derivation of a DNA clone corresponding to the viral agent of sheep-associated malignant catarrhal fever. *Res. Vet. Sci.* **50**:38-44

Brielmeier, M., Mautner, J., Laux, G., and Hammerschmidt, W., 1996, The latent membrane protein 2 gene of Epstein-Barr virus is important for efficient B cell immortalization. *J. Gen. Virol.* **77**:2807-2818

Bröker, B.M. and Fickenscher, H., 1999, Herpesvirus saimiri strategies for T cell stimulation and transformation. *Med. Microbiol. Immunol.* **187**:127-136

Bröker, B.M., Kraft, M.S., Klauenberg, U., Le Deist, F., de Villartay, J.P., Fleckenstein, B., Fleischer, B., and Meinl, E., 1997, Activation induces apoptosis in herpesvirus saimiri-transformed T cells independent of CD95 (Fas, APO-1). *Eur. J. Immunol.* **27**:2774-2780

Bröker, B.M., Tsygankov, A.Y., Fickenscher, H., Chitaev, N.A., Müller-Fleckenstein, I., Fleckenstein, B., Bolen, J.B., Emmrich, F. and Schulze-Koops, H., 1994, Engagement of the CD4 receptor inhibits the interleukin-2-dependent proliferation of human T cells transformed by herpesvirus saimiri. *Eur. J. Immunol.* **24**:843-850

Bröker, B.M., Tsygankov, A.Y., Müller-Fleckenstein, I., Guse, A.H., Chitaev, N.A., Biesinger, B., Fleckenstein, B., and Emmrich, F., 1993, Immortalization of human T cell clones by herpesvirus saimiri. Signal transduction analysis reveals functional CD3, CD4, and IL-2 receptors. *J. Immunol.* **151**:1184-1192

Brooks, J.W., Hamilton-Easton, A.M., Christensen, J.P., Cardin, R.D., Hardy, C.L., and Doherty, P.C., 1999, Requirement for CD40 ligand, CD4(+) T cells, and B cells in an infectious mononucleosis-like syndrome. *J. Virol.* **73**:9650-9654

Browning, G.F. and Studdert, M.J., 1989, Physical mapping of the genomic heterogeneity of isolates of equine herpesvirus 2 (equine cytomegalovirus). *Arch. Virol.* **104**:87-94

Bublot, M., Dubuisson, J., Van-Bressem, M.F., Danyi, S., Pastoret, P.P., and Thiry, E., 1991, Antigenic and genomic identity between simian herpesvirus aotus type 2 and bovine herpesvirus type 4. *J. Gen. Virol.* **72**:715-719

Bublot, M., Lomonte, P., Lequarre, A.S., Albrecht, J.C., Nicholas, J., Fleckenstein, B., Pastoret, P.P., and Thiry, E., 1992, Genetic relationships between bovine herpesvirus 4 and the gammaherpesviruses Epstein-Barr virus and herpesvirus saimiri. *Virology* **190**:654-665

Burger, R., Neipel, F., Fleckenstein, B., Savino, R., Ciliberto, G., Kalden, J.R., and Gramatzki, M., 1998, Human herpesvirus type 8 interleukin-6 homologue is functionally active on human myeloma cells. *Blood* **91**:1858-1863

Burysek, L., Yeow, W.S., Lubyowa, B., Kellum, M., Schafer, S.L., Huang, Y.Q., and Pitha, P.M., 1999, Functional analysis of human herpesvirus 8-encoded viral interferon regulatory factor 1 and its association with cellular interferon regulatory factors and p300. *J. Virol.* **73**:7334-7342

Buxton, D. and Reid, H.W., 1980, Transmission of malignant catarrhal fever to rabbits. *Vet. Rec.* **106**:243-245

Camarda, G., Spinetti, G., Bernardini, G., Mair, C., Davis, P.N., Capogrossi, M.C., and Napolitano, M., 1999, The equine herpesvirus 2 E1 open reading frame encodes a functional chemokine receptor. *J. Virol.* **73**:9843-9848

Card, G.L., Knowles, P., Laman, H., Jones, N., and McDonald, N.Q., 2000, Crystal structure of a gamma-herpesvirus cyclin-cdk complex. *EMBO J.* **19**:2877-2888

Cardin, R.D., Brooks, J.W., Sarawar, S.R., and Doherty, P.C., 1996, Progressive loss of CD8+T cell-mediated control of a gamma-herpesvirus in the absence of CD4+ T cells. *J.Exp.Med.* **184**:863-871

Castro, A.E., Daley, G.G., Zimmer, M.A., Whitenack, D.L., and Jensen, J., 1982, Malignant catarrhal fever in an Indian gaur and greater kudu: experimental transmission, isolation, and identification of a herpesvirus. *Am. J. Vet. Res.* **43**:5-11

Cesarman, E., Chang, Y., Moore, P.S., Said, J.W., and Knowles, D.M., 1995, Kaposi's sarcoma-associated herpesvirus-like DNA sequences in AIDS-related body-cavity-based lymphomas. *N. Engl. J. Med.* **332**:1186-1191

Chang, Y., Cesarman, E., Pessin, M.S., Lee, F., Culpepper, J., Knowles, D.M., and Moore, P.S., 1994, Identification of herpesvirus-like DNA sequences in AIDS-associated Kaposi's sarcoma. *Science* **266**:1865-1869

Chen, B.P., Wolfgang, C.D., and Hai, T., 1996, Analysis of ATF3, a transcription factor induced by physiological stresses and modulated by gadd153/Chop10. *Mol. Cell Biol.* **16**:1157-1168

Choi, J.K., Ishido, S., and Jung, J.U., 2000, The collagen repeat sequence is a determinant of the degree of herpesvirus saimiri STP transforming activity. *J. Virol.* **74**:8102-8110

Chou, C.S., Medveczky, M.M., Geck, P., Vercelli, D., and Medveczky, P.G., 1995, Expression of IL-2 and IL-4 in T lymphocytes transformed by herpesvirus saimiri. *Virology* **208**:418- 426

Chow, D., He, X., Snow, A.L., Rose-John, S., and Garcia, K.C., 2001, Structure of an extracellular gp130 cytokine receptor signaling complex. *Science* **291**:2150-2155

Ciufo, D.M., Cannon, J.S., Poole, L.J., Wu, F.Y., Murray, P., Ambinder, R.F., and Hayward, G.S., 2001, Spindle cell conversion by Kaposi's sarcoma-associated herpesvirus: formation of colonies and plaques with mixed lytic and latent gene expression in infected primary dermal microvascular endothelial cell cultures. *J. Virol.* **75**:5614-5626

Clambey, E.T., Virgin, H.W., and Speck, S.H., 2000, Disruption of the murine gammaherpesvirus 68 M1 open reading frame leads to enhanced reactivation from latency. *J. Virol.* **74**:1973-1984

Cooper, M., Goodwin, D.J., Hall, K.T., Stevenson, A.J., Meredith, D.M., Markham, A.F., and Whitehouse, A., 1999, The gene product encoded by ORF 57 of herpesvirus saimiri regulates the redistribution of the splicing factor SC-35. *J. Gen. Virol.* **80:**1311-1316

Copeland, K.F., McKay, P.J., and Rosenthal, K.L., 1995, Suppression of activation of the human immunodeficiency virus long terminal repeat by CD8+ T cells is not lentivirus specific. *AIDS Res. Hum. Retroviruses* **11:**1321-1326

Copeland, K.F., McKay, P.J., and Rosenthal, K.L., 1996, Suppression of the human immunodeficiency virus long terminal repeat by CD8+ T cells is dependent on the NFAT-1 element. *AIDS Res. Hum. Retroviruses* **12:**143-148

Damania, B., De Maria, M., Jung, J.U., and Desrosiers, R.C., 2000, Activation of lymphocyte signaling by the R1 protein of rhesus monkey rhadinovirus. *J. Virol.* **74:**2721-2730

Damania, B., Li, M., Choi, J.K., Alexander, L., Jung, J.U., and Desrosiers, R.C., 1999, Identification of the R1 oncogene and its protein product from the rhadinovirus of rhesus monkeys. *J. Virol.* **73:**5123-5131

Daniel, M.D., Silva, D., Jackman, D., Sehgal, P., Baggs, R.B., Hunt, R.D., King, N.W., and Melendez, L.V., 1976a, Reactivation of squirrel monkey heart isolate (herpesvirus saimiri strain) from latently infected human cell cultures and induction of malignant lymphoma in marmoset monkeys. *Bibl. Haematol.* **43:**392-395

Daniel, M.D., Silva, D., and Ma, N., 1976b, Establishment of owl monkey kidney 210 cell line for virological studies. *In Vitro* **12:**290

Davis, M.A., Stürzl, M., Blasig, C., Schreier, A., Guo, H.G., Reitz, M., Opalenik, S.R., and Browning, P.J., 1997, Expression of human herpesvirus 8-encoded cyclin D in Kaposi's sarcoma spindle cells. *J. Natl. Cancer Inst.* **89:**1868-1874

De Carli, M., Berthold, S., Fickenscher, H., Müller-Fleckenstein, I., D'Elios, M.M., Gao, Q., Biagiotti, R., Giudizi, M.G., Kalden, J.R., and Fleckenstein, B., 1993, Immortalization with herpesvirus saimiri modulates the cytokine secretion profile of established Th1 and Th2 human T cell clones. *J. Immunol.* **151:**5022-5030

Delabesse, E., Oksenhendler, E., Lebbe, C., Verola, O., Varet, B., and Turhan, A.G., 1997, Molecular analysis of clonality in Kaposi's sarcoma. *J. Clin. Pathol.* **50:**664-668

Del Prete, G., De Carli, M., D'Elios, M.M., Müller-Fleckenstein, I., Fickenscher, H., Fleckenstein, B., Almerigogna, F., and Romagnani, S., 1994, Polyclonal B cell activation induced by herpesvirus saimiri-transformed human CD4+ T cell clones. Role for membrane TNF-alpha/TNF-alpha receptors and CD2/CD58 interactions. *J. Immunol.* **153:**4872-4879

Derfuss, T., Fickenscher, H., Kraft, M.S., Henning, G., Lengenfelder, D., Fleckenstein, B., and Meinl, E., 1998, Antiapoptotic activity of the herpesvirus saimiri-encoded Bcl-2 homolog: stabilization of mitochondria and inhibition of caspase-3-like activity. *J. Virol.* **72:**5897-5904.

Desrosiers, R.C. and Falk, L.A., 1982, Herpesvirus saimiri strain variability. *J. Virol.* **43:**352-356

Desrosiers, R.C., Bakker, A., Kamine, J., Falk, L.A., Hunt, R.D., and King, N.W., 1985a, A region of the herpesvirus saimiri genome required for oncogenicity. *Science* **228:**184-187

Desrosiers, R.C., Burghoff, R.L., Bakker, A., and Kamine, J., 1984, Construction of replication-competent herpesvirus saimiri deletion mutants. *J. Virol.* **49:**343-348

Desrosiers, R.C., Kamine, J., Bakker, A., Silva, D., Woychik, R.P., Sakai, D.D., and Rottman, F.M., 1985b, Synthesis of bovine growth hormone in primates by using a herpesvirus vector. *Mol. Cell Biol.* **5:**2796-2803

Desrosiers, R.C., Mulder, C., and Fleckenstein, B., 1979, Methylation of herpesvirus saimiri DNA in lymphoid tumor cell lines. *Proc. Natl. Acad. Sci. USA* **76:**3839-3843

Desrosiers, R.C., Sasseville, V.G., Czajak, S.C., Zhang, X., Mansfield, K.G., Kaur, A., Johnson, R.P., Lackner, A.A., and Jung, J.U., 1997, A herpesvirus of rhesus monkeys related to the human Kaposi's sarcoma-associated herpesvirus. *J. Virol.* **71**:9764-9769

Desrosiers, R.C., Silva, D.P., Waldron, L.M., and Letvin, N.L., 1986, Nononcogenic deletion mutants of herpesvirus saimiri are defective for in vitro immortalization. *J. Virol.* **57**:701-705

Dittmer, D., Lagunoff, M., Renne, R., Staskus, K., Haase, A., and Ganem, D., 1998, A cluster of latently expressed genes in Kaposi's sarcoma-associated herpesvirus. *J. Virol.* **72**:8309-8315

Doherty, P.C., Tripp, R.A., Hamilton, E.A., Cardin, R.D., Woodland, D.L., and Blackman, M.A., 1997, Tuning into immunological dissonance: an experimental model for infectious mononucleosis. *Curr. Opin. Immunol.* **9**:477-483

Donofrio, G., Cavirani, S., and van Santen, V.L., 2000, Establishment of a cell line persistently infected with bovine herpesvirus-4 by use of a recombinant virus. *J. Gen. Virol.* **81**:1807-1814

Drummer, H.E., Reubel, G.H., and Studdert, M.J., 1996, Equine gammaherpesvirus 2 (EHV2) is latent in B lymphocytes. *Arch. Virol.* **141**:495-504

Duboise, S.M., Guo, J., Czajak, S., Desrosiers, R.C., and Jung, J.U., 1998a, STP and Tip are essential for herpesvirus saimiri oncogenicity. *J. Virol.* **72**:1308-1313

Duboise, S.M., Guo, J., Czajak, S., Lee, H., Veazey, R., Desrosiers, R.C., and Jung, J.U., 1998b, A role for herpesvirus saimiri orf14 in transformation and persistent infection. *J. Virol.* **72**:6770-6776

Duboise, S.M., Guo, J., Desrosiers, R.C., and Jung, J.U., 1996, Use of virion DNA as a cloning vector for the construction of mutant and recombinant herpesviruses. *Proc. Natl. Acad. Sci. USA* **93**:11389-11394

Dutia, B.M., Clarke, C.J., Allen, D.J., and Nash, A.A., 1997, Pathological changes in the spleens of gamma interferon receptor-deficient mice infected with murine gammaherpesvirus: a role for CD8 T cells. *J. Virol.* **71**:4278-4283

Dutia, B.M., Stewart, J.P., Clayton, R.A., Dyson, H., and Nash, A.A., 1999, Kinetic and phenotypic changes in murine lymphocytes infected with murine gammaherpesvirus-68 in vitro. *J. Gen. Virol.* **80**:2729-2736

Edington, N., Welch, H.M., and Griffiths, L., 1994, The prevalence of latent Equid herpesviruses in the tissues of 40 abattoir horses. *Equine Vet. J.* **26**:140-142

Egyed, L., 1998, Replication of bovine herpesvirus type 4 in human cells in vitro. *J. Clin. Microbiol.* **36**:2109-2111

Ehlers, B., Borchers, K., Grund, C., Frölich, K., Ludwig, H., and Buhk, H.J., 1999a, Detection of new DNA polymerase genes of known and potentially novel herpesviruses by PCR with degenerate and deoxyinosine-substituted primers. *Virus Genes* **18**:211-220

Ehlers, B., Ulrich, S., and Goltz, M., 1999b, Detection of two novel porcine herpesviruses with high similarity to gammaherpesviruses. *J. Gen. Virol.* **80**:971-978

Ellis, M., Chew, Y.P., Fallis, L., Freddersdorf, S., Boshoff, C., Weiss, R.A., Lu, X., and Mittnacht, S., 1999, Degradation of p27(Kip) cdk inhibitor triggered by Kaposi's sarcoma virus cyclin-cdk6 complex. *EMBO J.* **18**:644-653

Ensoli, B. and Stürzl, M., 1998, Kaposi's sarcoma: a result of the interplay among inflammatory cytokines, angiogenic factors and viral agents. *Cytokine Growth Factor Rev.* **9**:63-83

Ensser, A. and Fleckenstein, B., 1995, Alcelaphine herpesvirus type 1 has a semaphorin-like gene. *J. Gen. Virol.* **76**:1063-1067

Ensser, A., Glykofrydes, D., Niphuis, H., Kuhn, E.M., Rosenwirth, B., Heeney, J.L., Niedobitek, G., Müller-Fleckenstein, I., and Fleckenstein, B., 2001, Independence of herpesvirus induced T-cell lymphoma from viral cyclin D homologue. *J. Exp. Med.* **193**:637-642.

Ensser, A., Pfinder, A., Müller-Fleckenstein, I., and Fleckenstein, B., 1999, The URNA genes of herpesvirus saimiri (strain C488) are dispensable for transformation of human T cells in vitro. *J. Virol.* **73**:10551-10555

Ensser, A., Pflanz, R., and Fleckenstein, B., 1997, Primary structure of the alcelaphine herpesvirus 1 genome. *J. Virol.* **71**:6517-6525

Falk, L., Johnson, D., and Deinhardt, F., 1978, Transformation of marmoset lymphocytes in vitro with herpesvirus ateles. *Int. J. Cancer* **21**:652-657

Falk, L.A., Nigida, S.M., Deinhardt, F., Wolfe, L.G., Cooper, R.W., and Hernandez-Camacho, J.I., 1974, Herpesvirus ateles: properties of an oncogenic herpesvirus isolated from circulating lymphocytes of spider monkeys (Ateles sp.). *Int. J. Cancer* **14**:473-482

Falk, L.A., Wolfe, L.G., and Deinhardt, F., 1972a, Isolation of Herpesvirus saimiri from blood of squirrel monkeys (Saimiri sciureus). *J. Natl. Cancer Inst.* **48**:1499-1505

Falk, L.A., Wolfe, L.G., and Marczynska, F., 1972b, Characterization of lymphoid cell lines established from herpesvirus saimiri (HVS)-infected marmosets. *Bacteriol. Proc.* **38**:191

Feldmann, G., Fickenscher, H., Bodemer, W., Spring, M., Nisslein, T., Hunsmann, G., and Dittmer, U., 1997, Generation of herpes virus saimiri-transformed T-cell lines from macaques is restricted by reactivation of simian spuma viruses. *Virology* **229**:106-112

Fickenscher, H. and Fleckenstein, B., 1998, Growth-transformation of human T cells. *Meth. Microbiol.* **25**:573-603

Fickenscher, H. and Fleckenstein, B., 2001, Herpesvirus saimiri. *Philos. Trans. R. Soc. Lond. B Biol. Sci.* **356**:545-67.

Fickenscher, H., Biesinger, B., Knappe, A., Wittmann, S., and Fleckenstein, B., 1996a, Regulation of the herpesvirus saimiri oncogene stpC, similar to that of T-cell activation genes, in growth-transformed human T lymphocytes. *J. Virol.* **70**:6012-6019

Fickenscher, H., Bökel, C., Knappe, A., Biesinger, B., Meinl, E., Fleischer, B., Fleckenstein, B., and Bröker, B.M., 1997, Functional phenotype of transformed human alphabeta and gammadelta T cells determined by different subgroup C strains of herpesvirus saimiri. *J. Virol.* **71**:2252-2263

Fickenscher, H., Meinl, E., Knappe, A., Wittmann, S., and Fleckenstein, B., 1996b, TcR expression of herpesvirus saimiri immortalized human T cells. *Immunologist* **4**:41-43

Fischer, M., Goldschmitt, J., Peschel, C., Brakenhoff, J.P., Kallen, K.J., Wollmer, A., Grotzinger, J., and Rose-John, S., 1997, A bioactive designer cytokine for human hematopoietic progenitor cell expansion. *Nat. Biotechnol.* **15**:142-145

Fleckenstein, B. and Desrosiers, R.C., 1982, Herpesvirus saimiri and herpesvirus ateles. In The herpesviruses, Vol. 1 (B. Roizman, ed.), Plenum Press, New York, pp. 253-332.

Fleckenstein, B., Bornkamm, G.W., Mulder, C., Werner, F.J., Daniel, M.D., Falk, L.A., and Delius, H., 1978a, Herpesvirus ateles DNA and its homology with herpesvirus saimiri nucleic acid. *J. Virol.* **25**:361-373

Fleckenstein, B., Daniel, M.D., Hunt, R.D., Werner, J., Falk, L.A., and Mulder, C., 1978b, Tumour induction with DNA of oncogenic primate herpesviruses. *Nature* **274**:57-59

Fleckenstein, B., Müller, I., and Werner, J., 1977, The presence of herpesvirus saimiri genomes in virus-transformed cells. *Int. J. Cancer* **19**:546-554

Fodor, W.L., Rollins, S.A., Bianco-Caron, S., Rother, R.P., Guilmette, E.R., Burton, W.V., Albrecht, J.C., Fleckenstein, B., and Squinto, S.P., 1995, The complement control protein homolog of herpesvirus saimiri regulates serum complement by inhibiting C3 convertase activity. *J. Virol.* **69**:3889-3892

Fossiez, F., Djossou, O., Chomarat, P., Flores-Romo, L., Ait-Yahia, S., Maat, C., Pin, J.J., Garrone, P., Garcia, E., Saeland, S., Blanchard, D., Gaillard, C., Das, M.B., Rouvier, E.,

Golstein, P., Banchereau, J., and Lebecque, S., 1996, T cell interleukin-17 induces stromal cells to produce proinflammatory and hematopoietic cytokines. *J. Exp. Med.* **183**:2593-2603

Fossiez, F., Banchereau, J., Murray, R., van Kooten, C., Garrone, P., and Lebecque, S., 1998, Interleukin-17. *Int. Rev. Immunol.* **16**:541-551

Friborg, J., Kong, W., Hottiger, M.O., and Nabel, G.J., 1999, p53 inhibition by the LANA protein of KSHV protects against cell death. *Nature* **402**:889-894

Frölich, K., Li, H., and Müller-Doblies, U., 1998, Serosurvey for antibodies to malignant catarrhal fever-associated viruses in free-living and captive cervids in Germany. *J. Wildl. Dis.* **34**:777-782.

Fruehling, S. and Longnecker, R., 1997, The immunoreceptor tyrosine-based activation motif of Epstein-Barr virus LMP2A is essential for blocking BCR-mediated signal transduction. *Virology* **235**:241-251

Friedman-Kien, A.E., Laubenstein, L.J., Rubinstein, P., Buimovici-Klein, E., Marmor, M., Stahl, R., Spigland, I., Kim, K.S., Zolla-Pazner, S., 1982, Disseminated Kaposi's sarcoma in homosexual men. *Ann. Intern. Med.* **96**:693-700

Fuchs, P.G., Rüger, R., Pfister, H., and Fleckenstein, B., 1985, Genome organization of herpesvirus aotus type 2. *J. Virol.* **53**:13-18

Fujita, T., Sakakibara, J., Sudo, Y., Miyamoto, M., Kimura, Y., and Taniguchi, T., 1988, Evidence for a nuclear factor(s), IRF-1, mediating induction and silencing properties to human IFN-beta gene regulatory elements. *EMBO J.* **7**:3397-3405

Gallego, M.D., Santamaria, M., Pena, J., and Molina, I.J., 1997, Defective actin reorganization and polymerization of Wiskott-Aldrich T cells in response to CD3-mediated stimulation. *Blood* **90**:3089-3097

Ganem, D., 1997, KSHV and Kaposi's sarcoma: the end of the beginning? *Cell* **91**:157-160

Gao, S.J., Boshoff, C., Jayachandra, S., Weiss, R.A., Chang, Y., Moore, P.S., 1997, KSHV ORF K9 (vIRF) is an oncogene which inhibits the interferon signaling pathway. *Oncogene* **15**:1979-1985

Gao, S.J., Kingsley, L., Hoover, D.R., Spira, T.J., Rinaldo, C.R., Saah, A., Phair, J., Detels, R.P.-P., Chang, Y., and Moore, P.S., 1996a, Seroconversion to antibodies against Kaposi's sarcoma-associated herpesvirus-related latent nuclear antigens before the development of Kaposi's sarcoma. *N. Engl. J. Med.* **335**:233-241

Gao, S.J., Kingsley, L., Li, M., Zheng, W., Parravicini, C., Ziegler, J., Newton, R., Rinaldo, C.R., Saah, A., Phair, J., Detels, R., Chang, Y., and Moore, P.S., 1996b, KSHV antibodies among Americans, Italians and Ugandans with and without Kaposi's sarcoma. *Nat. Med.* **2**:925-928

Garber, A.C., Shu, M.A., Hu, J., and Renne, R., 2001, DNA binding and modulation of gene expression by the latency-associated nuclear antigen of Kaposi's sarcoma-associated herpesvirus. *J. Virol.* **75**:7882-7892

Giddens, W.E., Jr., Tsai, C.C., Morton, W.R., Ochs, H.D., Knitter, G.H., and Blakley, G.A., 1985, Retroperitoneal fibromatosis and acquired immunodeficiency syndrome in macaques. Pathologic observations and transmission studies. *Am.J. Pathol.* **119**:253-263

Gill, P., Tsai, Y., Rao, A.P., and Jones, P., 1997, Clonality in Kaposi's sarcoma. *N. Engl. J. Med.* **337**:570-571

Gill, P.S., Tsai, Y.C., Rao, A.P., Spruck, C.H., Zheng, T., Harrington, W.A.J., Cheung, T., Nathwani, B., and Jones, P.A., 1998, Evidence for multiclonality in multicentric Kaposi's sarcoma. *Proc. Natl. Acad. Sci. USA* **95**:8257-8261

Glykofrydes, D., Niphuis, H., Kuhn, E.M., Rosenwirth, B., Heeney, J.L., Bruder, J., Niedobitek, G., Müller-Fleckenstein, I., Fleckenstein, B., and Ensser, A., 2000, The vFLIP of herpesvirus saimiri provides an antiapoptotic function but is not essential for viral replication, transformation and pathogenicity. *J. Virol.* **74:**11919-11927

Godden-Kent, D., Talbot, S.J., Boshoff, C., Chang, Y., Moore, P., Weiss, R.A., and Mittnacht, S., 1997, The cyclin encoded by Kaposi's sarcoma-associated herpesvirus stimulates cdk6 to phosphorylate the retinoblastoma protein and histone H1. *J. Virol.* **71:**4193-4198

Goltz, M. and Ludwig, H., 1991, Biological and molecular aspects of bovine herpesvirus 4 (BHV-4). *Comp. Immunol. Microbiol. Infect. Dis.* **14:**187-195

Götze, R., 1930, Untersuchungen über das bösartige Katarrhalfieber des Rindes. *Dtsch. Tierärztl. Wochenschr.* **38:**487-491

Götze, R. and Liess, J., 1929, Erfolgreiche Übertragungsversuche des bösartigen Katarrhalfiebers von Rind von Rind. Identität mit der südafrikanischen Snotsiekte. *Dtsch. Tierärztl. Wochenschr.* **37:**433

Götze, R. and Liess, J., 1930, Untersuchungen über das bösartige Katarrhalfieber des Rindes. Schafe als Überträger. *Dtsch. Tierärztl. Wochenschr.* **38:**194-200

Gompels, U.A., Craxton, M.A., and Honess, R.W., 1988, Conservation of gene organization in the lymphotropic herpesviruses herpesvirus saimiri and Epstein-Barr virus. *J. Virol.* **62:**757-767

Goodwin, D.J., and Whitehouse, A., 2001, A gamma-2 herpesvirus nucleocytoplasmic shuttle protein interacts with importin alpha-1 and alpha-5. *J. Biol. Chem.* **276:**19905-19912

Goodwin, D.J., Hall, K.T., Giles, M.S., Calderwood, M.A., Markham, A.F., and Whitehouse, A., 2000, The carboxy terminus of the herpesvirus saimiri ORF 57 gene contains domains that are required for transactivation and transrepression. *J. Gen. Virol.* **81:**2253-2265

Goodwin, D.J., Hall, K.T., Stevenson, A.J., Markham, A.F., and Whitehouse, A., 1999, The open reading frame 57 gene product of herpesvirus saimiri shuttles between the nucleus and cytoplasm and is involved in viral RNA nuclear export. *J. Virol.* **73:**10519-10524

Grassmann, R. and Fleckenstein, B., 1989, Selectable recombinant herpesvirus saimiri is capable of persisting in a human T-cell line. *J. Virol.* **63:**1818-1821

Grassmann, R., Berchtold, S., Radant, I., Alt, M., Fleckenstein, B., Sodroski, J.G., Haseltine, W.A., and Ramstedt, U., 1992, Role of human T-cell leukemia virus type 1 X region proteins in immortalization of primary human lymphocytes in culture. *J. Virol.* **66:**4570-4575

Grassmann, R., Dengler, C., Müller-Fleckenstein, I., Fleckenstein, B., McGuire, K., Dokhelar, M.C., Sodroski, J.G., and Haseltine, W.A., 1989, Transformation to continuous growth of primary human T lymphocytes by human T-cell leukemia virus type I X-region genes transduced by a herpesvirus saimiri vector. *Proc. Natl. Acad. Sci. USA* **86:**3351-3355

Greensill, J., Sheldon, J.A., Murthy, K.K., Bessonette, J.S., Beer, B.E., and Schulz, T.F., 2000, A chimpanzee rhadinovirus sequence related to Kaposi's sarcoma-associated herpesvirus/human herpesvirus 8: increased detection after HIV-1 infection in the absence of disease. *AIDS* **14:**F129-F135

Greensill, J., Sheldon, J.A., Renwick, N.M., Beer, B.E., Norley, S., Goudsmit, J., and Schulz, T.F., 2000, Two distinct gamma-2 herpesviruses in African green monkeys: a second gamma-2 herpesvirus lineage among old world primates? *J. Virol.* **74:**1572-1577

Grüter, P., Tabernero, C., von Kobbe, C., Schmitt, C., Saavedra, C., Bachi, A., Wilm, M., Felber, B.K., and Izaurralde, E., 1998, TAP, the human homolog of Mex67p, mediates CTE-dependent RNA export from the nucleus. *Mol. Cell* **1:**649-659

Guo, J., Duboise, M., Lee, H., Li, M., Choi, J.K., Rosenzweig, M., and Jung, J.U., 1997, Enhanced downregulation of Lck-mediated signal transduction by a Y114 mutation of herpesvirus saimiri tip. *J. Virol.* **71:**7092-7096

Guo, J., Williams, K., Duboise, S.M., Alexander, L., Veazey, R., and Jung, J.U., 1998, Substitution of ras for the herpesvirus saimiri STP oncogene in lymphocyte transformation. *J. Virol.* **72**:3698-3704

Hall, K.T., Giles, M.S., Goodwin, D.J., Calderwood, M.A., Carr, I.M., Stevenson, A.J., Markham, A.F., and Whitehouse, A., 2000a, Analysis of gene expression in a human cell line stably transduced with herpesvirus saimiri. *J. Virol.* **74**:7331-7337

Hall, K.T., Giles, M.S., Goodwin, D.J., Calderwood, M.A., Markham, A.F., and Whitehouse, A., 2000b, Characterization of the herpesvirus saimiri ORF73 gene product. *J. Gen. Virol.* **81**:2653-2658

Hall, K.T., Stevenson, A.J., Goodwin, D.J., Gibson, P.C., Markham, A.F., and Whitehouse, A., 1999, The activation domain of herpesvirus saimiri R protein interacts with the TATA-binding protein. *J. Virol.* **73**:9756-9763

Handley, J.A., Sargan, D.R., Herring, A.J., and Reid, H.W., 1995, Identification of a region of the alcelaphine herpesvirus-1 genome associated with virulence for rabbits. *Vet. Microbiol.* **47**:167-181

Hartley, D.A. and Cooper, G.M., 2000, Direct binding and activation of STAT transcription factors by the herpesvirus saimiri protein Tip. *J. Biol. Chem.* **275**:16925-16932

Hartley, D.A., Amdjadi, K., Hurley, T.R., Lund, T.C., Medveczky, P.G., and Sefton, B.M., 2000, Activation of the Lck tyrosine protein kinase by the herpesvirus saimiri tip protein involves two binding interactions. *Virology* **276**:339-348

Hartley, D.A., Hurley, T.R., Hardwick, J.S., Lund, T.C., Medveczky, P.G., and Sefton, B.M., 1999, Activation of the Lck tyrosine-protein kinase by the binding of the tip protein of herpesvirus saimiri in the absence of regulatory tyrosine phosphorylation. *J. Biol. Chem.* **274**:20056-20059

Held, W., Acha-Orbea, H., MacDonald, H.R., and Waanders, G.A., 1994, Superantigens and retroviral infection: insights from mouse mammary tumor virus. *Immunol. Today* **15**:184-190

Held, W., Waanders, G.A., Shakhov, A.N., Scarpellino, L., Acha-Orbea, H., and MacDonald, H.R., 1993, Superantigen-induced immune stimulation amplifies mouse mammary tumor virus infection and allows virus transmission. *Cell* **74**:529-540

Hess, S., Kurrle, R., Lauffer, L., Riethmüller, G., and Engelmann, H., 1995, A cytotoxic CD40/p55 tumor necrosis factor receptor hybrid detects CD40 ligand on herpesvirus saimiri-transformed T cells. *Eur. J. Immunol.* **25**:80-86

Hiller, C., Wittmann, S., Slavin, S., and Fickenscher, H., 2000, Functional long-term thymidine kinase suicide gene expression in human T cells using a herpesvirus saimiri vector. *Gene Ther.* **7**:664-674

Hör, S., Ensser, A., Reiss, C., Ballmer-Hofer, K., and Biesinger, B., 2001, Herpesvirus saimiri protein StpB associates with cellular Src. *J. Gen. Virol.* **82**:339-344

Hoge, A.T., Hendrickson, S.B., and Burns, W.H., 2000, Murine gammaherpesvirus 68 cyclin D homologue is required for efficient reactivation from latency. *J. Virol.* **74**:7016-7023

Honess, R.W., Bodemer, W., Cameron, K.R., Niller, H.H., Fleckenstein, B., and Randall, R.E., 1986, The A+T-rich genome of herpesvirus saimiri contains a highly conserved gene for thymidylate synthase. *Proc. Natl. Acad. Sci. USA* **83**:3604-3608

Hoyle, J., Honess, R.W., and Randall, R.E., 1990, Differential expression of two immediate early genes of herpesvirus saimiri as detected by in situ hybridization. *J. Gen. Virol.* **71**:3005-3008

Hu, S., Vincenz, C., Buller, M., and Dixit, V.M., 1997, A novel family of viral death effector domain-containing molecules that inhibit both CD-95- and tumor necrosis factor receptor-1-induced apoptosis. *J. Biol. Chem.* **272**:9621-9624

Hunt, R.D., Melendez, L.V., Garcia, F.G., and Trum, B.F., 1972, Pathologic features of herpesvirus ateles lymphoma in cotton-topped marmosets (Saguinus oedipus). *J. Natl. Cancer Inst.* **49**:1631-1639

Huppes, W., Fickenscher, H., 't Hart, B.A., and Fleckenstein, B., 1994, Cytokine dependence of human to mouse graft-versus-host disease. *Scand. J. Immunol.* **40**:26-36

Husain, S.M., Usherwood, E.J., Dyson, H., Coleclough, C., Coppola, M.A., Woodland, D.L., Blackman, M.A., Stewart, J.P., and Sample, J.T., 1999, Murine gammaherpesvirus M2 gene is latency-associated and its protein a target for CD8(+) T lymphocytes. *Proc. Natl. Acad. Sci. USA* **96**:7508-7513

Hyun, T.S., Subramanian, C., Cotter, M.A., Thomas, R.A., and Robertson, E.S., 2001, Latency-associated nuclear antigen encoded by Kaposi's sarcoma- associated herpesvirus interacts with Tat and activates the long terminal repeat of human immunodeficiency virus type 1 in human cells. *J. Virol.* **75**:8761-8771

Inatsuki, A., Yasukawa, M., and Kobayashi, Y., 1989, Functional alterations of herpes simplex virus-specific CD4+ multifunctional T cell clones following infection with human T lymphotropic virus type I. *J. Immunol.* **143**:1327-1333

Itoh, S., Harada, H., Fujita, T., Mimura, T., and Taniguchi, T., 1989, Sequence of a cDNA coding for human IRF-2. *Nucl. Acids. Res.* **17**:8372

Jacoby, R.O., Reid, H.W., Buxton, D., and Pow, I., 1988, Transmission of wildebeest-associated and sheep-associated malignant catarrhal fever to hamsters, rats and guinea-pigs. *J. Comp. Pathol.* **98**:91-98

Jayachandra, S., Low, K.G., Thlick, A.E., Yu, J., Ling, P.D., Chang, Y., and Moore, P.S., 1999, Three unrelated viral transforming proteins (vIRF, EBNA2, and E1A) induce the MYC oncogene through the interferon-responsive PRF element by using different transcription coadaptors. *Proc. Natl. Acad. Sci. USA* **96**:11566-11571

Jenner, R.G., Alba, M.M., Boshoff, C., and Kellam, P., 2001, Kaposi's sarcoma-associated herpesvirus latent and lytic gene expression as revealed by DNA arrays. *J. Virol.* **75**:891-902

Johnson, D.R. and Jondal, M., 1981a, Herpesvirus ateles and herpesvirus saimiri transform marmoset T cells into continuously proliferating cell lines that can mediate natural killer cell-like cytotoxicity. *Proc. Natl. Acad. Sci. USA* **78**:6391-6395

Johnson, D.R. and Jondal, M., 1981b, Herpesvirus-transformed cytotoxic T-cell lines. *Nature* **291**:81-83

Jung, J.U. and Desrosiers, R.C., 1991, Identification and characterization of the herpesvirus saimiri oncoprotein STP-C488. *J. Virol.* **65**:6953-6960

Jung, J.U. and Desrosiers, R.C., 1992, Herpesvirus saimiri oncogene STP-C488 encodes a phosphoprotein. *J. Virol.* **66**:1777-1780

Jung, J.U., and Desrosiers, R.C., 1994, Distinct functional domains of STP-C488 of herpesvirus saimiri. *Virology* **204**:751-758

Jung, J.U. and Desrosiers, R.C., 1995, Association of the viral oncoprotein STP-C488 with cellular ras. *Mol. Cell Biol.* **15**:6506-6512

Jung, J.U., Choi, J.K., Ensser, A., and Biesinger, B., 1999, Herpesvirus saimiri as a model for gammaherpesvirus oncogenesis. *Semin. Cancer Biol.* **9**:231-239

Jung, J.U., Lang, S.M., Friedrich, U., Jun, T., Roberts, T.M., Desrosiers, R.C., and Biesinger, B., 1995a, Identification of Lck-binding elements in tip of herpesvirus saimiri. *J. Biol. Chem.* **270**:20660-20667

Jung, J.U., Lang, S.M., Jun, T., Roberts, T.M., Veillette, A., and Desrosiers, R.C., 1995b, Downregulation of Lck-mediated signal transduction by tip of herpesvirus saimiri. *J. Virol.* **69**:7814-7822

Jung, J.U., Stäger, M., and Desrosiers, R.C., 1994, Virus-encoded cyclin. *Mol. Cell Biol.*
14:7235-7244

Jung, J.U., Trimble, J.J., King, N.W., Biesinger, B., Fleckenstein, B.W., and Desrosiers, R.C.,
1991, Identification of transforming genes of subgroup A and C strains of herpesvirus
saimiri. *Proc. Natl. Acad. Sci. USA* 88:7051-7055

Jussila, L., Valtola, R., Partanen, T.A., Salven, P., Heikkila, P., Matikainen, M.T., Renkonen,
R., Kaipainen, A., Detmar, M., Tschachler, E., Alitalo, R., and Alitalo, K., 1998,
Lymphatic endothelium and Kaposi's sarcoma spindle cells detected by antibodies against
the vascular endothelial growth factor receptor-3. *Cancer Res.* 58:1599-1604

Kapadia, S.B., Molina, H., van Berkel, V., Speck, S.H., and Virgin, H.W., 1999, Murine
gammaherpesvirus 68 encodes a functional regulator of complement activation. *J. Virol.*
73:7658-7670

Kaschka-Dierich, C., Werner, F.J., Bauer, I., and Fleckenstein, B., 1982, Structure of
nonintegrated, circular herpesvirus saimiri and herpesvirus ateles genomes in tumor cell
lines and in vitro-transformed cells. *J. Virol.* 44:295-310

Kedes, D.H., Ganem, D., Ameli, N., Bacchetti, P., and Greenblatt, R., 1997, The prevalence
of serum antibody to human herpesvirus 8 (Kaposi sarcoma-associated herpesvirus)
among HIV-seropositive and high-risk HIV-seronegative women. *JAMA* 277:478-481

Kellam, P., Boshoff, C., Whitby, D., Matthews, S., Weiss, R.A., and Talbot, S.J., 1997,
Identification of a major latent nuclear antigen, LNA-1, in the human herpesvirus 8
genome. *J. Hum. Virol.* 1:19-29

Kempf, W., Adams, V., Hassam, S., Schmid, M., Moos, R., Briner, J., and Pfaltz, M., 1994,
[Detection of human herpesvirus type 6, human herpesvirus type 7, cytomegalovirus and
human papillomavirus in cutaneous AIDS- associated Kaposi's sarcoma]. *Verh. Dtsch.
Ges. Pathol.* 78:260-264

Kempf, W., Adams, V., Pfaltz, M., Briner, J., Schmid, M., Moos, R., and Hassam, S., 1995,
Human herpesvirus type 6 and cytomegalovirus in AIDS-associated Kaposi's sarcoma: no
evidence for an etiological association. *Hum. Pathol.* 26:914-919

Kempf, W., Adams, V., Wey, N., Moos, R., Schmid, M., Avitabile, E., and Campadelli-
Fiume, G., 1997, CD68+ cells of monocyte/macrophage lineage in the environment of
AIDS-associated and classic-sporadic Kaposi sarcoma are singly or doubly infected with
human herpesviruses 7 and 8. *Proc. Natl. Acad. Sci. USA* 94:7600-7605

Kennedy, J., Rossi, D.L., Zurawski, S.M., Vega, F. Jr., Kastelein, R.A., Wagner, J.L.,
Hannum, C.H., and Zlotnik, A., 1996, Mouse IL-17: a cytokine preferentially expressed by
alpha beta TCR + CD4-CD8-T cells. *J. Interferon Cytokine Res.* 16:611-617

Kennedy, M.M., Cooper, K., Howells, D.D., Picton, S., Biddolph, S., Lucas, S.B., McGee,
J.O., and O'Leary, J.J., 1998, Identification of HHV8 in early Kaposi's sarcoma:
implications for Kaposi's sarcoma pathogenesis. *Mol. Pathol.* 51:14-20

Kiyotaki, M., Desrosiers, R.C., and Letvin, N.L., 1986, Herpesvirus saimiri strain 11
immortalizes a restricted marmoset T8 lymphocyte subpopulation in vitro. *J. Exp. Med.*
164:926-931

Kiyotaki, M., Solomon, K.R., and Letvin, N.L., 1988, Herpesvirus ateles immortalizes in vitro
a CD3+CD4+CD8+ marmoset lymphocyte with NK function. *J. Immunol.* 140:730-736

Klein, J.L., Fickenscher, H., Holliday, J.E., Biesinger, B., and Fleckenstein, B., 1996,
Herpesvirus saimiri immortalized gamma delta T cell line activated by IL-12. *J. Immunol.*
156:2754-2760

Kliche, S., Kremmer, E., Hammerschmidt, W., Koszinowski, U., and Haas, J., 1998,
Persistent infection of epstein-barr virus-positive B lymphocytes by human herpesvirus 8.
J. Virol. 72:8143-8149

Knappe, A., Feldmann, G., Dittmer, U., Meinl, E., Nisslein, T., Wittmann, S., Mätz-Rensing, K., Kirchner, T., Bodemer, W., and Fickenscher, H., 2000a, Herpesvirus saimiri-transformed macaque T cells are tolerated and do not cause lymphoma after autologous reinfusion. *Blood* **95**:3256-3261

Knappe, A., Hiller, C., Niphuis, H., Fossiez, F., Thurau, M., Wittmann, S., Kuhn, E.M., Lebecque, S., Banchereau, J., Rosenwirth, B., Fleckenstein, B., Heeney, J., and Fickenscher, H., 1998a, The interleukin-17 gene of herpesvirus saimiri. *J. Virol.* **72**:5797-5801

Knappe, A., Hiller, C., Thurau, M., Wittmann, S., Hofmann, H., Fleckenstein, B., and Fickenscher, H., 1997, The superantigen-homologous viral immediate-early gene ie14/vsag in herpesvirus saimiri-transformed human T cells. *J. Virol.* **71**:9124-9133

Knappe, A., Hör, S., Wittmann, S., and Fickenscher, H., 2000b, Induction of a novel cellular homolog of interleukin-10, AK155, by transformation of T lymphocytes with herpesvirus saimiri. *J. Virol.* **74**:3881-3887

Knappe, A., Thurau, M., Niphuis, H., Hiller, C., Wittmann, S., Kuhn, E.M., Rosenwirth, B., Fleckenstein, B., Heeney, J., and Fickenscher, H., 1998b, T-cell lymphoma caused by herpesvirus saimiri C488 independently of ie14/vsag, a viral gene with superantigen homology. *J. Virol.* **72**:3469-3471

Koomey, J.M., Mulder, C., Burghoff, R.L., Fleckenstein, B., and Desrosiers, R.C., 1984, Deletion of DNA sequence in a nononcogenic variant of herpesvirus saimiri. *J. Virol.* **50**:662-665

Kraft, M.S., Henning, G., Fickenscher, H., Lengenfelder, D., Tschopp, J., Fleckenstein, B., and Meinl, E., 1998, Herpesvirus saimiri transforms human T-cell clones to stable growth without inducing resistance to apoptosis. *J. Virol.* **72**:3138-3145

Kretschmer, C., Murphy, C., Biesinger, B., Beckers, J., Fickenscher, H., Kirchner, T., Fleckenstein, B., and Rüther, U., 1996, A herpes saimiri oncogene causing peripheral T-cell lymphoma in transgenic mice. *Oncogene* **12**:1609-1616

Kulkarni, A.B., Holmes, K.L., Fredrickson, T.N., Hartley, J.W., and Morse, H.C., 1997, Characteristics of a murine gammaherpesvirus infection immunocompromised mice. *In Vivo* **11**:281-291

Kumanogoh, A., Watanabe, C., Lee, I., Wang, X., Shi, W., Araki, H., Hirata, H., Iwahori, K., Uchida, J., Yasui, T., Matsumoto, M., Yoshida, K., Yakura, H., Pan, C., Parnes, J.R., and Kikutani, H., 2000, Identification of CD72 as a lymphocyte receptor for the class IV semaphorin CD100. A novel mechanism for regulating B cell signaling. *Immunity* **13**:621-631

Kung, S.H. and Medveczky, P.G., 1996, Identification of a herpesvirus saimiri cis-acting DNA fragment that permits stable replication of episomes in transformed T cells. *J. Virol.* **70**:1738-1744

Lacey, S.F., Weinhold, K.J., Chen, C.H., McDanal, C., Oei, C., and Greenberg, M.L., 1998, Herpesvirus saimiri transformation of HIV type 1 suppressive CD8+ lymphocytes from an HIV type 1-infected asymptomatic individual. *AIDS Res. Hum. Retroviruses* **14**:521-531

Lacoste, V., Mauclere, P., Dubreuil, G., Lewis, J., Georges-Courbot, M.C., and Gessain, A., 2000a, KSHV-like herpesviruses in chimps and gorillas. *Nature* **407**:151-152

Lacoste, V., Mauclere, P., Dubreuil, G., Lewis, J., Georges-Courbot, M.C., Rigoulet, J., Petit, T., and Gessain, A., 2000b, Simian homologues of human gamma-2 and betaherpesviruses in mandrill and drill monkeys. *J. Virol.* **74**:11993-11999

Lacoste, V., Mauclere, P., Dubreuil, G., Lewis, J., Georges-Courbot, M. C., and Gessain, A., 2001, A novel gamma 2-herpesvirus of the Rhadinovirus 2 lineage in chimpanzees. *Genome Res.* **11**:1511-1519

Lagunoff, M. and Ganem, D., 1997, The structure and coding organization of the genomic termini of Kaposi's sarcoma-associated herpesvirus. *Virology* **236**:147-154

Lang, G. and Fleckenstein, B., 1990, Trans activation of the thymidylate synthase promoter of herpesvirus saimiri. *J. Virol.* **64**:5333-5341

Lange, C., Liehr, T., Goen, M., Gebhart, E., Fleckenstein, B., and Ensser, A., 1998, New eukaryotic semaphorins with close homology to semaphorins of DNA viruses. *Genomics* **51**:340-350

Lebbe, C., Blum, L., Pellet, C., Blanchard, G., Verola, O., Morel, P., Danne, O., and Calvo, F., 1998, Clinical and biological impact of antiretroviral therapy with protease inhibitors on HIV-related Kaposi's sarcoma. *AIDS* **12**:F45-F49

Lee, B. S., Alvarez, X., Ishido, S., Lackner, A.A., and Jung, J.U., 2000, Inhibition of intracellular transport of B cell antigen receptor complexes by Kaposi's sarcoma-associated herpesvirus K1. *J. Exp. Med.* **192**:11-21

Lee, H., Choi, J.K., Li, M., Kaye, K., Kieff, E., and Jung, J.U., 1999, Role of cellular tumor necrosis factor receptor-associated factors in NF-kappaB activation and lymphocyte transformation by herpesvirus saimiri STP. *J. Virol.* **73**:3913-3919

Lee, H., Guo, J., Li, M., Choi, J.K., DeMaria, M., Rosenzweig, M., and Jung, J.U., 1998a, Identification of an immunoreceptor tyrosine-based activation motif of K1 transforming protein of Kaposi's sarcoma-associated herpesvirus. *Mol. Cell Biol.* **18**:5219-5228

Lee, H., Trimble, J.J., Yoon, D.W., Regier, D., Desrosiers, R.C., and Jung, J.U., 1997, Genetic variation of herpesvirus saimiri subgroup A transforming protein and its association with cellular src. *J. Virol.* **71**:3817-3825

Lee, H., Veazey, R., Williams, K., Li, M., Guo, J., Neipel, F., Fleckenstein, B., Lackner, A., Desrosiers, R.C., and Jung, J.U., 1998b, Deregulation of cell growth by the K1 gene of Kaposi's sarcoma-associated herpesvirus. *Nat. Med.* **4**:435-440

Leith, J.G., Copeland, K.F., McKay, P.J., Bienzle, D., Richards, C.D., and Rosenthal, K.L., 1999, T cell-derived suppressive activity: evidence of autocrine noncytolytic control of HIV type 1 transcription and replication. *AIDS Res. Hum. Retroviruses* **15**:1553-1561

Li, H., Dyer, N., Keller, J., and Crawford, T.B., 2000a, Newly recognized herpesvirus causing malignant catarrhal fever in white- tailed deer (Odocoileus virginianus). *J. Clin. Microbiol.* **38**:1313-1318

Li, H., Keller, J., Knowles, D.P., and Crawford, T.B., 2001, Recognition of another member of the malignant catarrhal fever virus group: an endemic gammaherpesvirus in domestic goats. *J. Gen. Virol.* **82**:227-232

Li, H., Shen, D.T., Knowles, D.P., Gorham, J.R., and Crawford, T.B., 1994, Competitive inhibition enzyme-linked immunosorbent assay for antibody in sheep and other ruminants to a conserved epitope of malignant catarrhal fever virus. *J. Clin. Microbiol.* **32**:1674-1679

Li, H., Shen, D.T., Otoole, D., Knowles, D.P., Gorham, J.R., and Crawford, T.B., 1995, Investigation of sheep-associated malignant catarrhal fever virus-infection in ruminants by PCR and competitive-inhibition enzyme-linked-immunosorbent-assay. *J. Clin. Microbiol.* **33**:2048-2053

Li, H., Snowder, G., and Crawford, T.B., 1999a, Production of malignant catarrhal fever virus-free sheep. *Vet. Microbiol.* **65**:167-172

Li, H., Snowder, G., O'Toole, D., and Crawford, T.B., 1998a, Transmission of ovine herpesvirus 2 in lambs. *J. Clin. Microbiol.* **36**:223-226

Li, H., Snowder, G., O'Toole, D., and Crawford, T.B., 2000b, Transmission of ovine herpesvirus 2 among adult sheep. *Vet. Microbiol.* **71**:27-35

Li, H., Wang, H., and Nicholas, J., 2001, Detection of direct binding of human herpesvirus 8-encoded interleukin- 6 (vIL-6) to both gp130 and IL-6 receptor (IL-6R) and identification of amino acid residues of vIL-6 important for IL-6R-dependent and -independent signaling. *J. Virol.* **75**:3325-3334

Li, H., Westover, W.C., and Crawford, T.B., 1999b, Sheep-associated malignant catarrhal fever in a petting zoo. *J. Zoo. Wildl. Med.* **30**:408-412

Li, M., Lee, H., Guo, J., Neipel, F., Fleckenstein, B., Ozato, K., and Jung, J.U., 1998b, Kaposi's sarcoma-associated herpesvirus viral interferon regulatory factor. *J. Virol.* **72**:5433-5440

Li, M., Lee, H., Yoon, D.-W., Albrecht, J.C., Fleckenstein, B., Neipel, F., and Jung, J.U., 1997, Kaposi's sarcoma-associated herpesvirus encodes a functional cyclin. *J. Virol.* **71**:1984-1991

Lifson, A.R., Darrow, W.W., Hessol, N.A., O'Malley, P.M., Barnhart, L., Jaffe, H.W., and Rutherford, G.W., 1990, Kaposi's sarcoma among homosexual and bisexual men enrolled in the San Francisco City Clinic Cohort Study. *J. Acquir. Immune Defic. Syndr.* **3**:S32-S37

Lim, C., Gwack, Y., Hwang, S., Kim, S., and Choe, J., 2001, The transcriptional activity of cAMP response element-binding protein-binding protein is modulated by the latency associated nuclear antigen of Kaposi's sarcoma-associated herpesvirus. *J. Biol. Chem.* **276**:31016-31022

Lin, R., Genin, P., Mamane, Y., Sgarbanti, M., Battistini, A., Harrington, W.J., Jr., Barber, G.N., and Hiscott, J., 2001, HHV-8 encoded vIRF-1 represses the interferon antiviral response by blocking IRF-3 recruitment of the CBP/p300 coactivators. *Oncogene* **20**:800-811

Lin, R., Mustafa, A., Nguyen, H., Gewert, D., and Hiscott, J., 1994, Mutational analysis of interferon (IFN) regulatory factors 1 and 2. Effects on the induction of IFN-beta gene expression. *J. Biol. Chem.* **269**:17542-17549

Lisitsyn, N.A., Lisitsyn, N.M., and Wigler, M., 1993, Cloning the differences between to complex genomes. *Science* **259**:946-951

Loken, T., Aleksandersen, M., Reid, H., and Pow, I., 1998, Malignant catarrhal fever caused by ovine herpesvirus-2 in pigs in Norway. *Vet. Rec.* **143**: 464-467

Lomonte, P., Bublot, M., van Santen, V., Keil, G., Pastoret, P.P., and Thiry, E., 1996, Bovine herpesvirus 4: genomic organization and relationship with two other gammaherpesviruses, Epstein-Barr virus and herpesvirus saimiri. *Vet. Microbiol.* **53**:79-89

Lukac, D.M., Kirshner, J.R., and Ganem, D., 1999, Transcriptional activation by the product of open reading frame 50 of Kaposi's sarcoma-associated herpesvirus is required for lytic viral reactivation in B cells. *J. Virol.* **73**:9348-9361

Lukac, D.M., Renne, R., Kirshner, J.R., and Ganem, D., 1998, Reactivation of Kaposi's sarcoma-associated herpesvirus infection from latency by expression of the ORF 50 transactivator, a homolog of the EBV R protein. *Virology* **252**:304-312

Lund, T., Garcia, R., Medveczky, M.M., Jove, R., and Medveczky, P.G., 1997a, Activation of STAT transcription factors by herpesvirus saimiri Tip-484 requires p56lck. *J. Virol.* **71**:6677-6682

Lund, T., Medveczky, M.M., Geck, P., and Medveczky, P.G., 1995, A herpesvirus saimiri protein required for interleukin-2 independence is associated with membranes of transformed T cells. *J. Virol.* **69**:4495-4499

Lund, T., Medveczky, M.M., and Medveczky, P.G., 1997b, Herpesvirus saimiri Tip-484 membrane protein markedly increases p56lck activity in T cells. *J. Virol.* **71**:378-382

Lund, T., Medveczky, M.M., Neame, P.J., and Medveczky, P.G., 1996, A herpesvirus saimiri membrane protein required for interleukin-2 independence forms a stable complex with p56lck. *J. Virol.* **70**:600-606

Lund, T., Prator, P.C., Medveczky, M.M., and Medveczky, P.G., 1999, The Lck binding domain of herpesvirus saimiri tip-484 constitutively activates Lck and STAT3 in T cells. *J. Virol.* **73**:1689-1694

Mackewicz, C.E., Orque, R., Jung, J., and Levy, J.A., 1997, Derivation of herpesvirus saimiri-transformed CD8+ T cell lines with noncytotoxic anti-HIV activity. *Clin. Immunol. Immunopathol.* **82**:274-281

Mann, D.J., Child, E.S., Swanton, C., Laman, H., and Jones, N., 1999, Modulation of p27(Kip1) levels by the cyclin encoded by Kaposi's sarcoma-associated herpesvirus. *EMBO J.* **18**:654-663

Marczynska, B., Falk, L., Wolfe, L., and Deinhardt, F., 1973, Transplantation and cytogenetic studies of herpesvirus saimiri-induced disease in Marmoset monkeys. *J. Natl. Cancer Inst.* **50**:331-337

Martin, J.N., Ganem, D.E., Osmond, D.H., Page-Shafer, K.A., Macrae, D., and Kedes, D.H., 1998, Sexual transmission and the natural history of human herpesvirus 8 infection. *N. Engl. J. Med.* **338**:948-954

Masood, R., Lunardi-Iskandar, Y., Jean, L.F., Murphy, J.R., Waters, C., Gallo, R.C., and Gill, P., 1994, Inhibition of AIDS-associated Kaposi's sarcoma cell growth by DAB389-interleukin 6. *AIDS Res. Hum. Retroviruses* **10**:969-975

Medveczky, M.M., Geck, P., Clarke, C., Byrnes, J., Sullivan, J.L., and Medveczky, P.G., 1989a, Arrangement of repetitive sequences in the genome of herpesvirus sylvilagus. *J. Virol.* **63**:1010-1014

Medveczky, M.M., Geck, P., Sullivan, J.L., Serbousek, D., Djeu, J.Y., and Medveczky, P. G., 1993, IL-2 independent growth and cytotoxicity of herpesvirus saimiri- infected human CD8 cells and involvement of two open reading frame sequences of the virus. *Virology* **196**:402-412

Medveczky, M.M., Szomolanyi, E., Hesselton, R., DeGrand, D., Geck, P., and Medveczky, P.G., 1989b, Herpesvirus saimiri strains from three DNA subgroups have different oncogenic potentials in New Zealand white rabbits. *J. Virol.* **63**:3601-3611

Medveczky, P., Szomolanyi, E., Desrosiers, R.C., and Mulder, C., 1984, Classification of herpesvirus saimiri into three groups based on extreme variation in a DNA region required for oncogenicity. *J. Virol.* **52**:938-944

Meinl, E. and Fickenscher, H., 2000, Viral transformation of lymphocytes. In Lymphocytes. A practical approach (S. L. Rowland-Jones and A. J. McMichael, eds.), Oxford University Press, Oxford, pp. 55-74.

Meinl, E., Fickenscher, H., Hoch, R.M., Malefyt, R.D., de Waal Malefyt, R., Hart, B.A., Wekerle, H., Hohlfeld, R., and Fleckenstein, B., 1997, Growth transformation of antigen-specific T cell lines from rhesus monkeys by herpesvirus saimiri. *Virology* **229**:175-182

Meinl, E., Fickenscher, H., Thome, M., Tschopp, J., and Fleckenstein, B., 1998, Anti-apoptotic strategies of lymphotropic viruses. *Immunol. Today* **19**:474-479

Meinl, E., 't Hart, B.A., Bontrop, R.E., Hoch, R.M., Iglesias, A., de Waal, M.R., Fickenscher, H., Müller-Fleckenstein, I., Fleckenstein, B., and Wekerle, H., 1995a, Activation of a myelin basic protein-specific human T cell clone by antigen-presenting cells from rhesus monkeys. *Int. Immunol.* **7**:1489-1495

Meinl, E., Hohlfeld, R., Wekerle, H., and Fleckenstein, B., 1995b, Immortalization of human T cells by Herpesvirus saimiri. *Immunol. Today* **16**:55-58

Melendez, L.V., Daniel, M.D., Hunt, R.D., and Garcia, F.G., 1968, An apparently new herpesvirus from primary kidney cultures of the squirrel monkey (Saimiri sciureus). *Lab. Anim. Care* **18**:374-381

Melendez, L.V., Hunt, R.D., Daniel, M.D., Garcia, F.G., and Fraser, C.E., 1969, Herpesvirus saimiri. II. Experimentally induced malignant lymphoma in primates. *Lab. Anim. Care* **19**:378-386

Melendez, L.V., Hunt, R.D., King, N.W., Barahona, H.H., Daniel, M.D., Fraser, C.E., and Garcia, F.G., 1972, Herpesvirus ateles, a new lymphoma virus of monkeys. *Nat. New Biol.* **235**:182-184

Merlo, J.J. and Tsygankov, A.Y., 2001, Herpesvirus saimiri oncoproteins Tip and StpC synergistically stimulate NF-kappaB activity and interleukin-2 gene expression. *Virology* **279**:325-338

Meuer, S.C., Hussey, R.E., Fabbi, M., Fox, D., Acuto, O., Fitzgerald, K.A., Hodgdon, J.C., Protentis, J.P., Schlossman, S.F., and Reinherz, E.L., 1984, An alternative pathway of T-cell activation: a functional role for the 50 kd T11 sheep erythrocyte receptor protein. *Cell* **36**:897-906

Miles, S.A., Martinez-Maza, O., Rezai, A., Magpantay, L., Kishimoto, T., Nakamura, S., Radka, S.F., and Linsley, P.S., 1992, Oncostatin M as a potent mitogen for AIDS-Kaposi's sarcoma-derived cells. *Science* **255**:1432-1434

Miller, C.L., Lee, J.H., Kieff, E., and Longnecker, R., 1994, An integral membrane protein (LMP2) blocks reactivation of Epstein-Barr virus from latency following surface immunoglobulin crosslinking. *Proc. Natl. Acad. Sci. USA* **91**:772-776

Miller, G., Heston, L., Grogan, E., Gradoville, L., Rigsby, M., Sun, R., Shedd, D., Kushnaryov, V.M., Grossberg, S., and Chang, Y., 1997, Selective switch between latency and lytic replication of Kaposi's sarcoma herpesvirus and Epstein-Barr virus in dually infected body cavity lymphoma cells. *J. Virol.* **71**:314-324

Miller, G., Rigsby, M.O., Heston, L., Grogan, E., Sun, R., Metroka, C., Levy, J.A., Gao, S.J.C.-Y., and Moore, P., 1996, Antibodies to butyrate-inducible antigens of Kaposi's sarcoma-associated herpesvirus inpatients with HIV-1 infection. *N. Engl. J. Med.* **334**:1292-1297

Mine, T., Harada, K., Matsumoto, T., Yamana, H., Shirouzu, K., Itoh, K., and Yamada, A., 2000, CDw108 expression during T-cell development. *Tissue Antigens* **55**:429-436

Mittrücker, H.W., Müller-Fleckenstein, I., Fleckenstein, B., and Fleischer, B., 1992, CD2-mediated autocrine growth of herpesvirus saimiri-transformed human T lymphocytes. *J. Exp. Med.* **176**:909-913

Mittrücker, H.W., Müller-Fleckenstein, I., Fleckenstein, B., and Fleischer, B., 1993, Herpes virus saimiri-transformed human T lymphocytes: normal functional phenotype and preserved T cell receptor signalling. *Int. Immunol.* **5**:985-990

Miyoshi, I., Kubonishi, I., Yoshimoto, S., Akagi, T., Ohtsuki, Y., Shiraishi, Y., Nagata, K., and Hinuma, Y., 1981, Type C virus particles in a cord T-cell line derived by co-cultivating normal human cord leukocytes and human leukaemic T cells. *Nature* **294**:770-771

Molden, J., Chang, Y., You, Y., Moore, P.S., and Goldsmith, M.A., 1997, A Kaposi's sarcoma-associated herpesvirus-encoded cytokine homolog (vIL-6) activates signaling through the shared gp130 receptor subunit. *J. Biol. Chem.* **272**:19625-19631

Moore, P.S., Boshoff, C., Weiss, R.A., and Chang, Y., 1996, Molecular mimicry of human cytokine and cytokine response pathway genes by KSHV. *Science* **274**:1739-1744

Muralidhar, S., Pumfery, A.M., Hassani, M., Sadaie, M.R., Azumi, N., Kishishita, M., Brady, J.N., Doniger, J., Medveczky, P., and Rosenthal, L.J., 1998, Identification of kaposin (open reading frame K12) as a human herpesvirus 8 (Kaposi's sarcoma-associated herpesvirus) transforming gene. *J. Virol.* **72**:4980-498

Murakami-Mori, K., Taga, T., Kishimoto, T., and Nakamura, S., 1995, AIDS-associated Kaposi's sarcoma (KS) cells express oncostatin M (OM)-specific receptor but not leukemia inhibitory factor/OM receptor or interleukin-6 receptor. Complete block of OM-induced KS cell growth and OM binding by anti-gp130 antibodies. *J. Clin. Invest.* **96**:1319-1327

Müller-Doblies, U.U., Li, H., Hauser, B., Adler, H., and Ackermann, M., 1998, Field validation of laboratory tests for clinical diagnosis of sheep-associated malignant catarrhal fever. *J. Clin. Microbiol.* **36**:2970-2972

Murphy, C., Kretschmer, C., Biesinger, B., Beckers, J., Jung, J., Desrosiers, R.C., Müller-Hermelink, H.K., Fleckenstein, B.W., and Rüther, U., 1994, Epithelial tumours induced by a herpesvirus oncogene in transgenic mice. *Oncogene* **9**:221-226

Murphy, P.M., 1994, The molecular biology of leukocyte chemoattractant receptors. *Annu. Rev. Immunol.* **12**:593-633

Murray, M.J., Eichorn, E.S., Dubovi, E.J., Ley, W.B., and Cavey, D.M., 1996, Equine herpesvirus type 2: prevalence and seroepidemiology in foals. *Equine Vet. J.* **28**:432-436

Murthy, S.C., Trimble, J.J., and Desrosiers, R.C., 1989, Deletion mutants of herpesvirus saimiri define an open reading frame necessary for transformation. *J. Virol.* **63**:3307-3314

Nakamura, S., Salahuddin, S.Z., Biberfeld, P., Ensoli, B., Markham, P.D., Wong-Staal, F., and Gallo, R.C., 1988, Kaposi's sarcoma cells: long-term culture with growth factor from retrovirus-infected CD4+ T cells. *Science* **242**:426-430

Nakamura, H., Li, M., Zarycki, J., and Jung, J.U., 2001, Inhibition of p53 tumor suppressor by viral interferon regulatory factor. *J. Virol.* **75**:7572-7582

Nava, V.E., Cheng, E.H., Veliuona, M., Zou, S., Clem, R.J., Mayer, M.L., and Hardwick, J.M., 1997, Herpesvirus saimiri encodes a functional homolog of the human bcl-2 oncogene. *J. Virol.* **71**:4118-4122

Neipel, F., Albrecht, J.C., Ensser, A., Huang, Y.Q., Li, J.J., Friedman-Kien, A.E., and Fleckenstein, B., 1997a, Human herpesvirus 8 encodes a homolog of interleukin-6. *J. Virol.* **71**:839-842

Neipel, F., Albrecht, J.C., and Fleckenstein, B., 1997b, Cell-homologous genes in the Kaposi's sarcoma-associated rhadinovirus human herpesvirus 8: determinants of its pathogenicity? *J. Virol.* **71**:4187-4192

Neipel, F., Albrecht, J.C., Ensser, A., Huang, Y.Q., Li, J.J., Friedman-Kien, A.E., and Fleckenstein, B., 1997c, The genome of human herpesvirus 8 cloned from Kaposi's sarcoma. Genbank accession number U93872

Nguyen, H., Mustafa, A., Hiscott, J., and Lin, R., 1995, Transcription factor IRF-2 exerts its oncogenic phenotype through the DNA binding/transcription repression domain. *Oncogene* **11**:537-544

Nicholas, J., Cameron, K.R., Coleman, H., Newman, C., and Honess, R.W., 1992a, Analysis of nucleotide sequence of the rightmost 43 kbp of herpesvirus saimiri (HVS) L-DNA: general conservation of genetic organization between HVS and Epstein-Barr virus. *Virology* **188**:296-310

Nicholas, J., Cameron, K.R., and Honess, R.W., 1992b, Herpesvirus saimiri encodes homologues of G protein-coupled receptors and cyclins. *Nature* **355**:362-365

Nicholas, J., Coles, L.S., Newman, C., and Honess, R.W., 1991, Regulation of the herpesvirus saimiri (HVS) delayed-early 110-kilodalton promoter by HVS immediate-early gene products and a homolog of the Epstein-Barr virus R trans activator. *J. Virol.* **65**:2457-2466

Nicholas, J., Gompels, U.A., Craxton, M.A., and Honess, R.W., 1988, Conservation of sequence and function between the product of the 52- kilodalton immediate-early gene of herpesvirus saimiri and the BMLF1- encoded transcriptional effector (EB2) of Epstein-Barr virus. *J. Virol.* **62**:3250-3257

Nicholas, J., Ruvolo, V.R., Burns, W.H., Sandford, G., Wan, X., Ciufo, D., Hendrickson, S.B., Guo, H.G., Hayward, G.S., and Reitz, M.S., 1997, Kaposi's sarcoma-associated human herpesvirus-8 encodes homologues of macrophage inflammatory protein-1 and interleukin-6. *Nat. Med.* **3**:287-292

Nicholas, J., Smith, E.P., Coles, L., and Honess, R., 1990, Gene expression in cells infected with gammaherpesvirus saimiri: properties of transcripts from two immediate-early genes. *Virology* **179**:189-200

Nick, S., Fickenscher, H., Biesinger, B., Born, G., Jahn, G., and Fleckenstein, B., 1993, Herpesvirus saimiri transformed human T cell lines: a permissive system for human immunodeficiency viruses. *Virology* **194**:875-877

Nordengrahn, A., Rusvai, M., Merza, M., Ekstrom, J., Morein, B., and Belak, S., 1996, Equine herpesvirus type 2 (EHV-2) as a predisposing factor for Rhodococcus equi pneumonia in foals: prevention of the bifactorial disease with EHV-2 immunostimulating complexes. *Vet. Microbiol.* **51**:55-68

Oksenhendler, E., Cazals-Hatem, D., Schulz, T.F., Barateau, V., Grollet, L., Sheldon, J., Clauvel, J.P., Sigaux, F., and Agbalika, F., 1998, Transient angiolymphoid hyperplasia and Kaposi's sarcoma after primary infection with human herpesvirus 8 in a patient with human immunodeficiency virus infection. *N. Engl. J. Med.* **338**:1585-1590

O'Toole, D., Li, H., Miller, D., Williams, W.R., and Crawford, T.B., 1997, Chronic and recovered cases of sheep-associated malignant catarrhal fever in cattle. *Vet. Rec.* **140**:519-524

O'Toole, D., Li, H., Roberts, S., Rovnak, J., DeMartini, J., Cavender, J., Williams, B., and Crawford, T., 1995, Chronic generalized obliterative arteriopathy in cattle: a sequel to sheep-associated malignant catarrhal fever. *J. Vet. Diagn. Invest.* **7**:108-121

Pacheco-Castro, A., Marquez, C., Toribio, M.L., Ramiro, A.R., Trigueros, C., and Regueiro, J.R., 1996, Herpesvirus saimiri immortalization of alpha beta and gamma delta human T-lineage cells derived from CD34+ intrathymic precursors in vitro. *Int. Immunol.* **8**:1797-1805

Parry, B.C., Simas, J.P., Smith, V.P., Stewart, C.A., Minson, A.C., Efstathiou, S., and Alcami, A., 2000, A broad spectrum secreted chemokine binding protein encoded by a herpesvirus. *J. Exp. Med.* **191**:573-578

Peters, M., Müller, A.M., and Rose-John, S., 1998, Interleukin-6 and soluble interleukin-6 receptor: direct stimulation of gp130 and hematopoiesis. *Blood* **92**:3495-3504

Pines, J., 1995, Cyclins, CDKs and cancer. *Semin. Cancer Biol.* **6**:63-72

Plowright, W., Ferris, R.D., and Scott, G.R., 1960, Blue wildebeest and the aetiological agent of bovine malignant catarrhal fever. *Nature* **188**:1167-1169

Plummer, G., Bowling, C.P., and Goodheart, C.R., 1969, Comparison of four horse herpesviruses. *J. Virol.* **4**:738-741

Popovic, M., Lange-Wantzin, G., Sarin, P.S., Mann, D., and Gallo, R.C., 1983, Transformation of human umbilical cord blood T cells by human T-cell leukemia/lymphoma virus. *Proc. Natl. Acad. Sci. USA* **80**:5402-5406

Raab, M.S., Albrecht, J.C., Birkmann, A., Yaguboglu, S., Lang, D., Fleckenstein, B., and Neipel, F., 1998, The immunogenic glycoprotein gp35-37 of human herpesvirus 8 is encoded by open reading frame K8.1. *J. Virol.* **72**:6725-6731

Rabkin, C.S., Janz, S., Lash, A., Coleman, A.E., Musaba, E., Liotta, L., Biggar, R.J., and Zhuang, Z., 1997, Monoclonal origin of multicentric Kaposi's sarcoma lesions. *N. Engl. J. Med.* **336**:988-993

Rabkin, C.S., Schulz, T.F., Whitby, D., Lennette, E.T., Magpantay, L.I., Chatlynne, L., and Biggar, R.J., 1998, Interassay correlation of human herpesvirus 8 serologic tests. HHV-8 Interlaboratory Collaborative Group. *J. Infect. Dis.* **178**:304-309

Radkov, S.A., Kellam, P., and Boshoff, C., 2000, The latent nuclear antigen of kaposi sarcoma-associated herpesvirus targets the retinoblastoma-E2F pathway and with the oncogene hras transforms primary rat cells. *Nat. Med.* **6**:1121-1127

Rainbow, L., Platt, G.M., Simpson, G.R., Sarid, R., Gao, S.J., Stoiber, H., Herrington, C.S., Moore, P.S., and Schulz, T.F., 1997, The 222- to 234-kilodalton latent nuclear protein (LNA) of Kaposi's sarcoma-associated herpesvirus (human herpesvirus 8) is encoded by orf73 and is a component of the latency-associated nuclear antigen. *J. Virol.* **71**:5915-5921

Rajcani, J., Blaskovic, D., Svobodova, J., Ciampor, F., Huckova, D., and Stanekova, D., 1985, Pathogenesis of acute and persistent murine herpesvirus infection in mice. *Acta Virol.* **29**:51-60

Randall, R.E., Newman, C., and Honess, R.W., 1984, A single major immediate-early virus gene product is synthesized in cells productively infected with herpesvirus saimiri. *J. Gen. Virol.* **65**:1215-1219

Randall, R.E., Newman, C., and Honess, R.W., 1985, Asynchronous expression of the immediate-early protein of herpesvirus saimiri in populations of productively infected cells. *J. Gen. Virol.* **66**:2199-2213

Reichel, M., Matis, J., Lesso, J., and Stancekova, M., 1991, Polypeptides synthesized in rabbit cells infected with murine herpesvirus (MHV): a comparison of proteins specified by various MHV strains. *Acta Virol.* **35**:268-275

Reid, H.W. and Bridgen, A., 1991, Recovery of a herpesvirus from a roan antelope (Hippotragus equinus). *Vet. Microbiol.* **28**:269-278

Reid, H.W. and Buxton, D., 1989, Malignant catarrhal fever and the gammaherpesviruses of bovidae. In Herpesvirus Diseases of Cattle, Horses, and Pigs (G. Wittman, ed.), Kluwer, Boston, pp. 116-162.

Renne, R., Barry, C., Dittmer, D., Compitello, N., Brown, P.O., and Ganem, D., 2001, Modulation of cellular and viral gene expression by the latency-associated nuclear antigen of Kaposi's sarcoma-associated herpesvirus. *J. Virol.* **75**:458-468

Renne, R., Zhong, W., Herndier, B., McGrath, M.S., Abbey, N., Kedes, D.H., and Ganem, D.E., 1996, Lytic growth of Kaposi's sarcoma-associated herpesvirus (human herpesvirus 8) in culture. *Nat. Med.* **2**:342-346

Rizvi, S.M., Slater, J.D., Wolfinger, U., Borchers, K., Field, H.J., and Slade, A.J., 1997, Detection and distribution of equine herpesvirus 2 DNA in the central and peripheral nervous systems of ponies. *J. Gen. Virol.* **78**:1115-1118

Rodriguez-Gallego, C., Corell, A., Pacheco, A., Timon, M., Regueiro, J.R., Allende, L.M., Madrono, A., and Arnaiz-Villena, A., 1996, Herpesvirus saimiri transformation of T cells in CD3 gamma immunodeficiency: phenotypic and functional characterization. *J. Immunol. Methods* **198**:177-186

Roizman, B., Desrosiers, R.C., Fleckenstein, B., Lopez, C., Minson, A.C., and Studdert, M.J., The Herpesvirus Study Group of the International Committee on Taxonomy of Viruses, 1992, The family Herpesviridae: an update. *Arch. Virol.* **123**:425-449

Rose, T.M., Strand, K.B., Schultz, E.R., Schaefer, G., Rankin, G.W., Thouless, M.E., Tsai, C.C., and Bosch, M.L., 1997, Identification of two homologs of the Kaposi's sarcoma-associated herpesvirus (human herpesvirus 8) in retroperitoneal fibromatosis of different macaque species. *J. Virol.* **71**:4138-4144

Rother, R.P., Rollins, S.A., Fodor, W.L., Albrecht, J.C., Setter, E., Fleckenstein, B., and Squinto, S.P., 1994, Inhibition of complement-mediated cytolysis by the terminal complement inhibitor of herpesvirus saimiri. *J. Virol.* **68**:730-737

Rouvier, E., Luciani, M.F., Mattei, M.G., Denizot, F., and Golstein, P., 1993, CTLA-8, cloned from an activated T cell, bearing AU-rich messenger RNA instability sequences, and homologous to a herpesvirus saimiri gene. *J. Immunol.* **150**:5445-5456

Rovnak, J., Quackenbush, S.L., Reyes, R.A., Baines, J.D., Parrish, C.R., and Casey, J.W., 1998, Detection of a novel bovine lymphotropic herpesvirus. *J. Virol.* **72**:4237-4242

Russo, J.J., Bohenzky, R.A., Chien, M.C., Chen, J., Yan, M., Maddalena, D., Parry, J.P., Peruzzi, D., Edelman, I.S., Chang, Y., and Moore, P.S., 1996, Nucleotide sequence of the Kaposi sarcoma-associated herpesvirus (HHV8). *Proc. Natl. Acad. Sci. USA* **93**:14862-14867

Ruszczak, Z., Mayer da Silva, A., and Orfanos, C.E., 1987a, Angioproliferative changes in clinically noninvolved, perilesional skin in AIDS-associated Kaposi's sarcoma. *Dermatologica* **175**:270-279

Ruszczak, Z., Mayer da Silva, A., and Orfanos, C.E., 1987b, Kaposi's sarcoma in AIDS. Multicentric angioneoplasia in early skin lesions. *Am. J. Dermatopathol.* **9**:388-398

Saha, K., Bentsman, G., Chess, L., and Volsky, D. J., 1998, Endogenous production of beta-chemokines by CD4+, but not CD8+, T-cell clones correlates with the clinical state of human immunodeficiency virus type 1 (HIV-1)-infected individuals and may be responsible for blocking infection with non-syncytium-inducing HIV-1 in vitro. *J. Virol.* **72**:876-881

Saha, K., Ware, R., Yellin, M.J., Chess, L., and Lowy, I., 1996, Herpesvirus saimiri-transformed human CD4+ T cells can provide polyclonal B cell help via the CD40 ligand as well as the TNF-alpha pathway and through release of lymphokines. *J. Immunol.* **157**:3876-3885

Sakurada, S., Katano, H., Sata, T., Ohkuni, H., Watanabe, T., and Mori, S., 2001, Effective human herpesvirus 8 infection of human umbilical vein endothelial cells by cell-mediated transmission. *J. Virol.* **75**:7717-7722

Salahuddin, S.Z., Nakamura, S., Biberfeld, P., Kaplan, M,H., Markham, P.D., Larsson, L., and Gallo, R.C., 1988, Angiogenic properties of Kaposi's sarcoma-derived cells after long-term culture in vitro. *Science* **242**:430-433

Samaniego, F., Markham, P.D., Gendelman, R., Watanabe, Y., Kao, V., Kowalski, K., Sonnabend, J.A., Pintus, A., Gallo, R.C., and Ensoli, B., 1998, Vascular endothelial growth factor and basic fibroblast growth factor present in Kaposi's sarcoma (KS) are induced by inflammatory cytokines and synergize to promote vascular permeability and KS lesion development. *Am. J. Pathol.* **152**:1433-1443

Sangster, M.Y., Topham, D.J., D'Costa, S., Cardin, R.D., Marion, T.N., Myers, L.K., and Doherty, P.C., 2000, Analysis of the virus-specific and nonspecific B cell response to a persistent B-lymphotropic gammaherpesvirus. *J. Immunol.* **164**:1820-1828

Sarawar, S.R., Brooks, J.W., Cardin, R.D., Mehrpooya, M., and Doherty, P.C., 1998, Pathogenesis of murine gammaherpesvirus-68 infection in interleukin-6-deficient mice. *Virology* **249**:359-366

Sarid, R., Flore, O., Bohenzky, R.A., Chang, Y., and Moore, P.S., 1998, Transcription mapping of the Kaposi's sarcoma-associated herpesvirus (human herpesvirus 8) genome in a body cavity-based lymphoma cell line (BC-1). *J. Virol.* **72**:1005-1012

Sarid, R., Wiezorek, J.S., Moore, P.S., and Chang, Y., 1999, Characterization and cell cycle regulation of the major Kaposi's sarcoma-associated herpesvirus (human herpesvirus 8) latent genes and their promoter. *J. Virol.* **73**:1438-1446

Schirm, S., Müller, I., Desrosiers, R.C., and Fleckenstein, B., 1984, Herpesvirus saimiri DNA in a lymphoid cell line established by in vitro transformation. *J. Virol.* **49**:938-946

Schneider, U., Schwenk, H.U., and Bornkamm, G., 1977, Characterization of EBV-genome negative "null" and "T" cell lines derived from children with acute lymphoblastic leukemia and leukemic transformed non-Hodgkin lymphoma. *Int. J. Cancer* **19**:521-526

Schock, A. and Reid, H.W., 1996, Characterisation of the lymphoproliferation in rabbits experimentally affected with malignant catarrhal fever. *Vet. Microbiol.* **53**:111-119

Schock, A., Collins, R.A., and Reid, H.W., 1998, Phenotype, growth regulation and cytokine transcription in Ovine herpesvirus-2 (OHV-2)-infected bovine T-cell lines. *Vet. Immunol. Immunopathol.* **66**:67-81

Schofield, A., 1994, PhD Thesis Investigation of the origins of replication of herpesvirus saimiri. The Open University, London, England.

Schrum, S., Probst, P., Fleischer, B., and Zipfel, P.F., 1996, Synthesis of the CC-chemokines MIP-1alpha, MIP-1beta, and RANTES is associated with a type 1 immune response. *J. Immunol.* **157**:3598-3604

Schultz, E.R., Rankin, G.W., Blanc, M.P., Raden, B.W., Tsai, C.C., and Rose, T.M., 2000, Characterization of two divergent lineages of macaque rhadinoviruses related to Kaposi's sarcoma-associated herpesvirus. *J. Virol.* **74**:4919-4928

Sciortino, M.T., Perri, D., Medici, M.A., Foti, M., Orlandella, B.M., and Mastino, A., 2000, The gamma-2 herpesvirus Bovine herpesvirus 4 causes apoptotic infection in permissive cell lines. *Virology* **277**:27-39

Seal, B.S., Heuschele, W.P., and Klieforth, R.B., 1989a, Prevalence of antibodies to alcelaphine herpesvirus-1 and nucleic acid hybridization analysis of viruses isolated from captive exotic ruminants. *Am. J. Vet. Res.* **50**:1447-1453

Seal, B.S., Klieforth, R.B., Welch, W.H., and Heuschele, W.P., 1989b, Alcelaphine herpesviruses 1 and 2 SDS-PAGE analysis of virion polypeptides, restriction endonuclease analysis of genomic DNA, and virus replication restriction in different cell types. *Arch. Virol.* **106**:301-320

Searles, R.P., Bergquam, E.P., Axthelm, M.K., and Wong, S.W., 1999, Sequence and genomic analysis of a Rhesus macaque rhadinovirus with similarity to Kaposi's sarcoma-associated herpesvirus/human herpesvirus 8. *J. Virol.* **73**:3040-3053

Shi, W., Kumanogoh, A., Watanabe, C., Uchida, J., Wang, X., Yasui, T., Yukawa, K., Ikawa, M., Okabe, M., Parnes, J.R., Yoshida, K., and Kikutani, H., 2000, The class IV semaphorin CD100 plays nonredundant roles in the immune system. Defective B and T cell activation in CD100-deficient mice. *Immunity* **13**:633-642

Siegal, B., Levinton-Kriss, S., Schiffer, A., Sayar, J., Engelberg, I., Vonsover, A., and Ramon, Y.R.-E., 1990, Kaposi's sarcoma in immunosuppression. Possibly the result of a dual viral infection. *Cancer* **65**:492-498

Simas, J.P., Bowden, R.J., Paige, V., and Efstathiou, S., 1998, Four tRNA-like sequences and a serpin homologue encoded by murine gammaherpesvirus 68 are dispensable for lytic replication in vitro and latency in vivo. *J. Gen. Virol.* **79**:149-153

Simmer, B., Alt, M., Buckreus, I., Berthold, S., Fleckenstein, B., Platzer, E., and Grassmann, R., 1991, Persistence of selectable herpesvirus saimiri in various human haematopoietic and epithelial cell lines. *J. Gen. Virol.* **72**:1953-1958

Simpson, G.R., Schulz, T.F., Whitby, D., Cook, P.M., Boshoff, C., Rainbow, L., Howard, M.R., Gao, S.J., Bohenzky, R.A., Simmonds, P., Lee, C., de Ruiter, A., Hatzakis, A., Tedder, R.S., Weller, I.V.D., Weiss, R.A., and Moore, P.S., 1996, Prevalence of Kaposi's sarcoma associated herpesvirus infection measured by antibodies to recombinant capsid protein and latent immunofluorescence antigen. *Lancet* **348**:1133-1138

Spriggs, M.K., 1997, Interleukin-17 and its receptor. *J. Clin. Immunol.* **17**:366-369

Staskus, K.A., Zhong, W., Gebhard, K., Herndier, B., Wang, H., Renne, R., Beneke, J., Pudney, J., Anderson, D.J., Ganem, D., and Haase, A.T., 1997, Kaposi's sarcoma-associated herpesvirus gene expression in endothelial (spindle) tumor cells. *J. Virol.* **71**:715-719

Stephan, V., Wahn, V., Le Deist, F., Dirksen, U., Bröker, B., Müller-Fleckenstein, I., Horneff, G., Schroten, H., Fischer, A., and de Saint Basile, G., 1996, Atypical X-linked severe combined immunodeficiency due to possible spontaneous reversion of the genetic defect in T cells. *N. Engl. J. Med.* **335**:1563-1567

Stevenson, A.J., Cooper, M., Griffiths, J.C., Gibson, P.C., Whitehouse, A., Jones, E.F., Markham, A.F., Kinsey, S.E., and Meredith, D.M., 1999, Assessment of herpesvirus saimiri as a potential human gene therapy vector. *J. Med. Virol.* **57**:269-277

Stevenson, A.J., Clarke, D., Meredith, D.M., Kinsey, S.E., Whitehouse, A., and Bonifer, C., 2000a, Herpesvirus saimiri-based gene delivery vectors maintain heterologous expression throughout mouse embryonic stem cell differentiation in vitro. *Gene Ther.* **7**:464-471

Stevenson, A.J., Frolova-Jones, E., Hall, K.T., Kinsey, S.E., Markham, A.F., Whitehouse, A., and Meredith, D.M., 2000b, A herpesvirus saimiri-based gene therapy vector with potential for use in cancer immunotherapy. *Cancer Gene Ther.* **7**:1077-1085

Stevenson, A.J., Giles, M.S., Hall, K.T., Goodwin, D.J., Calderwood, M.A., Markham, A.F., and Whitehouse, A., 2000c, Specific oncolytic activity of herpesvirus saimiri in pancreatic cancer cells. *Br. J. Cancer* **83**:329-332

Strand, K., Harper, E., Thormahlen, S., Thouless, M.E., Tsai, C., Rose, T., and Bosch, M.L., 2000, Two distinct lineages of macaque gamma herpesviruses related to the Kaposi's sarcoma associated herpesvirus. *J. Clin. Virol.* **16**:253-269

Stürzl, M., Blasig, C., Schreier, A., Neipel, F., Hohenadl, C., Cornali, E., Ascherl, G., Esser, S., Brockmeyer, N.H., Ekman, M., Kaaya, E.E., Tschachler, E., and Biberfeld, P., 1997, Expression of HHV-8 latency-associated T0.7 RNA in spindle cells and endothelial cells of AIDS-associated, classical and African Kaposi's sarcoma. *Int. J. Cancer* **72**:68-71

Stürzl, M., Hohenadl, C., Zietz, C., Castanos-Velez, E., Wunderlich, A., Ascherl, G., Biberfeld, P., Monini, P., Browning, P.J., and Ensoli, B., 1999a, Expression of K13/v-FLIP gene of human herpesvirus 8 and apoptosis in Kaposi's Sarcoma spindle cells. *J. Natl. Cancer Inst.* **91**:1725-1733

Stürzl, M., Wunderlich, A., Ascherl, G., Hohenadl, C., Monini, P., Zietz, C., Browning, P.J., Neipel, F., Biberfeld, P., and Ensoli, B., 1999b, Human herpesvirus-8 (HHV-8) gene expression in Kaposi's sarcoma (KS) primary lesions: an in situ hybridization study. *Leukemia* **13**:S110-S112

Sun, R., Lin, S.F., Staskus, K., Gradoville, L., Grogan, E., Haase, A., and Miller, G., 1999, Kinetics of Kaposi's sarcoma-associated herpesvirus gene expression. *J. Virol.* **73**:2232-2242

Sunil-Chandra, N.P., Arno, J., Fazakerley, J., and Nash, A.A., 1994, Lymphoproliferative disease in mice infected with murine gammaherpesvirus 68. *Am. J. Pathol.* **145**:818-826

Sunil-Chandra, N.P., Efstathiou, S., Arno, J., and Nash, A.A., 1992, Virological and pathological features of mice infected with murine gamma-herpesvirus 68. *J. Gen. Virol.* **73**:2347-2356

Swanton, C., Mann, D.J., Fleckenstein, B., Neipel, F., Peters, G., and Jones, N., 1997, Herpesviral cyclin/Cdk6 complexes evade inhibition by cdk inhibitor proteins. *Nature* **390**:184-187

Szomolanyi, E., Medveczky, P., and Mulder, C., 1987, In vitro immortalization of marmoset cells with three subgroups of herpesvirus saimiri. *J. Virol.* **61**:3485-3490

Telford, E.A., Studdert, M.J., Agius, C.T., Watson, M.S., Aird, H.C., and Davison, A.J., 1993, Equine herpesviruses 2 and 5 are gamma-herpesviruses. *Virology* **195**:492-499

Telford, E.A., Watson, M.S., Aird, H.C., Perry, J., and Davison, A.J., 1995, The DNA sequence of equine herpesvirus 2. *J. Mol. Biol.* **249**:520-528

Thome, M., Martinon, F., Hofmann, K., Rubio, V., Steiner, V., Schneider, P., Mattmann, C., and Tschopp, J., 1999, Equine herpesvirus-2 E10 gene product, but not its cellular homologue, activates NF-kappaB transcription factor and c-Jun N-terminal kinase. *J. Biol. Chem.* **274**:9962-9968

Thome, M., Schneider, P., Hofmann, K., Fickenscher, H., Meinl, E., Neipel, F., Mattmann, C., Burns, K., Bodmer, J.L., Schroter, M., Scaffidi, C., Krammer, P.H., Peter, M.E., and Tschopp, J., 1997, Viral FLICE-inhibitory proteins (FLIPs) prevent apoptosis induced by death receptors. *Nature* **386**:517-521

Thomson, B. J. and Nicholas, J., 1991, Superantigen function. *Nature* **351**:530

Thurau, M., Whitehouse, A., Wittmann, S., Meredith, D., and Fickenscher, H., 2000, Distinct transcriptional and functional properties of the R transactivator gene orf50 of the transforming herpesvirus saimiri strain C488. *Virology* **268**:167-177

Trimble, J.J., Murthy, S.C., Bakker, A., Grassmann, R., and Desrosiers, R.C., 1988, A gene for dihydrofolate reductase in a herpesvirus. *Science* **239**:1145-1147

Troidl, B., Simmer, B., Fickenscher, H., Müller-Fleckenstein, I., Emmrich, F., Fleckenstein, B., and Gebhart, E., 1994, Karyotypic characterization of human T-cell lines immortalized by herpesvirus saimiri. *Int. J. Cancer* **56**:433-438.

Tsygankov, A.Y., Bröker, B.M., Fargnoli, J., Ledbetter, J.A., and Bolen, J.B., 1992, Activation of tyrosine kinase p60fyn following T cell antigen receptor cross-linking. *J. Biol. Chem.* **267**:18259-18262

Ulrich, S., Goltz, M., and Ehlers, B., 1999, Characterization of the DNA polymerase loci of the novel porcine lymphotropic herpesviruses 1 and 2 in domestic and feral pigs. *J. Gen. Virol.* **80**:3199-3205

Usherwood, E.J., Stewart, J.P., and Nash, A.A., 1996a, Characterization of tumor cell lines derived from murine gammaherpesvirus-68-infected mice. *J. Virol.* **70**:6516-6518

Usherwood, E.J., Stewart, J.P., Robertson, K., Allen, D.J., and Nash, A.A., 1996b, Absence of splenic latency in murine gammaherpesvirus 68-infected B cell-deficient mice. *J. Gen. Virol.* **77**:2819-2825

van Dyk, L.F., Hess, J.L., Katz, J.D., Jacoby, M., Speck, S.H., and Virgin, H.W. IV, 1999, The murine gammaherpesvirus 68 v-cyclin gene is an oncogene that promotes cell cycle progression in primary lymphocytes. *J. Virol.* **73**:5110-5122

van Dyk, L.F., Virgin, H.W. IV, and Speck, S.H., 2000, The murine gammaherpesvirus 68 v-cyclin is a critical regulator of reactivation from latency. *J. Virol.* **74**:7451-7461

Vanderplasschen, A., Markine-Goriaynoff, N., Lomonte, P., Suzuki, M., Hiraoka, N., Yeh, J.C., Bureau, F., Willems, L., Thiry, E., Fukuda, M., and Pastoret, P.P., 2000, A multipotential beta-1,6-N-acetylglucosaminyl-transferase is encoded by bovine herpesvirus type 4. *Proc. Natl. Acad. Sci. USA* **97**:5756-5761

Varga, J., Fodor, L., Rusvai, M., Soos, I., and Makrai, L., 1997, Prevention of Rhodococcus equi pneumonia of foals using two different inactivated vaccines. *Vet. Microbiol.* **56**:205-212

Vella, C., Gregory, J., Bristow, R., Troop, M., Easterbrook, P., Zheng, N., and Daniels, R., 1999a, Isolation of HIV type 1 from long-term nonprogressors in herpesvirus saimiri-immortalized T cells. *AIDS Res. Hum. Retroviruses* **15**:1145-1147

Vella, C., Fickenscher, H., Atkins, C., Penny, M., and Daniels, R., 1997, Herpesvirus saimiri-immortalized human T-cells support long-term, high titred replication of human immunodeficiency virus types 1 and 2. *J. Gen. Virol.* **78**:1405-1409

Vella, C., Zheng, N.N., Vella, G., Atkins, C., Bristow, R.G., Fickenscher, H., and Daniels, R.S., 1999b, Enhanced replication of M-tropic HIV-1 strains in herpesvirus saimiri immortalised T-cells which express CCR5. *J. Virol. Methods* **79**:51-63

Virgin, H.W., Latreille, P., Wamsley, P., Hallsworth, K., Weck, K.E., Dal Canto, A.J., and Speck, S.H., 1997, Complete sequence and genomic analysis of murine gammaherpesvirus 68. *J. Virol.* **71**:5894-5904

Virgin, H.W., Presti, R.M., Li, X.Y., Liu, C., and Speck, S.H., 1999, Three distinct regions of the murine gammaherpesvirus 68 genome are transcriptionally active in latently infected mice. *J. Virol.* **73**:2321-2332

Wang, G.H., Bertin, J., Wang, Y., Martin, D.A., Wang, J., Tomaselli, K.J., Armstrong, R.C., and Cohen, J.I., 1997, Bovine herpesvirus 4 BORFE2 protein inhibits Fas- and tumor necrosis factor receptor 1-induced apoptosis and contains death effector domains shared with other gamma-2 herpesviruses. *J. Virol.* **71**:8928-8932

Wang, G.H., Garvey, T.L., and Cohen, J.I., 1999, The murine gammaherpesvirus-68 M11 protein inhibits Fas- and TNF-induced apoptosis. *J. Gen. Virol.* **80**:2737-2740

Weber, F., Meinl, E., Drexler, K., Czlonkowska, A., Huber, S., Fickenscher, H., Müller-Fleckenstein, I., Fleckenstein, B., Wekerle, H., and Hohlfeld, R., 1993, Transformation of human T-cell clones by herpesvirus saimiri: intact antigen recognition by autonomously growing myelin basic protein-specific T cells. *Proc. Natl. Acad. Sci. USA* **90**:11049-11053

Weck, K.E., Barkon, M.L., Yoo, L.I., Speck, S.H., and Virgin, H.W. IV, 1996, Mature B cells are required for acute splenic infection, but not for establishment of latency, by murine gammaherpesvirus 68. *J. Virol.* **70**:6775-6780

Wehner, L.E., Schröder, N., Kamino, K., Friedrich, U., Biesinger, B., and Rüther, U., 2001, Herpesvirus saimiri Tip gene causes T-cell lymphomas in transgenic mice. *DNA Cell Biol.* **20**:81-88.

Weninger, W., Partanen, T.A., Breiteneder-Geleff, S., Mayer, C., Kowalski, H., Mildner, M., Pammer, J., Stürzl, M., Kerjaschki, D., Alitalo, K., and Tschachler, E., 1999, Expression of vascular endothelial growth factor receptor-3 and podoplanin suggests a lymphatic endothelial cell origin of Kaposi's sarcoma tumor cells. *Lab. Invest.* **79**:243-251

Whitehouse, A., Carr, I.M., Griffiths, J.C., and Meredith, D.M., 1997a, The herpesvirus saimiri ORF50 gene, encoding a transcriptional activator homologous to the Epstein-Barr virus R protein, is transcribed from two distinct promoters of different temporal phases. *J. Virol.* **71**:2550-2554

Whitehouse, A., Cooper, M., Hall, K.T., and Meredith, D.M., 1998a, The open reading frame (ORF) 50a gene product regulates ORF 57 gene expression in herpesvirus saimiri. *J. Virol.* **72**:1967-1973

Whitehouse, A., Cooper, M., and Meredith, D.M., 1998b, The immediate-early gene product encoded by open reading frame 57 of herpesvirus saimiri modulates gene expression at a posttranscriptional level. *J. Virol.* **72**:857-861

Whitehouse, A., Stevenson, A.J., Cooper, M., and Meredith, D.M., 1997b, Identification of a cis-acting element within the herpesvirus saimiri ORF 6 promoter that is responsive to the HVS.R transactivator. *J. Gen. Virol.* **78**:1411-1415

Wiese, N., Tsygankov, A.Y., Klauenberg, U., Bolen, J.B., Fleischer, B., and Bröker, B.M., 1996, Selective activation of T cell kinase p56lck by herpesvirus saimiri protein tip. *J. Biol. Chem.* **271**:847-852

Wiyono, A., Baxter, S.I., Saepulloh, M., Damayanti, R., Daniels, P., and Reid, H.W., 1994, PCR detection of ovine herpesvirus-2 DNA in Indonesian ruminants - normal sheep and clinical cases of malignant catarrhal fever. *Vet. Microbiol.* **42**:45-52

Wong, S.W., Bergquam, E.P., Swanson, R.M., Lee, F.W., Shiigi, S.M., Avery, N.A., Fanton, J.W., and Axthelm, M.K., 1999, Induction of B cell hyperplasia in simian immunodeficiency virus-infected rhesus macaques with the simian homologue of Kaposi's sarcoma-associated herpesvirus. *J. Exp. Med.* **190**:827-840

Wright, J., Falk, L.A., Collins, D., and Deinhardt, F., 1976, Mononuclear cell fraction carrying herpesvirus saimiri in persistently infected squirrel monkeys. *J. Natl. Cancer Inst.* **57**:959-962

Yamamoto, N., Okada, M., Koyanagi, Y., Kannagi, M., and Hinuma, Y., 1982, Transformation of human leukocytes by cocultivation with an adult T cell leukemia virus producer cell line. *Science* **217**:737-739

Yamanashi, Y., Mori, S., Yoshida, M., Kishimoto, T., Inoue, K., Yamamoto, T., and Toyoshima, K., 1989, Selective expression of a protein-tyrosine kinase, p56lyn, in hematopoietic cells and association with production of human T-cell lymphotropic virus type I. *Proc. Natl. Acad. Sci. USA* **86**:6538-6542

Yan, M., Lee, J., Schilbach, S., Goddard, A., and Dixit, V., 1999, mE10, a novel caspase recruitment domain-containing proapoptotic molecule. *J. Biol. Chem.* **274**:10287-10292

Yang, B.T., Chen, S.C., Leach, M.W., Manfra, D., Homey, B., Wiekowski, M., Sullivan, L., Jenh, C.H., Narula, S.K., Chensue, S.W., and Lira, S.A., 2000, Transgenic expression of the chemokine receptor encoded by human herpesvirus 8 induces an angioproliferative disease resembling Kaposi's sarcoma. *J. Exp. Med.* **191**:445-454

Yao, Z., Fanslow, W.C., Seldin, M.F., Rousseau, A.M., Painter, S.L., Comeau, M.R., Cohen, J.I., and Spriggs, M.K., 1995a, Herpesvirus saimiri encodes a new cytokine, IL-17, which binds to a novel cytokine receptor. *Immunity* **3**:811-821

Yao, Z., Maraskovsky, E., Spriggs, M.K., Cohen, J.I., Armitage, R.J., and Alderson, M.R., 1996a, Herpesvirus saimiri open reading frame 14, a protein encoded by T lymphotropic herpesvirus, binds to MHC class II molecules and stimulates T cell proliferation. *J. Immunol.* **156**:3260-3266

Yao, Z., Painter, S.L., Fanslow, W.C., Ulrich, D., Macduff, B.M., Spriggs, M.K., and Armitage, R.J., 1995b, Human IL-17: a novel cytokine derived from T cells. *J. Immunol.* **155**:5483-5486

Yao, Z., Timour, M., Painter, S., Fanslow, W., and Spriggs, M., 1996b, Complete nucleotide sequence of the mouse CTLA8 gene. *Gene* **168**:223-225

Yasukawa, M., Inoue, Y., Kimura, N., and Fujita, S., 1995, Immortalization of human T cells expressing T-cell receptor gamma delta by herpesvirus saimiri. *J. Virol.* **69**:8114-8117

Yoon, D.W., Lee, H., Seol, W., DeMaria, M., Rosenzweig, M., and Jung, J.U., 1997, Tap: a novel cellular protein that interacts with tip of herpesvirus saimiri and induces lymphocyte aggregation. *Immunity* **6**:571-582

Yssel, H., de Waal Malefyt, R., Duc, D., Blanchard, D., Gazzolo, L., de Vries, J.E., and Spits, H., 1989, Human T cell leukemia/lymphoma virus type I infection of a CD4+ proliferative/cytotoxic T cell clone progresses in at least two distinct phases based on changes in function and phenotype of the infected cells. *J. Immunol.* **142**:2279-2289

Zapata, D.A., Pacheco-Castro, A., Torres, P.S., Ramiro, A.R., Jose, E.S., Alarcon, B., Alibaud, L., Rubin, B., Toribio, M.L., and Regueiro, J.R., 1999, Conformational and biochemical differences in the TCR.CD3 complex of CD8(+) versus CD4(+) mature lymphocytes revealed in the absence of CD3gamma. *J. Biol. Chem.* **274**:35119-35128

Zimmer, M.A., McCoy, C.P., and Jensen, J.M., 1981, Comparative pathology of the African form of malignant catarrhal fever in captive Indian gaur and domestic cattle. *J. Am. Vet. Med. Assoc.* **179**:1130-1135.

Zimmermann, W., Broll, H., Ehlers, B., Buhk, H.J., Rosenthal, A., and Goltz, M., 2001, Genome sequence of bovine herpesvirus 4, a bovine rhadinovirus, and identification of an origin of DNA replication. *J. Virol.* **75**:1186-1194

Zimring, J.C., Goodbourn, S., and Offermann, M.K., 1998, Human herpesvirus 8 encodes an interferon regulatory factor (IRF) homolog that represses IRF-1-mediated transcription. *J. Virol.* **72**:701-707

Chapter 6.2

Interaction of Papillomaviral Oncoproteins with Cellular Factors

SIGRUN SMOLA-HESS AND HERBERT J. PFISTER
Department of Virology, University of Cologne, Fürst-Pückler-Str. 56, Cologne, Germany

1. INTRODUCTION

1.1 Classification of human papillomaviruses

Human papillomaviruses (HPV) are small, non-enveloped double-stranded DNA viruses. They belong to the *papillomaviridae* family, members of which infect squamous epithelia and cause proliferative diseases (papillomas) in a number of vertebrates, including man, non-human primates, cattle, rabbits and dogs, in a highly species-specific manner. Papillomaviruses are classified according to their host species and their genetic relatedness. Based on their nucleotide sequence, HPV separate into distinct types, when they are less than 90% homologous in the open reading frame of the late gene L1.

To date more than 85 different HPV types are sequenced and more than 120 putatively novel types have been partially characterised (zur Hausen, 2000). As a consequence of infection and viral persistence in the epithelium, a variety of proliferative lesions may arise in the skin or mucosa according to their preferential sites of infection. These include benign warts and condylomas but also dysplastic lesions, which may further de-differentiate and progress to cancer. Distinct clinical entities have been found to be associated with distinct HPV types. Based on the biological or clinical properties, HPV types have therefore been categorised into "low risk" types mainly associated with benign disease only and "high risk" types also associated with malignant disease (Table 1).

Table 1. Papillomavirus types and associated disease

Papillomavirus (most common types)		Associated disease
Cutaneous low risk	HPV 1, 2, 3, 10, 27	Skin warts
Cutaneous high risk	HPV 5, 8, 17, 20	EV-specific skin tumors
Mucosal low risk	HPV 6, 11	Condylomata acuminata
		Laryngeal and conjunctival papilloma
Mucosal high risk	HPV 16, 18, 31, 45	Cervical dysplasia and cancer

In the past decade it has been a major challenge in papillomavirus research to understand the differences between high risk and low risk types and to unravel mechanisms that lead to malignant transformation.

Irrespective of the HPV type viral genome replication and production of viral offspring are the obvious major goals that have to be reached with a limited number of viral proteins expressed (see chapter 1.2.). This is achieved via a strong adaptation to and interaction with the host cell, the keratinocyte. Thus, the viral life cycle (see chapter 1.3.) is closely linked to the keratinocyte's life cycle.

1.2   HPV genome organisation

The icosahedral capsids of papillomaviruses enclose a double-stranded circular DNA genome of approximately 8,000 bp. Only one strand is transcribed into mRNA. The genomes of all known papillomaviruses are similarly organised (figure 1).

A long control region (LCR) can be separated from the "early region" encoding regulatory genes (E) and the "late region", which encompasses genes (L) coding for the capsid proteins. These are only expressed in productively infected cells.

The main role of the early proteins E1 and E2 is seen in the regulation of viral replication and transcription. The early proteins E5, E6 and E7 alter host cell conditions, e.g. cellular proliferation and differentiation, in favour of the viral life cycle. As a "side effect" of this function, the respective proteins encoded by high risk viruses may also lead to cellular immortalisation and transformation and are therefore called viral "oncoproteins".

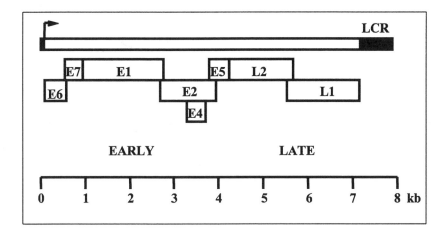

Figure 1. Genome organisation of HPV16. The circular genome is shown in a linearised form. Open reading frames are indicated by open boxes. The black boxes of the genome bar represent the non-coding regions. The scale is in kilo base pairs (kb).

1.3   Viral life cycle

Both cellular factors involved in proliferation and the cellular differentiation program are essential for the virus to replicate and to mature, respectively.

According to the current model, PV infect keratinocytes of the basal cell layers via a microtrauma of the skin or mucosa. In basal cells the early genes start to be weakly expressed and a maintenance copy number of the viral genome is established. With keratinocyte differentiation the virus reaches suprabasal and upper layers of the epithelium. In the spinous layer vegetative replication to high copy numbers and in upper epithelial strata viral maturation and assembly take place. Keratinocytes have to exit the cell cycle in order to differentiate. However, as HPV are lacking a DNA-dependent DNA polymerase they depend on the activity of cellular enzymes. In order to solve that problem, a major task of early viral proteins is to overcome this cell cycle arrest and keep cells cycling in the suprabasal epithelial layers.

Keratinocyte differentiation is regulated by a variety of transcription factors derived from the AP-1 and AP-2 families, and the c/EBP, STAT and POU domain families of transcription factors (for review see Eckert *et al.*, 1997). Parameters that influence differentiation include the distribution pattern of distinct members of these families, their amount of expression and their phosphorylation state. These parameters and their regulation by exogenous factors determine their activity in the epithelium. Interestingly, different studies implicate that the same factors are involved in the regulation of HPV early promoter activity (Desaintes and Demeret, 1996). This may explain the close association of HPV biology with keratinocyte differentiation.

Progression of an infected cell to malignancy is associated with integration of HPV into the host genome in about 88 % (Klaes *et al.*, 1999). From the viral point of view, integration into the host genome, deregulation of viral gene expression and malignant transformation, which finally abolishes the keratinocyte's capacity to differentiate, appear rather to be an "accident". Nevertheless, from the clinical point of view, mechanisms that lead to this "accident" are of major importance. In the following chapter we will give an overview on the viral oncoproteins E5, E6 and E7 that are causally associated with growth stimulation and transformation. We will focus on the interactions of these proteins with cellular factors and the functional consequences that arise thereof.

2.   PAPILLOMAVIRAL ONCOPROTEINS

The E6 and E7 proteins are the major viral oncoproteins. A predominant role of both these proteins in the viral life cycle is to overcome negative cell cycle control checkpoints in order to allow viral replication in differentiating cells. As a consequence, the oncoproteins do not only drive the host cell into abnormal proliferation but may also cause genomic instability. The accumulation of genetic damage will finally promote malignant progression, a function, which has come into the focus of research. Furthermore, the E6 oncoprotein promotes the DNA integration into the host genome (Kessis *et al.*, 1996) and has the capacity to activate telomerase activity (Klingelhutz *et al.*, 1996). Under certain conditions, the E5 protein may also lead to growth stimulation and transformation.

Whereas none of these oncoproteins has intrinsic enzymatic properties, there is increasing evidence that they directly or indirectly alter the activity of cellular factors by protein-protein-interactions.

2.1   The E5 oncoprotein

2.1.1   E5 protein characteristics and structure

The papillomavirus E5 gene products represent small, extremely hydrophobic polypeptides consisting of 44-90 amino acids. They are located in the Golgi apparatus or anchored in the plasma and nuclear membranes (Burkhardt *et al.*, 1989; Conrad *et al.*, 1993). The E5 proteins (approximately 44 amino acids) are highly conserved in animal papillomaviruses leading to fibropapillomas in vivo. Among those, the BPV-1 E5 protein is best characterised. A carboxy terminal C-X-C motif is involved in dimer (and possibly also oligomer) formation. E5 appears to play an important role in cellular transformation by BPV-1. The E5 ORF is not conserved between these species and HPVs and even not among the HPVs themselves. Nevertheless the HPV E5 proteins are predicted to share a common structure. They are nearly double in size compared to BPV-1 E5 and reveal three highly hydrophobic regions supposed to form α-helices which serve as separate transmembrane domains (figure 2) (Bubb *et al.*, 1988; Ullman *et al.*, 1994).

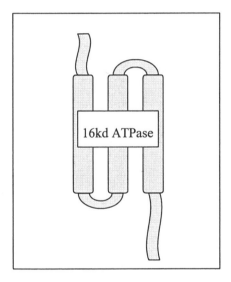

Figure 2. Schematic presentation of the HPV16 E5 protein. From sequence analysis a trimodal hydrophobic structure is predicted. In contrast to the BPV-1 E5 protein, HPV16 E5 does not interact with cellular growth factor receptors but with the 16 kDa subunit of the H^+-dependent vacuolar ATPase (Conrad *et al.*, 1993).

The E5 proteins of HPV 6 and 11 share some structural properties with regions important for transformation in the BPV-1 E5 protein, as they contain a C-X-C motif near the carboxy terminus and a glutamine residue within the hydrophobic domain. However, they do not form dimers (Conrad *et al.*, 1993). For HPV16 this has been controversially discussed. While dimers were observed in Western blot analysis from exfoliated cervical cells (Kell *et al.*, 1994), the E5 protein of HPV16 failed to form dimers when overexpressed in COS cells (Conrad *et al.*, 1993). The E5 protein of human genital high risk viruses is expressed in productive infections, whereas a role in keratinocyte transformation is questioned (Conrad-Stöppler *et al.*, 1996; Swan *et al.*, 1994). In cervical carcinoma, due to viral integration, the E5 ORF is often uncoupled from the early promoter resulting in the loss of E5 expression. In the EV-associated, cutaneous, high risk HPV8 an E5 ORF is even lacking (Fuchs *et al.*, 1986). The role of HPV E5 in the viral life cycle and in human diseases still has to be defined.

2.1.2   Interactions of E5 with cellular factors

BPV-1 E5 forms complexes with different transmembrane proteins. Thus, E5 directly binds to the transmembrane domain of the platelet-derived growth factor receptor ß (PDGFR) and functionally interacts with the epidermal growth factor receptor (EGFR) and the colony-stimulating factor-1 receptor (CSF-1R), (Cohen *et al.*, 1993; Goldstein *et al.*, 1994; Petti and DiMaio, 1992). Through these interactions, the receptors may be activated in a ligand-independent manner. Moreover, their signals, e.g. increased receptor phosphorylation, mitogen activated protein kinase activity and phospholipase C-γ-1 activity are enhanced even in the absence of a physiological ligand (Crusius *et al.*, 1997; Gu and Matlashewski, 1995). This may result from interference with receptor degradation and internalisation (Martin *et al.*, 1989; Waters *et al.*, 1992).

Also HPV16 and HPV6 E5 have been shown to co-operate with the EGFR and PDGFR (Leechanachai *et al.*, 1992; Pim *et al.*, 1992; Straight *et al.*, 1993). Moreover, HPV16 E5 enhances endothelin receptor signalling (Venuti *et al.*, 1998). Whereas HPV6 E5 also associates with the EGFR, the related erbB2 receptor and the PDGFR, HPV16 E5 does not bind to cellular growth factor receptors (Conrad *et al.*, 1994). Another common target of both BPV and HPV E5 proteins is a 16 kDa membrane pore protein representing a subunit of the H^+-dependent vacuolar ATPase (Conrad *et al.*, 1993; Goldstein and Schlegel, 1990). For ATPase binding, the glutamine residue within the hydrophobic domain seems to play an important role (Andresson *et al.*, 1995). As a result from this interaction, acidification of endosomes is inhibited. This has been suggested to be responsible for the prolonged retention and reduced degradation of the EGFR in the presence of the E5 protein (Straight *et al.*, 1995). However, the binding of HPV16 E5 to the 16kDa ATPase subunit could be dissociated from the E5-mediated EGFR overactivation (Rodriguez *et al.*, 2000).

2.2   The E6 oncoprotein

2.2.1   E6 protein characteristics and structure

The HPVE6 proteins are predominantly located in the nucleus and can also be detected in non-nuclear membranes (Chen *et al.*, 1995; Grossman *et al.*, 1989; Kanda *et al.*, 1991). They are approximately 150 amino acids in size resulting in a molecular mass of 18 kDa. Although the E6 sequences differ between HPV types and intratypic variants, the overall structure of E6 proteins is well conserved. They comprise four C-X-X-C motifs with a spacing of 29-30 amino acids (Barbosa *et al.*, 1989; Cole and Danos, 1987). Formation of two zinc fingers is important for protein stability and activity (Kanda *et al.*, 1991). They play a role in protein-protein interactions, cellular transformation and transcriptional activation. A second structural motif, X-S/T-X-V/L has been found in the carboxy terminus of the high-risk but not low-risk E6 proteins binding to so-called PDZ domain-containing proteins (Kiyono *et al.*, 1997) (figure 3). At least HPV16 E6 is phosphorylated by PKN, a fatty acid- and Rho small G protein-activated serine/threonine kinase (Gao *et al.*, 2000).

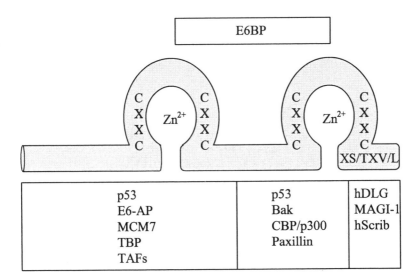

Figure 3. Schematic presentation of the HPV E6 protein containing two zinc fingers. Interacting proteins whose binding regions have been mapped in HPV16 or HPV18 E6 are indicated.

Four different splice variants of E6 messenger-RNA have been described (E6*I-IV) coding for rather unstable proteins. HPV18 E6*I has been shown to bind to the full length E6 protein. As it may interfere with E6 functions, it has been postulated to constitute a viral control for E6 (Pim et al., 1997).

2.2.2   Interactions of E6 with cellular factors

E6 proteins have no intrinsic enzymatic activities. They exert their functions mainly via interactions with cellular proteins. Interactions with high-risk E6 proteins that have been elucidated so far are summarised in figure 3 and will be discussed in detail below.

2.2.2.1   The tumor suppressor protein p53

Most knowledge has been accumulated on the interaction of HPV E6 and the tumor suppressor protein p53. The latter had originally been described in the context of the SV40 large T antigen and the adenovirus E1B 55k, two viral oncoproteins binding to p53 (Lane and Crawford, 1979; Linzer and Levine, 1979; Sarnow et al., 1982). Responding to a variety of stimuli resulting in DNA damage, the sequence-specific transcriptional activator p53 is well known as a "guardian of the cell cycle". It activates genes, e.g. p21, an inhibitor of cyclin dependent kinases (El-Deiry et al., 1993; Xiong et al., 1993) inducing cell cycle arrest in the G1 phase and thus allowing DNA repair. P53 may also induce apoptosis in order to remove the damaged cells (for review of p53 functions see Somasundaram, 2000).

E6 proteins, predominantly of the mucosal high risk types, may bind to p53 and promote their degradation (Scheffner et al., 1990; Werness et al., 1990). As a consequence, E6 does not only functionally block p53's transcriptional activity but strongly reduces its half-life (Crook et al., 1994; Hoppe-Seyler and Butz, 1993; Hubbert et al., 1992). Within the p53 gene, polymorphism occurs at codon 72 resulting either in the amino acid arginine or proline. Storey et al. (1998) showed that p53 is more susceptible to HPV16 E6-mediated degradation in the arginine form than in the proline form. It was further speculated that arginine homozygotes were over-represented within a certain collective of cervical carcinoma patients. Most of the subsequent studies failed to reproduce this finding (Helland et al., 1998; Hildesheim et al., 1998; Josefsson et al., 1998; Minaguchi et al., 1998). However, a trend for increased frequency of p53 arginine homozygotes was again confirmed among cervical carcinoma patients infected with a HPV16 variant encoding an E6 protein with a leucine at position 83 (Brady et al., 1999). In line with this observation, the 83L variant was significantly over-represented in p53 arginine homozygous women with cervical cancer in comparison to 83V variants (van Duin *et al.*, 2000). This may point to a specific risk of the combination E6 83L and p53 arginine.

The interaction between E6 and p53 has long been regarded as the central mechanism of how E6 proteins from high risk types exert their oncogenic function. Interestingly, E6 proteins from low risk types may also bind to p53 but fail to degrade it (Crook *et al.*, 1991; Elbel *et al.*, 1997; Lechner and Laimins, 1994) with the exception of HPV11 E6 which has been reported to degrade the arginine form of p53 (Storey *et al.*, 1998). E6 proteins from cutaneous high risk types (HPV5, 8 and 47) (Elbel *et al.*, 1997; Kiyono *et al.*, 1994; Steger and Pfister, 1992) even fail to bind p53, suggesting other modes of E6 action in the host cell.

Viral induction of p53 degradation is only observed in the context of HPVE6 but neither of SV40 large T antigen, nor of adenovirus E1B. It requires complex formation with a 100 kDa protein called E6-associated protein (E6-AP), representing an ubiquitin-protein ligase. E6-AP belongs to the HECT E3 proteins directly transferring ubiquitin to their substrates (Huibregtse *et al.*, 1991, 1993a,b; Scheffner *et al.*, 1993). It stably interacts with E6 proteins derived from mucosal "high risk" but not "low risk" HPV types. For HPV16 E6, domains interacting with either p53 or E6-AP have been extensively studied. (Crook *et al.*, 1991; Dalal *et al.*, 1996; Elbel *et al.*, 1997; Foster *et al.*, 1994; Nakagawa *et al.*, 1995; Pim *et al.*, 1994; Sherman *et al.*, 1997; Waddell and Jenkins, 1998). In summary, a carboxy terminal domain of HPV16 E6 (residues 106-115) and an independent amino terminal moiety are involved in p53 binding. The amino terminal portion also comprises residues/domains involved in E6-AP binding as well as in p53 degradation. Of note, the E6 binding and DNA binding domains in the interaction partner p53 overlap, possibly explaining the degradation-independent inhibitory effect of E6 on p53 (Elbel *et al.*, 1997; Li and Coffino, 1996; Mansur *et al.*, 1995).

Complex formation of E6 and E6-AP with p53 leads to ubiquitination of p53 and subsequent degradation (Huibregtse *et al.*, 1991; Scheffner *et al.*, 1993). In fact, blocking E6-AP expression in an antisense approach increased p53 levels only in HPV-positive cells (Beer-Romero *et al.*, 1997). Although E6-AP may lead to ubiquitination of other proteins in the absence of E6, in the case of p53, complex formation via E6 is required for E6-AP-mediated p53 degradation. Binding of E6 to E6-AP is, however, not sufficient for the recruitment of p53 as demonstrated with BPV-1 E6 (Chen *et al.*, 1995). Of note, HPV16 E6 even induces self-ubiquitination and degradation of the E6-AP ubiquitin-protein ligase (Kao *et al.*, 2000).

Blocking p53 functions by interaction with E6 leads to the abrogation of cell cycle checkpoints and inhibits p53-dependent apoptosis. In terms of the viral life cycle this permits the virus to exploit host cell factors for its own

DNA replication and may help virally infected cells to survive. As a "side effect" of neutralising or eliminating the important control molecule p53, however, chromosomal instability may arise leading to the accumulation of genetic mutations, a prerequisite for oncogenic transformation.

2.2.2.2   Cellular transcription factors and co-activators

The papillomaviral oncoproteins E6 and E7 have been shown to modulate transcription of a variety of heterologous promoters (Desaintes *et al.*, 1992). As described above, an important mechanism of how E6 regulates transcription is the abrogation of p53 function. This may be accomplished by binding and degrading p53 in an E6-AP dependent manner. On the other hand, it has recently been reported that the HPV16 E6 oncoprotein may also downregulate p53 activity by targeting the transcriptional co-activator p300/CBP (Patel *et al.*, 1999; Zimmermann *et al.*, 1999). The transcriptional co-activators p300 or CBP are recruited by p53 and synergize with p53 in transcriptional activation. These co-activators have been shown to play a role in many cellular functions including the regulation of cellular differentiation (Missero *et al.*, 1995). Intriguingly, the adenovirus E1A oncoprotein is able to block p53-mediated transactivation in a CBP-dependent manner (Lill *et al.*, 1997; Sang *et al.*, 1997; Somasundaram and El-Deiry, 1997). CBP/p300 activates transcription by modifying histones and non-histone transcription factors via acetylation and by linking DNA-bound transcription factors to components of the basal transcription complex (Nakajima *et al.*, 1997). E6 directly interacts with CBP/p300 through its carboxy terminal zinc finger amino acids (100-147) (Zimmermann *et al.*, 1999). The CBP/p300-dependent effect could be clearly separated from degradation-dependent events. A HPV16 E6 mutant (L50G) that binds CBP/p300, but not E6-AP, was still able to inhibit p53-mediated transcriptional activation (Zimmermann *et al.*, 1999). Thus, p53 function is targeted by HPV E6 proteins in two different ways, (i) by binding and blocking the transcriptional co-activator p300/CBP and (ii) by E6-AP-dependent degradation. The interaction of E6 with p300/CBP may, however, not only influence p53 activity. Complex functional interactions with E6 are also expected from the papillomaviral E2 protein which has recently been shown to bind also to p53 (Massimi *et al.*, 1999) as well as p300/CBP (Lee *et al.*, 2000; Peng *et al.*, 2000).

Besides p53 p300/CBP also co-activates other cellular transcription factors including the c-AMP-regulated enhancer binding protein (CREB) (Chrivia *et al.*, 1993), c-myb (Dai *et al.*, 1996; Oelgeschlager *et al.*, 1996) and YY-1 (Lee *et al.*, 1995). Moreover, p300/CBP has been shown to bind to cytokine-inducible transcription factors. These include members of the activator protein-1 (AP-1) family, signal transducers and activators of transcription (STATs) and nuclear factor-kappa B (NF-κB) (Horvai *et al.*, 1997; Kamei *et al.*, 1996; Perkins *et al.*, 1997), which are involved in cellular differentiation. In fact Patel *et al.* (Patel *et al.*, 1999) could show that HPV16 E6 decreased the ability of p300 to activate a NF-κB-responsive reporter construct. In general, E6-mediated blocking of p300/CBP function may not only have implications on the action of p53 but also on cytokine-mediated signalling and differentiation of the host keratinocyte.

In addition, mucosal as well as cutaneous E6 proteins have been shown to bind to Gps2 or AML-1 (Degenhardt and Silverstein, 2001). This protein has previously been shown to form a complex with the PV E2 protein and CBP/p300 (Breiding *et al.*, 1997; Peng *et al.*, 2000). Binding sites have been mapped for HPV6 E6 involving amino acids 1-36 and 94-131. E6 proteins derived from HPV6 and 18 target Gps2 degradation in vivo and HPV18 E6 suppresses Gps2-mediated transactivation. This might represent a second mechanism of how E6 proteins attenuate the function of CBP/p300.

The E6 protein may also directly bind to components of the host cell transcription machinery and to regulatory factors of specific signalling pathways.

Thus, different E6 proteins were shown to directly bind to the TATA-binding protein (TBP) (Massimi *et al.*, 1997) and to TBP associated factors (TAFs). A detailed in vitro interaction analysis has been performed with HPV8 E6, demonstrating that its amino terminal portion interacts with TBP, $TAF_{II}28$, a N-terminally truncated form of $TAF_{II}135$ and weakly with $TAF_{II}20$ (Enzenauer *et al.*, 1998). The functional significance of these findings, however, still has to be determined.

Interferons (IFNs) are known as potent antiproliferative and anti-viral cytokines (Biron, 1998). Two types of IFNs exist, type I IFN comprising IFN-α and –ß, and type II IFN (IFN-γ). Many viruses have acquired mechanisms to interfere with the IFN regulatory pathways. While some poxviruses for example encode IFN-binding proteins to block the action of type I IFN, human papillomaviruses chose a different mode of action. HPV16 E6 and to a lesser extent HPV18 E6 bind to the so-called interferon regulatory factor-3 (IRF-3) and thereby inhibit the induction of IFN-ß mRNA (Ronco *et al.*, 1998). Moreover, the HPV18 E6 protein has been shown to impair IFN-α but not IFN-γ signalling (Li *et al.*, 1999). In part, this could be explained by physical interaction of E6 with the tyrosine kinase tyk-2. This member of the Janus kinase (JAK) family associates with the cytoplasmic tail of the IFN receptor. After receptor activation JAKs will normally be autophosphorylated and thereby activated to phosphorylate the receptor. The tyrosine-phosphorylated receptor will then be able to bind the src homology 2 (SH2) domains of STATs forming the DNA-binding complex called ISGF-3 (IFN-stimulated gene factor-3). In the presence of HPV18 E6, which is supposed to impair the binding of tyk2 to the IFN-α receptor chain I, tyk-2, STAT1 and STAT2 showed decreased tyrosine phosphorylation after IFN-α treatment. However, these proteins are not degraded as seen with p53. As a consequence, DNA-binding and transactivation capacities of the transcription factor ISGF-3 are reduced.

2.2.2.3   Other E6-interacting proteins

Although the interactions of the E6 oncoprotein especially with the tumor suppressor protein p53 and with other cellular transcriptional regulators may account for many of the E6-mediated alterations in the host cell, they may not explain all observed E6 functions. One example is the fact that E6 may modulate apoptosis not only in a p53- dependent but also in a p53-independent fashion. Therefore, additional cellular targets for E6 interactions were searched and found that might also have important roles in the viral life cycle and/or HPV-induced pathogenesis.

To start with the capacity of the E6 protein to protect cells from undergoing apoptosis, a second cellular protein, Bak, was shown to interact with the E6 protein of HPV18, 16 and 11 (Thomas and Banks, 1998, 1999). For HPV18 E6 the interaction could be mapped to its carboxy terminal part. Bak belongs to the bcl-2 family of apoptosis-regulating proteins, which may form homo- and heterodimers. Whereas bcl-2 and bcl-X_L are anti-apoptotic representatives of this family, bax and bak, for example, display pro-apoptotic qualities. Bak is physiologically expressed in differentiating keratinocytes where papillomaviruses replicate (Krajewski et al., 1996). In vitro, it was observed that E6 derived from HPV18 and HPV16, and to a lesser extent also from HPV11, protected cells from Bak-induced apoptosis. Interestingly, not only E6 binds to Bak, but also E6-AP, even in the absence of E6. Thus, Bak represents the first natural target of E6-AP, which alone or in complex with E6 leads to the ubiquitin/proteasome-mediated degradation of Bak.

Besides playing a role in cellular proliferation and transformation, myc proteins have also been implicated in the induction of cellular apoptosis (Evan et al., 1992). In analogy to Bak and p53, an ubiquitin/proteasome targeting complex is formed between E6, E6-AP and cellular myc proteins (c-myc, N-myc) which are subsequently degraded (Gross-Mesilaty et al., 1998).

Four additional molecules, hDLG (the human homologue of the drosophila discs large protein), MAGI-1 (membrane-associated guanylate kinase with an inverted arrangement of protein-protein interaction domains), hScrib (Vartul, a human homologue of the Drosophila tumor suppressor Scribble) and E6TP1 (E6-targeted protein-1) bind to high risk E6 proteins and are degraded thereafter (Gao *et al.*, 1999; Glaunsinger *et al.*, 2000; Kiyono *et al.*, 1997; Nakagawa and Huibregtse, 2000). These proteins have in common that they encompass PDZ domains (PSD-95/disc large/ZO-1). This is a 80-90 amino acids structural motif involved in protein-protein interactions. HDLG and MAGI-1, members of a protein family called MAGUKs (membrane associated guanylate kinase homologues) as well as hScrib bind at least through one of its PDZ domains to the carboxy terminus of high risk E6 proteins. HPV18 E6 binds to hDLG more efficiently than HPV16 E6 (Gardiol *et al.*, 1999; Kiyono *et al.*, 1997; Lee *et al.*, 1997). Of note, this E6 motif (X-S/T-X-V/L), which is very similar to the recently defined PDZ binding motif X-S/T-X-V, is not conserved in E6 proteins derived from low risk HPV types and appears to be critical for the transforming activity in rodent cells. Transfer of this motif to a low risk E6 protein confers hDLG binding and allows degradation mediated by the respective low risk E6 protein. In contrast to other E6-associated proteins mentioned above, E6-AP does not seem to play a role in hDLG degradation (Pim *et al.*, 2000).

The E6TP1 protein displays homology to GAP-proteins (GTPase-activating proteins). Within this family of proteins several have been identified as tumor suppressor proteins. E6TP-1 is also degraded by high risk, but not by low risk E6 proteins. Although containing a PDZ domain, this domain does not seem to contribute to E6 binding. Nevertheless, E6TP1 was preferentially degraded by E6 mutants that are competent for cellular immortalisation (Gao *et al.*, 1999), further labelling it as a potential transformation-relevant target of E6.

A structurally different protein (E6BP or ERC-55) binding to E6 from BPV-1 or the cancer related HPV types HPV16, HPV18 and HPV31, but not from the low risk types HPV6 and 11, turned out to be a calcium-binding protein of the endoplasmic reticulum (Chen *et al.*, 1995). Its murine homologue, VAF1, specifically interacts with vitamin D receptor (Imai *et al.*, 1997). Binding of E6 to ERC-55 correlated with the transforming activity of BPV-1 E6 mutants suggesting a role in E6-dependent oncogenicity. The binding domain in ERC-55 falls within an EF-hand motif, in which a α-helical motif is of importance for E6 binding. A similar motif predicted to form an α-helix has been identified in the E6-binding region of E6-AP as well as another HPV16 and BPV-1 E6-associated protein paxillin implicated in integrin signalling (Tong and Howley, 1997; Vande Pol *et al.*, 1998). Binding of both, E6-AP and paxillin, to E6 seems to be mutually exclusive. This peptide sequence forming a α-helix may therefore be regarded as an E6-binding motif (Chen *et al.*, 1998). In paxillin, this protein sequence has previously been called LD motif important for the association with vinculin and focal adhesion kinase. For E6 binding the first LD motif appears to be most important (Tong *et al.*, 1997). Mutational analysis of BPV-1 E6 correlates its ability to bind paxillin to anchorage independent growth induced by BPV-1 E6. BPV-1 E6 moreover associates with the trans-Golgi network-specific clathrin adaptor (AP-1) (Tong *et al.*, 1998). The role of this association in the papillomaviral life cycle, however, is unclear at the moment and warrants resolution.

In a yeast two hybrid system screen the minichromosome maintenance protein 7 (MCM7) was found to be associated with the amino terminus of E6 proteins, linking the papillomaviral oncoprotein to proteins involved in the control of cellular DNA replication (Kukimoto *et al.*, 1998). In yeast, E6 proteins derived from cancer-associated HPVs (HPV16, 18 and 58) associate more strongly with MCM7 than the E6 proteins from low risk viruses (HPV6 and 11). It is tempting to speculate that E6 might interfere with pRb binding by this protein as their binding sites in MCM7 overlap, yet such an E6 function has still to be determined. Interestingly, MCM7 may be degraded by HPV18 E6 in an E6-AP-ubiquitin dependant way (Kuhne and Banks, 1998).

2.3   The E7 oncoprotein

2.3.1   E7 protein characteristics and structure

The papillomaviral E7 ORF encodes a zinc-binding protein, which comprises about 100 amino acids and is about 11 kDa in size. It is predominantly located in the nucleus but may also be found in the cytoplasm (Sato *et al.*, 1989; Smotkin and Wettstein, 1987). The advances in understanding of the E7 oncoprotein have profited from the fact that E7 displays similarities with the adenovirus E1A and the simian virus large T antigen. These similarities are not only found on a functional but also on a structural level. All three viral oncoproteins contain two conserved domains or regions (CR1 and CR2). The CR2 regions are substrates of casein kinase II (CKII) whose phosphorylation may led to the enhancement of protein function (Barbosa *et al.*, 1990; Firzlaff *et al.*, 1989; Massimi *et al.*, 1996; Phelps *et al.*, 1992). Also in CR2 of most E7 proteins a motif L-X-C-X-E is found, which is involved in a variety of protein-protein interactions (Chellappan *et al.*, 1992). Besides CR1 and CR2, two C-X-X-C motifs forming a zinc finger (also referred to as CR3) in the carboxy terminus of the E7 protein contribute to its transforming activity (Jewers *et al.*, 1992;

Watanabe *et al.*, 1990). Although the carboxy terminal sequence of E7 is largely unrelated to the E1A and large T antigen sequences, the C-X-X-C motif is also found in E1A (Phelps *et al.*, 1988). The distance between these two motifs (29 or 30 amino acids) is highly conserved among E7 proteins and confers stability to the E7 protein (Watanabe *et al.*, 1992). In this regard, the metal binding domain could be functionally replaced by homologous sequences of the E6 protein (Mavromatis *et al.*, 1997). Besides its role in stabilisation, the carboxy terminal part of E7 also serves as an interaction domain for E7 itself leading to dimerisation (Clemens *et al.*, 1995; McIntyre *et al.*, 1993; Zwerschke *et al.*, 1996) and for cellular factors as described below.

2.3.2   Interactions of E7 with cellular factors

Figure 4 summarises the interactions between HPV16 E7 and cellular factors and illustrates the respective docking sites in E7 characterised so far.

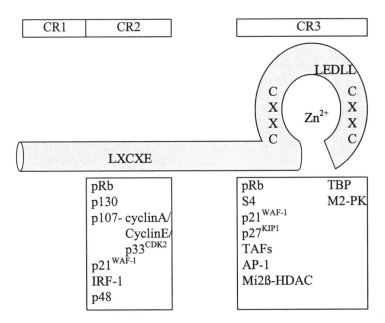

Figure 4. Schematic presentation of the HPV16 E7 protein containing one zinc finger. Interacting proteins whose binding regions have been mapped to HPV16 E7 are indicated.

2.3.2.1   The retinoblastoma family and other cell cycle regulatory proteins

For the HPV life cycle to take place, the host cell must be kept in cycle in order to provide all necessary cellular proteins required for viral replication. However, this is a delicate challenge for the virus: HPV naturally infects keratinocytes which physiologically stop cycling in order to differentiate. This functional state of the keratinocytes is exploited by the virus to express late genes and to mature. Thus, viral functions are required that permit the replication in suprabasal keratinocytes but nevertheless finally allow terminal differentiation of the infected epithelium.

The E7 protein has in fact been shown to enhance cell proliferation (Foster and Galloway, 1996) and to overcome arrest in the G1-phase of the cell cycle induced by a variety of stimuli (Banks et al., 1990; Slebos et al., 1994; Vousden et al., 1993). A number of these functions are executed by the HPV E7 protein via interactions with cell cycle regulatory proteins. The most prominent finding was the direct association of E7 with the retinoblastoma tumor suppressor protein (pRb) (Dyson et al., 1989; Münger et al., 1989). Two further proteins related to pRb, p130 and p107, all containing a related "pocket" domain, were found to bind to E7 (Dyson et al., 1992; Stirdivant et al., 1992). The conserved motif L-X-C-X-E in the CR2 of E7 and structurally related oncoproteins is mainly responsible for binding to these pocket proteins (Barbosa et al., 1990; Lee et al., 1998). In addition, pRb associates also with E7 in its carboxy terminal portion (Patrick et al., 1994). As a consequence of E7 binding, pRb is phosphorylated and degraded via the ubiquitin-proteasome pathway (Boyer et al., 1996) potentially involving the binding of E7 to the S4 subunit of the proteasome complex (Berezutskaya and Bagchi, 1997). This releases transcription factors of the E2F/DP family, which are bound to hypophosphorylated pRb, however at a site which is separate from the E7 binding site (Wu et al., 1993). Free E2F factors subsequently activate cell cycle regulatory genes acting at the G1/S transition, e.g. cyclin E and cyclin A (Botz et al., 1996; Ohtani et al., 1995; Schulze et al., 1995). HPV16 E7 has also been shown to associate with both cyclins A and E and cyclin-dependent kinase 2 ($p33^{CDK2}$) displaying histone H1 kinase activity. The association is most likely mediated via p107 (Davies et al., 1993; Tommasino et al., 1993). The p107 protein is not degraded after E7 binding, but E2F may be physically sequestered from the E2F/p107 complex in the G1 phase (Zerfass et al., 1995) also leading to E2F-dependent gene activation. During S-phase E7 does not disrupt the E2F-p107-cyclinA/$p33^{CDK2}$ complex but seems to associate with it (Arroyo et al., 1993).

A fatal consequence of the binding to and inactivation of pocket proteins was first demonstrated in the developing lens of transgenic mice showing inhibition of differentiation and the ultimate loss of HPV16 E7 expressing cells by p53-dependent apoptosis (Howes *et al.*, 1994; Pan and Griep, 1994). This was not the case if p53 was inactivated underlining the necessity for the virus of also targeting the tumor suppressor protein p53 via E6 co-expression.

Comparing different HPV types, it had been observed that E7 binding to pocket proteins corresponds to its transforming capacity (Heck *et al.*, 1992; Phelps *et al.*, 1992). However, wart induction (as shown for cottontail rabbit PV) and immortalisation capacity has also been observed for E7 mutants with defective pRb binding pointing to the relevance of additional mechanisms (Defeo-Jones *et al.*, 1993; Jewers *et al.*, 1992).

Cyclin dependent kinases (cdks) are key enzymes controlling different cell cycle checkpoints as they inactivate pocket proteins via phosphorylation. This activity is blocked by cyclin dependent kinase inhibitors (cki), e.g. $p21^{WAF-1}$, $p27^{KIP1}$, $p15^{INK4B}$ and $p16^{INK4}$. It could be demonstrated that the E7 protein also binds to the p53-inducible ckis $p21^{WAF-1}$ and $p27^{KIP1}$ (Funk *et al.*, 1997; Jones *et al.*, 1997; Zerfass-Thome *et al.*, 1996). $p21^{WAF-1}$ binding involves the pRb binding site and the carboxy terminus of E7 (Jones *et al.*, 1997). Thereby, the physiological cki function to inactivate cyclin /CDK2 complexes is blocked. As cki binding, and in part also pRB binding, is mediated through the carboxy terminus of E7 it is understandable that both the amino and carboxy terminal portions of the E7 oncogene play an important role in overcoming the cell cycle control and also in cellular transformation.

2.3.2.2 Cellular factors involved in transcriptional regulation

The interactions of the E7 protein with pRb implicate a modulatory role of the viral oncoprotein in gene transcription. As described above, released E2F certainly plays a major role in transactivation. In addition, pRb may directly impair RNA polymerase III transcription (Sutcliffe *et al.*, 2000; Sutcliffe *et al.*, 1999) and block to the polI activator upstream binding factor UBF (Cavanaugh *et al.*, 1995). This inhibition could be reverted by the L-X-C-X-E motif, which is present in E7 proteins.

Apart from these pRb-dependent effects, E7 may regulate transcription through the association with other general or sequence-specific transcription factors. HPV16 and HPV8 E7 were reported to complex with the core component of the TFIID, the TATA Box binding protein TBP (Enzenauer *et al.*, 1998; Massimi *et al.*, 1997; Phillips and Vousden, 1997). Four amino acids in the zinc finger (amino acids 79-83) were shown to be essential for TBP binding (Massimi *et al.*, 1997). The binding affinity, however, is increased by CKII-mediated phosphorylation in the amino terminal portion of HPV16 E7 (Massimi *et al.*, 1997). Although a clear functional consequence has not yet been demonstrated, HPV16 E7 and HPV8 E7 interact with various TBP associated factors (Enzenauer *et al.*, 1998; Mazzarelli *et al.*, 1995). The E7 zinc finger moreover mediates binding to Mi2ß, a component of the NURD histone deacetylase complex, containing also HDAC, a histone deacetylase (Brehm *et al.*, 1999). Histone deacetylation favors the formation of compact, repressive chromatin. The association between E7 and Mi2ß may result in the deregulation of gene expression and cell cycle control. Although pRb may play a role in this regard as it recruits HDAC and thereby represses E2F-regulated S-phase specific genes (Magnaghi-Jaulin *et al.*, 1998), the E7-mediated effect on histone deacetylase is pRb independent. In fact an E7 point mutant still binding pRb but failing to bind Mi2ß/HDAC also fails to overcome cell cycle arrest.

Besides these more general E7-driven mechanisms to regulate gene transcription, the oncoprotein has also been implicated in the modulation of specific signalling pathways, e.g. of the cytokine IFN-α, which is also targeted by the viral E6 protein. E7, however, seems to employ a different mechanism. It impairs the formation of the IFN-α-inducible transcription complex ISGF3 by binding to the DNA-binding component p48 (Barnard and McMillan, 1999). The binding site in HPV16 E7 was mapped to amino acids 17-37, which also comprise the pRb binding site. In this study no effect was observed on IFN-γ-mediated signalling, which does not involve p48. In contrast, it was reported that E7 may also inhibit IFN-γ function by impairing IRF-1 DNA binding activity (Perea *et al.*, 2000). IFN-γ-induced STAT-1 binding activity is not altered by E7. The amino terminal domain of E7 interacts with IRF-1, but it does not target it for degradation. IRF-1 is not only a mediator of IFN-signalling but is also involved in the regulation of the IFN-ß promoter itself (Park *et al.*, 2000). Thus E7 inhibits the IRF-1-mediated activation of the IFN-ß promoter. The negative regulatory activity of E7 does not only require the amino terminal domain of E7 interacting with IRF-1 but also its carboxy terminus involved in recruitment of HDAC independent of pRb.

A controversial discussion has been raised on the effect of the E7 oncoprotein on the function of the activator protein-1 (AP-1). The family of AP-1 transcription factors does not only play a role in transcriptional regulation of a variety of cellular effector molecules, including cytokines and matrix-metalloproteinases, but is also involved in the regulation of keratinocyte differentiation (Benbow and Brinckerhoff, 1997; Eckert and

Welter, 1996). On the one hand, E7 has been described to bind AP-1 factors composed of the family members jun and fos via its zinc finger and to transactivate these factors (Antinore *et al.*, 1996). On the other hand, pRb binds to AP-1 and transactivates gene expression through the AP-1 binding site (Nead *et al.*, 1998; Nishitani *et al.*, 1999). Here, activation of c-jun-mediated transcription by pRb could be relieved by the expression of adenovirus E1A, as well as HPV E7. This result is in concordance with the suggestion that AP-1-mediated transactivation by pRb is important for cell growth control and with the role of the AP-1 family of transcription factors in keratinocyte differentiation.

2.3.2.3   Other E7-interacting proteins

Additional targets interacting with the E7 protein have been defined. One molecule which has been found to interact with the metal binding domain of E7 represents the hTid-1 protein, the human homologue of the Drosophila tumor suppressor protein Tid56 (Schilling *et al.*, 1998). However, the role of this interaction in the viral life cycle is still unclear.

With few exceptions, e.g. the S4 subunit of the proteasome complex (Berezutskaya and Bagchi, 1997), most cellular factors interacting with E7 are found in the nucleus. Nevertheless, cytoplasmic targets of E7, which itself is detectable in the cytoplasm of cells (Sato *et al.*, 1989; Smotkin and Wettstein, 1987) have also been described. As such, E7 has been reported to bind and influence the function of a cytoplasmic enzyme, the M2 pyruvate kinase (M2-PK) (Zwerschke *et al.*, 1999). The C-terminally located amino acid motif L-E-D-L-L in HPV16 E7 (positions 79-83), previously suggested to mediate the interaction with TBP and to play a role in E7-mediated transformation (Massimi *et al.*, 1997), is essential for this interaction. This finding is of particular interest as many tumor cells display an altered carbohydrate metabolism. Binding of E7 results in a conformation-dependent change in the affinity for its substrate phosphoenol-pyruvate. This favors an increase of phosphometabolites then available for biosynthetic processes, which are required in proliferating cells.

E7 has also been shown to form a complex with F-actin, one component of the cellular cytoskeleton (Rey *et al.*, 2000). This interaction seems to be specific as it may be abolished if CKII-dependent E7 phosphorylation is prevented. Additionally, in cells expressing E7, a defect in actin polymerisation was observed.

3.   SUMMARY

Papillomavirus oncoproteins play a major role in the PV life cycle. Their main task is to overcome cell cycle control and to counteract immune surveillance. As they do not display any intrinsic enzymatic activity, they exert their function via protein-protein interactions. Many of these interactions are highly conserved among different HPV types, especially between the "high risk" types, and can be reduced to the conserved overall structure of the oncoproteins. As outlined above these complex interactions may have profound consequences for the host keratinocyte. With respect to the high risk PV, these alterations may even lead beyond the goal of PV amplification, namely the oncogenic transformation of the host cell.

REFERENCES

Andresson, T., Sparkowski, J., Goldstein, D. J., and Schlegel, R., 1995, Vacuolar H(+)-ATPase mutants transform cells and define a binding site for the papillomavirus E5 oncoprotein. *J. Biol. Chem.* **270**: 6830-6837.

Antinore, M. J., Birrer, M. J., Patel, D., Nader, L., and McCance, D. J., 1996, The human papillomavirus type 16 E7 gene product interacts with and trans-activates the AP1 family of transcription factors. *Embo J.* **15**: 1950-1960.

Arroyo, M., Bagchi, S., and Raychaudhuri, P., 1993, Association of the human papillomavirus type 16 E7 protein with the S-phase-specific E2F-cyclin A complex. *Mol. Cell. Biol.* **13**: 6537-6546.

Banks, L., Barnett, S. C., and Crook, T., 1990, HPV-16 E7 functions at the G1 to S phase transition in the cell cycle. *Oncogene* **5**: 833-837.

Barbosa, M. S., Edmonds, C., Fisher, C., Schiller, J. T., Lowy, D. R., and Vousden, K. H., 1990, The region of the HPV E7 oncoprotein homologous to adenovirus E1a and Sv40 large T antigen contains separate domains for Rb binding and casein kinase II phosphorylation. *Embo J.* **9**: 153-160.

Barbosa, M. S., Lowy, D. R., and Schiller, J. T., 1989, Papillomavirus polypeptides E6 and E7 are zinc-binding proteins. *J. Virol.* **63**: 1404-1407.

Barnard, P. and McMillan, N. A., 1999, The human papillomavirus E7 oncoprotein abrogates signaling mediated by interferon-alpha. *Virology* **259**: 305-313.

Beer-Romero, P., Glass, S., and Rolfe, M., 1997, Antisense targeting of E6AP elevates p53 in HPV-infected cells but not in normal cells. *Oncogene* **14**: 595-602.

Benbow, U. and Brinckerhoff, C. E., 1997, The AP-1 site and MMP gene regulation: what is all the fuss about? *Matrix Biol.* **15**: 519-526.

Berezutskaya, E. and Bagchi, S., 1997, The human papillomavirus E7 oncoprotein functionally interacts with the S4 subunit of the 26 S proteasome. *J. Biol. Chem.* **272**: 30135-30140.

Biron, C. A., 1998, Role of early cytokines, including alpha and beta interferons (IFN-alpha/beta), in innate and adaptive immune responses to viral infections. *Semin. Immunol.* **10**: 383-390.

Botz, J., Zerfass-Thome, K., Spitkovsky, D., Delius, H., Vogt, B., Eilers, M., Hatzigeorgiou, A., and Jansen-Durr, P., 1996, Cell cycle regulation of the murine cyclin E gene depends on an E2F binding site in the promoter. *Mol. Cell. Biol.* **16:** 3401-3409.

Boyer, S. N., Wazer, D. E., and Band, V., 1996, E7 protein of human papilloma virus-16 induces degradation of retinoblastoma protein through the ubiquitin-proteasome pathway. *Cancer Res.* **56:** 4620-4624.

Brady, C. S., Duggan-Keen, M. F., Davidson, J. A., Varley, J. M., and Stern, P. L., 1999, Human papillomavirus type 16 E6 variants in cervical carcinoma: relationship to host genetic factors and clinical parameters. *J. Gen. Virol.* **80:** 3233-3240.

Brehm, A., Nielsen, S. J., Miska, E. A., McCance, D. J., Reid, J. L., Bannister, A. J., and Kouzarides, T., 1999, The E7 oncoprotein associates with Mi2 and histone deacetylase activity to promote cell growth. *Embo J.* **18:** 2449-2458.

Breiding, D. E., Sverdrup, F., Grossel, M. J., Moscufo, N., Boonchai, W., and Androphy, E. J., 1997, Functional interaction of a novel cellular protein with the papillomavirus E2 transactivation domain. *Mol. Cell. Biol.* **17:** 7208-7219.

Bubb, V., McCance, D. J., and Schlegel, R., 1988, DNA sequence of the HPV-16 E5 ORF and the structural conservation of its encoded protein. *Virology* **163:** 243-246.

Burkhardt, A., Willingham, M., Gay, C., Jeang, K. T., and Schlegel, R., 1989, The E5 oncoprotein of bovine papillomavirus is oriented asymmetrically in Golgi and plasma membranes. *Virology* **170:** 334-339.

Cavanaugh, A. H., Hempel, W. M., Taylor, L. J., Rogalsky, V., Todorov, G., and Rothblum, L. I., 1995, Activity of RNA polymerase I transcription factor UBF blocked by Rb gene product. *Nature* **374:** 177-180.

Chellappan, S., Kraus, V. B., Kroger, B., Münger, K., Howley, P. M., Phelps, W. C., and Nevins, J. R., 1992, Adenovirus E1A, simian virus 40 tumor antigen, and human papillomavirus E7 protein share the capacity to disrupt the interaction between transcription factor E2F and the retinoblastoma gene product. *Proc. Natl. Acad. Sci U S A* **89:** 4549-4553.

Chen, J. J., Hong, Y., Rustamzadeh, E., Baleja, J. D., and Androphy, E. J., 1998, Identification of an alpha helical motif sufficient for association with papillomavirus E6. *J. Biol. Chem.* **273:** 13537-13544.

Chen, J. J., Reid, C. E., Band, V., and Androphy, E. J., 1995, Interaction of papillomavirus E6 oncoproteins with a putative calcium-binding protein. *Science* **269:** 529-531.

Chrivia, J. C., Kwok, R. P., Lamb, N., Hagiwara, M., Montminy, M. R., and Goodman, R. H., 1993, Phosphorylated CREB binds specifically to the nuclear protein CBP. *Nature* **365:** 855-859.

Clemens, K. E., Brent, R., Gyuris, J., and Münger, K., 1995, Dimerization of the human papillomavirus E7 oncoprotein in vivo. *Virology* **214:** 289-293.

Cohen, B. D., Goldstein, D. J., Rutledge, L., Vass, W. C., Lowy, D. R., Schlegel, R., and Schiller, J. T., 1993, Transformation-specific interaction of the bovine papillomavirus E5 oncoprotein with the platelet-derived growth factor receptor transmembrane domain and the epidermal growth factor receptor cytoplasmic domain. *J. Virol.* **67:** 5303-5311.

Cole, S. T. and Danos, O., 1987, Nucleotide sequence and comparative analysis of the human papillomavirus type 18 genome. Phylogeny of papillomaviruses and repeated structure of the E6 and E7 gene products. *J. Mol. Biol.* **193:** 599-608.

Conrad, M., Bubb, V. J., and Schlegel, R., 1993, The human papillomavirus type 6 and 16 E5 proteins are membrane-associated proteins which associate with the 16-kilodalton pore-forming protein. *J. Virol.* **67:** 6170-6178.

Conrad, M., Goldstein, D., Andresson, T., and Schlegel, R., 1994, The E5 protein of HPV-6, but not HPV-16, associates efficiently with cellular growth factor receptors. *Virology* **200:** 796-800.

Conrad-Stöppler, M., Straight, S. W., Tsao, G., Schlegel, R., and McCance, D. J., 1996, The E5 gene of HPV-16 enhances keratinocyte immortalization by full-length DNA. *Virology* **223:** 251-254.

Crook, T., Fisher, C., Masterson, P. J., and Vousden, K. H., 1994, Modulation of transcriptional regulatory properties of p53 by HPV E6. *Oncogene* **9:** 1225-1230.

Crook, T., Tidy, J. A., and Vousden, K. H., 1991, Degradation of p53 can be targeted by HPV E6 sequences distinct from those required for p53 binding and trans-activation. *Cell* **67:** 547-556.

Crusius, K., Auvinen, E., and Alonso, A., 1997, Enhancement of EGF- and PMA-mediated MAP kinase activation in cells expressing the human papillomavirus type 16 E5 protein. *Oncogene* **15:** 1437-1444.

Dai, P., Akimaru, H., Tanaka, Y., Hou, D. X., Yasukawa, T., Kanei-Ishii, C., Takahashi, T., and Ishii, S., 1996, CBP as a transcriptional coactivator of c-Myb. *Genes Dev.* **10:** 528-540.

Dalal, S., Gao, Q., Androphy, E. J., and Band, V., 1996, Mutational analysis of human papillomavirus type 16 E6 demonstrates that p53 degradation is necessary for immortalization of mammary epithelial cells. *J. Virol.* **70:** 683-688.

Davies, R., Hicks, R., Crook, T., Morris, J., and Vousden, K., 1993, Human papillomavirus type 16 E7 associates with a histone H1 kinase and with p107 through sequences necessary for transformation. *J. Virol.* **67:** 2521-2528.

Defeo-Jones, D., Vuocolo, G. A., Haskell, K. M., Hanobik, M. G., Kiefer, D. M., McAvoy, E. M., Ivey-Hoyle, M., Brandsma, J. L., Oliff, A., and Jones, R. E., 1993, Papillomavirus E7 protein binding to the retinoblastoma protein is not required for viral induction of warts. *J. Virol.* **67:** 716-725.

Degenhardt, Y. Y. and Silverstein, S. J., 2001, Gps2, a protein partner for human papillomavirus E6 proteins [In Process Citation]. *J. Virol.* **75:** 151-160.

Desaintes, C. and Demeret, C., 1996, Control of papillomavirus DNA replication and transcription. *Semin. Cancer Biol.* **7:** 339-347.

Desaintes, C., Hallez, S., Van Alphen, P., and Burny, A., 1992, Transcriptional activation of several heterologous promoters by the E6 protein of human papillomavirus type 16. *J. Virol.* **66:** 325-333.

Dyson, N., Guida, P., Münger, K., and Harlow, E., 1992, Homologous sequences in adenovirus E1A and human papillomavirus E7 proteins mediate interaction with the same set of cellular proteins. *J. Virol.* **66:** 6893-6902.

Dyson, N., Howley, P. M., Münger, K., and Harlow, E., 1989, The human papilloma virus-16 E7 oncoprotein is able to bind to the retinoblastoma gene product. *Science* **243:** 934-937.

Eckert, R. L., Crish, J. F., and Robinson, N. A., 1997, The epidermal keratinocyte as a model for the study of gene regulation and cell differentiation. *Physiol. Rev.* **77:** 397-424.

Eckert, R. L. and Welter, J. F., 1996, Transcription factor regulation of epidermal keratinocyte gene expression. *Mol. Biol. Rep.* **23:** 59-70.

El-Deiry, W. S., Tokino, T., Velculescu, V. E., Levy, D. B., Parsons, R., Trent, J. M., Lin, D., Mercer, W. E., Kinzler, K. W., and Vogelstein, B., 1993, WAF1, a potential mediator of p53 tumor suppression. *Cell* **75:** 817-825.

Elbel, M., Carl, S., Spaderna, S., and Iftner, T., 1997, A comparative analysis of the interactions of the E6 proteins from cutaneous and genital papillomaviruses with p53 and E6AP in correlation to their transforming potential. *Virology* **239:** 132-149.

Enzenauer, C., Mengus, G., Lavigne, A., Davidson, I., Pfister, H., and May, M., 1998, Interaction of human papillomavirus 8 regulatory proteins E2, E6 and E7 with components of the TFIID complex. *Intervirology* **41**: 80-90.

Evan, G. I., Wyllie, A. H., Gilbert, C. S., Littlewood, T. D., Land, H., Brooks, M., Waters, C. M., Penn, L. Z., and Hancock, D. C., 1992, Induction of apoptosis in fibroblasts by c-myc protein. *Cell* **69**: 119-128.

Firzlaff, J. M., Galloway, D. A., Eisenman, R. N., and Luscher, B., 1989, The E7 protein of human papillomavirus type 16 is phosphorylated by casein kinase II. *New Biol.* **1**: 44-53.

Foster, S. A., Demers, G. W., Etscheid, B. G., and Galloway, D. A., 1994, The ability of human papillomavirus E6 proteins to target p53 for degradation in vivo correlates with their ability to abrogate actinomycin D-induced growth arrest. *J. Virol.* **68**: 5698-5705.

Foster, S. A. and Galloway, D. A., 1996, Human papillomavirus type 16 E7 alleviates a proliferation block in early passage human mammary epithelial cells. *Oncogene* **12**: 1773-1779.

Fuchs, P. G., Iftner, T., Weninger, J., and Pfister, H., 1986, Epidermodysplasia verruciformis-associated human papillomavirus 8: genomic sequence and comparative analysis. *J. Virol.* **58**: 626-634.

Funk, J. O., Waga, S., Harry, J. B., Espling, E., Stillman, B., and Galloway, D. A., 1997, Inhibition of CDK activity and PCNA-dependent DNA replication by p21 is blocked by interaction with the HPV-16 E7 oncoprotein. *Genes Dev.* **11**: 2090-2100.

Gao, Q., Kumar, A., Srinivasan, S., Singh, L., Mukai, H., Ono, Y., Wazer, D. E., and Band, V., 2000, PKN binds and phosphorylates human papillomavirus E6 oncoprotein. *J. Biol. Chem.* **275**: 14824-14830.

Gao, Q., Srinivasan, S., Boyer, S. N., Wazer, D. E., and Band, V., 1999, The E6 oncoproteins of high-risk papillomaviruses bind to a novel putative GAP protein, E6TP1, and target it for degradation. *Mol. Cell. Biol.* **19**: 733-744.

Gardiol, D., Kuhne, C., Glaunsinger, B., Lee, S. S., Javier, R., and Banks, L., 1999, Oncogenic human papillomavirus E6 proteins target the discs large tumour suppressor for proteasome-mediated degradation. *Oncogene* **18**: 5487-5496.

Glaunsinger, B. A., Lee, S. S., Thomas, M., Banks, L., and Javier, R., 2000, Interactions of the PDZ-protein MAGI-1 with adenovirus E4-ORF1 and high-risk papillomavirus E6 oncoproteins [In Process Citation]. *Oncogene* **19**: 5270-5280.

Goldstein, D. J., Li, W., Wang, L. M., Heidaran, M. A., Aaronson, S., Shinn, R., Schlegel, R., and Pierce, J. H., 1994, The bovine papillomavirus type 1 E5 transforming protein specifically binds and activates the beta-type receptor for the platelet-derived growth factor but not other related tyrosine kinase-containing receptors to induce cellular transformation. *J. Virol.* **68**: 4432-4441.

Goldstein, D. J. and Schlegel, R., 1990, The E5 oncoprotein of bovine papillomavirus binds to a 16 kd cellular protein. *Embo J.* **9**: 137-145.

Gross-Mesilaty, S., Reinstein, E., Bercovich, B., Tobias, K. E., Schwartz, A. L., Kahana, C., and Ciechanover, A., 1998, Basal and human papillomavirus E6 oncoprotein-induced degradation of Myc proteins by the ubiquitin pathway. *Proc. Natl. Acad. Sci U S A* **95**: 8058-8063.

Grossman, S. R., Mora, R., and Laimins, L. A., 1989, Intracellular localization and DNA-binding properties of human papillomavirus type 18 E6 protein expressed with a baculovirus vector. *J. Virol.* **63**: 366-374.

Gu, Z. and Matlashewski, G., 1995, Effect of human papillomavirus type 16 oncogenes on MAP kinase activity. *J. Virol.* **69**: 8051-8056.

Heck, D. V., Yee, C. L., Howley, P. M., and Münger, K., 1992, Efficiency of binding the retinoblastoma protein correlates with the transforming capacity of the E7 oncoproteins of the human papillomaviruses. *Proc. Natl. Acad. Sci U S A* **89**: 4442-4446.

Helland, A., Langerod, A., Johnsen, H., Olsen, A. O., Skovlund, E., and Borresen-Dale, A. L., 1998, p53 polymorphism and risk of cervical cancer. *Nature* **396**: 530-531; discussion 532.

Hildesheim, A., Schiffman, M., Brinton, L. A., Fraumeni, J. F., Jr., Herrero, R., Bratti, M. C., Schwartz, P., Mortel, R., Barnes, W., Greenberg, M., McGowan, L., Scott, D. R., Martin, M., Herrera, J. E., and Carrington, M., 1998, p53 polymorphism and risk of cervical cancer [letter; comment]. *Nature* **396**: 531-532.

Hoppe-Seyler, F. and Butz, K., 1993, Repression of endogenous p53 transactivation function in HeLa cervical carcinoma cells by human papillomavirus type 16 E6, human mdm-2, and mutant p53. *J. Virol.* **67**: 3111-3117.

Horvai, A. E., Xu, L., Korzus, E., Brard, G., Kalafus, D., Mullen, T. M., Rose, D. W., Rosenfeld, M. G., and Glass, C. K., 1997, Nuclear integration of JAK/STAT and Ras/AP-1 signaling by CBP and p300. *Proc. Natl. Acad. Sci U S A* **94**: 1074-1079.

Howes, K. A., Ransom, N., Papermaster, D. S., Lasudry, J. G., Albert, D. M., and Windle, J. J., 1994, Apoptosis or retinoblastoma: alternative fates of photoreceptors expressing the HPV-16 E7 gene in the presence or absence of p53 [published erratum appears in Genes Dev 1994 Jul 15;8(14):1738]. *Genes Dev.* **8**: 1300-1310.

Hubbert, N. L., Sedman, S. A., and Schiller, J. T., 1992, Human papillomavirus type 16 E6 increases the degradation rate of p53 in human keratinocytes. *J. Virol.* **66**: 6237-6241.

Huibregtse, J. M., Scheffner, M., and Howley, P. M., 1991, A cellular protein mediates association of p53 with the E6 oncoprotein of human papillomavirus types 16 or 18. *Embo J.* **10**: 4129-4135.

Huibregtse, J. M., Scheffner, M., and Howley, P. M., 1993a, Cloning and expression of the cDNA for E6-AP, a protein that mediates the interaction of the human papillomavirus E6 oncoprotein with p53. *Mol. Cell. Biol.* **13**: 775-784.

Huibregtse, J. M., Scheffner, M., and Howley, P. M., 1993b, Localization of the E6-AP regions that direct human papillomavirus E6 binding, association with p53, and ubiquitination of associated proteins. *Mol. Cell. Biol.* **13**: 4918-4927.

Imai, T., Matsuda, K., Shimojima, T., Hashimoto, T., Masuhiro, Y., Kitamoto, T., Sugita, A., Suzuki, K., Matsumoto, H., Masushige, S., Nogi, Y., Muramatsu, M., Handa, H., and Kato, S., 1997, ERC-55, a binding protein for the papilloma virus E6 oncoprotein, specifically interacts with vitamin D receptor among nuclear receptors. *Biochem. Biophys. Res. Commun.* **233**: 765-769.

Jewers, R. J., Hildebrandt, P., Ludlow, J. W., Kell, B., and McCance, D. J., 1992, Regions of human papillomavirus type 16 E7 oncoprotein required for immortalization of human keratinocytes. *J. Virol.* **66**: 1329-1335.

Jones, D. L., Alani, R. M., and Münger, K., 1997, The human papillomavirus E7 oncoprotein can uncouple cellular differentiation and proliferation in human keratinocytes by abrogating p21Cip1-mediated inhibition of cdk2. *Genes Dev.* **11**: 2101-2111.

Josefsson, A. M., Magnusson, P. K., Ylitalo, N., Quarforth-Tubbin, P., Ponten, J., Adami, H. O., and Gyllensten, U. B., 1998, p53 polymorphism and risk of cervical cancer. *Nature* **396**: 531; discussion 532.

Kamei, Y., Xu, L., Heinzel, T., Torchia, J., Kurokawa, R., Gloss, B., Lin, S. C., Heyman, R. A., Rose, D. W., Glass, C. K., and Rosenfeld, M. G., 1996, A CBP integrator complex mediates transcriptional activation and AP-1 inhibition by nuclear receptors. *Cell* **85**: 403-414.

Kanda, T., Watanabe, S., Zanma, S., Sato, H., Furuno, A., and Yoshiike, K., 1991, Human papillomavirus type 16 E6 proteins with glycine substitution for cysteine in the metal-binding motif. *Virology* **185**: 536-543.

Kao, W. H., Beaudenon, S. L., Talis, A. L., Huibregtse, J. M., and Howley, P. M., 2000, Human papillomavirus type 16 E6 induces self-ubiquitination of the E6AP ubiquitin-protein ligase. *J. Virol.* **74**: 6408-6417.

Kell, B., Jewers, R. J., Cason, J., Pakarian, F., Kaye, J. N., and Best, J. M., 1994, Detection of E5 oncoprotein in human papillomavirus type 16-positive cervical scrapes using antibodies raised to synthetic peptides. *J. Gen. Virol.* **75**: 2451-2456.

Kessis, T. D., Connolly, D. C., Hedrick, L., and Cho, K. R., 1996, Expression of HPV16 E6 or E7 increases integration of foreign DNA. *Oncogene* **13**: 427-431.

Kiyono, T., Hiraiwa, A., Fujita, M., Hayashi, Y., Akiyama, T., and Ishibashi, M., 1997, Binding of high-risk human papillomavirus E6 oncoproteins to the human homologue of the Drosophila discs large tumor suppressor protein. *Proc. Natl. Acad. Sci U S A* **94**: 11612-11616.

Kiyono, T., Hiraiwa, A., Ishii, S., Takahashi, T., and Ishibashi, M., 1994, Inhibition of p53-mediated transactivation by E6 of type 1, but not type 5, 8, or 47, human papillomavirus of cutaneous origin. *J. Virol.* **68**: 4656-4661.

Klaes, R., Woerner, S. M., Ridder, R., Wentzensen, N., Duerst, M., Schneider, A., Lotz, B., Melsheimer, P., and von Knebel Doeberitz, M., 1999, Detection of high-risk cervical intraepithelial neoplasia and cervical cancer by amplification of transcripts derived from integrated papillomavirus oncogenes. *Cancer Res.* **59**: 6132-6136.

Klingelhutz, A. J., Foster, S. A., and McDougall, J. K., 1996, Telomerase activation by the E6 gene product of human papillomavirus type 16. *Nature* **380**: 79-82.

Krajewski, S., Krajewska, M., and Reed, J. C., 1996, Immunohistochemical analysis of in vivo patterns of Bak expression, a proapoptotic member of the Bcl-2 protein family. *Cancer Res.* **56**: 2849-2855.

Kuhne, C. and Banks, L., 1998, E3-ubiquitin ligase/E6-AP links multicopy maintenance protein 7 to the ubiquitination pathway by a novel motif, the L2G box. *J. Biol. Chem.* **273**: 34302-34309.

Kukimoto, I., Aihara, S., Yoshiike, K., and Kanda, T., 1998, Human papillomavirus oncoprotein E6 binds to the C-terminal region of human minichromosome maintenance 7 protein. *Biochem. Biophys. Res. Commun.* **249**: 258-262.

Lane, D. P. and Crawford, L. V., 1979, T antigen is bound to a host protein in SV40-transformed cells. *Nature* **278**: 261-263.

Lechner, M. S. and Laimins, L. A., 1994, Inhibition of p53 DNA binding by human papillomavirus E6 proteins. *J. Virol.* **68**: 4262-4273.

Lee, D., Lee, B., Kim, J., Kim, D. W., and Choe, J., 2000, cAMP response element-binding protein-binding protein binds to human papillomavirus E2 protein and activates E2-dependent transcription. *J. Biol. Chem.* **275**: 7045-7051.

Lee, J. O., Russo, A. A., and Pavletich, N. P., 1998, Structure of the retinoblastoma tumour-suppressor pocket domain bound to a peptide from HPV E7. *Nature* **391**: 859-65.

Lee, J. S., Galvin, K. M., See, R. H., Eckner, R., Livingston, D., Moran, E., and Shi, Y., 1995, Relief of YY1 transcriptional repression by adenovirus E1A is mediated by E1A-associated protein p300 [published erratum appears in Genes Dev 1995 Aug 1;9(15):1948-9]. *Genes Dev.* **9**: 1188-1198.

Lee, S. S., Weiss, R. S., and Javier, R. T., 1997, Binding of human virus oncoproteins to hDlg/SAP97, a mammalian homolog of the Drosophila discs large tumor suppressor protein. *Proc. Natl. Acad. Sci U S A* **94**: 6670-6675.

Leechanachai, P., Banks, L., Moreau, F., and Matlashewski, G., 1992, The E5 gene from human papillomavirus type 16 is an oncogene which enhances growth factor-mediated signal transduction to the nucleus. *Oncogene* **7:** 19-25.

Li, S., Labrecque, S., Gauzzi, M. C., Cuddihy, A. R., Wong, A. H., Pellegrini, S., Matlashewski, G. J., and Koromilas, A. E., 1999, The human papilloma virus (HPV)-18 E6 oncoprotein physically associates with Tyk2 and impairs Jak-STAT activation by interferon-alpha. *Oncogene* **18:** 5727-5737.

Li, X. and Coffino, P., 1996, High-risk human papillomavirus E6 protein has two distinct binding sites within p53, of which only one determines degradation. *J. Virol.* **70:** 4509-4516.

Lill, N. L., Grossman, S. R., Ginsberg, D., DeCaprio, J., and Livingston, D. M., 1997, Binding and modulation of p53 by p300/CBP coactivators. *Nature* **387:** 823-827.

Linzer, D. I. and Levine, A. J., 1979, Characterization of a 54K dalton cellular SV40 tumor antigen present in SV40-transformed cells and uninfected embryonal carcinoma cells. *Cell* **17:** 43-52.

Magnaghi-Jaulin, L., Groisman, R., Naguibneva, I., Robin, P., Lorain, S., Le Villain, J. P., Troalen, F., Trouche, D., and Harel-Bellan, A., 1998, Retinoblastoma protein represses transcription by recruiting a histone deacetylase. *Nature* **391:** 601-605.

Mansur, C. P., Marcus, B., Dalal, S., and Androphy, E. J., 1995, The domain of p53 required for binding HPV 16 E6 is separable from the degradation domain. *Oncogene* **10:** 457-465.

Martin, P., Vass, W. C., Schiller, J. T., Lowy, D. R., and Velu, T. J., 1989, The bovine papillomavirus E5 transforming protein can stimulate the transforming activity of EGF and CSF-1 receptors. *Cell* **59:** 21-32.

Massimi, P., Pim, D., and Banks, L., 1997, Human papillomavirus type 16 E7 binds to the conserved carboxy-terminal region of the TATA box binding protein and this contributes to E7 transforming activity. *J. Gen. Virol.* **78:** 2607-2613.

Massimi, P., Pim, D., Bertoli, C., Bouvard, V., and Banks, L., 1999, Interaction between the HPV-16 E2 transcriptional activator and p53. *Oncogene* **18:** 7748-7754.

Massimi, P., Pim, D., Storey, A., and Banks, L., 1996, HPV-16 E7 and adenovirus E1a complex formation with TATA box binding protein is enhanced by casein kinase II phosphorylation. *Oncogene* **12:** 2325-2330.

Mavromatis, K. O., Jones, D. L., Mukherjee, R., Yee, C., Grace, M., and Münger, K., 1997, The carboxyl-terminal zinc-binding domain of the human papillomavirus E7 protein can be functionally replaced by the homologous sequences of the E6 protein. *Virus Res.* **52:** 109-118.

Mazzarelli, J. M., Atkins, G. B., Geisberg, J. V., and Ricciardi, R. P., 1995, The viral oncoproteins Ad5 E1A, HPV16 E7 and SV40 TAg bind a common region of the TBP-associated factor-110. *Oncogene* **11:** 1859-1864.

McIntyre, M. C., Frattini, M. G., Grossman, S. R., and Laimins, L. A., 1993, Human papillomavirus type 18 E7 protein requires intact Cys-X-X-Cys motifs for zinc binding, dimerization, and transformation but not for Rb binding. *J. Virol.* **67:** 3142-3150.

Minaguchi, T., Kanamori, Y., Matsushima, M., Yoshikawa, H., Taketani, Y., and Nakamura, Y., 1998, No evidence of correlation between polymorphism at codon 72 of p53 and risk of cervical cancer in Japanese patients with human papillomavirus 16/18 infection. *Cancer Res.* **58:** 4585-4586.

Missero, C., Calautti, E., Eckner, R., Chin, J., Tsai, L. H., Livingston, D. M., and Dotto, G. P., 1995, Involvement of the cell-cycle inhibitor Cip1/WAF1 and the E1A-associated p300 protein in terminal differentiation. *Proc. Natl. Acad. Sci U S A* **92:** 5451-5455.

Münger, K., Werness, B. A., Dyson, N., Phelps, W. C., Harlow, E., and Howley, P. M., 1989, Complex formation of human papillomavirus E7 proteins with the retinoblastoma tumor suppressor gene product. *Embo J.* **8**: 4099-4105.

Nakagawa, S. and Huibregtse, J. M., 2000, Human scribble (Vartul) is targeted for ubiquitin-mediated degradation by the high-risk papillomavirus E6 proteins and the E6AP ubiquitin-protein ligase [In Process Citation]. *Mol. Cell. Biol.* **20**: 8244-8253.

Nakagawa, S., Watanabe, S., Yoshikawa, H., Taketani, Y., Yoshiike, K., and Kanda, T., 1995, Mutational analysis of human papillomavirus type 16 E6 protein: transforming function for human cells and degradation of p53 in vitro. *Virology* **212**: 535-542.

Nakajima, T., Uchida, C., Anderson, S. F., Lee, C. G., Hurwitz, J., Parvin, J. D., and Montminy, M., 1997, RNA helicase A mediates association of CBP with RNA polymerase II. *Cell* **90**: 1107-1112.

Nead, M. A., Baglia, L. A., Antinore, M. J., Ludlow, J. W., and McCance, D. J., 1998, Rb binds c-Jun and activates transcription. *Embo J.* **17**: 2342-2352.

Nishitani, J., Nishinaka, T., Cheng, C. H., Rong, W., Yokoyama, K. K., and Chiu, R., 1999, Recruitment of the retinoblastoma protein to c-Jun enhances transcription activity mediated through the AP-1 binding site. *J. Biol. Chem.* **274**: 5454-5461.

Oelgeschlager, M., Janknecht, R., Krieg, J., Schreek, S., and Luscher, B., 1996, Interaction of the co-activator CBP with Myb proteins: effects on Myb-specific transactivation and on the cooperativity with NF-M. *Embo J.* **15**: 2771-2780.

Ohtani, K., DeGregori, J., and Nevins, J. R., 1995, Regulation of the cyclin E gene by transcription factor E2F1. *Proc. Natl. Acad. Sci U S A* **92**: 12146-12150.

Pan, H. and Griep, A. E., 1994, Altered cell cycle regulation in the lens of HPV-16 E6 or E7 transgenic mice: implications for tumor suppressor gene function in development. *Genes Dev.* **8**: 1285-1299.

Park, J. S., Kim, E. J., Kwon, H. J., Hwang, E. S., Namkoong, S. E., and Um, S. J., 2000, Inactivation of interferon regulatory factor-1 tumor suppressor protein by HPV E7 oncoprotein. Implication for the E7-mediated immune evasion mechanism in cervical carcinogenesis. *J. Biol. Chem.* **275**: 6764-6769.

Patel, D., Huang, S. M., Baglia, L. A., and McCance, D. J., 1999, The E6 protein of human papillomavirus type 16 binds to and inhibits co-activation by CBP and p300. *Embo J.* **18**: 5061-5072.

Patrick, D. R., Oliff, A., and Heimbrook, D. C., 1994, Identification of a novel retinoblastoma gene product binding site on human papillomavirus type 16 E7 protein. *J. Biol. Chem.* **269**: 6842-6850.

Peng, Y. C., Breiding, D. E., Sverdrup, F., Richard, J., and Androphy, E. J., 2000, AMF-1/Gps2 binds p300 and enhances its interaction with papillomavirus E2 proteins. *J. Virol.* **74**: 5872-5879.

Perea, S. E., Massimi, P., and Banks, L., 2000, Human papillomavirus type 16 E7 impairs the activation of the interferon regulatory factor-1. *Int. J. Mol. Med.* **5**: 661-666.

Perkins, N. D., Felzien, L. K., Betts, J. C., Leung, K., Beach, D. H., and Nabel, G. J., 1997, Regulation of NF-kappaB by cyclin-dependent kinases associated with the p300 coactivator. *Science* **275**: 523-527.

Petti, L. and DiMaio, D., 1992, Stable association between the bovine papillomavirus E5 transforming protein and activated platelet-derived growth factor receptor in transformed mouse cells. *Proc. Natl. Acad. Sci U S A* **89**: 6736-6740.

Phelps, W. C., Münger, K., Yee, C. L., Barnes, J. A., and Howley, P. M., 1992, Structure-function analysis of the human papillomavirus type 16 E7 oncoprotein. *J. Virol.* **66**: 2418-2427.

Phelps, W. C., Yee, C. L., Münger, K., and Howley, P. M., 1988, The human papillomavirus type 16 E7 gene encodes transactivation and transformation functions similar to those of adenovirus E1A. *Cell* **53**: 539-547.

Phillips, A. C. and Vousden, K. H., 1997, Analysis of the interaction between human papillomavirus type 16 E7 and the TATA-binding protein, TBP. *J. Gen. Virol.* **78**: 905-909.

Pim, D., Collins, M., and Banks, L., 1992, Human papillomavirus type 16 E5 gene stimulates the transforming activity of the epidermal growth factor receptor. *Oncogene* **7**: 27-32.

Pim, D., Massimi, P., and Banks, L., 1997, Alternatively spliced HPV-18 E6* protein inhibits E6 mediated degradation of p53 and suppresses transformed cell growth. *Oncogene* **15**: 257-264.

Pim, D., Storey, A., Thomas, M., Massimi, P., and Banks, L., 1994, Mutational analysis of HPV-18 E6 identifies domains required for p53 degradation in vitro, abolition of p53 transactivation in vivo and immortalisation of primary BMK cells. *Oncogene* **9**: 1869-1876.

Pim, D., Thomas, M., Javier, R., Gardiol, D., and Banks, L., 2000, HPV E6 targeted degradation of the discs large protein: evidence for the involvement of a novel ubiquitin ligase. *Oncogene* **19**: 719-725.

Rey, O., Lee, S., Baluda, M. A., Swee, J., Ackerson, B., Chiu, R., and Park, N. H., 2000, The E7 oncoprotein of human papillomavirus type 16 interacts with F-actin in vitro and in vivo. *Virology* **268**: 372-381.

Rodriguez, M. I., Finbow, M. E., and Alonso, A., 2000, Binding of human papillomavirus 16 E5 to the 16 kDa subunit c (proteolipid) of the vacuolar H+-ATPase can be dissociated from the E5-mediated epidermal growth factor receptor overactivation. *Oncogene* **19**: 3727-3732.

Ronco, L. V., Karpova, A. Y., Vidal, M., and Howley, P. M., 1998, Human papillomavirus 16 E6 oncoprotein binds to interferon regulatory factor-3 and inhibits its transcriptional activity. *Genes Dev.* **12**: 2061-2072.

Sang, N., Avantaggiati, M. L., and Giordano, A., 1997, Roles of p300, pocket proteins, and hTBP in E1A-mediated transcriptional regulation and inhibition of p53 transactivation activity. *J. Cell Biochem.* **66**: 277-285.

Sarnow, P., Ho, Y. S., Williams, J., and Levine, A. J., 1982, Adenovirus E1b-58kd tumor antigen and SV40 large tumor antigen are physically associated with the same 54 kd cellular protein in transformed cells. *Cell* **28**: 387-394.

Sato, H., Watanabe, S., Furuno, A., and Yoshiike, K., 1989, Human papillomavirus type 16 E7 protein expressed in Escherichia coli and monkey COS-1 cells: immunofluorescence detection of the nuclear E7 protein. *Virology* **170**: 311-315.

Scheffner, M., Huibregtse, J. M., Vierstra, R. D., and Howley, P. M., 1993, The HPV-16 E6 and E6-AP complex functions as a ubiquitin-protein ligase in the ubiquitination of p53. *Cell* **75**: 495-505.

Scheffner, M., Werness, B. A., Huibregtse, J. M., Levine, A. J., and Howley, P. M., 1990, The E6 oncoprotein encoded by human papillomavirus types 16 and 18 promotes the degradation of p53. *Cell* **63**: 1129-1136.

Schilling, B., De-Medina, T., Syken, J., Vidal, M., and Münger, K., 1998, A novel human DnaJ protein, hTid-1, a homolog of the Drosophila tumor suppressor protein Tid56, can interact with the human papillomavirus type 16 E7 oncoprotein. *Virology* **247**: 74-85.

Schulze, A., Zerfass, K., Spitkovsky, D., Middendorp, S., Berges, J., Helin, K., Jansen-Durr, P., and Henglein, B., 1995, Cell cycle regulation of the cyclin A gene promoter is mediated by a variant E2F site. *Proc. Natl. Acad. Sci U S A* **92**: 11264-11268.

Sherman, L., Jackman, A., Itzhaki, H., Conrad-Stöppler, M., Koval, D., and Schlegel, R., 1997, Inhibition of serum- and calcium-induced differentiation of human keratinocytes by HPV16 E6 oncoprotein: role of p53 inactivation. *Virology* **237**: 296-306.

Slebos, R. J., Lee, M. H., Plunkett, B. S., Kessis, T. D., Williams, B. O., Jacks, T., Hedrick, L., Kastan, M. B., and Cho, K. R., 1994, p53-dependent G1 arrest involves pRB-related proteins and is disrupted by the human papillomavirus 16 E7 oncoprotein. *Proc. Natl. Acad. Sci U S A* **91**: 5320-5324.

Smotkin, D. and Wettstein, F. O., 1987, The major human papillomavirus protein in cervical cancers is a cytoplasmic phosphoprotein. *J. Virol.* **61**: 1686-1689.

Somasundaram, K., 2000, Tumor suppressor p53: regulation and function. *Front. Biosci.* **5**: D424-D437.

Somasundaram, K. and El-Deiry, W. S., 1997, Inhibition of p53-mediated transactivation and cell cycle arrest by E1A through its p300/CBP-interacting region. *Oncogene* **14**: 1047-1057.

Steger, G. and Pfister, H., 1992, In vitro expressed HPV 8 E6 protein does not bind p53. *Arch. Virol.* **125**: 355-360.

Stirdivant, S. M., Ahern, J. D., Oliff, A., and Heimbrook, D. C., 1992, Retinoblastoma protein binding properties are dependent on 4 cysteine residues in the protein binding pocket. *J. Biol. Chem.* **267**: 14846-14851.

Storey, A., Thomas, M., Kalita, A., Harwood, C., Gardiol, D., Mantovani, F., Breuer, J., Leigh, I. M., Matlashewski, G., and Banks, L., 1998, Role of a p53 polymorphism in the development of human papillomavirus-associated cancer. *Nature* **393**: 229-234.

Straight, S. W., Herman, B., and McCance, D. J., 1995, The E5 oncoprotein of human papillomavirus type 16 inhibits the acidification of endosomes in human keratinocytes. *J. Virol.* **69**: 3185-3192.

Straight, S. W., Hinkle, P. M., Jewers, R. J., and McCance, D. J., 1993, The E5 oncoprotein of human papillomavirus type 16 transforms fibroblasts and effects the downregulation of the epidermal growth factor receptor in keratinocytes. *J. Virol.* **67**: 4521-4532.

Sutcliffe, J. E., Brown, T. R., Allison, S. J., Scott, P. H., and White, R. J., 2000, Retinoblastoma protein disrupts interactions required for RNA polymerase III transcription [In Process Citation]. *Mol. Cell. Biol.* **20**: 9192-9202.

Sutcliffe, J. E., Cairns, C. A., McLees, A., Allison, S. J., Tosh, K., and White, R. J., 1999, RNA polymerase III transcription factor IIIB is a target for repression by pocket proteins p107 and p130. *Mol. Cell. Biol.* **19**: 4255-4261.

Swan, D. C., Vernon, S. D., and Icenogle, J. P., 1994, Cellular proteins involved in papillomavirus-induced transformation. *Arch. Virol.* **138**: 105-115.

Thomas, M. and Banks, L., 1999, Human papillomavirus (HPV) E6 interactions with Bak are conserved amongst E6 proteins from high and low risk HPV types. *J. Gen. Virol.* **80**: 1513-1517.

Thomas, M. and Banks, L., 1998, Inhibition of Bak-induced apoptosis by HPV-18 E6. *Oncogene* **17**: 2943-2954.

Tommasino, M., Adamczewski, J. P., Carlotti, F., Barth, C. F., Manetti, R., Contorni, M., Cavalieri, F., Hunt, T., and Crawford, L., 1993, HPV16 E7 protein associates with the protein kinase p33CDK2 and cyclin A. *Oncogene* **8**: 195-202.

Tong, X., Boll, W., Kirchhausen, T., and Howley, P. M., 1998, Interaction of the bovine papillomavirus E6 protein with the clathrin adaptor complex AP-1. *J. Virol.* **72**: 476-482.

Tong, X. and Howley, P. M., 1997, The bovine papillomavirus E6 oncoprotein interacts with paxillin and disrupts the actin cytoskeleton. *Proc. Natl. Acad. Sci U S A* **94**: 4412-4417.

Tong, X., Salgia, R., Li, J. L., Griffin, J. D., and Howley, P. M., 1997, The bovine papillomavirus E6 protein binds to the LD motif repeats of paxillin and blocks its interaction with vinculin and the focal adhesion kinase. *J. Biol. Chem.* **272:** 33373-33376.

Ullman, C. G., Haris, P. I., Kell, B., Cason, J., Jewers, R. J., Best, J. M., Emery, V. C., and Perkins, S. J., 1994, Hypothetical structure of the membrane-associated E5 oncoprotein of human papillomavirus type 16. *Biochem. Soc. Trans.* **22:** 439S.

van Duin, M., Snijders, P. J., Vossen, M. T., Klaassen, E., Voorhorst, F., Verheijen, R. H., Helmerhorst, T. J., Meijer, C. J., and Walboomers, J. M., 2000, Analysis of human papillomavirus type 16 E6 variants in relation to p53 codon 72 polymorphism genotypes in cervical carcinogenesis. *J. Gen. Virol.* **81 Pt 2:** 317-325.

Vande Pol, S. B., Brown, M. C., and Turner, C. E., 1998, Association of Bovine Papillomavirus Type 1 E6 oncoprotein with the focal adhesion protein paxillin through a conserved protein interaction motif. *Oncogene* **16:** 43-52.

Venuti, A., Salani, D., Poggiali, F., Manni, V., and Bagnato, A., 1998, The E5 oncoprotein of human papillomavirus type 16 enhances endothelin-1-induced keratinocyte growth. *Virology* **248:** 1-5.

Vousden, K. H., Vojtesek, B., Fisher, C., and Lane, D., 1993, HPV-16 E7 or adenovirus E1A can overcome the growth arrest of cells immortalized with a temperature-sensitive p53. *Oncogene* **8:** 1697-1702.

Waddell, S. and Jenkins, J. R., 1998, Defining the minimal requirements for papilloma viral E6-mediated inhibition of human p53 activity in fission yeast. *Oncogene* **16:** 1759-1765.

Watanabe, S., Kanda, T., Sato, H., Furuno, A., and Yoshiike, K., 1990, Mutational analysis of human papillomavirus type 16 E7 functions. *J. Virol.* **64:** 207-214.

Watanabe, S., Sato, H., Furuno, A., and Yoshiike, K., 1992, Changing the spacing between metal-binding motifs decreases stability and transforming activity of the human papillomavirus type 18 E7 oncoprotein. *Virology* **190:** 872-875.

Waters, C. M., Overholser, K. A., Sorkin, A., and Carpenter, G., 1992, Analysis of the influences of the E5 transforming protein on kinetic parameters of epidermal growth factor binding and metabolism. *J. Cell. Physiol.* **152:** 253-263.

Werness, B. A., Levine, A. J., and Howley, P. M., 1990, Association of human papillomavirus types 16 and 18 E6 proteins with p53. *Science* **248:** 76-79.

Wu, E. W., Clemens, K. E., Heck, D. V., and Münger, K., 1993, The human papillomavirus E7 oncoprotein and the cellular transcription factor E2F bind to separate sites on the retinoblastoma tumor suppressor protein. *J. Virol.* **67:** 2402-2407.

Xiong, Y., Hannon, G. J., Zhang, H., Casso, D., Kobayashi, R., and Beach, D., 1993, p21 is a universal inhibitor of cyclin kinases. *Nature* **366:** 701-704.

Zerfass, K., Levy, L. M., Cremonesi, C., Ciccolini, F., Jansen-Durr, P., Crawford, L., Ralston, R., and Tommasino, M., 1995, Cell cycle-dependent disruption of E2F-p107 complexes by human papillomavirus type 16 E7. *J. Gen. Virol.* **76:** 1815-1820.

Zerfass-Thome, K., Zwerschke, W., Mannhardt, B., Tindle, R., Botz, J. W., and Jansen-Durr, P., 1996, Inactivation of the cdk inhibitor p27KIP1 by the human papillomavirus type 16 E7 oncoprotein. *Oncogene* **13:** 2323-2330.

Zimmermann, H., Degenkolbe, R., Bernard, H. U., and O'Connor, M. J., 1999, The human papillomavirus type 16 E6 oncoprotein can down-regulate p53 activity by targeting the transcriptional coactivator CBP/p300. *J. Virol.* **73:** 6209-6219.

zur Hausen, H., 2000, Papillomaviruses causing cancer: evasion from host-cell control in early events in carcinogenesis. *J. Natl. Cancer Inst.* **92:** 690-698.

Zwerschke, W., Joswig, S., and Jansen-Durr, P., 1996, Identification of domains required for transcriptional activation and protein dimerization in the human papillomavirus type-16 E7 protein. *Oncogene* **12:** 213-220.

Zwerschke, W., Mazurek, S., Massimi, P., Banks, L., Eigenbrodt, E., and Jansen-Durr, P., 1999, Modulation of type M2 pyruvate kinase activity by the human papillomavirus type 16 E7 oncoprotein. *Proc. Natl. Acad. Sci U S A* **96:** 1291-1296.

7. Defense mechanisms

Chapter 7.1

Cytomegalovirus Glycoproteins Interacting with MHC Class I Molecules and the MHC-encoded Peptide Transporter

ALBERT ZIMMERMANN, ANNE HALENIUS, AND HARTMUT HENGEL
Robert Koch-Institut, Division of Viral Infections, Nordufer 20, 13353 Berlin, Germany

1. INTRODUCTION

1.1 The MHC class I pathway of antigen presentation to CD8[+] T lymphocytes

After successful entry into a susceptible host the major function of the immune system is to detect and eliminate invading viruses or at least to control their replication. As a first line of defence, natural killer (NK) cells, type I interferons and cytokines contribute to immediately activating innate immune responses and occur at times preceding adaptive immunity. Next, naive T and B lymphocytes are concentrated to sites of antigen presentation and primary stimulation in lymphoid organs where they interact with professional antigen presenting cells like dendritic cells to be clonally expanded. Preactivated effector cells are localized to sites of infection to execute their effector functions. Antigen-specific lymphocytes generated during acute viral infection enter the memory pool and provide host protection on reexposure to the virus.

CD8[+] T lymphocytes represent a major constituent of antiviral defence mechanisms (Ahmed and Biron, 1999). CD8[+] T cells recognize non-self peptides of viral origin bound to major histocompatibility complex (MHC) class I molecules on the surface of infected cells (Fig.1) (Heemels and Ploegh, 1995). MHC class I molecules are type I transmembrane proteins

Structure-Function Relationships of Human Pathogenic Viruses, Edited by
Holzenburg and Bogner, Kluwer Academic/Plenum Publishers, New York, 2002

binding peptide ligands that provide an extracellular representation of intracellular antigen content. Peptides are inspected by the CD8$^+$ T lymphocyte subset through interactions of their clonotypic T cell receptors with the MHC class I-peptide complex. In this pathway, viral proteins are cleaved into peptides by the ubiquitin/proteasome system in the cytosol (Rock and Goldberg, 2000). Due to their cytosolic origin, peptides have to be translocated across the membrane barrier of the endoplasmic reticulum (ER) to encounter the peptide binding site of MHC class I molecules in the ER lumen. To this end, peptides are translocated by a specific peptide transporter, the transporter associated with antigen processing (TAP), which comprises two homologous subunits, TAP1 and TAP2 (Momburg and Hämmerling, 1998).

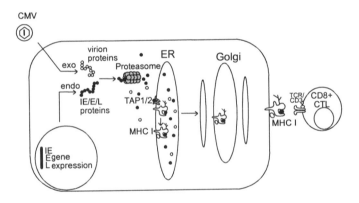

Figure 1. The major histocompatibility complex (MHC) class I pathway of antigen processing and presentation. Viral proteins *de novo* synthesized in the infected cell (endogenous proteins) or structural proteins of the virion (exogenous proteins) are processed into peptides (black and unfilled circles) by the proteasome. Peptides are translocated by TAP1/2 (shown in grey) into the lumen of the endoplasmic reticulum (ER) for binding to MHC class I/β$_2$m heterodimers (white). Timeric MHC class I complexes are transported via the Golgi to the cell surface for presentation of peptides to CD8$^+$ cytotoxic T lymphocytes (CTL). Abbreviations: IE, immediate early; E, early; L, late; TCR/CD3: T cell receptor complex.

Stable binding of peptides to MHC I molecules in the ER requires correct folding and association of the MHC I heavy chain with the soluble β$_2$-

microglobulin (β_2m) light chain. Several ER chaperones participate in this process, including calnexin, calreticulin, thiol-dependent reductase ER-60 and tapasin (Suh *et al.,* 1994; Sadasivan *et al.,* 1996; Lindquist *et al.,* 1998). Tapasin mediates the association of MHC I/$\beta_2$microglobulin/calreticulin/ER-60 complexes with the TAP1/2 transporter resulting in a transient TAP1/2 - MHC class I heavy chain - $\beta_2$microglobulin - calreticulin - ER-60 - tapasin assembly complex (Ortmann *et al.,* 1997). Following peptide binding, heterotrimeric MHC I/$\beta_2$microglobulin/peptide complexes dissociate from TAP1/2 and exit from the ER for their transport via the constitutive secretory route to the plasma membrane to present the peptide to the T cell receptor.

1.2　Cytomegalovirus infection and immune control

Cytomegaloviruses (CMVs) represent prototypes of the β-subfamily of the herpesvirus group widely distributed among mammals (Mocarski, 1996). CMVs are highly species specific documenting their intimate relationship with the host. Based on genomic sequences of several CMVs infecting rodents and primates, respectively, it is generally thought that CMVs have cospeciated with their hosts during protracted periods of evolution. CMVs are characterized by a slow replication rate and a relatively broad cell-type specific tropism which includes different epithelial, endothelial, myelo-monocytic and smooth muscle cells as potential targets (Sinzger *et al.,* 1995). The antiviral defence against CMVs both during primary as well as latent and recurrent infection is critically dependent on hierarchical but also redundant effector functions mediated mainly by T lymphocytes (Reddehase *et al.,* 1985; Polic *et al.,* 1998). As a consequence, CMVs can cause severe disease upon immunosuppression or in the immunologically immature host. *In vivo* depletion experiments and adoptive transfer studies in mouse CMV (MCMV)-infected mice identified the CD8[+] rather than the CD4[+] T cell subset as essential and sufficient to clear acute infection in visceral organs and to mediate protection from an otherwise lethal disease (Reddehase *et al.,* 1985, Reddehase *et al.,* 1987; Jonjic *et al.,* 1988). The efficacy of adoptively transferred CMV-specific CD8[+] T cells for controlling CMV infection was confirmed in humans (Riddell *et al.,* 1992). A significant role for CD4[+] T lymphocytes was demonstrated for controlling CMV infection in the salivary gland (Jonjic *et al.,* 1989). A strong antiviral effect operating in *vivo* was documented for certain cytokines. In particular, depletion of endogenous IFNγ and TNFα resulted in an increase in MCMV replication (Lucin *et al.,* 1992; Pavic *et al.,* 1993), CD8[+] T cell recognition and effector function (Hengel *et al.,* 1994) as well as viral peptide processing (Geginat *et al.,* 1997). Natural killer (NK) cells and virus-specific antibodies can significantly contribute to virus control but fail to protect against lethal CMV infections (Shellam *et al.*, 1981; Welsh *et al.*, 1991; Jonjic *et al.*, 1994).

1.3 Establishment and maintenance of cytomegaloviral persistance

As with other herpesviruses, the primary CMV infection is followed by a lifelong persistent infection. Despite efficient clearance of viral replication after primary infection by the host's immune system, the CMV genome persists in a latent form associated with minimal viral gene expression without virus production. Recurrence of productive infection occurs periodically resulting in virus shedding and transmission to non-immune hosts. Thus recurrence of a productive infection requires successful CMV replication in the face of repeatedly primed antiviral immune responses. The CMV dsDNA genome of approx. 235 kbp provides the luxury of a very large protein-coding capacity comprising about 200 separate open reading frames (Chee *et al.*, 1990, Rawlinson *et al.*, 1996). This permits the exposure of a multitude of potential antigens to the host immune system and allows an elaborate array of proteins designed purely to manipulate immune recognition and effector functions (Hengel *et al.*, 1998). During the last few years quite a number of CMV gene products have been identified which subvert MHC class I-restricted antigen presentation of infected cells to CD8$^+$ T lymphocytes. The CMV-encoded MHC I inhibitors established a paradigm for immune receptor modulation by pathogens and represent one of the best characterized immune evasion strategies known to date.

2.   MCMV ENCODED INHIBITORS OF MHC CLASS I MOLECULES

2.1   MCMV *m04*/gp34 associates with MHC class I complexes for transport to the cell surface

The *m04* open reading frame belongs to the MCMV *m02* gene family of type I transmembrane glycoprotein comprising 15 members which lie in tandem on the top strand at the left end of the prototype MCMV genome. The 34 kDa MCMV *m04*-encoded glycoprotein (gp) was identified as a protein coprecipitating with MHC class I molecules in MCMV infected cells (Kleinen *et al.*, 1997). Expressed during the early and late phase of MCMV infection gp34 is abundantly synthesized to form tight complexes with folded and β_2m-associated MHC class I molecules in the ER. Most of the gp34 molecules found in association with MHC I contain two endoglycosidase H-resistant mature oligosaccharide chains and can be subjected to surface iodination, indicating that gp34/MHC class I complexes readily leave the ER and reach the cell surface (Fig. 2) (Kleinen *et al.,* 1997). This observation suggested that gp34/MHC I complexes may escape the *m152*/gp40-mediated transport block of MHC I molecules formed in the endoplamic reticulum Golgi intermediate compartment (ERGIC). The display of MHC class I complexes bound to gp34 on the cell surface may prevent recognition of the CMV-infected target cell by cytotoxic effector cells, either CD8+ MHC class I-restricted CTL, NK cells, or both (Kleinen *et al.,* 1997).

2.2   MCMV *m06*/gp48 associates with MHC class I complexes for re-routing to lysosomes for degradation

Analysis of MCMV deletion mutant viruses identified another *m02* family member, the *m06*-encoded 48 kDa glycoprotein, to be involved in the rapid loss of MHC class I surface expression of MCMV-infected fibroblasts and impaired peptide presentation to CD8+ CTL (Reusch *et al.,* 1999; Hengel *et al.,* 1999). During the MCMV replication cycle, the expression of gp48 starts later than the transcription of gene *m152* but is maintained throughout the late phase of infection. Like *m04*/gp34, gp48 binds directly to the MHC class I complex in the ER. This association is mediated by the luminal/transmembrane part of the protein. The complex leaves from the ER, passes the Golgi, but instead of being expressed on the cell surface it is targeted to a lysosomal compartment where MHC I and gp48 are degraded by Leupeptin and Concanamycin A sensitive-mechanisms (Fig.2). In the presence of these inhibitors gp48 accumulates and can be co-localized with LAMP-1, a marker of the late endosome and lysosome (Reusch *et al.,* 1999). The cytoplasmic tail of gp48 contains two di-leucine motifs, the membrane-proximal one of both is required to re-route MHC class I/gp48 complexes to the lysosome.

2.3   MCMV *m152*/gp40 accumulates MHC class I complexes in the ER-Golgi intermediate compartment (ERGIC)

MCMV was the first herpesvirus to suggest a direct interference with the MHC class I pathway of antigen presentation (Del Val *et al.,* 1989). When the fate of MHC class I molecules was studied in MCMV-infected cells, the formation and stability of class I complexes was found intact, whereas the maturation of the MHC I glycoprotein and its transport into the *medial*-Golgi compartment was inhibited (Del Val *et al.*, 1992). A combined approach of constructing MCMV deletion mutants and injecting cloned MCMV DNA fragments into cells and screening them for MHC class I molecule retention by immunofluorescence microscopy allowed the identification of the *m152* gene that mediates this effect (Thäle *et al.*, 1995, Ziegler *et al.*, 1997). *m152* belongs to the *m145* gene family of MCMV and is transcribed within a first set of early genes. *m152* encodes a type I transmembrane glycoprotein of 37 and 40 kDa, respectively, that arrests the export of mouse class I complexes from the ERGIC (see Fig. 2), resulting in a drastically reduced density of membrane-residing MHC class I molecules and impaired peptide presentation to CD8$^+$ T cells (Ziegler *et al.*, 1997). A MCMV recombinant in which the *m152* gene was deleted has been used to demonstrate the *in vivo* significance of this factor (Krmpotic *et al.*, 1999). The *m152*-deficient virus replicates in vitro as efficient as wild type virus, but to significantly lower titers in multiple organs of infected mice. This effect was abrogated in mice deficient in β_2m or CD8, indicating that *m152*/gp40 protects the virus against CD8$^+$ T cell control *in vivo* (Krmpotic *et al.*, 1999; Hengel et al., 1999). Unlike *m04*/gp34 and *m06*/gp48, gp40 has not been shown to form stable complexes with MHC I. The retained MHC I molecules consist of the complete trimolecular complex of heavy chain, β_2m and peptide and show an extended half-life. gp40 affects the transport of a large array of mouse class I alleles, while human class I alleles are largely resistant to the gp40-mediated effect. Deletion of the cytoplasmic tail did not lift its effect on MHC I retention (Ziegler *et al.*, 1997), indicating that gp40 differs in its function from the prototypic inhibitor of MHC I transport, the E3/19K protein of adenovirus (Jackson *et al.*, 1990). In contrast to E3/19K, a secreted *m152*/gp40 variant protein consisting of only the luminal part of gp40 still has MHC class I retention capacity. Moreover, continuos expression of gp40 is not required for MHC class I retention, since MHC I molecules are still excluded from anterograde transport when gp40 has left the ERGIC and reached an endosomal/lysosomal compartment (Ziegler *et al.*, 2000). Hypothetically, retention of MHC I molecules could be triggered by a gp40 induced alteration of MHC I complexes in the ERGIC excluding MHC I from the interaction with obligate cargo receptors. Alternatively, the hypothesized modification of MHC I molecules might serve as a retention signal which connects proteins to the retrograde transport machinery for permanent retrieval.

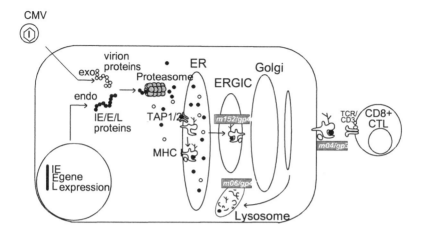

Figure 2. Mechanisms of Mouse cytomegalovirus (MCMV) to evade MHC class I restricted immunity. MCMV gene functions responsible for interference with the MHC class I presentation pathway are indicated as explained in the text and summarized in Table 1. Abbreviation: ERGIC, endoplasmic reticulum Golgi intermediate compartment.

3. HCMV ENCODED INHIBITORS OF MHC CLASS I MOLECULE EXRESSION

3.1 HCMV *US3*/gp23 accumulates MHC class I complexes in the ER-Golgi intermediate compartment

US3 represents the only immediate early (IE) gene within the US segment of the HCMV genome. US3 is expressed abundantly immediately after infection but US3 transcription is eventually abolished 6 to 8 hrs p.i.*US3* codes for products of 23 and 17 kDa, the larger of which binds to β_2m-associated MHC class I molecules in the ER (Ahn *et al.*, 1996b; Jones *et al.*, 1996). Since gp23-associated MHC class I molecules are stable at

37°C in NP40-lysates, they are likely to be loaded with peptide. In *US3*-transfected cells MHC class I/β_2m heterodimers are stably retained in the ER, reflected by an endoglycosidase H-sensitive phenotype (Fig. 3) (Ahn *et al.*, 1996b; Jones *et al.*, 1996). Mutational analysis of *US3*/gp23 using a panel of CD4/US3 chimeras revealed that the luminal domain of *US3*/gp23 is sufficient for ER retention (Lee *et al.*, 2000). Unlike the ER retention signal, the ability to associate with MHC class I required not only the luminal domain but in addition the transmembrane segment of the *US3*/gp23 molecule (Lee *et al.*, 2000). Following US3/gp23 another MHC class I inhibitor is expressed during the early phase of the HCMV replication cycle, *US11*/gp33, which also affects β_2m-associated MHC class I molecules (Ahn *et al.*, 1996b; Wiertz *et al.*, 1996a). It is conceivable that HCMV inhibits antigen presentation by a sequential multistep process which is initiated by *US3*/gp23-mediated MHC I retention followed by subsequent degradation due to *US11*/gp33 (Ahn *et al.*, 1996b). Moreover, polymorphic MHC class I /β_2m heterodimers are not the only target that are retained by *US3*/gp23 since nonpolymorphic HLA-G molecules are retained as well (Jun *et al.*, 2000). Amazingly, MHC class II molecules are also blocked in their transport from the ER via Golgi to the endocytic pathway (Tomazin *et al.*, 2001).

3.2   HCMV *US11*/gp33 mediates retrograde transport of MHC class I heavy chains to the cytosol

During the early phase of HCMV replication, HCMV gene expression results in a rapid decline of MHC class I surface expression and antigen presentation to CD8$^+$ T cells (Warren *et al.*, 1994; Hengel *et al.*, 1995). Pulse chase analysis of MHC molecules after metabolic labelling demonstrated that MHC class I synthesis was unaffected but the half-life of unassembled MHC class I heavy chains was drastically reduced resulting in a subsequent decrease of MHC class I complex formation in the ER (Beersma *et al.*, 1993; Yamashita *et al.*, 1993). By screening HCMV deletion mutants lacking open reading frames in the short component of the HCMV genome, Jones and co-workers identified two independent loci associated with down-regulating MHC class I heavy chains, one of which was shown to be *US11* (Jones *et al.*, 1995). During HCMV infection in fibroblasts and epithelial cells the *US11* encoded 33 kDa glycoprotein is synthesized at maximum levels 8-24 hours p.i. in the early phase (Jones and Muzithras, 1991; Benz *et al.*, 2001). In *US11*-transfected cells the half-life of free class I heavy chains was less than 1 minute (Wiertz *et al.*, 1996a). The breakdown of heavy chains was sensitive to proteasome inhibitors. Proteasome inhibition led to the accumulation of a deglycosylated lower molecular mass form of MHC I molecules. Cell fractionation experiments located the heavy chain intermediate in the cytosol (Fig. 3). Since MHC class I molecules become core-glycosylated in the presence of *US11*/gp33, they must have been inserted into the ER membrane before being exported to the cytosol. Further analysis of the *US11*/gp33-dependent degradation process showed that heavy chains become ubiquitinated before they are degraded. Ubiquitination of the cytosolic tail of heavy chains is not required for its dislocation and degradation, and ubiquitination of the heavy chain does not appear to be the signal to start dislocation (Shamu *et al.*, 1999). The majority of ubiquitinated heavy chains are deglycosylated and associated with membrane fractions, suggesting that ubiquitination occurs while the heavy chain is still bound to the ER membrane. Taken together, it appears that *US11*/gp33 accelerates a process that normally occurs with misfolded proteins that do not pass ER quality control.

3.3 HCMV *US2*/gp24 transfers MHC class I heavy chains via Sec61 to the cytosol for degradation

Similarly to *US11*/gp33, the 24 kDa glycoprotein encoded by *US2* targets ER resident MHC class I heavy chains for rapid degradation by the proteasome (Wiertz *et al.*, 1996b; Jones and Sun, 1997). The ER lumenal domain of *US2*/gp24 is sufficient to allow tight interaction with HLA-A molecules in the ER (Gewurz *et al.*, 2001). The physicochemical properties of the gp24/MHC class I complex suggest a 1:1 stoichiometry. The function of *US11*/gp33 and *US2*/gp24 differs in that different murine and human MHC class I alleles are recognized, e.g. *US2*/gp24 binds HLA-A2 but not HLA-B7, B27, Cw4 or HLA-E (Gewurz *et al.*, 2001, Machold *et al.*, 1997). Remarkably, studies of the *US2*/gp24-mediated mechanism revealed that dislocation of the glycosylated MHC class I heavy chain is mediated by the translocon/Sec61 complex (Wiertz *et al.*, 1996b). The translocon is composed of 3-4 copies of the Sec61 heterotrimeric complex forming a cylindrical structure with a central pore diameter of 40-60Å. The translocon mediates the co- or posttranslational entry of secreted proteins from the cytosol into the secretory pathway. The reverse translocation of the MHC class I heavy chain includes *US2*/gp24 itself which becomes deglycosylated in the cytosol to a 21 kDa protein. Dislocation of the MHC class I heavy chain requires its cytosolic tail, since a truncated tailless HLA-A2 mutant with preserved binding by *US11*/gp33 and *US2*/gp24 is stable in the presence of the viral proteins (Story *et al.*, 1999). Therefore, the viral proteins per se do not have the capability to remove heavy chains from the ER. One may speculate whether the proteasome could be the factor that binds the cytosolic tail of MHC class I heavy chains once bound by *US11*/gp33 and *US2*/gp24, respectively, and whether the proteasome extracts the molecule from the ER membrane (Story *et al.,* 1999).

3.4   HCMV *US6*/gp21 inhibits peptide translocation by the peptide transporter TAP

The supply of peptides is critical and limiting for the assembly of stable trimeric MHC class I complexes in the ER (Heemels and Ploegh, 1995; Momburg and Hämmerling, 1998). The import of peptides into the ER lumen of HCMV-infected fibroblasts was found to be continuously decreased during the early and late phase of replication until eventually abolished at 96 hrs p.i. (Hengel *et al.,* 1996). The inactivation of peptide transport in HCMV-infected cells is established despite augmented TAP transcription and synthesis (Warren *et al.,* 1994; Hengel *et al.,* 1996). The inhibition of peptide translocation is caused by HCMV *US6*/gp21 (Fig.3) (Hengel *et al.,* 1997; Ahn *et al.,* 1997; Lehner *et al.,* 1997) which is expressed with early/late kinetics peaking 72 hrs p.i. in the late phase of the HCMV replication cylce (Jones and Muzithras, 1991). The subcellular distribution of *US6*/gp21 is ER-restricted and identical with that of TAP. The viral protein is found associated with the transient TAP/MHC class I peptide loading complex consisting of TAP1/TAP2/MHC/β_2m/calreticulin/ tapasin/ ERp57 and with calnexin (Hengel *et al.,* 1997), an ER chaperone. In TAP-deficient T2 cells *US6*/gp21 fails to downregulate MHC-I indicating that the MHC-I subversive function of gpUS6 is exclusively dependent on TAP1/2 expression (Momburg and Hengel, 2001). Several findings indicate that *US6*/gp21 regulates the translocation of peptides by interactions with the lumenal surface of the TAP1/TAP2 heterodimer. Unlike the Herpes simplex virus inhibitor of TAP, ICP47 (Früh *et al.,* 1995; Hill *et al.,* 1995; Tomazin *et al,* 1996; Ahn *et al.,* 1996a), *US6*/gp21 does not interfere with peptide binding to the cytosolic face of TAP whereas ICP47 prevents peptide binding to TAP in the presence of *US6*/gp21 (Ahn *et al.,* 1997; Hengel *et al.,* 1997). Moreover, expression of a truncated mutant of *US6*/gp21 lacking the transmembrane and cytoplasmic domains resulted in a MHC-I-downregulated phenotype (Ahn *et al.,* 1997). Akin to the findings with ICP47, *US6*/gp21-mediated inhibition of TAP is restricted to certain species. Peptide translocation assays using cells infected with a *US6*/gp21-recombinant vaccinia virus showed that TAP in human, monkey, and rabbit cells could be inhibited whereas mouse and rat TAP were unaffected (Momburg and Hengel, 2001). Isolated expression of TAP and *US6*/gp21 indicates that the viral protein directly interacts with both subunits of TAP in a species-restricted manner (Momburg and Hengel, 2001). In a recent report, Neefjes and coworkers have used lateral mobility measurements of green fluorescent protein (GFP)-tagged TAP1 molecules to assess conformational changes of TAP induced by manipulation of energy and peptide supply (Reits *et al.*, 2000). Microinjection of synthetic peptides into cells resulted in a decreased lateral diffusion of TAP1-GFP that was interpreted as an opening of an otherwise closed translocation pore. This peptide-induced effect was dependent on the presence of ATP. Surprisingly, the presence of a soluble ER form of *US6*/gp21 resulted in a lower mobility of TAP similar to the reduction found after the injection of synthetic peptide (Reits *et al.*, 2000). This finding suggests that *US6*/gp21 induces profound changes in the conformation of TAP, possibly freezing an "open" conformation. Another argument comes from the observation that *US6*/gp21 inhibits ATP binding to TAP1 (Hewitt *et al.*, 2001). This is a conformational effect since the ER lumenal domain of *US6*/gp21is sufficient to inhibit ATP binding by the cytosolic nucleotide binding domain of TAP. These findings fit into a model in which the association of *US6*/gp21 with TAP induces a conformation of the transporter that prevents ATP binding and subsequent peptide translocation.

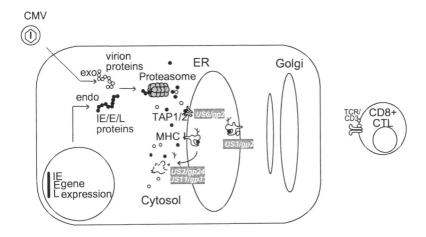

Figure 3. Mechanisms of human cytomegalovirus (HCMV) to evade MHC class I restricted immunity. HCMV gene functions responsible for interference with the MHC class I presentation pathway are indicated as explained in the text and summarized in Table 1.

4.   HCMV GLYCOPROTEINS INTERFERING WITH NATURAL KILLER CELL RECOGNITION

In contrast to CD8$^+$ CTL activity, NK cell activation is inhibited by surface expressed MHC class I molecules that are sensed by killer inhibitory receptors (KIR) on NK cells (Kärre, 1995). According to the "missing self" hypothesis of NK cell recognition, cytomegaloviral downregulation of MHC class I surface expression (stealth strategy) should lead to a higher susceptibility of CMV-infected cells to NK cell activity. Studies by Huard and Früh using NK cell clones to analyze recognition of CMV-infected targets as well as gene transfectants support this notion (Huard and Früh, 2000).

Strikingly, there is growing evidence that HCMV is able to counteract increased NK cell recognition by specific glycoproteins. The inhibitory receptor CD94/NKG2A is broadly distributed on NK cells and recognizes the ubiquitously expressed nonpolymorphic MHC class Ib molecule HLA-E. Plasma membrane expression of HLA-E requires the TAP-dependent binding of conserved peptides corresponding to residues 3–11 of the signal sequences from various HLA-A, HLA-C and some HLA-B heavy chains as well as from HLA-G (Fig. 4) (Braud *et al.,* 1998a, 1998b; Borrego *et al.* 1998; Lee *et al.,* 1998). Remarkably, this motif of HLA-E ligands is present in the leader sequence of the HCMV *UL40* ORF (Chee *et al.,* 1990). Expression of *UL40* conferred resistance to NK cell lysis via the CD94/NKG2A receptor by inducing membrane expression of HLA-E (Tomasec *et al.,* 2000; Ulbrecht *et al.,* 2000). In the context of HCMV infection exploiting the HLA-E-mediated inhibition of NK cells appears to be a perfect strategy to counteract NK attack of MHC-I deficient target cells, provided that the gpUL40-derived HLA-E ligand is able to bypass the TAP block established by the simultaneously expressed inhibitor *US6*/gp21. Expression of gpUL40 in TAP-deficient RMA-S cells efficiently induced surface expression of cotransfected HLA-E molecules indicating that binding of the gpUL40-derived peptide ligand is indeed TAP-independent (Ulbrecht *et al,.* 2000).

Another promising candidate factor of HCMV likely to interfere with NK recognition of HCMV-infected cells is the *UL16*-encoded glycoprotein (Fig. 4) (Cosman *et al.,* 2001) gpUL16, which is not required for HCMV replication *in vitro* (Kaye *et al.,* 1992). It is predicted to be a type I transmembrane protein and expressed on the surface of transfected cells. It binds to two members of ULBPs, a novel family of GPI-linked glycoproteins belonging to the extended MHC class I family, and to MICB (Cosman *et al.,* 2001). MICB is a nonclassical MHC class I antigen (Bahram *et al.,* 1994). It has the same basic structure as classical MHC I molecules but does not associate with β₂m (Groh *et al.*, 1996). Both ULBP and MICB are ligands for an activating receptor on NK cells, NKG2D/DAP10 and stimulate NK cells to produce chemokines, cytokines and trigger NK cytotoxicity. This interaction is blocked by soluble gpUL16, rising the possibility that the viral protein may represent a decoy receptor for NK cell recognition which masks the recognition of ULBPs and MICB (Cosman *et al.*, 2001). Alternatively, binding of gpUL16 to ULBPs and MICB could form a complex on the cell surface or retain these molecules within HCMV-infected cells.

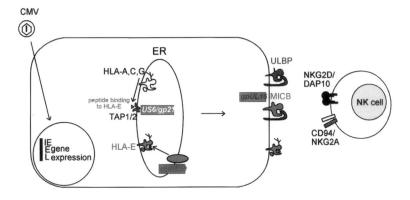

Fig. 4 HCMV gene functions which interfere with recognition of natural killer (NK) cells. The HCMV factors are indicated as explained in the text. Abbreviations: ULBP, UL16-binding proteins; MICB, MHC class I related protein B.

5.   PERSPECTIVE

CMVs which represent slowly replicating viruses with a large DNA genome and an extended exposure to immune attack are under particular pressure to acquire defence strategies against MHC class I-restricted immunity. This is reflected by the bewildering array of gene functions identified so far (see Table I). It appears that CMVs have evolved the most extensive genetic repertoire to evade CD8$^+$ T cell responses among viruses (Hengel and Koszinowski, 1997; Ploegh, 1998).

Table 1. MCMV and HCMV glycoproteins that interfere with the MHC class I pathway of antigen presentation to CD8$^+$ T cells

CMV ORF/gp	Target molecule within the MHC class I pathway	Compartment of interaction with target molecule	Supposed immune evasive mechanism	Destination of viral glycoprotein
MCMV m04/gp34	MHC class I complexes	ER	Disturbance of MHC I dependent NK and CD8$^+$ T cell recognition	Cell surface
MCMV m06/gp48	MHC class I complexes	ER	Proteolytic degradation	Endolysosome
MCMV m152/gp40	MHC class I complexes	ERGIC	Transient interaction inducing retention	Endolysosome
HCMV US3/gp23	MHC class I complexes	ER	Retention	ER
HCMV US2/gp21	MHC class I heavy chains	ER	Proteolytic degradation	Cytosol
HCMV US11/gp33	MHC class I heavy chains	ER	Proteolytic degradation	Cytosol
HCMV US6/gp21	TAP1/2	ER	Alteration of ATP binding to TAP; inactivation of peptide translocation	ER

HCMV and MCMV have found sophisticated means to interfere with distinct steps of cellular functions like protein transport or peptide translocation alongside the steps of MHC class I biogenesis and function. Since the MHC class I pathway of antigen presentation is part of the constitutive secretory pathway, CMV-encoded glycoproteins are predestined to act as MHC I inhibitors. Typically, a complete glycoprotein inhibitor is built by several distinct functional domains. Each submolecular module may execute at least one essential interaction with a specific cellular target molecule, e.g. intracellular re-routing to certain compartments by cytoplasmic targeting motifs or MHC class I binding by the lumenal domain. As sensors of cellular functions, the CMV inhibitors have contributed essentially not only to our current knowledge of antigen presentation but also to the elucidation of general pathways for intracellular protein transport.

All of the CMV genes that control MHC I expression are dispensable for virus replication *in vitro*. Except for MCMV *m152*/gp40 (Krmpotic *et al.*, 1999), little is known about their role during CMV infection *in vivo*. Recent reports suggest that at least some of the MHC I inhibitors of HCMV, i.e. *US2*/gp24 and *US3*/gp23 may be even multifunctional in that components of the MHC class II pathway are affected as well (Tomazin *et al.,* 1999; Johnson *et al.,* 2001). Another surprising but still preliminary observation comes from the *in vivo* analysis of MCMV mutants lacking MHC I inhibitors which relates to the consequence of increased susceptibility to NK cell attack due to MHC class I downregulation: Apparently, *m04*/gp34 and *m152*/gp40 found elegant means to overcome this dilemma since they are able to inhibit simultaneously both NK as well as CD8$^+$-mediated immunity *in vivo* (Krmpotic *et al.,* 2001). This finding illustrates that multiple immunological forces contributed to the perfect design of cytomegaloviral MHC I regulators.

Despite the redundant molecular mechanisms of CMVs to limit recognition of CD8$^+$ CTL and to avoid its elimination, MHC class I restricted T cell responses are readily generated in the immunocompetent host and required to hold CMV replication in check. This seeming paradox indicates that the viral immune evasion mechanisms operate *in vivo* with only a limited efficacy. MCMV-infected macrophages which are a major source of MHC class I peptides processed *in vivo* have been shown to resist MHC I downregulation and inhibition of peptide presentation (Hengel *et al.,* 2000). In addition, cytokines like Interferon-γ or TNF-α can counteract CMV evasion by restoring MHC class I presentation (Hengel *et al.,* 1994; Hengel *et al.,* 1995; Geginat *et al.,* 1997). Of course, CMVs have also found means against the threat of interferons (Miller *et al.,* 1998; Heise *et al.,* 1998), another exciting game of chess between virus and host (Benz and Hengel, 2000).

ACKNOWLEDGMENTS

Our work is supported by the Deutsche Forschungsgemeinschaft through grant He 2526/3-2, SFB 455 (A6), SFB 421 (A8) and the German Ministry of Research and Technology.

REFERENCES

Ahmed, R. and Biron, C.A., 1999, Immunity to viruses. In: *Fundamental Immunology*, (W.E. Paul, ed.) Lippincott-Raven, Philadelphia; pp. 1295-1334

Ahn, K., Angulo, A., Ghazal, P., Peterson, P. A., Yang, .Y., and Früh, K., 1996b, Human cytomegalovirus inhibits antigen presentation by a sequential multistep process., *Proc. Natl. Acad. Sci. USA* **93**:10990-10995

Ahn, K., Gruhler, A., Galocha, B., Jones, T.R., Wiertz, E.J.H.J., Ploegh, H. L., Peterson, P. A., and Früh, K., 1997, The ER-luminal domain of the HCMV glycoprotein US6 inhibits peptide translocation by TAP. *Immunity* **6:** 613-621

Ahn, K., Meyer, T.H., Uebel, S., Sempe, P., Djaballah, H., Yang, Y., Peterson, P.A., Früh, K., Tampe, R., 1996a, Molecular mechanism and species specificity of TAP inhibition by herpes simplex virus ICP47. *EMBO J* **15:**3247–3255

Bahram, S., Bresnahan, M., Geraghty, D.E. and Spies, T., 1994, A second lineage of mammalian major histocompatibility complex class I genes. *Proc. Natl. Acad. Sci. USA* **91:**6259-6263

Beersma, M.F.C., Bijlmakers, M.J.E., and Ploegh, H.L., 1993, Cytomegalovirus down-regulates HLA class I expression by reducing the stability of class I H chains. *J. Immunol.* **151:**4455-4464

Benz, C. and Hengel, H., 2000, MHC class I-subversive gene functions of cytomegalovirus and their regulation by interferons - an intricate balance. *Virus Genes* **21:**39-47

Benz, C., Reusch, U., Muranyi, W., Brune, W., Atalay, R., and Hengel, H., 2001, Efficient downregulation of major histocompatibilty complex class I molecules in human epithelial cells infected with cytomegalovirus. *J. Gen. Virol.* **82:** 2061-2070

Borrego, F., Ulbrecht, M., Weiss, E.H., Coligan, J.E., Brooks, A.G., 1998, Recognition of human histocompatibility leukocyte antigen (HLA)-E complexed with HLA class I signal sequence-derived peptides by CD94/NKG2 confers protection from natural killer cell-mediated lysis. *J. Exp. Med.* **187:**813–818

Braud, V.M., Allan, D.S., Wilson, D., McMichael, A.J., 1998a, TAP- and tapasin-dependent HLA-E surface expression correlates with the binding of an MHC class I leader peptide. *Curr. Biol.* **8:**1–10

Braud, V.M., Allan, D.S., O'Callaghan, C.A., Soderstrom, K., D'Andrea, A., Ogg, G.S., Lazetic, S., Young, N.T., Bell, J.I., Phillips, J.H., Lanier, L.L., McMichael, A.J., 1998b, HLA-E binds to natural killer cell receptors CD94/NKG2A, B and C. *Nature* **391:**795–799

Chee, M. S,, Bankier, A.T., Beck, S., Bohni, R., Brown, C.M., Cerny, R., Horsnell, T., Hutchison, C.A. 3d, Kouzarides, T., Martignetti, J.A., Preddi, E., Satchwell, S.C., Tomlinson, P., Weston, K.M., Barrell, B.G., 1990, Analysis of the protein-coding content of the sequence of human cytomegalovirus strain AD169. *Curr. Top. Microbiol. Immunol.* **154:**125–169

Cosman, D., Müllberg, J., Sutherland, C.L., Chin, W., Armitage, R., Fanslow, W., Kubin, M., and Chalupny, n.J., 2001, ULBPs, novel MHC class I-related molelcules, bind to CMV glycoprotein UL16 and stimulate NK cytotoxicity through the NKG2D receptor. *Immunity* **14:**123-133

Del Val, M., Hengel, H., Häcker, H., Hartlaub, U., Ruppert, T., Lucin, P. and Koszinowski, U. H., 1992, Cytomegalovirus prevents antigen presentation by blocking the transport of peptide-loaded major histocompatibility complex class I molecules into the medial-Golgi compartment. *J. Exp. Med.* **172:**729-738

Del Val, M., Münch, K., Reddehase, M.J., Koszinowski, U.H., 1989, Presentation of cytomegalovirus immediate-early antigens to cytolytic T lymphocytes is selectively blocked by viral genes expressed in the early phase. *Cell* **58:**305-315

Früh, K., Ahn, K., Djaballah, H., Sempe, P., van Endert, P.M., Tampe, R., Peterson, P.A., Yang, Y. 1995, A viral inhibitor of peptide transporters for antigen presentation. *Nature* **375:**415–418

Geginat, G., Ruppert, T., Hengel, H., Holtappels, R., and Koszinowski, U.H., 1997, IFN-γ is a prerequisite for optimal antigen processing of viral peptides in vivo. *J. Immunol.* **158:**3303-3310

Gewurz, B.E., Wang, E.W., Tortorella, D., Schust, D.J., and Ploegh, H.L., 2001, Human cytomegalovirus US2 endoplasmic reticulum domain dictates association with major histocompatibility complex class I in a locus-specific manner. *J. Virol.* **75:**5197-5204

Groh, V., Bahram, S., Bauer, S., Herman, A., Beauchamp, M., and Spies, T., 1996, Cell stress-regulated human major histocompatibility complex class I gene expressed in gastrointestinal epithelium. *Proc. Natl. Acad. Sci. USA* **93**:12445-12450

Heemels, M. T., and H. L. Ploegh. Generation, translocation, and presentation of MHC class I-restricted peptides. 1995. Annu. Rev. Biochem. **64**:463-491.

Heise, M., Connick, M., and Virgin, H.W., 1998, Murine cytomegalovirus inhibits interferon γ-induced antigen presentation to CD4 T cells by macrophages via regulation of expression of major histocompatibility complex class II-associated genes. *J. Exp. Med.* **187**:1037-1046

Hengel, H., Brune, W., and Koszinowski, U.H. 1998, Immune evasion by cytomegalovirus - survival strategies of a highly adapted opportunist. *Trends Microbiol.* **6**:190-197

Hengel, H., Eßlinger, C., Pool, J., Goulmy, E., and Koszinowski, U.H., 1995, Cytokines restore MHC class I complex formation and control antigen presentation in human cytomegalovirus-infected cells. *J. Gen. Virol.* **76**:2987-2997

Hengel, H., Flohr, T., Hämmerling, G.H., Koszinowski, U.H., and Momburg, F., 1996, Human cytomegalovirus inhibits peptide translocation into the endoplasmic reticulum for MHC class I assembly. *J. Gen. Virol.* **77**:2287-2296

Hengel, H., Koopmann, J.O., Flohr, T., Muranyi, W., Goulmy, E., Hämmerling, G.J., Koszinowski, U.H., and Momburg, F., 1997. A viral ER-resident glycoprotein inactivates the MHC-encoded peptide transporter. *Immunity* **6**:623-632

Hengel, H., and Koszinowski, U.H., 1997, Interference with antigen processing by viruses. *Cur. Opin. Immunol.*, **9**:470-476

Hengel, H., Lucin, P., Jonjic, S., Ruppert, T., and Koszinowski, U.H., 1994, Restoration of cytomegalovirus antigen presentation by gamma interferon combats viral escape. *J. Virol.* **68**:289-297

Hengel, H., Reusch, U., Geginat, G., Holtappels, R., Ruppert, T., Hellebrand, E., and Koszinowski, U.H., 2000, Macrophages escape inhibition of major histocompatibility complex class I dependent antigen presentation by cytomegalovirus. *J. Virol.* **74**:7861-7868

Hengel, H., Reusch, U., Gutermann, A., Ziegler, H., Jonjic, S., Lucin, P., and Koszinowski U.H., 1999, Cytomegaloviral control of MHC class I function in the mouse. *Immunol. Rev.* **168**:167-176

Hewitt, E.W., Sen Gupta, S. and Lehner P.J., 2001, The human cytomegalovirus gene product US6 inhibits ATP binding by TAP. *EMBO J.* **20**:387-396

Hill, A., Jugovic, P., York, I., Russ, G., Bennink, J., Yewdell, J., Ploegh, H., Johnson, D. 1995, Herpes simplex virus turns off the TAP to evade host immunity. *Nature* **375**:411–415

Huard, B., Früh, K., 2000, A role for MHC class I down-regulation in NK cell lysis of herpes virus-infected cells. *Eur. J. Immunol.* **30**:509–515

Jones, T.R., Hanson, L.K., Sun, L., Slater, J.S., Stenberg, R.S. and Campbell, A.E., 1995, Multiple independent loci within the human cytomegalovirus unique short region down-regulate expression of major histocompatibility complex class I heavy chains. *J. Virol.* **69**:4830-4841

Jones, T. R. and Muzithras, V. P., 1991, Fine mapping of transcripts expressed from the US6 gene family of human cytomegalovirus strain AD169. *J. Virol.* **65**:2024-2036

Jones, T.R., and Sun, L., 1997, Human cytomegalovirus US2 destabilizes major histocompatibility complex class I heavy chains. *J. Virol.* **71**:2970-2979

Jones, T.R., Wiertz, E.J.H.J., Sun, L., Fish, K., Nelson, J.A. and Ploegh, H.L., 1996, Human cytomegalovirus US3 impairs transport and maturation of major histocompatibility complex class I heavy chains. *Proc. Natl. Acad. Sci. USA* **93**:11327-11333

Jonjic, S., Del Val, M., Keil, G.M., Reddehase, M.J. and Koszinowski, U.H., 1988, A nonstructural viral protein expressed by a recombinant vaccinia virus protects against lethal cytomegalovirus infection. *J. Virol.* **62**:1653-1658

Jonjic, S., Mutter, W., Weiland, F., Reddehase, M.J., and Koszinowski, U.H., 1989, Site-restricted persistent cytomegalovirus infection after selective long-term depletion of CD4[+] T lymphocytes. *J. Exp. Med.* **169**:1199-1212

Jonjic, S., Pavic, I., Polic, B., Crnkovic, I., Lucin, P., and Koszinowski, U.H., 1994, Antibodies are not essential for the resolution of primary cytomegalovirus infection but limit dissemination of recurrent virus. *J. Exp. Med.* **179**:1713-1717

Jun, Y., Kim, E., Jin, M., Sung, H.C., Han, H., Geraghty, D.E., Ahn, K., 2000, Human cytomegalovirus gene products US3 and US6 down-regulate trophoblast class I MHC molecules. *J. Immunol.* **164**:805–811

Kärre, K., 1995, Express yourself or die: peptides, MHC molecules, and NK cells. *Science* **267**:978-979

Kaye, J., Browne, H., Stoffel, M., and Minson, T., 1992, The UL16 gene of human cytomegalovirus encodes a glycoprotein that is dispensable for growth in vitro. *J. Virol.* **66**:6609-6615

Kleijnen M., Huppa, J.B., Lucin, P., Mukherjee, P., Farrell, H., Campbell, A., Koszinowski, U.H., Hill, A.B. and Ploegh, H.L., 1997, A mouse cytomegalovirus glycoprotein, gp34, forms a complex with folded class I MHC molecules in the ER which is not retained but transported to the cell surface. *EMBO J.* **16**:685-694

Krmpotic, A., Bubic, I., Hengel, H., Scalzo, A., Koszinowski, U.H., and Jonjic, S., 2001, The MCMV m152/gp40 glycoprotein achieves simultaneous avoidance of CD8+ and NK cell-mediated immune control. Abstract. International Cytomegalovirus Workshop, Asilomar, CA, USA

Krmpotic, A., Messerle, M., Crnkovic-Mertens, I., Polic, B., Jonjic, S., and Koszinowski, U. H., 1999, The immunoevasive function encoded by the mouse cytomegalovirus gene m152 protects the virus against T cell conctrol *in vivo. J. Exp. Med.* **190**:1285-1295

Lee, N., Goodlett, D.R., Ishitani, A., Marquardt, H., Geraghty, D.E., 1998b, HLA-E surface expression depends on binding of TAP-dependent peptides derived from certain HLA class I signal sequences. *J. Immunol.* **160**:4951–4960

Lehner, P.J., Karttunen, J.T., Wilkinson, G.W.G. and Cresswell, P., 1997, The human cytomegalovirus US6 glycoprotein inhibits transporter associated with antigen processing-dependent peptide translocation. *Proc. Natl. Acad. Sci. USA* **94**:6904-6909

Lindquist J.A., Jensen, O.N., Mann, M., and Hämmerling, G.J., 1998, ER-60, a chaperone with thiol-dependent reductase activity involved in MHC class I assembly. *EMBO J.* **17**:2186-2195

Lucin, P., Pavic, I., Polic, B., Jonjic, S., and Koszinowski, U.H., 1992, Gamma interferon-dependent clearance of cytomegalovirus infection in salivary glands. *J. Virol.* **66**:1977-1984

Miller, D.M., Rahill, B.M., Boss, J.M., Lairmore, M.D., Durbin, J.E., Waldman, W.J., Sedmak, D.D., 1998, Human cytomegalovirus inhibits major histocompatibility complex class II expression by disruption of the Jak/Stat pathway. *J. Exp. Med.* **187**:675-683

Mocarski, E.S,. 1996, Cytomegaloviruses and their replication. In: *Virology*, (B.N. Fields, D.M. Knipe, and P.M. Howley, eds.) Volume 2. Lippincott-Raven, Philadelphia; pp. 2447-2492

Momburg, F. and Hämmerling, G.J., 1998, Generation and TAP-mediated transport of peptides for major histocompatibility complex class I molecules. *Adv. Immunol.* **68**:191–256

Momburg F. and Hengel, H., 2001, Corking the bottle neck: the transporter associated with antigen processing (TAP) as a target for immune subversion by viruses. *Curr. Top. Microbiol. Immunol., in press*

Ortmann, B., Androlewicz, M., and Cresswell, P., 1994, MHC class I / β_2-microglobulin complexes associate with TAP transporters before peptide binding. *Nature* **368**:864-867

Pavic, I., Polic, B., Crnkovic, I., Lucin, P., Jonjic, S., and Koszinowski, U.H., 1993, Participation of endogenous tumour necrosis factor α in host resistance to cytomegalovirus infection. *J. Gen. Virol.,* **74**:2215-2223

Ploegh, H.L., 1998, Viral Strategies of immune evasion. *Science* **280**:248-253.

Polic, B., Hengel, H., Krmpotic, A., Trgovcich, J., Pavic, I., Lucin, P., Jonjic, S., Koszinowski, U.H., 1988, Hierarchical and redundant lymphocyte subset control precludes cytomegalovirus replication during latent infection. *J. Exp. Med.* **188**:1047-1054

Rawlinson, W.D., Farrell, H.E., Barrell, B.G., 1996, Analysis of the complete DNA sequence of murine cytomegalovirus. *J. Virol.* **70**:8833-8849

Reddehase, M.J., Mutter, W., Münch, K., Bühring, H.J., and Koszinowski, U.H., 1987, CD8 positive T lymphocytes specific for murine cytomegalovirus immediate-early antigens mediate protective immunity. *J. Virol.* **61**:3102-3108

Reddehase, M.J., Weiland, F., Münch, K., Jonjic, S., Lüske, A., and Koszinowki, U.H., 1985, Interstitial murine cytomegalovirus pneumonia after irradiation: characterization of cells that limit viral replication during established infection of the lungs. *J. Virol.* **55**:264-273

Reits, E.A., Vos, J.C., Gromme, M., Neefjes, J., 2000, The major substrates for TAP in vivo are derived from newly synthesized proteins. *Nature* **404**:774–778

Reusch, U., Muranyi, W., Lucin, P., Burgert, H.G., Hengel, H., and Koszinowski, U.H., 1999, A cytomegalovirus glycoprotein reroutes MHC class I complexes to lysosomes for degradation. *EMBO J.* **18**:1081-1091

Riddell, S.R., Watanabe, K.S., Goodrich, J.M., Li, C.R., Agha, M.E., and Greenberg, P.D., 1992, Restoration of viral immunity in immunodeficient humans by the adoptive transfer of T cell clones. *Science* **257**:238-241

Rock, K.L., and Goldberg, A.L., 1999, Degradation of cell proteins and the generation of MHC class I-presented peptides. *Annu. Rev. Immunol.* **17**:739–779

Sadasivan, B., Lehner, P.J., Ortmann B, Spies, T., and Cresswell P., 1996, Roles for calreticulin and a novel glycoprotein, tapasin, in the interaction of MHC class I molecules with TAP. *Immunity* **5**:103-114

Schust, D.J., Tortorella, D., Seebach, J., Phan, C., and Ploegh, H.L., 1998, Trophoblast class I major histocompatibility complex (MHC) products are resistant to rapid degradation imposed by the human cytomegalovirus (HCMV) gene products US2 and US11. *J. Exp. Med.* **188**:497-503

Shamu, C.E., Story, C.M., Rapoport, T.A., and Ploegh, H.L., 1999, The pathway of US11-dependent degradation of MHC class I heavy chains involves a ubiquitin-conjugated intermediate. *J. Cell. Biol.* **147**:45-57

Shellam, G.R., Allan, J.E., Papadimitriou, J.M., and Bancroft, G.J., 1981, Increased susceptibility to cytomegalovirus infection in beige mutant mice. *Proc. Natl. Acad. Sci. USA* **78**:5104-5108

Sinzger, C., Grefte, A., Plachter, B., Gouw, A.S.H., The, T.H. and Jahn, G., 1995, Fibroblasts, epithelial cells, endothelial cells and smooth muscle cells are major targets of human cytomegalovirus infection in lung and gastrointestinal tissues. *J. Gen. Virol.* **76**:741-750

Story, C.M., Furman, M.H., and Ploegh, H.L., 1999, The cytosolic tail of class I MHC heavy chain is required for its dislocation by the human cytomegalovirus US2 and US11 gene products. *Proc. Natl. Acad. Sci. USA* **96**:8516-8521

Suh, W.K., Cohen-Doyle, M.F., Früh, K., Wang, K., Peterson, P.A., Williams, D.B., 1994, Interaction of MHC class I molecules with the transporter associated with antigen processing. *Science* **264**:1322-1326

Thäle, R., Szepan, U., Hengel, H., Geginat, G., Lucin, P., Koszinowski, U.H., 1995, Identification of the mouse cytomegalovirus genomic region affecting major histocompatibility complex class I molecule transport. *J. Virol.* **69**:6098-6105

Tomasec, P., Braud, V.M., Rickards, C., Powell, M.B., McSharry, B.P., Gadola, S., Cerundolo. V., Borysiewicz, L.K., McMichael, A.J., Wilkinson, G.W., 2000, Surface expression of HLA-E, an inhibitor of natural killer cells, enhanced by human cytomegalovirus gpUL40. *Science* **287**:1031–1033

Tomazin, R., Boname, J., Hegde, N., Lewinsohn, D.M., Altschuler, Y., Jones, T.R., Cresswell, P., Nelson, J.A., Riddell, S., and Johnson, D.C., 1999, Cytomegalovirus US2 destroys two components of the MHC class II pathway, preventing recognition by CD4+ T cells. *Nat. Med.* **9**:1039-1043.

Tomazin, R., Hegde, N., and Johnson, D.C., 2001, Inhibition of MHC class I and II antigen presentation by herpesviruses. Abstract. Keystone Symposium on Molecular Aspects of Viral Immunity, Keystone, Co, USA.

Tomazin, R., Hill, A.B., Jugovic, P., York, I., van Endert, P., Ploegh, H.L., Andrews, D.W., Johnson, D.C., 1996, Stable binding of the herpes simplex virus ICP47 protein to the peptide binding site of TAP. *EMBO J* **15**:3256–3266

Ulbrecht, M., Martinozzi, S., Grzeschik, M., Hengel, H., Ellwart, J.W., Pla, M., Weiss, E.H., 2000, Cutting edge: the human cytomegalovirus UL40 gene product contains a ligand for HLA-E and prevents NK cell-mediated lysis. *J. Immunol.* **164**:5019–5022

Warren, A.P, Ducroq, D.H,, Lehner, P.J. and Borysiewicz, L.K., 1994, Human-cytomegalovirus-infected cells have unstable assembly of major histocompatibility complex class I complexes and are resistant to lysis by cytotoxic T lymphocytes. *J. Virol.* **68**:2822-2829.

Welsh, R.M., Brubaker, J.O., Vargas-Cortes, M., and O'Donnell, C.L., 1991, Natural Killer (NK) cell responses to virus infections in mice with severe combined immunodeficiency. The stimulation of NK cells and the NK-dependent control of virus infections occur independently of T and B cell funtion. *J. Exp. Med.* **173**:1053-1063

Wiertz, E.J H.J., Jones, T.R., Sun, L., Bogyo, M., Geuze, H.J. and Ploegh, H.L., 1996a, The human cytomegalovirus US11 gene product dislocates MHC class I heavy chains from the endoplasmic reticulum to the cytosol. *Cell* **84**:769-779

Wiertz, E.J.H.J., Tortorella, D., Bogyo, M., Yu, J., Mothes, W., Jones, T.R., Rapoport, T.A. and Ploegh, H.L., 1996b, Sec61-mediated transfer of a membrane protein from the endoplasmic reticulum to the proteasome for destruction. *Nature* **384**:432-438

Yamashita, Y., Shimokata, K., Mizuno, S., Yamaguchi, H. and Nishiyama, Y., 1993, Down-regulation of the surface expression of class I MHC antigens by human cytomegalovirus. *Virology* **193**:727-736

Ziegler, H., Muranyi, W., Burgert, H.-G., Kremmer, E., and Koszinowski, H.U., 2000, The luminal part of the murine cytomegalovirus glycoprotein gp40 catlayzes the retention of MHC class I molecules, *EMBO J.* **19**:870-881

Ziegler, H., R. Thäle, P. Lucin, W. Muranyi, T. Flohr, H. Hengel, H. Farell, W. Rawlinson, and U.H. Koszinowski, 1997, A mouse cytomegalovirus glycoprotein retains MHC class I complexes in the ERGIC/cis-Golgi. *Immunity* **6**:57-66

Chapter 7.2

Regulation of Cellular Genes by Cytomegalovirus

MARK F. STINSKI AND YOON-JAE SONG
Department of Microbiology, School of Medicine, University of Iowa, Iowa City, Iowa 52242, USA

1.   INTRODUCTION

Cytomegaloviruses (CMVs) are members of the betaherpesviruses with the prototype of this classification group being human CMV. Animal CMVs, such as simian, guinea pig, rat, and murine CMV, have also been used to study betaherpesvirus replication and pathogenesis. Human and animal CMVs have many genes in common. For example, murine CMV, the most widely used model for CMV pathogenesis, has 78 open reading frames (ORFs) in common with the greater than 200 ORFs of human CMV. Two human viruses that are also genetically related to human CMV are human herpesvirus-6 (HHV-6) and HHV-7 which are associated with a disseminated infection in children called roseola.

Depending on location and economic status, between 40 and 90% of the population is infected with human CMV. The virus is the leading cause of congenital birth defects which include brain damage and sight and hearing defects. The virus also establishes disseminated infections in immunosuppressed individuals that can result in pneumonitis, retinitis, hepatitis, and gastroenteritis (Alford and Britt, 1990; Ho, 1991). The stromal and epithelial tissues are frequently damaged by CMV infection. In addition, both human and animal CMVs infect vascular endothelial and smooth muscle cells. CMV DNAs are detectable in lesions of the blood vessel (Melnick *et al.*, 1994). A correlation between CMV infection of smooth muscle cells and migration of these cells has been shown, and it is

Structure-Function Relationships of Human Pathogenic Viruses, Edited by
Holzenburg and Bogner, Kluwer Academic/Plenum Publishers, New York, 2002

possible that this migration of infected cells to sites of vascular wall inflammation is due to cytokine release from inflammatory cells (Streblow *et al.*, 1999). These observations and others implicate CMV infection as a potential factor involved in atherosclerosis and coronary restenosis (Bruggeman and van Dam-Mieras, 1991; Melnick, Adam, and Debakey, 1993; Melnick *et al.*, 1994; Persoons *et al.*, 1994; Span *et al.*, 1992; Speir *et al.*, 1994).

The betaherpesviruses typically establish life-long latent-recurrent infections. The latent human CMV genome has been detected in experimentally infected granulocyte-macrophage progenitors, naturally infected hematopoietic stem cells of the bone marrow and blood monocytes (Fish, Britt, and Nelson, 1996; Hahn, Jores, and Mocarski, 1998; Kondo, Kaneshima, and Mocarski, 1994; Rice, Schrier, and Oldstone, 1884; Soderberg-Naucler, Fish, and Nelson, 1997a; Soderberg-Naucler, Fish, and Nelson, 1997b; Taylor-Wiedeman *et al.*, 1991). Although murine CMV also infects cells of the myeloid lineage, these cells are regarded as temporary sites of latency and the stromal cells of various organs as enduring sites (Kurz *et al.*, 1997; Kurz *et al.*, 1999).

Human CMV latent genomes are very difficult to detect, with 0.01% or less of the naturally infected cells of the myeloid lineage carrying a latent viral genome (Slobedman and Mocarski, 1999). Maintenance of the viral DNA in these latently infected cells is not understood. Viral proteins are presumably not required because only 2% of the latently infected cells have viral specific transcripts. Although latency appears to proceed without latent viral transcription, reactivation from latency is a poorly understood event in the life cycle of the virus. Nevertheless, reactivation occurs periodically, and viral replication can result in contagion in bodily secretions which serves to spread the virus. Reactivation of the virus occurs when the latently infected cell is stimulated and differentiates into a macrophage. Cytokine secretion from T-cells has a role in this reactivation (Soderberg-Naucler, Fish, and Nelson, 1997a; Soderberg-Naucler, Fish, and Nelson, 1997b). One of the key genetic targets for reactivation from latency is the viral major immediate early (MIE) enhancer and promoter located in the unique long component of the viral genome. The MIE enhancer has multiple cis-acting sites for the binding of cellular transcription factors (reviewed in Meier and Stinski, 1996; Stinski, 1999). When the MIE enhancer is deleted, murine CMV can replicate in cell culture, but the mutant virus fails to cause disease in the mouse (Angulo *et al.*, 2000). Therefore, it is assumed that the MIE enhancer of human CMV is necessary for reactivation from latency and for productive replication in the host.

During reactivation from latency or primary infection, the MIE enhancer responds to stimulatory events that activate cellular transcription factors. The MIE promoter drives the expression of two viral regulatory proteins. These viral genes designated IE1 (UL123) and IE2 (UL122) encode viral proteins of 72,000 (IE72) and 86,000 (IE86) kDa of apparent molecular weight. Several other isomers are also encoded by the IE2 gene (reviewed in Stenberg, 1996). The functions of these viral gene products and their critical domains have been reviewed (Spector, 1996; Stenberg, 1996). The IE viral proteins are multifunctional proteins that affect cellular transcription and the cascade of viral gene expression. This temporal cascade of CMV gene expression is designated IE (alpha), early (beta) and late (gamma). IE gene transcription does not require *de novo* viral protein synthesis. Early genes require IE gene expression, and late genes require viral DNA replication. The IE proteins are either viral regulatory proteins or viral proteins involved in immune evasion. The early viral proteins, which are also involved in immune evasion, are required for viral DNA replication. The late viral proteins are mainly involved in virion assembly.

This chapter will focus on the effects of CMV and CMV-specified proteins on host cell gene expression. The effects of CMV on host cell signal transduction events and on the phases of the cell cycle will be reviewed. These virus-specified effects on the host cell that establish a favorable environment for viral replication are induced by either virion-associated proteins or viral proteins synthesized early after infection.

2. EFFECTS OF VIRUS ATTACHMENT TO THE HOST CELL

Virions and viral glycoproteins. After human CMV infection, there is an increase in the transcription of cellular genes such as c-myc, c-fos, and c-jun (Boldogh, AbuBakar, and Albrecht, 1990; Boldogh *et al.*, 1991). In addition, there is signal transduction via the phosphoinositide-specific phospholipase C pathway and an increase in secondary messengers like phosphotidyl inositol and diacylglycerol (Albrecht *et al.*, 1990; Albrecht *et al.*, 1991). The intracellular stores of calcium and cyclic AMP and GMP also increase. Drugs like papaverine that inhibit the above cellular response decrease the efficiency of CMV replication (Albrecht *et al.*, 1990), indicating that these cellular responses are important for the virus. The virion of human CMV also has phosphatases such as PP1 and PP2A that are delivered to the host cell and cause dephosphorylation of cellular proteins immediately after infection (Michelson *et al.*, 1996). However, it is not understood how transient dephosphorylation of cellular proteins affects viral replication. The above effects on the host cell do not require *de novo* viral protein synthesis, and they are also demonstrable with UV-inactivated virus (Michelson *et al.*, 1996).

It has recently been discovered that CMV also delivers to the host cell viral mRNAs as part of the virion (Bresnahan and Shenk, 2000a). The viral mRNAs can be translated early after infection, but the effect of the viral proteins coded by the mRNAs is currently not understood. After CMV infection, there are changes in cellular gene expression. Gene microarray assays demonstrated a four-fold or greater increase of at least 124 cellular mRNAs and a decrease of at least 134 cellular mRNAs (Zhu *et al.*, 1998). Cellular gene expression affected by human CMV infection included cell structural and matrix genes, cell cycle genes, cellular ligands and receptors, transcription factors, translation factors, prostaglandins, interferon response genes, and miscellaneous genes. Approximately 15 interferon-inducible mRNAs were increased four-fold or greater early after human CMV infection (Zhu *et al.*, 1998; Zhu, Cong, and Shenk, 1997).

The activation of interferon-responsive genes is related to virion attachment. Virion-associated glycoproteins required for attachment and entry into host cells are gB, gH, which complexes with gL and gO, and gN, which complexes with gM (review in Britt and Mach, 1996; Mach *et al.*, 2000). After first stage attachment to a cellular proteoglycan, cellular receptors facilitate second stage attachment of CMV. These second stage attachment receptors are currently unknown. Human CMV glycoprotein gB and gH stimulate cellular transcription factor activity. The transcription factor NF-κB is activated presumably by a signal transduction event that releases the inhibitor I-κB and allows for translocation of NF-κB to the nucleus. There is also an increase in the binding of NF-κB to its cognate site on DNA and an increase in cellular mRNAs from genes regulated by NF-κB (Yurochko *et al.*, 1997; Yurochko *et al.*, 1995). There is also an increase in transcription factor SP-1 and the binding of SP-1 to its cognate site on DNA (Yurochko *et al.*, 1997). SP-1 activity affects NF-κB activity and vice-versa. Whether these effects are exclusive to human CMV glycoproteins gB and gH is currently not known. Human CMV soluble gB induces expression of the interferon response genes and an interferon response element binding factor (Navarro *et al.*, 1998). After phosphorylation, the factor is translocated to the nucleus and can complex with other cellular transcription factors such as CREB binding protein (CBP) which is a co-activator of transcription with histone acetylase activity. CBP contributes to the opening of the chromatin for activation of transcription. There is also an increase in the RNA steady-state level of 2'-5' oligoadenylate synthetase and interferon-stimulated gene 54 (Boyle, Pietropaolo, and Compton, 1999). Activation of interferon response gene expression is assumed to be linked to a viral ligand-cellular receptor-mediated signal transduction event. Although activation of interferon response genes appears counter to successful CMV replication, the signal transduction event may be necessary for virion entry and endocytosis. Human CMV may have a virus-specified mechanism to neutralize the effects of an interferon response.

3. OTHER SIGNAL TRANSDUCTION EVENTS ACTIVATED BY CMV

CMVs normally infect terminally differentiated cells and protein kinase pathways are normally suppressed in these cells. There is robust activation of the MAPK/ERK signal transduction pathway at approximately 4 hours after infection with human CMV (Rodems and Spector, 1998). MAPK/ERK are cellular signaling kinase pathways which are activated by phosphorylation of tyrosine and threonine residues in response to various external stimuli. Transcription factors at the end of the MAPK/ERK signal transduction pathway, such as SRF/Ets, CREB/ATF, and AP-1, are activated by phosphorylation. These transcription factors can induce transcription from both the viral and the cellular DNAs. CMV may specify viral proteins that induce an activation of the MAPK/ERK pathway and the timing of this activation implies that viral proteins synthesized *de novo* are involved. The activation of the pathway may be due to an increase in the rate of phosphorylation or to an inactivation of phosphatase activity. Inhibitors of the MAPK/ERK pathway affect the early viral promoter activity and presumably the efficiency of infectious virus production (Rodems and Spector, 1998). Therefore, it is assumed that human CMV requires an activation of the MAPK/ERK pathway for efficient viral replication.

Upstream of CMV early promoters are putative cis-acting sites for the binding of cellular transcription factors activated by the MAPK/ERK pathway. For example, early viral promoters upstream of viral genes, such as UL4 and UL112-113, have SRF/Ets and CREB/ATF cis-acting sites, respectively. Mutation of these cis-acting sites effects the level of promoter activity in transient transfection assays and in the context of the viral genome (Chen and Stinski, 2000; Rodems, Clark, and Spector, 1998). Likewise, inhibitors of the MAPK/ERK pathway effect early viral gene expression (Rodems and Spector, 1998).

Viral regulatory proteins are presumably also affected by the MAPK/ERK pathway. The human CMV IE86 protein encoded by the IE2 gene (UL122) is phosphorylated *in vitro* and *in vivo* by ERK (Harel and Alwine, 1998). The IE86 protein has a region rich in serine and threonine residues that is phosphorylated during infection.

There is also an activation of the p38 signal transduction pathway at approximately 8 hours after infection (Johnson, Huong, and Huang, 2000). Again, the timing implies that viral proteins synthesized *de novo* after infection activate this pathway. While mitogen-activated kinase kinase six (MKK6) appears to be necessary for a CMV induction of the p38 pathway at early times after infection, MKK3 and MKK4 are not activated (Johnson, Huong, and Huang, 2000). Early activation of the p38 pathway may be maintained by virus-induced inhibition of dephosphorylation allowing the p38 pathway to be sustained into the late stages of the viral replication cycle (Johnson, Huong, and Huang, 2000). The activity of the p38 pathway is also maintained at late times after infection by an increase in upstream kinases MKK3 and 6. Preliminary data suggest that these pathways might not be induced by the IE72 or IE86 viral proteins (Johnson, Huong, and Huang, 2000), but may be induced by other IE or early viral proteins. Like the MAPK/ERK, the p38 kinase can phosphorylate transcription factors, such as SRF/Ets and CREB/ATF, for activation of early viral and cellular promoters resulting in a more efficient viral replication. Although herpes simplex virus activates the JNK pathway, this pathway is not reported to be activated by human CMV infection.

4.   EFFECTS OF CMV ON THE CELL CYCLE

Terminally differentiated cells withdraw from the cell cycle and are in G_0 arrest or G_1 phase. These cells have low levels of dideoxynucleotide triphosphates and the biosynthetic enzymes to make precursors for DNA synthesis. A cell widely used for permissive human CMV infection is the human fibroblast (HF). These cells withdraw from the cell cycle by contact inhibition and can be induced to withdraw by serum deprivation. Since CMVs do not encode many of the biosynthetic enzymes for DNA precursor synthesis, the virus must have a mechanism to overcome G_1 arrest. After human CMV infection of HF cells, there is approximately a 30-fold increase in the size of the thymidylate triphosphate pool (Biron *et al.*, 1986). Human CMV infection of HF cells stimulates cellular enzymes such as thymidine kinase, dihydrofolate reductase, thymidylate synthetase, and ribonucleotide reductase which provide precursors for DNA synthesis (Gribaudo *et al.*, 2000). In addition, human CMV primes the cell for cellular DNA synthesis by inducing the synthesis of cellular cyclins and cyclin-dependent kinases (Cdks). Cyclin E and Cdk2 are highly activated after CMV infection of HF cells (Bresnahan, Albrecht, and Thompson, 1998; Bresnahan *et al.*, 1997; Jault *et al.*, 1995; Salvant, Fortunato, and Spector, 1998). The Cdks phosphorylate tumor suppressor proteins like retinoblastoma (Rb) family members for activation of the E2F family of transcription factors (Nevins, 1992; Weinberg, 1995). CMV infection also increases transcription factor E2F expression (Salvant, Fortunato, and Spector, 1998) which along with cyclin E comprises a feed-forward loop allowing amplification of signals that promote cell cycle progression from G_1 to S phase. An increase in cyclins, Cdks, and E2F is indicative of cell cycle progression from G_1 to S phase. Cdk activation is important for CMV replication because cells transfected with a dominant negative mutant of Cdk2 inhibit CMV replication (Bresnahan *et al.*, 1997). With the CMV-infected HF cell, there is progression from G_0 to G_1, but the majority of the cells are blocked at the G_1/S transition point. There is some progression into S phase where another block in cell cycle progression occurs (Lu and Shenk, 1996). In contrast, other permissive cell lines for human CMV infection like the astrocytoma cell line (U373) and the differentiated embryonal carcinoma cells (NT2D) are associated with more progression into S phase after infection (Murphy *et al.*, 2000; Sinclair *et al.*, 2000). Many of the infected U373 and differentiated carcinoma cells are blocked at a later stage in their cell cycle. *De novo* viral protein synthesis is required for cell cycle progression in the terminally differentiated cell and UV-inactivated virus has no effect. The differences between HF cell strains and U373 and NT2D cell lines is not fully understood, but it may be related to virion-associated proteins such as tegument protein UL69. This viral protein, which is discussed in more detail below, is translocated to the nucleus of HF cells and blocks G_1/S transition (Lu and Shenk, 1999). In both cell types, the cellular biosynthesis pathways are sustained to insure the presence of the necessary precursors for viral DNA replication. Thus, human CMV encodes gene products that can both stimulate and block cell cycle progression. The block may be related to direct virion-associated protein effects, indirect effects such as induction of interferon responsive genes, or viral IE protein effects.

After human CMV infection of HF cells, the mRNA levels of cyclin E and Cdk2 increase significantly (Bresnahan *et al.*, 1997; Jault *et al.*, 1995; Salvant, Fortunato, and Spector, 1998). In addition, there is an increase in Cdk2 translocation from the cytoplasm to the nucleus (Bresnahan, Thompson, and Albrecht, 1997). The effect on cyclin A, which is necessary for initiation of the S phase, is marginal as is the effect on cyclin B, which is necessary for late S through G_2 (Jault *et al.*, 1995). Inhibitors of cell cycle progression, such as p21 and p27, are expressed at reduced levels (Bresnahan *et al.*, 1996). In addition, human CMV induces degradation of p21 (Chen et al., 2001). These events are considered critical for CMV replication because inhibitors of Cdk activity like roscovitine and olomoucine, inhibit infectious virus production (Bresnahan *et al.*, 1997). In addition, an inhibitor of biosynthetic enzyme activity, such as ZD1694, a quinazoline-based folate analog that inhibits thymidylate synthetase, blocks CMV replication, and the addition of 20 μM thymidine can abrogate the effect of the drug (Gribaudo *et al.*, 2000). The pathway that leads to an absence or a reduction of cellular DNA synthesis in the human CMV-infected cell is not understood. Nevertheless, the virus uses the activated cell for viral DNA synthesis.

The Rb pocket protein family is associated with histone deacetylase activity (Brehm *et al.*, 1998; Luo, Postigo, and Dean, 1998; Magnaghi-Jaulin *et al.*, 1998). The Rb cellular proteins, when hypophosphorylated, prevent cell cycle progression into S phase by inhibiting the activity of transcription factor E2F. Human CMV infection causes hyperphosphorylation of Rb (Jault *et al.*, 1995; Sinclair *et al.*, 2000) which in-turn dissociates Rb protein from E2F and activates the E2F family of transcription factors. The E2F transcription factor recognizes cis-sites upstream of cellular promoters and associates with histone acetylase activity (Trouche and Kouzarides, 1996) and thereby induces expression of cell cycle regulators, enzymatic machinery for DNA precursor and DNA synthesis, and regulatory factors for initiation of cellular DNA synthesis. Rb remains highly phosphorylated into the late stages of human CMV infection thus keeping E2F transcription factors activated.

Human CMV infection of HF cells also elevates the level of tumor suppressor protein p53 (Bonin and McDougall, 1997; Francoise *et al.*, 1995; Jault *et al.*, 1995). The induction of p53 would normally lead to cell cycle arrest or apoptosis (Clarke *et al.*, 1993), but these pathways are inhibited by CMV. CMV infection of HF cells also stimulates TGF-β1 expression (Michelson *et al.*, 1994) with evidence pointing to activation by an IE protein (Yoo *et al.*, 1996). TGF-β1 normally inhibits progression at the G_1 phase of the cell cycle (Massague, 1990). Both p53- and TGF-β1-mediated cell cycle arrest is probably overcome by the phosphorylation of Rb and the effects of E2F activated transcription on cellular genes for cell cycle progression in CMV-infected cells. The induction of TGF-β1 in the host could be beneficial to CMV infection because TGF-β1 also down regulates the host immune response.

CMVs characteristically have a long delay between entry into the host cell and the initiation of viral DNA replication. Viral DNA replication initiates at approximately 18 hours post infection (h.p.i), and reaches peak levels by 72 h.p.i. (Stinski, 1978). The slow accumulation of viral progeny may be related to the slow rate of viral DNA replication which is determined by the phase of the cell cycle. Cells infected in G_0/G_1 proceed to viral DNA synthesis in the absence of cellular DNA synthesis. In contrast, cells infected in S phase proceed through mitosis and delay viral gene expression until G_0/G_1 (Salvant, Fortunato, and Spector, 1998). After expression of viral IE proteins, the cells are induced towards G_1/S. However further cell cycle progression is blocked, and the cell specializes in viral DNA synthesis.

5. EFFECT OF INDIVIDUAL CMV PROTEINS ON VIRAL REPLICATION AND CELL CYCLE PROGRESSION

From attachment and entry through the temporal phases of viral gene expression, CMV is affecting cell cycle progression. To better understand the multiple effects of the virus on the host cell, it is useful to assay the role of individual viral proteins.

5.1 Viral tegument proteins

The viral tegument proteins are located between the viral envelope and the nucleocapsid. Although the number of different CMV tegument proteins and their biological activity has not been determined, tegument proteins of herpesviruses have important biological activity that determines the efficiency of viral replication. Two tegument proteins of human CMV encoded by the UL82 and UL69 genes affect replication efficiency.

After attachment and entry, the viral tegument protein pp71, which is encoded by the UL82 gene, is transported to the nucleus of the infected cell (Hensel *et al.*, 1996). In the nucleus, the viral tegument protein can be found associated with small subnuclear punctuate structures or nuclear domains referred to as PODS or ND-10 domains. It is also conjugated to ubiquitin-like protein hDAAX (Hofmann, Sindre, and Stamminger, 2000). Whether or not this event is linked to virus replication efficiency is currently not understood. However, deletion of the UL82 gene significantly affects virus replication efficiency (Bresnahan and Shenk, 2000b).

We found that the pp71 tegument protein activated cellular transcription factors such as CREB/ATF and AP-1 (Liu and Stinski, 1992) which in turn had positive effects on the MIE and early viral promoters (Baldick *et al.*, 1997; Bresnahan and Shenk, 2000b; Liu and Stinski, 1992). Others found that pp71 affected transcription from heterologous viral promoters like the herpes simplex virus (HSV) IE promoters (Homer *et al.*, 1999). These effects may explain why pp71 increases plaque formation by infectious human CMV DNA (Baldick *et al.*, 1997).

One potential effect of pp71 on HF cells is cell cycle progression. The pp71 tegument protein appears to push HF cells from G_0 to G_1 (T. Shenk, Personal communication). This would increase the level of cyclins and cyclin dependent kinases and the level of biosynthetic enzymes for DNA synthesis which in turn would increase viral replication efficiency.

The UL69 gene also affects the efficiency of CMV replication in HF cells. At low MOI, viruses mutated in the UL69 gene grow slower and take longer to reach peak levels of infectious virus (Lu Hayashi, Blankenship, and Shenk, 2000). The UL69 tegument protein of human CMV, a homologue of HSV ICP27, also enhances transcription from the MIE promoter (Winkler, Rice, and Stamminger, 1994) and may also affect mRNA transport. In contrast to pp71, the viral tegument protein does not enhance plaque formation by infectious viral DNA (Baldick *et al.*, 1997). As mentioned previously, the UL69 protein can independently block the G_1/S transition (Lu Hayashi, Blankenship, and Shenk, 2000). Recombinant viruses lacking the UL69 gene fail to induce this block in cell cycle progression (Lu Hayashi, Blankenship, and Shenk, 2000) and consequently, there is a delay in viral DNA replication and late gene expression which can be rescued by UL69 tegument protein provided in trans (Lu Hayashi, Blankenship, and Shenk, 2000).

5.2   Viral immediate early proteins

The IE1 (UL123) gene of human CMV that encodes the IE72 protein is dispensable for viral replication after high MOI, but viral replication is inefficient after low MOI (0.01 PFU/cell) (Greaves and Mocarski, 1998; Mocarski, 1996). The IE72 protein also localizes to ND-10s in the nucleus and is conjugated by ubiquitin-like proteins SUMO-1 and SUMO-2 (Xu *et al.*, 2000). The IE72 protein can disperse the promyelocytic leukemia protein (PML) and the Sp100 protein associated with ND-10s (Ahn and Hayward, 1997; Ahn, Brignole, and Hayward, 1998; Muller and Dejean, 1999; Wilkinson *et al.*, 1998). Since the IE72 protein also has a positive effect on the efficiency of viral replication, it is possible that localization to the ND-10s and conjugation by ubiquitin-like proteins has an effect on the host cell that favors viral replication. One hypothesis is that herpesvirus proteins that localize the ND-10s and are conjugated by ubiquitin-like proteins induce degradation of cellular proteins which might interfere with the viral replication process (Everett, 2000).

It is proposed that the IE72 protein pushes the cell into S phase by countering the repressive activity of Rb family member p107 (Poma *et al.*, 1996). The viral protein binds *in vitro* to the N-terminus of p107, but it does not bind to the C-terminus containing the Rb-like pocket region. The N-terminus of Rb is different from p107 which might explain why IE72 protein does not bind to Rb. By binding to tumor suppressor protein p107, IE72 protein may free E2F transcription factor. The viral IE72 protein activates transcription from E2F containing promoters such as the dihydrofolate reductase promoter (Margolis *et al.*, 1995). It is also proposed that the IE72 protein has protein kinase activity and phosphorylates the Rb family members p107 and p130 as well as itself (Pajovic *et al.*, 1997). Phosphorylation causes release of E2F from p107 and p130 thus activating E2F containing promoters. The IE72 protein also phosphorylates E2F-1, -2, and –3, but not E2F-4 or –5 (Pajovic et al., 1997). Although we (Murphy *et al.*, 2000) and others (Castillo, Yurochko, and Kowalik, 2000) find the IE72 protein to push cells into the S phase, it is a mild effect relative to the effect of the viral IE86 protein. In addition, the IE72 protein does not delay the G_2/M phase of the cell cycle.

The IE72 protein does not dramatically affect the cell cycle profile of nonpermissive cells like normal rat or mouse fibroblasts (Castillo, Yurochko, and Kowalik, 2000). In contrast, in p53 negative cells IE72 protein can push cells into the S phase. Therefore, p53 may negate the proliferative effects of the viral IE72 protein. In permissive cells, IE72 protein increases the level of p53 protein localized to the nucleus (Castillo, Yurochko, and Kowalik, 2000) which may mask the effect of IE72 on promoting cell cycle progression. Nevertheless, the effects of IE72 protein on the cell cycle are considered real, because transfection of the viral gene expressing the IE72 protein can reverse the enlarged flat cell phenotype of human osteosarcoma Saos-2 cells.

The IE86 viral protein encoded by the IE2 gene (UL122) also localizes within or adjacent to ND-10s and is modified by ubiquitin-like SUMO proteins (Ahn et al., 2001; Hofmann, Floss, and Stamminger, 2000). Unlike IE72 protein, IE86 protein does not disperse PML or Sp100 protein. The wild type IE86 protein pushes as many as four-fold more cells into the S phase compared to a mutant IE86 protein with site-specific mutations in a putative zinc-finger motif (Murphy *et al.*, 2000). Exit from the S phase is delayed in cells expressing wild type IE86 protein. In addition, a significant amount of BrdU incorporation into cellular DNA is detected in IE86 protein expressing cells (Castillo, Yurochko, and Kowalik, 2000; Wiebusch and Hagemeier, 2001). Nevertheless, cell division is blocked by the wild type IE86 protein, but not by a mutant IE86 protein (Murphy *et al.*, 2000; Weibusch and Hagemeier, 1999).

The viral IE86 protein is reported to interact with a number of cellular proteins that affect the cell cycle. GST-IE86 fusion protein in which the amino end of the IE86 protein is truncated interacts *in vitro* with p53, Rb, and p21 (Hagemeier *et al.*, 1994; Sinclair *et al.*, 2000; Sommer, Scully, and Spector, 1994; Speir *et al.*, 1994). The dimerization domain of the IE86 protein must be present for these interactions. Amino acids between 290 and 390 are critical for binding to Rb even though the viral IE86 protein does not have a LXCXE amino acid motif typical of proteins that interact with Rb. Others find that the IE86 protein region between amino acids 241 to 369 is the major Rb-binding domain while a secondary region maps to amino acids 1 to 85 which is common also to the IE72 protein (Fortunato *et al.*, 1997). The IE86 protein with amino acids 136 to 290, 519 to 579, or 544 to 579 deleted interacts better with Rb than wild type IE86 protein (Sommer, Scully, and Spector, 1994). The pocket domain and the carboxyl end of Rb are necessary for interaction with IE86 protein (Fortunato *et al.*, 1997). Phosphorylation of Rb by cyclin-dependent kinases diminishes the interaction with IE86 protein. The following biological assays support the notion that IE86 protein interacts with Rb *in vivo*: (i) The viral IE86 protein alleviates an Rb-mediated repression of a cellular promoter bearing an E2F responsive element in transient transfection assays. (ii) Rb alleviates viral IE86 protein-mediated repression of the human CMV MIE promoter in transient transfection assays. (iii) The IE86 protein counters the Rb-mediated enlarged flat cell phenotype of Saos-2 cells. (iv) IE86 protein-Rb complexes can be immunoprecipitated from human CMV-infected cells (Fortunato *et al.*, 1997; Hagemeier *et al.*, 1994). It is likely that freeing Rb from E2F activates the E2F transcription factor which is one mechanism by which the viral IE86 protein pushes cells towards the S phase. As discussed below, E2F activity is controlled by multiple mechanisms and IE86 protein association with Rb provides only one of the possible mechanisms for activation of E2F.

The IE86 protein also interacts with p53 and p21 (Bonin and McDougall, 1997; Sinclair *et al.*, 2000; Speir *et al.*, 1994). p53 restricts cell cycle progression and p21 is a cyclin-dependent kinase inhibitor which also restricts cell cycle progression. The IE86 protein increases the level of p53 localized to the nucleus of the cell (Bonin and McDougall, 1997; Muganda *et al.*, 1994; Speir *et al.*, 1994). The interaction of IE86 protein with p53 does not interfere with p53 binding to its cognate site on DNA, but it does interfere with transactivation of a p53 responsive promoter (Tsai *et al.*, 1996). Therefore, it is possible that IE86 protein interferes with the expression of the inhibitors of cyclin dependent kinases. Interaction of IE86 protein with p21 may also prevent p21 repression of cyclin E-dependent kinase activity and consequently, prevent growth arrest. It is also proposed that the IE86 protein activates the cyclin E promoter through direct binding to a site between +35 and +66 relative to the transcription start site (Bresnahan, Albrecht, and Thompson, 1998). These results from transient transfection assays and *in vitro* DNA binding assays have not been confirmed in the virus-infected cell.

Using gene microarray analysis, we have found that the viral IE86 protein expressed by an adenovirus vector significantly induces the expression of E2F and E2F-regulated cellular promoters. There is a 4-fold or greater increase in the steady-state level of mRNAs of the cell cycle regulatory genes, the cellular biosynthetic and DNA synthesis genes, and the genes that regulate initiation of cellular DNA synthesis (Song and Stinski, 2001). These viral and cellular gene products are sufficient for cell cycle progression towards the S phase in contact-inhibited HF cells or in serum-starved cells that are in G_0. These results indicate that the effects of inhibiting cell cycle progression by p53, p16/p21/p27, and TGF-β1 are negated by IE86 protein expression. While it was proposed that the viral IE86 protein alleviates the p53 G_1 checkpoint function and allows for cellular proliferation (Speir *et al.*, 1994), the human CMV IE86 protein is different from other DNA virus proteins such as adenovirus E1a/E1b, simian virus 40 T-antigen, and papilloma virus E6/E7 proteins. These DNA virus proteins can independently induce cellular transformation while the IE86 protein independently blocks cell division without inducing apoptosis (Murphy *et al.*, 2000; Weibusch and Hagemeier, 1999). While the IE86 protein may impair p53's ability to block cell cycle progression, the DNA damage checkpoint function of p53 is apparently not impaired (Bonin and McDougall, 1997). The IE86 protein efficiently drives quiescence cells into S phase, delays exit from the cell cycle, and delays apoptosis by possibly inhibiting p53-mediated apoptosis. Both the IE72 and IE86 proteins can block induction of apoptosis by tumor necrosis factor alpha (TNF-α) (Zhu, Shen, and Shenk, 1995). This allows IE86 protein expressing cells to remain viable, yet they do not divide (Murphy *et al.*, 2000). How the IE86 protein triggers a check point in cell cycle progression in the S phase is currently not understood.

Another human CMV IE protein with anti-apoptotic activity is the UL37 exon 1 protein. The UL37 gene is a part of an IE transcription unit between the UL36 and UL38 genes (Colberg-Poley, 1996). Exon 1 of UL37 codes for a 163 amino acid protein that localizes to the outer mitochondrial membrane. It suppresses apoptosis downstream of the caspase-8 activation pathway and upstream of cytochrome c release. It blocks TNF-α or Fas-mediated apoptosis (Goldmacher *et al.*, 1999) and presumably plays an important role in the human CMV infectious process. Murine CMV, which efficiently infects vascular endothelial cells, may also block apoptosis. One viral gene involved in blocking apoptosis is M45 which encodes a viral protein with homology to ribonucleotide reductase (RR1) (Brune *et al.*, 2001). Human CMV has a gene homologous to M45.

6.   SUMMARY

Figure 1 depicts the time course of signal transduction pathway activation after infection of HF cells with human CMV. Activation of the cellular MAPK/ERK and p38 signal transduction pathways are important events for the successful and efficient replication of the virus. Early after infection with human CMV, when viral proteins have been synthesized *de novo*, there is a robust stimulation of the MAPK/ERK and p38 signal transduction pathways. At the end of both of these signal transduction pathways are cellular transcription factors, such as SRF/Ets, CREB/ATF, and AP-1, that are activated by phosphorylation. These cellular transcription factors in combination with viral IE proteins synthesized *de novo* enhance transcription from both viral and cellular promoters. The IE proteins of human CMV, such as IE72, IE86, and UL37, are anti-apoptotic and stimulatory for viral and cellular transcription. Evidence suggests that the murine CMV RR1 homologue is anti-apoptotic in virus-infected vascular endothelial cells. There is a significant enhancement in the expression of cellular genes involved in the regulation of the cell cycle, the biosynthesis of cellular DNA and DNA precursors, and the initiation of cellular DNA synthesis. However, the expression of the viral IE proteins also prevents exit from the cell cycle and cell division. In doing so, the virus primes the cell for viral DNA replication. These events are important to the biology of the virus because CMVs normally infect differentiated cells that have withdrawn from the cell cycle and are in the G_0 or G_1 phase of the cell cycle. Since human CMV only encodes one of the biosynthetic enzymes for making precursors for DNA synthesis, it must have a mechanism to stimulate cell cycle progression. Figure 2 depicts the virus-specified proteins involved in pushing the cell towards the S phase. The tegument protein pp71 pushes the differentiated cell that is normally in G_0 to the G_1 phase. The viral IE72 and IE86 proteins push the cell towards S phase so that all the cellular genes with E2F-regulated promoters are activated such as the cyclins and cyclin dependent kinases, E2F transcription factors, the biosynthetic enzymes for DNA synthesis, and the regulatory proteins for initiation of cellular DNA synthesis. Cellular gene products highly activated by human CMV or by the IE86 protein of human CMV are depicted in Figure 2. While it is not to the advantage of the virus to compete with the cell, the viral tegument protein encoded by the UL69 gene efficiently blocks the S phase in HF cells, which are the most permissive cell for human CMV replication. In addition, the viral IE86 protein blocks cell division in permissive cells such as HF, smooth muscle, and endothelial cells. The virus establishes a "license for replication" and it does this best in the HF cells. However, in the host, CMV

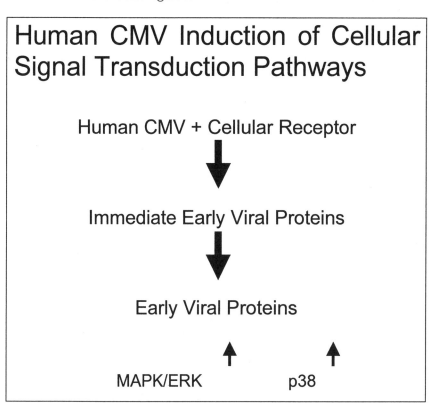

Figure 1. CMV effects on cellular signal transduction pathways.

replication is slower and less efficient in many cell types like macrophage and endothelial cells and consequently, the virus may not establish all the parameters for efficient replication. Cell types that permit fast or slow replication of CMV are important in understanding the biology of a virus that induces chronic and persistent infection.

As discussed previously, human CMV may be involved in restenosis following angioplasty. Smooth muscle cells localize to the site of injury and migrate into the vessel intima and sub-intima causing restenosis. A human CMV-specified cytokine receptor encoded by the viral US28 gene is proposed to be involved in the migration of smooth muscle cells (Streblow *et al.*, 1999). The infected smooth muscle cells may either release cytokines that attract inflammatory cells or respond to cytokines released by inflammatory cells and accumulate at the site of injury.

The IE proteins of human CMV, such as IE72, IE86, and UL37 exon 1, may induce the cell to remain viable by pushing the cell towards S phase, delaying exit from the cell cycle, and inhibiting the apoptotic pathway. The cells could potentially remain in a proliferative like state until viral DNA replication is initiated. CMV infection and progression to disease relates to alterations in the regulation of cellular genes by this ubiquitous virus.

Human CMV Induction of Cell Cycle Progression

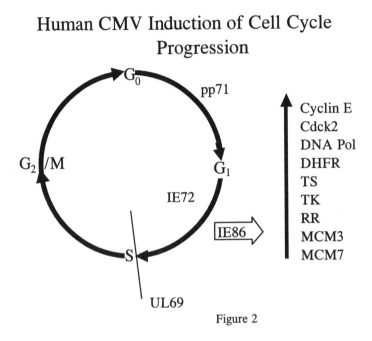

Figure 2

Figure 2. The effects of human CMV proteins on the cell cycle.

ACKNOWLEDGMENTS

We thank Dr. Al Klingelhutz for a critical reading of the manuscript and members of the laboratory for helpful discussion. Our work was supported by grant AI-13562 from the National Institutes of Health.

REFERENCES

Ahn, J.-H., and Hayward, G. S. (1997). The major immediate-early proteins IE1 and IE2 of human cytomegalovirus colocalize with and disrupt PML-associated nuclear bodies at very early times in infected permissive cells. *J. Virol.* **71,** 4599-4613.

Ahn, J.-H., Xu, Y., Jang, W.-J., Matunis, M., and Hayward, G. S. (2001). Evaluation of interactions of human cytomegalovirus immediate-early IE2 regulatory protein with small ubiquitin-like modifiers and their conjugation enzyme Ubc9. J. Virol. **75,** 3859-3872

Ahn, J. H., Brignole, E., and Hayward, G. S. (1998). Disruption of PML subnuclear domains by the acidic IE1 protein of human cytomegalovirus is mediated through interaction with PML and may modulate a RING finger-dependent cryptic transactivator function of PML. *Mol. Cell. Biol.* **18,** 4899-4913.

Albrecht, T., Boldogh, I., Fons, M., Abubakar, S., and Deng, C. Z. (1990). Cell activation signals and the pathogenesis of human cytomegalovirus infection. *Intervirology* **31,** 68-75.

Albrecht, T., Fons, M. P., Boldogh, I., Abubakar, I., Deng, S., and Millinoff, D. (1991). Metabolic and cellular effects of human cytomegalovirus infection. *Transplant. Proc.* **23,** 48-55.

Alford, C. A., and Britt, W. J. (1990). Cytomegalovirus. *In B. N. Fields, D. M. Knipe, et al (ed.), Raven Press, Ltd., New York,* 1981-2010.

Angulo, A., Messerle, M., Griffiths, M., and Ghazal, P. (2000). In vitro and in vivo characterization of an enhancerless murine cytomegalovirus. 25th International Herpesvirus Workshop, Portland, OR.

Baldick, C. J., Marchini, A., Patterson, C. E., and Shenk, T. (1997). Human cytomegalovirus tegument protein pp71 (ppUL82) enhances the infectivity of viral DNA and accelerates the infectious cycle. *J. Virol.* **71,** 4400-4408.

Biron, K. K., Fyfe, J. A., Stanat, S. C., Leslie, K., Sorrell, J. A., and Lambe, C. U. (1986). A human cytomegalovirus mutant resistant to the nucleoside analog 9-[2-hydroxy-1-(hydroxymethyl)ethoxy]methylguanine (BW B759U) induces reduced levels of BW B759U triphosphate. *Proc. Natl. Acad. Sci. USA* **83,** 8769-8773.

Boldogh, I., AbuBakar, S., and Albrecht, T. (1990). Activation of proto-oncogenes: an immediate early event in human cytomegalovirus infection. *Science* **247,** 561-564.

Boldogh, I., AbuBakar, S., Deng, C. Z., and Albrecht, T. (1991). Transcriptional activation of cellular oncogenes fos, jun, and myc by human cytomegalovirus. *J. Virol.* **65,** 1568-1571.

Bonin, L. R., and McDougall, J. K. (1997). Human cytomegalovirus IE2 86-kilodalton protein binds p53 but does not abrogate G1 checkpoint function. *J. Virol.* **71,** 5861-5870.

Boyle, K. A., Pietropaolo, R. L., and Compton, T. (1999). Engagement of the cellular receptor for glycoprotein B of human cytomegalovirus activates the interferon-responsive pathway. *Mol. Cell. Biol.* **19,** 3607-3613.

Brehm, A., Miska, E. A., McCance, D. J., Reid, J. L., Bannister, A. J., and Kouzarides, T. (1998). Retinoblastoma protein recruits histone deacetylase to repress transcription. *Nature* **239,** 597-601.

Bresnahan, W. A., Albrecht, T., and Thompson, E. A. (1998). The cyclin E promoter is activated by human cytomegalovirus 86-kDa immediate early protein. *J. Biol. Chem.* **273,** 22075-22082.

Bresnahan, W. A., Boldogh, I., Chi, P., Thompson, E. A., and Albrecht, T. (1997). Inhibition of cellular Cdk2 activity blocks human cytomegalovirus replication. *Virology* **231,** 239-247.

Bresnahan, W. A., Boldogh, I., Thompson, E. A., and Albrecht, T. (1996). Human cytomegalovirus inhibits cellular DNA synthesis and arrests productively infected cells in late G1. *Virology* **224,** 150-160.

Bresnahan, W. A., and Shenk, T. (2000a). Subset of viral transcripts packaged within human cytomegalovirus particles. *Science* **288,** 2373-2376.

Bresnahan, W. A., and Shenk, T. E. (2000b). UL82 virion protein activates expression of immediate early viral genes in human cytomegalovirus-infected cells. *Proc. Natl. Acad. of Sci. USA* **97,** 14506-14511.

Bresnahan, W. A., Thompson, E. A., and Albrecht, T. (1997). Human cytomegalovirus infection results in altered Cdk2 subcellular localization. *J. Gen. Virol.* **78,** 1993-1997.

Britt, W. J., and Mach, M. (1996). Human cytomegalovirus glycoproteins. *Intervirology* **39,** 401-412.

Bruggeman, C. A., and van Dam-Mieras, M. C. E. (1991). The possible role of cytomegalovirus in atherogenesis. *Progr. Med. Virol.* **38,** 1-26.

Brune, W., Menard, C., Heesemann, J., and Koszinowski, U. H. (2001). A ribonucleotide reductase homolog of cytomegalovirus and endothelial cell tropism. *Science* **291,** 303-305.

Castillo, J. P., Yurochko, A. D., and Kowalik, T. F. (2000). Role of human cytomegalovirus immediate-early proteins in cell growth control. *J. Virol.* **74,** 8028-8037.

Chen, J., and Stinski, M. F. (2000). Activation of transcription of the human cytomegalovirus early UL4 promoter by the Ets transcription factor binding element. *J. Virol.* **74,** 9845-9857.

Chen, Z., Knutson, E., Kurosky, A., and Albrecht, T. (2001). Degradation of p21[Cip1] in cells productively infected with human cytomegalovirus. J. Virol. **75,** 3613-3625.

Clarke, A. R., Purdie, C. A., Harrison, D. J., Morris, R. G., Bird, C. C., Hooper, M. L., and Wyllie, A. H. (1993). Tymocyte apoptosis induced by p53-dependent and independent pathways. *Nature* **362,** 849-852.

Colberg-Poley, A. M. (1996). Functional roles of immediate early proteins encoded by the human cytomegalovirus UL36-38, UL115-119, TRS1/IRS1 and US3 loci. *Intervirology* **39,** 350-360.

Everett, R. D. (2000). ICP0 induces the accumulation of colocalizing conjugated ubiquitin. *J. Virol.* **74,** 9994-10005.

Fish, K. N., Britt, W., and Nelson, J. A. (1996). A novel mechanism for persistence of human cytomegalovirus in macrophages. *J. Virol.* **70,** 1855-1862.

Fortunato, E. A., Sommer, M. H., Yoder, K., and Spector, D. H. (1997). Identification of domains within the human cytomegalovirus major immediate-early 86-kilodalton protein and the retinoblastoma protein required for physical and functional interaction with each other. *J. Virol.* **71,** 8176-8185.

Francoise, M. J., Jault, J., Ruchti, F., Fortunato, A., Clark, C., Corbeil, J., Richman, D. D., and Spector, D. H. (1995). Cytomegalovirus infection induces high levels of cyclins, phosphorylated Rb, and p53, leading to cell cycle arrest. *J. Virol.* **69,** 6697-6704.

Goldmacher, V. S., Bartle, L. M., Skaletskaya, A., Dionne, C. A., Nedersha, N. L., Vater, C. A., Han, J.-W., Lutz, R. J., Watanabe, S., McFarland, E. D. C., Kieff, E. D., Mocarski, E. S., and Chittenden, T. (1999). A cytomegalovirus-encoded mitochondria-localized inhibitor of apoptosis structurally unrelated to Bcl-2. *Proc. Natl. Acad. Sci. USA* **96**, 12536-12541.

Greaves, R. F., and Mocarski, E. S. (1998). Defective growth correlates with reduced accumulation of viral DNA replication protein after low-multiplicity infection by a human cytomegalovirus ie1 mutant. *J. Virol.* **72**, 366-379.

Gribaudo, G., Riera, L., Lembo, D., De Andrea, M., Gariglio, M., Rudge, T. L., Johnson, L. F., and Landolfo, S. (2000). Murine cytomegalovirus stimulates cellular thymidylate synthase gene expression in quiescent cells and requires the enzyme for replication. *J. Virol.* **74**, 4979-4987.

Hagemeier, C., Caswell, R., Hayhurst, G., Sinclair, J., and Kouzarides, T. (1994). Functional interaction between the HCMV IE2 transactivator and the retinoblastoma protein. *EMBO J.* **13**, 2897-2903.

Hahn, G., Jores, R., and Mocarski, E. S. (1998). Cytomegalovirus remains latent in a common precursor of dendritic and myeloid cells. *Proc. Natl. Acad. Sci.* **95**, 3937-3942.

Harel, N. Y., and Alwine, J. C. (1998). Phosphorylation of the human cytomegalovirus 86-kilodalton immediate-early protein IE2. *J. Virol.* **72**, 5481-5492.

Hensel, G. M., Meyer, H. H., Buchman, I., Pommerehne, D., Schmolke, B., Plachter, B., Radsak, K., and Kern, H. F. (1996). Intracellular localization and expression of the human cytomegalovirus matrix protein pp71 (UL82): evidence for its translocation to the nucleus. *J. Gen. Virol.* **77**, 3087-3097.

Ho, M. (1991). Cytomegalovirus: Biology and Infection. *Plenum Publishing Corp., New York.*

Hofman, H., Floss, S., and Stamminger, T. (2000). Covalent modification of the transactivator protein IE2-86 of human cytomegalovirus by conjugation to the ubiquitin-homologous proteins SUM0-1 and hSMT3b. J. Virol. **74**, 2510-2524.

Hofmann, H., Sindre, H., and Stamminger, T. (2000). Functional interaction between the pp71 protein of human cytomegalovirus and the PML-interacting protein hDAXX. *25th Internaltional Herpesvirus Workshop, Portland, OR.*

Homer, E. G., Rinaldi, A., Nicholi, M. J., and Preston, C. M. (1999). Activation of herpesvirus gene expression by the human cytomegalovirus protein pp71. *J. Virol.* **73**, 8512-8518.

Jault, F. M., Jault, J.-M., Ruchti, F., Fortunato, E. A., Clark, C., Corbeil, J., Richman, D. D., and Spector, D. H. (1995). Cytomegalovirus infection induces high levels of cyclins, phosphorylated RB, and p53, leading to cell cycle arrest. *J. Virol.* **69**, 6697-6704.

Johnson, R. A., Huong, S. M., and Huang, E. S. (2000). Activation of the mitogen-activated protein kinase p38 by human cytomegalovirus infection through two distinct pathways: a novel mechanism for activation of p38. *J. Virol.* **74**, 1158-1167.

Kondo, K., Kaneshima, H., and Mocarski, E. S. (1994). Human cytomegalovirus latent infection of granulocyte-macrophage progenitors. *Proc. Natl. Acad. Sci. USA* **91**, 11879-11883.

Kurz, S., Steffens, H.-P., Mayer, A., Harris, J. R., and Reddehase, M. J. (1997). Latency versus persistence or intermittent recurrences: Evidence for a latent state of murine cytomegalovirus in the lungs. *J. Virol.* **71**, 2980-2987.

Kurz, S. K., Rapp, M., Steffens, H.-P., Grzimek, N. K. A., Schmalz, S., and Reddehase, M. J. (1999). Focal transcriptional activity of murine cytomegalovirus during latency in the lungs. *J. Virol.* **73**, 482-494.

Liu, B., and Stinski, M. F. (1992). Human cytomegalovirus contains a tegument protein that enhances transcription from promoters with upstream ATF and AP-1 cis-acting elements. *J. Virol.* **66**, 4434-4444.

Lu Hayashi, M., Blankenship, C., and Shenk, T. (2000). Human cytomegalovirus UL69 protein is required for efficient accumulation of infected cells in the G1 phase of the cell cycle. *Proc. Natl. Acad. Sci. USA* **97**, 2692-2696.

Lu, M., and Shenk, T. (1996). Human cytomegalovirus infection inhibits cell cycle progression at multiple points including the transition from G_1 to S. *J. Virol.* **70**, 8850-8857.

Lu, M., and Shenk, T. (1999). Human cytomegalovirus UL69 protein induces cells to accumulate in G_1 phase of the cell cycle. *J. Virol.* **73**, 676-683.

Luo, R. X., Postigo, A. A., and Dean, D. C. (1998). Rb interacts with histone deacetylase to repress transcription. *Cell* **92**, 463-473.

Mach, M., Kropff, B., Dal Monte, P., and Britt, W. (2000). Complex formation by human cytomegalovirus glycoproteins M (gpUL100) and N (gpUL73). *J. Virol.* **74**, 11881-11892.

Magnaghi-Jaulin, L., Groisman, R., Naguibneva, I., Robin, P., Lorain, S., Le Villain, J. P., Troalen, F., Trouche, D., and Harel-Bellan, A. (1998). Retinoblastoma protein represses transcription by recruiting a histone deacetylase. *Nature* **391**, 601-604.

Margolis, M. J., Pajovic, S., Wong, E. L., Wade, M., Jupp, R., Nelson, J. A., and Azizkhan, J. C. (1995). Interaction of the 72-kilodalton human cytomegalvirus IE1 gene product with E2F1 coincides with E2F-dependent activation of dihydrofolate reductase transcription. *J. Virol.* **69**, 7759-7767.

Massague, J. (1990). The transforming growth factor-beta family. *Annu. Rev. Cell Biol.* **6**, 597-641.

Meier, J. L., and Stinski, M. F. (1996). Regulation of human cytomegalovirus immediate-early gene expression. *Intervirology* **39**, 331-342.

Melnick, J. L., Adam, E., and Debakey, M. E. (1993). Cytomegalovirus and atherosclerosis. *Eur. Heart J.* **14**, 30-38.

Melnick, J. L., Hu, C., Burek, J., Adam, E., and DeBakey, M. E. (1994). Cytomegalovirus DNA in arterial walls of patients with atherosclerosis. *J. Med. Virol.* **42**, 170-174.

Michelson, S., Alcami, J., Kim, S.-J., Danielpour, D., Licard, L., Bessia, C., Paya, C., and Virelizier, J.-L. (1994). Human cytomegalovirus infection induces production of transforming growth factor beta 1. *J. Virol.* **68**, 5730-5737.

Michelson, S., Turowski, P., Picard, L., Goris, J., Landini, M. P., Topilko, A., Hemmings, B., Bessia, C., Garcia, A., and Virelizier, J. L. (1996). Human cytomegalovirus carries serine/threonine protein phosphatases PP1 and a host-cell drived PP2A. *J. Virol.* **70**, 1415-1423.

Mocarski, E. S., Kemble, G. Lyle, J., Greaves, R. F. (1996). A deletion mutant in the human cytomegalovirus gene encoding IE1 491aa is replication defective due to a failure in autoregulation. *Proc. Natl. Acad. Sci. USA* **93**, 11321-11326.

Muganda, P., Mendoza, O., Hernandez, J., and Qian, Q. (1994). Human cytomegalovirus elevates levels of the cellular protein p53 in infected fibroblasts. *J. Virol.* **68**, 8028-8034.

Muller, S., and Dejean, A. (1999). Viral immediate-early proteins abrogate the modification by SUMO-1 of PML and Sp100 proteins correlating with nuclear body disruption. *J. Virol.* **73**, 5139-5143.

Murphy, E. A., Streblow, D. N., Nelson, J. A., and Stinski, M. F. (2000). The human cytomegalovirus IE86 protein can block cell cycle progression after inducing transition into the S-phase of permissive cells. *J. Virol.* **74**, 7108-7118.

Navarro, L., Mowen, K., Rodems, S., Weaver, B., Reich, N., Spector, D., and David, M. (1998). Cytomegalovirus activates interferon immediate-early response gene expression and an interferon regulatory factor 3-containing interferon-stimulated response element-binding complex. *Mol. Cell. Biol.* **18**, 3796-3802.

Nevins, J. R. (1992). E2F: a link between the Rb tumor suppressor and viral oncoproteins. *Science* **258**, 424-429.

Pajovic, S., Wong, E. L., Black, A. R., and Azizkhan, J. C. (1997). Identification of a viral kinase that phosphorylates specific E2Fs and pocket proteins. *Mol. and Cell. Biol.* **17**, 6459-6464.

Persoons, M. C. J., Daemen, M. J. A. P., Bruning, J. H., and Bruggeman, C. A. (1994). Active cytomegalovirus infection of arterial smooth muscle cells in immunocompromised rats: a clue to herpesvirus-associated atherogenesis? *Cric. Res.* **72**, 214-220.

Poma, E. E., Kowalik, T. F., Zhu, L., Sinclair, J. H., and Huang, E.-S. (1996). The human cytomegalovirus IE1-72 protein interacts with the cellular p107 protein and relieves p107-mediated transcriptional repression of an E2F-responsive promoter. *J. Virol.* **70**, 7867-7877.

Rice, G. P. A., Schrier, R. D., and Oldstone, M. B. A. (1984). Cytomegalovirus infects human lymphocytes and monocytes: virus expression is restricted to immediate-early gene products. *Proc. Natl. Acad. Sci. USA* **81**, 6134-6138.

Rodems, S. M., Clark, C. L., and Spector, D. H. (1998). Separate DNA elements containing ATF/CREB and IE86 binding sites differentially regulate the human cytomegalovirus UL112-113 promoter at early and late times in the infection. *J. Virol.* **72**, 2697-2707.

Rodems, S. M., and Spector, D. H. (1998). Extracellular signal-regulated kinase activity is sustained early during human cytomegalovirus infection. *J. Virol.* **72**, 9173-9180.

Salvant, B. S., Fortunato, E. A., and Spector, D. H. (1998). Cell cycle dysregulation by human cytomegalovirus: Influence of the cell cycle phase at the time of infection and effects on cyclin transcription. *J. Virol.* **72**, 3729-3741.

Sinclair, J., Baillie, J., Bryant, L., and Caswell, R. (2000). Human cytomegalovirus mediates cell cycle progression through G_1 into early S phase in terminally differentiated cells. *J. Gen. Virol.* **81**, 1553-1565.

Slobedman, B., and Mocarski, E. S. (1999). Quantitative analysis of latent human cytomegalovirus. *J. Virol.* **73**, 4806-4812.

Soderberg-Naucler, C., Fish, K. N., and Nelson, J. A. (1997a). Interferon-gamma and tumor necrosis factor-alpha specifically induce formation of cytomegalovirus-permissive monocyte-derived macrophages that are refractory to the antiviral activity of these cytokines. *J. Clin. Invest.* **100**, 3154-3163.

Soderberg-Naucler, C., Fish, K. N., and Nelson, J. A. (1997b). Reactivation of latent human cytomegalovirus by allogeneic stimulation of blood cells from healthy donors. *Cell* **91**, 119-126.

Sommer, M. H., Scully, A. L., and Spector, D. H. (1994). Transactivation by the human cytomegalovirus IE2 86-kilodalton protein requires a domain that binds to both the TATA box-binding protein and the retinoblastoma protein. *J. Virol.* **68,** 6223-6231.

Song, Y.-J., and Stinski, M. F. (2001). Effect of the human cytomegalovirus IE86 protein on expression of cellular genes that regulate the cell cycle, biosynthetic enzymes for DNA synthesis, and the initiation of DNA synthesis: A gene microarray analysis. (Manuscript in preparation).

Span, A. H. M., Grauls, G., Bosman, F., Van Boven, C. P. A., and Bruggeman, C. A. (1992). Cytomegalovirus infection induces vascular injury in the rat. *Atherosclerosis* **93,** 41-52.

Spector, D. H. (1996). Activation and regulation of human cytomegalovirus early genes. *Intervirology* **39,** 361-377.

Speir, E., Modali, R., Huang, E.-S., Leon, M. B., Shawl, F., Finkel, T., and Epstein, S. E. (1994). Potential role of human cytomegalovirus and p53 interaction in coronary restenosis. *Science* **265,** 391-394.

Stenberg, R. M. (1996). The human cytomegalovirus major immediate-early gene. *Intervirology* **39,** 343-349.

Stinski, M. F. (1978). Sequence of protein synthesis in cells infected by human cytomegalovirus: early and late virus-induced polypeptides. *J. Virol.* **26,** 686-701.

Stinski, M. F. (1999). Cytomegalovirus promoter for expression in mammalian cells. *In Gene Expression Systems: Using Nature for the Art of Expression, eds. J. M. Ferandez and J. P. Hoeffler, Academic Press, San Diego, CA,* 211-233.

Streblow, D. N., Soderberg-Naucler, C., Vieiera, J., Smith, P., Wakabayashi, E., Ruchti, F., Mattison, K., Altschulaer, Y., and Nelson, J. A. (1999). The human cytomegalovirus chemokine receptor US28 mediates vascular smooth muscle cell migration. *Cell* **99,** 511-520.

Taylor-Wiedeman, J., Sissons, J. G., Borysiewicz, L. K., and Sinclair, J. H. (1991). Monocytes are a major site of persistence of human cytomegalovirus in peripheral blood mononuclear cells. *J. Gen. Virol.* **72,** 2059-2064.

Trouche, D., and Kouzarides, T. (1996). E2F1 and E1A 12S have a homologous activation domain regulated by Rb and CBP. *Proc. Natl. Acad. Sci.* **93,** 11268-11273.

Tsai, H. L., Kou, G. H., Chen, S. C., Wu, C. W., and Lin, Y. S. (1996). Human cytomegalovirus immediate-early protein IE2 tethers a transcriptional repression domain to p53. *J. Biol. Chem.* **271,** 3534-3540.

Weibusch, L., and Hagemeier, C. (2001). The human cytomegalovirus immediate early 2 protein dissociates cellular DNA synthesis from cyclin-dependent kinase activation. *EMBO J.* **20,** 1086-1098.

Weibusch, L., and Hagemeier, C. (1999). Human cytomegalovirus 86-kilodalton IE2 protein blocks cell cycle progression in G1. *J. Virol.* **73,** 9274-9283.

Weinberg, R. (1995). The retinoblastoma and cell cycle control. *Cell* **81,** 323-330.

Wilkinson, G. W., Kelly, C., Sinclair, J. H., and Richards, C. (1998). Disruption of PML-associated nuclear bodies mediated by the human cytomegalovirus major immediate early gene product. *J. Gen. Virol.* **79,** 1233-1245.

Winkler, M., Rice, S. A., and Stamminger, T. (1994). UL69 of human cytomegalovirus, an open reading frame with homology to ICP27 of herpes simplex virus, encodes a transactivator of gene expression. *J. Virol.* **68,** 3943-3954.

Xu, Y., Ahn, J.-H., Jang, W.-J., and Hayward, G. S. (2000). Covalent modification of human cytomegalovirus major immediate-early proteins IE1 and IE2 by small ubiquitin-related modifiers. 25th International Herpesvirus Workshop, Portland, OR.

Yoo, Y. D., Chiou, C.-J., Choi, K. S., Yi, Y., Michelson, S., Kim, S., Hayward, G. S., and Kim, S.-J. (1996). The IE2 regulatory protein of human cytomegalovirus induces expression of the human transforming growth factor beta 1 gene through an Egr-1 binding site. *J. Virol.* **70**, 7062-7070.

Yurochko, A. D., Hwang, E.-S., Rasmussen, L., Keay, S., Pereira, L., and Huang, E.-S. (1997). The human cytomegalovirus UL55 (gB) and UL75 (gH) glycoprotein ligands initiate the rapid activation of Sp1 and NF-kB during infection. *J. Virol.* **71**, 5051-5059.

Yurochko, A. D., Kowalik, T. F., Huong, S. M., and Huang, E. S. (1995). Human cytomegalovirus upregulates NF-kappa B activity by transactivating the NF-kappa B p105/p50 and p65 promoters. *J. Virol.* **69**, 5391-5400.

Zhu, H., Cong, J.-P., Mamtora, G., Gingeras, T., and Shenk, T. (1998). Cellular gene expression altered by human cytomegalovirus: Global monitoring with oligonucleotide arrays. *Proc. Natl. Acad. Sci. USA* **95**, 14470-14475.

Zhu, H., Cong, J. P., and Shenk, T. (1997). Use of differential display analysis to assess the effect of human cytomegalovirus infection on the accumulation of cellular RNAs: Induction of interferon-responsive RNAs. *Proc. Natl. Acad. Sci. U.S.A.* **94**, 13985-13990.

Zhu, H., Shen, Y., and Shenk, T. (1995). Human cytomegalovirus IE1 and IE2 proteins block apoptosis. *J. Virol.* **69**, 7960-7970.

Index